TECHNICAL ELECTRICITY AND ELECTRONICS

PETER BUBAN & MARSHALL SCHMITT

McGraw-Hill Book Company

New York   St. Louis   San Francisco   Dallas   Düsseldorf   London
Mexico   Panama   Rio de Janeiro   Singapore   Sydney   Toronto

McGRAW-HILL PUBLICATIONS IN INDUSTRIAL EDUCATION

CHRIS H. GRONEMAN, Consulting Editor

BOOKS IN SERIES

DRAWING AND BLUEPRINT READING—Coover
GENERAL DRAFTING: A COMPREHENSIVE EXAMINATION—Blum
GENERAL METALS—Feirer
GENERAL POWER MECHANICS—Worthington, Margules and Crouse
GENERAL SHOP—Groneman and Feirer
GENERAL WOODWORKING—Groneman
TECHNICAL ELECTRICITY AND ELECTRONICS—Buban and Schmitt
TECHNICAL WOODWORKING—Groneman and Glazener
UNDERSTANDING ELECTRICITY AND ELECTRONICS—Buban and Schmitt

ABOUT THE AUTHORS of TECHNICAL ELECTRICITY AND ELECTRONICS

PETER BUBAN teaches Electricity and Electronics at Quincy Senior High School, Quincy, Illinois.
He has been an electronics technician and Chief Radioman in the U.S. Navy.
He has worked as an electrician and in a radio manufacturing plant.
He holds an M.A. degree.

MARSHALL L. SCHMITT is Specialist, Industrial Arts, U.S. Office of Education,
Washington, D.C. In addition to teaching electricity and electronics courses in high school and
college, he has had extensive military experience in the electricity and electronics fields.
He holds an Ed.D. degree.

TECHNICAL ELECTRICITY AND ELECTRONICS

Copyright © 1972 by McGraw-Hill, Inc. All Rights Reserved. Printed in the United States of America.
No part of this publication may be reproduced, stored in a retrieval system,
or transmitted, in any form or by any means, electronic, mechanical, photocopying, recording,
or otherwise, without the prior written permission of the publisher.

07-008639-7

FOREWORD

The technology of electricity and electronics is expanding at a fantastic rate. This growth pattern will be even more impressive and challenging in the future. Virtually all segments of industry depend upon electrical energy. Living is made more comfortable and meaningful because of the everyday commodities and services that are the result of electrical and electronic research and practical application. The rapid development and successful activities of space exploration are possible because of a relatively new electronic device, the computer.

Technical Electricity and Electronics is an advanced text; it is a companion to *Understanding Electricity and Electronics* by the same authors. It presents many of the sophisticated concepts and systems functioning in modern technology. The introductory text has had wide acceptance. This advanced title should prove equally popular with students, teachers, and interested persons working in the area of this rapidly expanding industry.

This book is the result of many years of purposeful activities in teaching, supervising, and writing instructional materials. The content has been closely read and favorably evaluated by outstanding educators in the field of electricity and electronics.

The editor, the authors, and the publisher feel that *Technical Electricity and Electronics* will fill a basic need for technical courses in electricity and electronics. Many individuals interested, conversant, and trained in this complex area will be required in the future. Students now studying the fascinating field of electronic technology may be the ones who will envision and design the complex electronic systems that will make travel beyond the moon a reality.

Chris H. Groneman

PREFACE

Technical Electricity and Electronics is designed to provide an extended range of learning experiences in the field of electricity and electronics. The text includes both basic and advanced concepts. It can be effectively used by upper-level students who have had no prior background in these areas or by students who have had previous experiences in industrial arts electricity and electronics in the junior or senior high school.

The text presents a direct, well-illustrated approach to theoretical topics and related practical devices and circuit systems. A special effort has been made to correlate theory and practice so that the student can participate in a wide variety of technological learning experiences. In line with this purpose, the text contains a number of broad-range topics of particular interest to the aspiring technician. These include (1) fundamental theories, laws, and devices; (2) circuit analysis; (3) testing and measuring procedures; (4) troubleshooting procedures; (5) servicing procedures; (6) drafting techniques and diagram interpretation; (7) circuit assembly and wiring techniques; (8) electronic communications; and (9) sensing and control circuits. In addition, the projects section contains a wide variety of activities designed to challenge students at all levels of ability.

The mathematical content is both broad and varied. Every effort has been made to show the close relationship between mathematics and design and to present mathematics as a "tool" to be used for the solution of practical problems. In this connection, it is hoped that the presentation of the slide rule in Appendix C will stimulate interest in the use of what is an increasingly important item in the technician's "stock" of tools.

The authors and the publisher sincerely welcome your comments and suggestions with regard to this volume.

Peter Buban
Marshall L. Schmitt

CONTENTS

Foreword iii
Preface iv

Section 1. Fundamental Theories, Laws, and Devices 2

Unit 1. Electric Energy and Power 2
Unit 2. The Content of Matter 4
Unit 3. Electrons in Action 18
Unit 4. Semiconductors 26
Unit 5. Ohm's Law and Electrical Power 36
Unit 6. Resistance and Resistive Devices 41
Unit 7. Capacitance and Capacitors 51
Unit 8. Safety 64

Section 2. Magnetism and Electromagnetism 74

Unit 9. The Nature of Magnetism 74
Unit 10. Electromagnetism 82
Unit 11. Alternating Voltage and Current 89
Unit 12. Self-Inductance 101
Unit 13. Transformers 105
Unit 14. Miscellaneous Magnetic Devices 117

Section 3. Circuit Analysis 123

Unit 15. Resistive Circuits 123
Unit 16. Inductive Circuits 135
Unit 17. Capacitive Circuits 150
Unit 18. Resonance and Tuned Circuits 164

Section 4. Functions of Devices and Circuits 176

Unit 19. Rectifiers and Rectifier Circuits 176
Unit 20. Transistors 194
Unit 21. Transistor Amplifier Stages 205
Unit 22. Electron Tubes 221
Unit 23. Amplifier Systems 229
Unit 24. Oscillators 244

Section 5. Tests and Measurements 258

Unit 25. Meters 258
Unit 26. Measuring Current, Voltage, and Resistance 264
Unit 27. Other Test Instruments 278
Unit 28. Troubleshooting Procedures 296

Section 6. Drafting and Fabrication Techniques 309

Unit 29. Electronic Drafting 309
Unit 30. Assembly and Wiring Techniques 320
Unit 31. Printed Circuits 345
Unit 32. Integrated Circuits 357

Section 7. Electronic Communications 366

Unit 33. Transmitter Fundamentals 366
Unit 34. Modulation 377
Unit 35. Receiver Fundamentals 387
Unit 36. Transmitter and Receiver Systems 394
Unit 37. Television 410
Unit 38. Radar and Lasers 429

Section 8. Sensing and Control Systems 438

Unit 39. Sensing and Control Circuits 438

Section 9. Projects 446

Line Filter 447
Safety Flasher 448
Battery Charger 449
Transistor Tester 450
Photoelectric Controller 452
Motor Speed Controller 453
Intercom 454
Transistor Power Supply 456
Portable P.A. System 458
Engine Analyzer 460

Appendices 463

A. Trigonometric Functions 463
B. Logarithms 463
C. The Slide Rule 464
D. Plane Vectors 466
E. International System of Units 466
F. The Greek Alphabet 467
G. NPN and PNP Transistor Circuit Symbols 467
H. Chassis Wiring Color Code 468
I. Selected Frequency Allocations 468

Index 469

TECHNICAL ELECTRICITY AND ELECTRONICS

SECTION 1
FUNDAMENTAL THEORIES, LAWS, AND DEVICES

In this section, the most important fundamental theories and laws of electricity and electronics are presented in conjunction with their application to the operation of basic devices found in most circuit systems. These devices include a number of semiconductor products, which are so important in modern solid-state technology.

The topic of safety is also included in this section, since it is of prime importance in many electrical and electronic activities. A knowledge of safety practices and procedures will provide the student with general guidelines to be followed while performing a variety of activities that can be dangerous if safety is not considered.

UNIT 1. Electric Energy and Power

The study of electricity and electronics is, for the most part, the study of the energy provided by the flow of electrons in various conductors and devices that form complete circuit systems. This study is based on a number of theories and laws, some that have been known for centuries and some that are relatively new (Fig. 1-1). *Electricity* is the name given to the field of study relating to circuit systems in which *electrons* flow primarily through solid-wire conductors of various forms. The term *electronics* is applied to a study of circuit systems in which electron tubes and solid-state devices such as diodes and transistors control the flow of electrons.

POTENTIAL AND KINETIC ENERGY

Energy, which is the ability to do work, exists in two forms: potential energy and kinetic energy. *Potential energy* is defined as energy that is "stored" or as energy that is due to position. An example of potential energy is the energy existing across the terminals of a battery. This energy, in the form of a voltage, can do the work of causing electrons to move if a circuit is connected to the battery's terminals. The resulting flow of electrons can in turn cause work to be done. For example, it can cause a lamp to light or a bell to ring. Energy of motion is termed *kinetic energy.*

CONVERSION OF ENERGY

The law of conservation of energy states that energy cannot be created or destroyed. However, energy can be converted from one form into another. Electric energy in the form of a flow of electrons can, for example, be converted into heat energy, light energy, chemical energy, or magnetic energy.

Consider what happens when an ordinary incandescent lamp or light bulb is in operation. As electrons flow through the filament of the lamp, the resulting friction produces heat, causing the filament to incandesce, or become white-hot. Some of the electric energy is therefore converted into heat energy. At the same time, the incandescent filament produces light, which represents the conversion of some of the electric energy into light energy as well.

Fig. 1-1. Astronaut Aldrin and the Apollo 11 lunar module upon the surface of the moon represent centuries of man's efforts to understand and use scientific theories and natural laws. (*NASA*)

EFFICIENCY

The degree to which electric energy in any circuit system is converted into the desired output energy is referred to as the *efficiency* of the system. An incandescent lamp is, in most cases, operated to produce light. Since approximately one-half of the electric energy input to the lamp circuit is converted into heat energy during the process of producing light, the incandescent lamp is only about 50 percent efficient.

POWER

Power is the time rate of doing work, or the time required to convert one form of energy into another form of energy.

The power rating of devices and machines is a measure of the rate at which they are able to convert energy. If, for example, one motor is more powerful than another, it is because it is able to convert electric energy into magnetic energy at a faster rate and, as a result, it is able to do more work during any given period of time.

Electric energy is most commonly furnished by electrochemical cells and batteries and by electromagnetic generators. The amount of power available from such devices depends upon the rate at which they are capable of supplying energy.

ELECTRIC CIRCUITS

An *electric circuit* is a system of conductors and devices used for the purpose of converting electric energy into one or more different forms of energy. A functional electric circuit most often consists of four basic parts: an *energy source* such as a battery, the *conductors* or wires through which electrons flow, the *control device* such as a switch, and the *load*. The load is that device or machine within which the actual energy conversion takes place. The incandescent lamp and the electric motor previously mentioned are common examples of electric loads.

For Review and Discussion

1. Define energy.
2. Explain the difference between potential energy and kinetic energy.
3. Is an electric current an example of potential or kinetic energy? Why?
4. State the law of conservation of energy.
5. Define power.
6. What is an electric circuit?
7. Name the four basic parts of a functional electric circuit.
8. What is the function of the load in a circuit?

UNIT 2. The Content of Matter

Anything that occupies space and has weight is called *matter*. Matter is composed of extremely small particles called *atoms*. All matter can be categorized into either one of two groups: elements or compounds. In an element, all the atoms are the same. At the present time, more than 100 different elements are known to exist. These include substances found in nature, such as aluminum, carbon, copper, germanium, and silicon. A number of additional elements such as neptunium and plutonium can be artificially produced by a process of nuclear bombardment, whereby atomic particles such as neutrons or protons are propelled into the nuclei of atoms at tremendous velocities. The combination of components and equipment systems necessary to accelerate the particles to the necessary velocity is known as a particle accelerator (Fig. 2-1).

The structure of the atoms within any given element determines the characteristics of that element. Since each element has the same characteristics wherever it is found, we can assume that all the atoms of a given element are identical in nature. Since no two elements have exactly the same characteristics, we can also assume that different elements are composed of different kinds of atoms.

Most of the matter on earth is composed of combinations of elements called *compounds*. Water, for example, is a compound consisting of the elements hydrogen and oxygen. The smallest particle of any compound that retains the original characteristics of that compound is called a *molecule*.

ATOMIC STRUCTURE

The word *atom* is derived from the Greek word *atomos*, meaning indivisible. During the latter part of the nineteenth century, scientists began to discard the theory that atoms were indivisible. They conducted experiments which showed that the atom was composed of subparticles. The arrangement of these particles inside the atom is explained by the *electron theory*, which defines all electronic activity.

Fig. 2-1. Particle source and control components of a Van de Graaff accelerator.
(*High Voltage Engineering Corp.*)

The Atomic Solar System. In the study of electricity and electronics, one subatomic particle, the electron, is of primary importance. The electron is the fundamental negative charge of electricity. Within the atom, the electrons revolve about the nucleus or center portion in a definite pattern of concentric "shells" or orbits (Fig. 2-2). The proton, which is the fundamental positive charge of electricity, is found within the nucleus. The number of protons within the nucleus of any particular atom determines the atomic number of that atom.

Fig. 2-2. Electrons and nucleus of an oxygen atom.

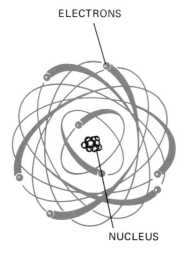
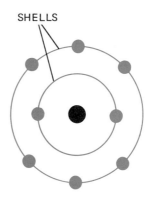

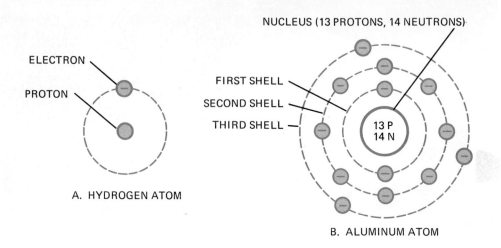

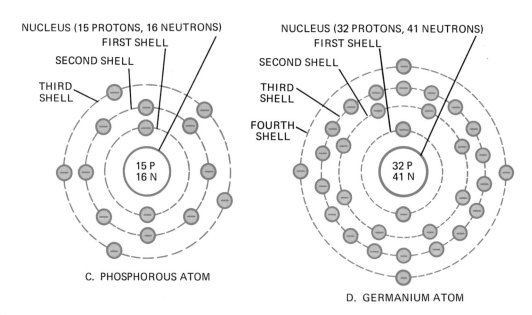

Fig. 2-3. Electron and proton content of the atoms of four common elements.

Atoms of different elements differ from one another in the number of electrons and protons they contain (Fig. 2-3). In its natural state, an atom of any element contains equal numbers of electrons and protons. Since the negative charge of each electron is equal in magnitude to the positive charge of each proton, the two opposite charges cancel. In this condition, the atom is electrically neutral, or in balance.

Energy Levels. A stable (in balance) atom has a certain amount of energy, which is equal to the sum of the energies of its electrons. Electrons, in turn, have different energies called *energy levels*. The energy level of an electron is proportional to its distance from the nucleus. Hence, the energy levels of electrons in shells (orbits) farther from the nucleus are higher than those of electrons in shells nearer the nucleus.

It is possible for an atom to absorb energy such as that possessed by heat and light. When this occurs, some electrons within the atom move to higher energy levels, and the energy of the entire atom is raised. The amount of energy necessary for an electron to be raised from one energy level to the next higher energy level is called a *quantum*. An atom in which this has occurred is said to be in an excited state.

An atom in the excited state is unstable. When the exciting (excitation) energy is removed, electrons that have been raised to higher energy levels tend to return to their original energy levels. As a result, the atom releases energy equal to one quantum, or photon, for each electron that has returned to a lower energy level. This energy is in the form of electromagnetic radiation such as light, heat, or radio waves.

Atomic Weight and Isotopes. The system of atomic weights is a relative measure. The standard for atomic weights is the carbon-12 atom, with an assigned atomic weight of 12.0000. Thus if another atom had an atomic weight of 24.0000, it would be exactly twice as heavy as an atom of carbon 12.

Many atoms of the same element can exhibit different atomic weights because of a difference in the number of neutrons (uncharged particles) in their nuclei but these different atoms have the same atomic number (number of protons). For example, the nuclei of ordinary hydrogen atoms contain only one proton. The nuclei of atoms of a second form of hydrogen, deuterium, contain one proton and one neutron. Elements which, like deuterium, have an atomic weight different from that of the "parent" element are called *isotopes*.

Binding Energy. Scientists have discovered that when protons and neutrons join to form atomic nuclei, a portion of their mass is changed into a form of energy that holds them together. This is called *binding energy*. While the nature of this energy is not yet thoroughly understood, we can assume that it overcomes the force of electrostatic repulsion between the protons (like charges tend to repel each other).

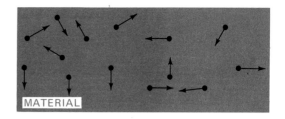

Fig. 2-4. Random motion of free electrons through a conductor.

FREE ELECTRONS

The planetary electrons that revolve around the nucleus of an atom possess energy. The amount of energy possessed by an individual electron depends on its distance from the nucleus. As the result of collisions with other atomic particles in a material, some electrons acquire energy that allows them to change orbits and move into a higher energy level farther from the nucleus. As the outer-shell electrons also collide with other particles, some of these may acquire enough energy to break away from the attraction of the nucleus and become free (mobile) electrons. Under this condition, the free electrons move about within the material in a random motion (Fig. 2-4).

Free electrons remain in the mobile state for only a comparatively short period of time. By means of a process known as *recombination*, they soon release the acquired energy and once again become part of an atom. In any given material, the process of electron release and recombination occurs continuously at a fairly constant rate. Because of this, there are approximately the same number of free electrons in a given volume of material at all times, provided that other factors such as temperature remain constant.

Atomic Quota. Each shell of an atom can accommodate a definite number of electrons. The maximum number of electrons each shell can accommodate is referred to as *quota* (Fig. 2-5). If the quota is filled in the outermost or high-energy shell, an atom will exhibit no tendency to enter into chemical combination with the atoms of any other element. Therefore, an ele-

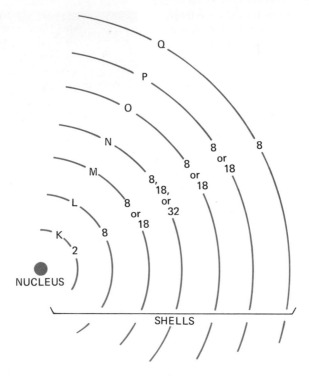

Fig. 2-5. Identification of atomic shells and the quota of electrons for each shell.

ment made up of such atoms is said to be *inert*. Some common inert elements are the gases argon, helium, and neon.

When the outer shell of an atom lacks its full quota of electrons, it is capable of gaining or losing electrons. Whether it will gain or lose depends upon how nearly filled the quota is and how closely the electrons are "bound" to their nuclei. The atoms of the majority of metallic elements, for example, have only one or two electrons in their outer shells. Thus, their quotas are comparatively far from being filled. As a result, the atoms tend to lose these electrons. On the other hand, the outer shells of the atoms of most nonmetallic elements are more nearly filled with their quotas of electrons. These atoms tend to gain electrons.

The electrical characteristics of an atom are directly related to the actions and the energy of planetary electrons. This is true primarily with regard to the electrons in the outer shell because it is these electrons that determine the stability of an atom or the extent to which it will enter into electronic or chemical activity. The outer-shell electrons are also commonly referred to as *valence electrons*.

Ions. If an atom has lost one or more electrons, the protons outnumber the electrons; thus, the atom carries a net electrical charge that is positive in nature. In this condition, the atom is called a *positive ion*. If an atom gains electrons, its net electrical charge becomes negative and the atom is referred to as a *negative ion* (Fig. 2-6).

THE ELECTRIC CHARGE

Since some atoms can lose electrons and other atoms can gain electrons, it is possible to cause a transfer of electrons from one material (object) to another. When this takes place, the normal balanced (equal) distribution of the positive and negative charges in each object no longer exists. Therefore, one of the materials will contain more than its usual number of electrons, while the other will contain less than its usual number of electrons. In this condition, the objects are said to be electrically charged.

A transfer of electrons between materials can be accomplished in a number of ways. One simple method is by means of friction. When, for example, a Bakelite rod is rubbed with flannel cloth, electrons move from the cloth to the rod, and thus the rod becomes negatively charged. If a glass rod is rubbed with a silk cloth, electrons move from the glass to the cloth, thereby producing a positive charge upon the rod.

The Coulomb. The magnitude of the electrical charge an object possesses is determined by the number of electrons within the object as compared with the number of protons. The unit by which this magnitude is indicated is the coulomb, which represents 6,280,000,000,000,000,000 (or 6,280 quadrillion) electrons. Thus, if an object contains 6,280 quadrillion more electrons than protons, its charge is one negative coulomb. Likewise, if an object contains 6,280 quadrillion more protons than electrons, its charge is equal to one positive coulomb.

The Electrostatic or Dielectric Field. The fundamental characteristic of an electrical charge is its ability to exert a force that attracts charges of opposite polarity and repels charges of like polarity. This force is present within the elec-

---

Mathematical Note: A number such as 6,280 quadrillion is often written (expressed) in terms of scientific notation or as the product of a number between 1 and 10 and a power of 10. Thus, the number 6,280,000,000,000,000,000 can also be written as 6.28×10^{18}. Examples of other numbers expressed in terms of scientific notation are:

$$0.000001 = 1 \times 10^{-6}$$
$$0.001 = 1 \times 10^{-3}$$
$$0.0746 = 7.46 \times 10^{-2}$$
$$100 = 1 \times 10^{2}$$
$$1,000 = 1 \times 10^{3}$$
$$270,000 = 2.7 \times 10^{5}$$
$$1,000,000 = 1 \times 10^{6}$$

Fig. 2-6. Positive and negative ions.

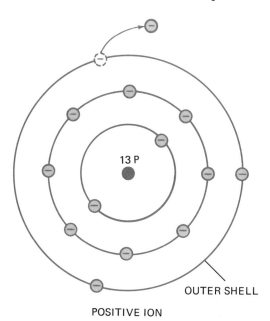

POSITIVE ION

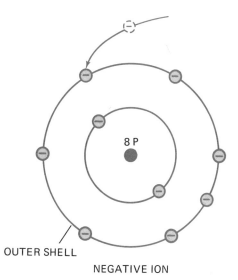

NEGATIVE ION

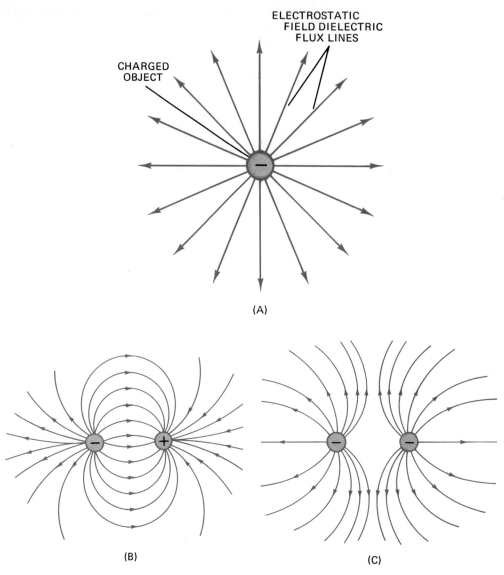

Fig. 2-7. The electrostatic field. (A) surrounding a charged object, (B) between two charges of opposite polarity, (C) between two charges of like polarity.

trostatic field surrounding every charged object (Fig. 2-7A). When two charges of opposite polarity (+ and −) are brought near each other, the electrostatic field is concentrated in the area between them (Fig. 2-7B). According to Coulomb's law, the force between these charges is directly proportional to the product of their magnitudes and inversely proportional to the square of the distance between them.

Electrostatic Induction. An electrostatic field has the ability to repel or to attract electrons, depending upon the polarity of the charge that produced it. As a result, the field can transfer

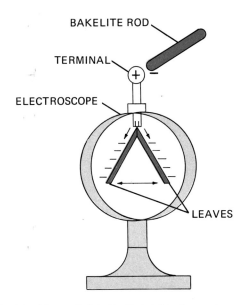

Fig. 2-8. Electrostatic induction illustrated by the use of an electroscope.

Discharge. An object that is positively charged will, under the proper conditions, accept from an available source the number of electrons that will cause it to return to the neutral or balanced state. An object that is negatively charged will pass its surplus electrons on to any material that will accept them. When a charged object has returned to its neutral state, the object is said to have become discharged. If two oppositely charged objects with charges of equal magnitude come into contact with each other, the surplus electrons from the negatively charged object will pass to the positively charged object until both have become discharged (Fig. 2-9).

The earth is considered to be an inexhaustible source of electrons that can either supply electrons or accept them without noticeable effect. Therefore, when any charged object, regardless of its polarity, is brought into contact with the earth, it will become discharged.

this charge to another object by electrostatic induction. For example, assume a negatively charged Bakelite rod is brought near the metallic terminal of an electroscope (Fig. 2-8). Since the rod is negatively charged, its electrostatic field repels electrons from the terminal, forcing them downward into the leaves of the electroscope. As a result, the terminal becomes positively charged while the leaves become negatively charged. The electrostatic fields of the charges that have been induced within the leaves, in turn, cause the leaves to repel each other because their charges are of like polarity.

Static Electricity. A charged object will retain its charge temporarily if there is no immediate transfer of electrons to or from it. In this condition, the charge is said to be at rest or in a stationary (static) state. Electricity at rest is called *static electricity*. While the effects of static electricity are generally considered to be a nuisance, the electrostatic field that accompanies a static charge is extremely useful in many applications. It is, for example, the basis of capacitor and electron tube action, which are discussed in later units. Electrostatic fields are also used in dielectric heating units, electrostatic spray painting, air cleaners, and in several manufacturing processes.

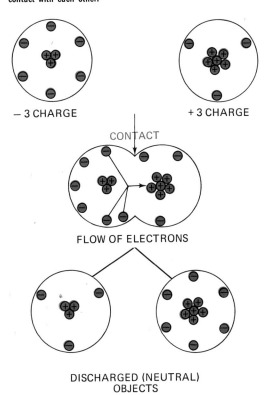

Fig. 2-9. The flow of electrons between two charged objects in contact with each other.

Fig. 2-10. Corona discharge from the terminal of a Van de Graaff generator operating at approximately 5 million volts. Such a discharge is a display of uncontrolled energy. (*High Voltage Engineering Corp.*)

The length of time a charged object in air will retain its charge depends upon the kind of material from which it is made and upon the humidity of the air. If the magnitude of the charge is sufficient, the object will discharge into the air with a visible spark or flash, which is known as a corona discharge (Fig. 2-10).

DIFFERENCE IN POTENTIAL

Because of the force of its electrostatic field, an electric charge has the ability to do the work of moving electrons (Fig. 2-11A). This ability of a charge to do work is referred to as its potential. Since a negative charge has the potential for repelling electrons and a positive charge has the potential for attracting electrons, there is a difference in potential between these charges (Fig. 2-11B).

When two charges of opposite polarity and equal magnitude are in contact with a material through which electrons will pass (a conductor), the difference in potential between them will result in a movement of electrons from the negative charge to the positive charge until both charges are neutralized (Fig. 2-12A). If, under the same condition, the two charges are of the same

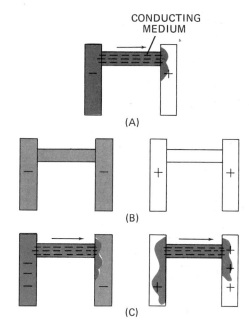

Fig. 2-12. Difference in potential existing between unlike charges and between like charges of unequal magnitude.

polarity and equal in magnitude, there will be no movement of electrons (Fig. 2-12B). In this case, there is no difference in potential between the charges, and each charge retains its individual potential.

In addition to the difference in potential that exists between charges of opposite polarity, a difference in potential also exists between two charges of the same polarity that are different in magnitude (Fig. 2-12C). In this case, the difference in potential is the result of a difference in the capabilities of the individual potentials to do work.

Zero Potential. The potential of charges is often compared to the potential of the earth, which is considered to be the zero reference potential level. With reference to this zero potential, a charge can be positive or negative, depending upon the nature of the charge. When any conductor, such as a metallic wire or other metallic object, is in contact with the earth, it must be at the same potential as the earth. Thus, there is no difference in potential between it and the earth. In this condition the object is said to be "grounded."

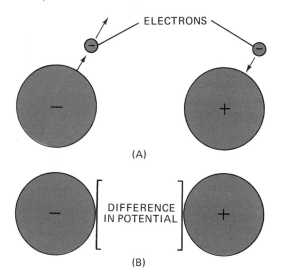

Fig. 2-11. Difference in potential: (electrostatic force between two oppositely charged objects which causes electrons to move).

Electromotive Force. A difference in potential can exist between any two of a number of points within an electrostatic field. The sum of these various differences in potential is commonly referred to as an electromotive force (emf). An electromotive force is the total force of an electrostatic field that causes electrons to move (Fig. 2-13). To produce a continuous movement of electrons through a conductor, the differences in potential between all points within the conductor must be maintained. To maintain the required differences in potential a device such as a battery or a generator is used.

CURRENT, VOLTAGE, AND RESISTANCE

The electrical content of matter provides us with two of the three factors that affect the operation of every electric-electronic circuit and device. These are electrons and the electromotive force that causes electrons to move. A third factor, resistance, is also present in every application of electricity-electronics. Resistance is not directly related to the electrical content of matter; it is the effect that particles of matter have upon the movement of electrons.

Current. The movement or the flow of electrons through a conductor is called a *current*. Current flow is represented by the letter symbol I. The basic unit in which current is measured is the ampere (amp). One ampere of current is defined as the movement of one coulomb (6.28×10^{18} electrons) past any point of a conductor during one second of time.

Table 2-1. Common metric prefixes and their numerical equivalents.

PREFIX AND SYMBOL	NUMERICAL EQUIVALENT
giga (G)	$1,000,000,000 = 10^9$
mega (M)	$1,000,000 = 10^6$
kilo (k)	$1,000 = 10^3$
milli (m)	$0.001 = 10^{-3}$
micro (μ)	$0.000001 = 10^{-6}$
nano (n)	$0.000000001 = 10^{-9}$
pico (p)	$0.000000000001 = 10^{-12}$

When it is desirable to express a magnitude of current smaller than the ampere, the milliampere (ma) and the microampere (μa) units are used. One milliampere is equivalent to one-thousandth (0.001) of an ampere, and one microampere is equivalent to one-millionth (0.000001) of an ampere.

Voltage. The term voltage (represented by the letter symbol E) is commonly used to indicate both a difference in potential and an electromotive force. The unit in which voltage is measured is the volt. One volt is defined as that magnitude of electromotive force required to cause a current of one ampere to pass through a conducting medium having a resistance of one ohm.

A magnitude of voltage less than one volt is expressed in terms of millivolts (mV) or microvolts (μV). Larger magnitudes of voltage are expressed in kilovolts (kV). One kilovolt equals one thousand volts.

Mathematical Note: The metric system (as opposed to the English system) is the universal system of measurement. In this system, a number of prefixes of Greek origin are combined with basic units of measurement such as the volt, ampere, and ohm. The prefixes most commonly used in electricity-electronics applications and their numerical equivalents are given in Table 2-1.

Fig. 2-13. Electromotive force: the sum of the various differences of potential between two points.

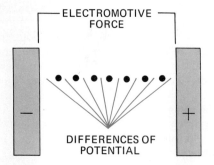

Table 2-2. Conversion of units.

TO CONVERT FROM:	TO:	MULTIPLY BY:
Units	Pico Units	1,000,000,000,000
Units	Micro Units	1,000,000
Units	Milli Units	1,000
Micro Units	Pico Units	1,000,000
Milli Units	Micro Units	1,000
Kilo Units	Units	1,000
Mega Units	Units	1,000,000
Mega Units	Kilo Units	1,000

TO CONVERT FROM:	TO:	DIVIDE BY:
Pico Units	Units	1,000,000,000,000
Micro Units	Units	1,000,000
Pico Units	Micro Units	1,000,000
Milli Units	Units	1,000
Micro Units	Milli Units	1,000
Units	Kilo Units	1,000
Units	Mega Units	1,000,000
Kilo Units	Mega Units	1,000

Resistance. As electrons move through a conductor, they collide with atoms, ions, and other particles within the material. These collisions retard or oppose the movement of electrons. The opposition to current is called *electrical resistance* and is represented by the letter symbol *R*. The unit of resistance is the ohm, a term that is often expressed by using the upper case of the Greek letter *omega* (Ω). Thus, 6 ohms may be written as 6 Ω, 325 ohms may be written as 325 Ω, and so forth.

One ohm is defined as that amount of resistance that will limit the current in a conductor to one ampere when the voltage applied to the conductor is one volt. Larger amounts of resistance are commonly expressed in kilohms (kΩ or k) and in megohms (MΩ or M). One kilohm equals one thousand ohms and one megohm equals one million ohms.

As an aid in the conversion of units, refer to Table 2-2. On studying this table you will notice that to convert a given unit of measure-

Mathematical Note (conversion of units): It is often necessary or desirable to convert one unit of measurement to another unit that may be larger or smaller. To do this, a system of multiplying and dividing by 10 or by a power of 10 is used.

When a number is to be multiplied by 10 or by a power of 10, move its decimal point to the right as many places as there are zeros following the 1 of the multiplier. Thus

$$0.0535 \times 10 = 0.535$$
$$43 \times 100 = 4,300$$
$$92.08 \times 1,000 = 92,080$$

When a number is to be divided by 10 or by a power of 10, move its decimal point to the left as many places as there are zeros following the 1 of the divisor. Thus

$$\frac{0.045}{10} = 0.0045$$

$$\frac{68}{100} = 0.68$$

$$\frac{7,380}{1,000} = 7.38$$

ment to a smaller unit, multiplication is used. To convert a given unit of measurement to a larger unit, division is used. These processes are illustrated by the following examples that are related to the conversion of current, voltage, and resistance units of measurement:

$$0.423 \text{ ampere} = 0.423 \times 1,000$$
$$= 423 \text{ milliamperes}$$

$$82 \text{ milliamperes} = \frac{82}{1,000} = 0.082 \text{ ampere}$$

$$0.0067 \text{ milliampere} = 0.0067 \times 1,000$$
$$= 6.7 \text{ microamperes}$$

$$7,440 \text{ volts} = \frac{7,440}{1,000} = 7.44 \text{ kilovolts}$$

$$220,000 \text{ ohms} = \frac{220,000}{1,000} = 220 \text{ kilohms}$$

$$4.7 \text{ kilohms} = 4.7 \times 1,000 = 4,700 \text{ ohms}$$

$$2.2 \text{ megohms} = 2.2 \times 1,000$$
$$= 2,200 \text{ kilohms}$$

OTHER CURRENTS

By definition, an electric current is a movement of any electrical charge regardless of its polarity. This, of course, includes electrons. However, there are also two other kinds of currents that are made use of in the field of electricity-electronics. These consist of the movement of ions and of holes.

Ionic Currents in Liquids. When certain compounds such as sodium chloride (NaCl or common salt) or sulfuric acid (H_2SO_4) are dissolved in water, their molecules ionize (break up) into charged particles. As a result, the solution is able to conduct an electric current. Any solution that exhibits this property of conduction is referred to as an *electrolyte*.

A basic example of ionic (ionization) current occurs when sodium chloride is dissolved in pure water. Before the sodium chloride is added to the water, the water will not conduct an electric current (Fig. 2-14A). When sodium chloride is added to the water, it ionizes into positive sodium (Na^+) ions and negative chloride (Cl^-) ions. The sodium ions are attracted to the plate (cathode) that is connected to the negative terminal of the battery. Likewise, the chloride ions are attracted to the plate (anode) that is connected to the positive terminal of the battery (Fig. 2-14B). This produces a movement of ionic charges (current) through the electrolyte in opposite directions.

When a chloride ion reaches the anode, the ion releases its excess electron, which then moves through the external circuit to the positive terminal of the battery. At the same time, a sodium ion reaches the cathode, where it combines with an electron that has passed through the external circuit from the negative terminal of the battery to the cathode. These

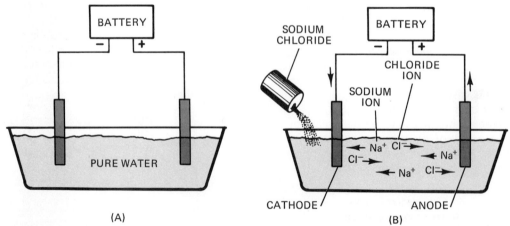

Fig. 2-14. Ionic currents in a solution consisting of sodium chloride (common salt) and water.

processes produce a continuous ionic current within the electrolyte and a continuous electronic current in the external circuit.

Chemical changes (reactions) also result from the conduction of ionic current. In the previous example, chlorine gas, hydrogen gas, and sodium hydroxide are produced as the current flows through the electrolyte. The chlorine gas is released at the anode, the hydrogen gas is released at the cathode, and the sodium hydroxide settles to the bottom of the solution container.

The process of producing chemical changes or variations in the properties of substances by ionic currents in liquids is called *electrolysis*. The most common form of electrolysis is that which occurs in electrochemical voltaic cells and batteries to produce electromotive force (voltage). Electrolysis is also widely used in processes that are related to electroplating, the production of chemicals, refining copper, and the extraction of metals such as aluminum and magnesium from their ores.

Ionic Currents in Gases. The atoms within a gas do not ionize to any significant extent. Because of this, a gas is not, under normal conditions, a very good conductor of an electric current. However, when a gas such as neon is subjected to a relatively high voltage, some of its outer shell (valence) electrons gain enough energy to escape from parent atoms, thus producing ionization. In this condition, the free electrons are attracted by the positive terminal of the voltage source, while the positive neon (Ne^+) ions are attracted by the negative terminal of the voltage source. As a result, there is a movement of both ionic and electron currents within the gas (Fig. 2-15).

In a neon lamp or a neon-sign tube, the ionization process results in the production of a glowing light. The ionization of gases is also made use of in the operation of certain types of gas-filled electron tubes.

The voltage required to ionize a gas under a given condition to a degree that will allow the conduction of current through the gas is referred to as the striking, starting, or ionization voltage. With small-size neon lamps such as those used in dial indicators, the ionization voltage is usually from 60 to 90 volts. The ionization voltage required for the operation of a rather long tube in a neon sign may be 10,000 volts or more.

Plasma. A mass of ionized gas in which there is an equal number of electrons and positive ions is known as *plasma*. As these charges move within any given volume of a gas, the smaller mass of the electrons causes them to move outward, thus producing what is referred to as a *sheath*. When the electrons and the ions are made to recombine, the plasma gives off a tremendous amount of energy in the form of heat, which is utilized in special kinds of torches.

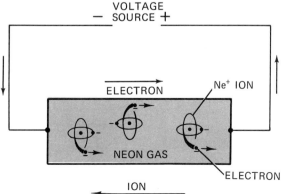

Fig. 2-15. Ionic and electronic currents produced as the result of the ionization of a gas.

Other possible applications of plasma include the production of the intense heat required for thermonuclear reactions, and the generation of ultrahigh-frequency microwaves, which are useful for radio communication purposes.

Holes. Holes are regions resulting from the absence of electrons in the atomic structure of certain semiconductor materials. Because a hole has a net positive charge, its movement has an effect similar to the movement of a positively charged particle. The theory of holes and their activity within semiconductor materials and devices is presented in Unit 4.

For Review and Discussion

1. Describe the basic structure of an atom.
2. Explain what is meant by atomic energy levels.
3. What are free electrons?
4. What is meant by atomic quota?
5. Define an ion.
6. Under what condition does an object acquire a negative charge? A positive charge?
7. Describe the characteristics of an electrostatic field.
8. Describe the process of electrostatic induction.
9. Define static electricity.
10. What is meant by a corona discharge?
11. Explain what is meant by a difference in potential.
12. Name and define the basic units of current, voltage, and resistance measurement.
13. What causes electrical resistance?
14. Define an electrolyte.
15. Describe the process of ionization in a liquid and the movement of ionic current through a liquid.
16. Explain the process of ionization in a gas such as neon.
17. Define plasma.
18. What is meant by a hole?

UNIT 3. Electrons in Action

In their natural state, electrons are in constant orbital motion around the nuclei of atoms. This motion is the basis of electrostatic and magnetic energy. To transmit electrical energy or convert it into other forms of energy, however, there must be a continuous flow of electrons through the conductors and components of a circuit system. When this occurs, the energy of moving electrons is converted into other forms of energy, resulting in the production of heat, light, magnetism, chemical action, and physiological (shock) effects.

CONDUCTION OF ELECTRONIC CURRENT

When a voltage is applied to the ends of a metallic conductor, differences in potential exist between all points along the conductor. As a result, a negative to positive direction is imposed upon the random movement of free electrons (Fig. 3-1). The rate of this movement or "drift" of electrons depends upon the magnitude of the applied voltage and the nature of the conducting material.

As electrons enter the positive terminal of the voltage source, an equal number of electrons leave or are "injected" into the conductor from

Fig. 3-1. Electrons move through a conductor in a negative to positive direction.

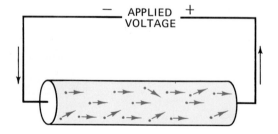

the negative terminal. In this connection, it is important to remember that the voltage source does not itself provide electrons; they are already present within the conductor material. The voltage merely provides the energy (force) that causes the electrons to move.

Velocity. As electrons move through a conductor under the influence of voltage, their velocity tends to increase. However, as the electrons collide with molecules and atomic particles within the material, their velocity is checked. As a result, they move through the material at a rather uniform velocity of approximately 0.428 inch per second.

The Electrical Impulse. Although individual electrons move through a conductor at a relatively slow speed, an electrical impulse or the action of an electronic drift through a conductor moves at the speed of light (186,000 miles per second). This condition can be better understood if we imagine a conductor 186,000 miles long connected to a battery through a switch. One second after the switch is closed, a drift of electrons (the electrical impulse) is established through all points along the entire length of the conductor.

DIRECT CURRENT

Direct current (dc) is current that moves through a conductor or circuit in one direction only. This direction is from a point of negative polarity to a point of positive polarity. The current results from the application of a voltage of constant polarity. Sources such as cells and batteries produce direct current. The voltage supplied by these sources is called direct-current voltage (dc voltage), or simply direct voltage.

Direct current may be either "pure," varying, or pulsating (Fig. 3–2). *Pure* direct current, which is not only unidirectional (one direction) but also of a constant magnitude, is characteristic of the current produced as the result of a cell or battery voltage. *Varying* direct current fluctuates or changes in magnitude; it is usually present in direct-current generator circuits. *Pulsating* direct current is conducted in short bursts or pulses whose duration is determined by the period of time during which a control device in a circuit is in the "on" condition. A common example is in the automobile ignition system, where breaker points rapidly open and close the circuit that connects the battery to the ignition coil (Fig. 3–3). Voltage and current pulses are also extremely important in the operation of several different types of electronic circuits, including those found in communications equipment, control devices, and computers.

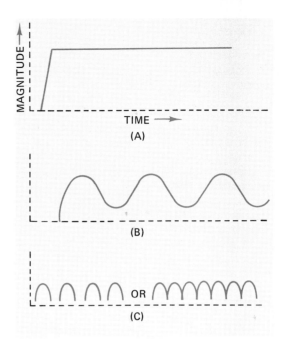

Fig. 3-2. Direct current: (A) "pure" direct current, (B) varying direct current, (C) pulsating direct current.

Fig. 3-3. Example of the waveform of the pulsating direct current produced by the action of the breaker points in an automobile ignition system.

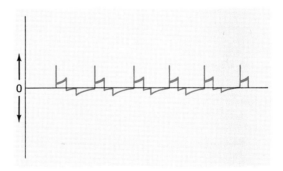

ALTERNATING CURRENT

The voltage present across the two lines (wires) of a typical circuit in the home changes periodically in polarity and constantly in magnitude. As a result, this voltage produces an alternating current (ac), which also periodically changes in direction and constantly changes in magnitude. The voltage supplied by an alternator (alternating-current generator) is referred to as an *alternating-current voltage* (ac voltage) or simply as an *alternating voltage*.

The Sine Waveform. The directional and fluctuating characteristics of the most common type of alternating current are described by a graph known as a *sine waveform* (Fig. 3–4). On this graph, the perpendicular distance from the horizontal base line AC to any point on the curve indicates the magnitude of the current at any instant of time. Beginning with point A, the current gradually rises to a maximum magnitude, then decreases to zero at point B. At this time, the current reverses its direction, again rises to a maximum magnitude, then decreases back to zero at point C.

Fig. 3-4. Sine waveform of alternating current.

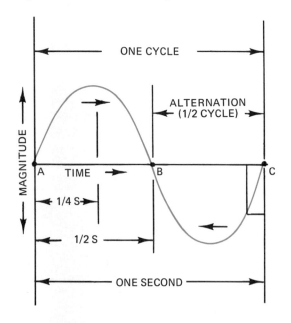

Frequency. The number of times that the variations of an alternating current represented by the sine waveform are repeated during one second of time is referred to as the *frequency* of the current. Frequency is expressed in hertz (Hz), which has replaced cycles per second (cps). Thus, one hertz equals one cycle per second. High frequencies are expressed in kilohertz (kHz) and megahertz (MHz). One kilohertz equals one thousand hertz and one megahertz equals one million hertz. Examples of designating frequency in hertz and in equivalent cycles per second are as follows:

60 hertz = 60 cycles per second
550 kilohertz = 550 kilocycles per second
1,220 megahertz = 1,220 megacycles per second

In the United States, the current used for the transmission of electrical energy and the operation of common products has a frequency of 60 hertz. This is a very low frequency compared with the extremely high frequencies of up to millions of hertz present in radio, television, radar, and other electronic circuits.

Alternating Current vs. Direct Current. Alternating current can be transmitted over long distances and at various voltages by the use of transformers, but direct current cannot. Therefore, alternating current is found in practically all power distribution systems. Since it is so readily available, alternating current is used for the operation of the vast majority of circuits. However, in many circuit applications such as general-purpose lighting and heating appliances, either alternating current or direct current can be used. In circuits containing components that depend for their operation on the magnetic effects produced by alternating current (as in some motors, transformers, and other devices equipped with large coils), only alternating current can be used. Other circuits and circuit processes are either designed for or must be operated with direct current. For example, direct current must be used in certain electrochemical processes such as battery charging, electroplating, and industrial electrolysis.

Rectification. When large amounts of direct current are required, direct-current generators or rectifier units that convert an alternating voltage

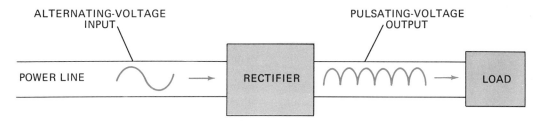

Fig. 3-5. Basic process of rectification.

to a direct voltage are used (Fig. 3-5). Most electronic circuits such as those found in radio and television receivers are equipped with rectifier circuits that produce the direct voltages necessary for the operation of transistors and electron tubes.

CONDUCTORS

A conductor is a material containing a large number of free electrons that can pass through the material quite easily under the influence of a voltage. While the quality of a conductor is most often stated in terms of its resistance, it is sometimes convenient to express this in terms of conductivity or conductance, G. Since conductance is the opposite of resistance, it can be expressed as the reciprocal of resistance, or $G = 1/R$. The unit of conductance is the mho (ohm spelled backward).

Heat has a definite effect on a conductor's resistance. For example, an increase in the operating temperature of a metallic conductor such as copper increases the velocity of atomic particles within the conductor. This in turn causes an increase in atomic collisions. As a result, the resistance also increases.

The measure of the effect of heat upon the resistance of a material is known as its *temperature coefficient*. This factor is usually expressed in terms of the increase or decrease in the resistance of the material, with temperature change given in degrees Celsius (°C) (formerly degrees Centigrade).

In general, the resistance of most commonly used metallic conductors increases as temperatures increase. These materials are said to have a positive temperature coefficient. The resistance of many nonmetallic conductors, such as carbon and graphite, decreases with an increase of temperature. These materials are said to have a negative temperature coefficient.

The resistances of conductors are often compared by referring to their relative resistivity. This is a measure of the resistance of a material as compared with the resistance of annealed copper when both materials are of equal length and cross-sectional area (Table 3-1).

Table 3-1. Relative resistivity of common metals and metallic alloys.

MATERIAL	RELATIVE RESISTANCE WITH COPPER = 1
Silver	0.95
Copper	1.00
Gold	1.32
Aluminum	1.64
Tungsten	3.16
Brass (Spring)	3.50
Zinc	3.52
Cadmium	4.34
Nickel	5.60
Pure Iron	5.90
Platinum	6.15
Lead	12.70
Mercury	13.68
Copper-Nickel (55%–45%)	28.40
Nickel-Chromium (80%–20%)	62.70
Carbon	2028.20

SKIN EFFECT

As the frequency of the current passing through a metallic conductor increases, the electrons tend to move only on or near the surface of the conductor. This phenomenon is known as *skin effect*. As a result of skin effect, the usable cross-sectional area of a conductor is effectively decreased, thus causing its resistance to increase at high frequencies.

In practice, skin effect is reduced by the use of larger wire and special kinds of conductors designed to have the largest possible surface area. These include hollow (tubular) conductors and Litz wire, which is made up of several fine strands of wire insulated from each other. Silver plating is also used to reduce the resistance of conductors at high frequencies.

SUPERCONDUCTIVITY

The fact that the electrical resistance of common conductors varies with changing temperatures has been a major factor in the development of the relatively new branch of physics called *cryogenics*, which is the study of the behavior of materials under conditions of ultralow temperatures. The extreme lower limit of these temperatures is called *absolute zero*, which is approximately -459.7 degrees Fahrenheit. To achieve temperatures approaching absolute zero, gases such as oxygen, hydrogen, and helium are subjected to a number of compression-expansion cycles until they reach their liquid or solid state.

As the temperatures of many elements and compounds approach absolute zero, the motion of the atomic particles within is gradually reduced. As a result, the resistance of the materials also decreases to a point that, for all practical purposes, is zero. This is referred to as a condition of *superconductivity*.

Metallic elements with excellent superconductive properties include molybdenum, rhenium, niobium, and zirconium. These metals, and alloys such as vanadium-silicon and niobium-zirconium, are in use or are being considered for use as conductors in electromagnet coils, transformer coils, and transmission lines. In these applications, the conductors are often maintained at superconductive temperatures by immersion in liquid helium or hydrogen.

INSULATORS

An *insulator* is a material which at ordinary room temperatures contains very few free electrons. For this reason, the conductivity of these materials is very low and, for all practical purposes, is usually considered to be zero.

There is no sharp dividing line between conductors and insulators. If, for example, a sufficiently high voltage is applied to a material ordinarily considered to be an excellent insulator, the voltage will cause electrons to be "torn" from their orbital bonds. As a result, the material will conduct current between the surfaces in contact with the source of the voltage. Under this condition, the material is said to be "punctured," and is often damaged beyond repair.

The voltage at which an insulator will conduct current is known as its *breakdown voltage*. This factor, of course, determines how effective a particular material is in actual use as an insulator. The magnitude of the voltage necessary to cause the flow of electrons through an insulator material is a measure of its dielectric strength. As might be expected, this factor is often used to compare the insulating qualities of different materials.

ELECTRON EMISSION

The flow of free electrons through a conductor, brought about by the application of voltage, is the principal method of utilizing electrical energy. However, in a number of electronic applications it is desirable to cause electrons to flow through a vacuum or through an air space under the influence of an electrostatic or magnetic field. The process by which these electrons are caused to move through a vacuum or gas is called *electron emission*. The word *emission* is derived from the Latin word *emittere* which means "to send out" or "to send forth."

Although free electrons move about within materials even when a voltage is not present, these electrons are not able to escape from the surface of a material under ordinary conditions. The reason for this is that their velocity is not great enough to overcome the attractive force within the material. The force, known as a *surface barrier*, is developed whenever an electron tends to leave the surface. When this occurs, a

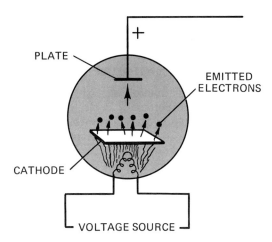

Fig. 3-6. Thermionic emission.

small positive charge is produced upon the surface of the material at the point from which the electron is tending to leave. This charge immediately attracts the electron and pulls it back to the material.

In order to make it possible for electrons to overcome the surface barrier and be emitted from a material, energy is required. This energy may be in the form of heat, light, or an electrostatic field. Because of variations in atomic structure, different amounts of energy must be expended to produce electron emission from different materials. The amount of energy necessary to accomplish this with any given material is referred to as the *work function* of that material.

Thermionic Emission. When heat is applied to most materials, the velocity of the orbital electrons is increased. As the temperature reaches a certain point, the electrons are accelerated to an extent that allows them to overcome the surface barrier. Under this condition, valence electrons are emitted from the surface of a material by a process referred to as *thermionic emission*. The materials most useful in producing such emission are tungsten and nickel alloys, often coated with thorium and/or metallic oxides.

A common application of thermionic emission is found in the operation of a typical electron tube, where an electrode known as a cathode is heated. As electrons are emitted from the cathode, they are attracted by the positive voltage applied to the plate of the tube (Fig. 3-6).

Photoelectric Emission. Photoelectric emission is the process by which electrons are emitted from materials such as zinc, potassium, cesium oxide, and other alkaline metals when they are exposed to light. In this process, the energy of the light rays is transferred to valence electrons within the materials, thus enabling them to overcome the surface barrier.

The velocity at which electrons are emitted from the surface of a material as a result of photoelectric emission is determined by the wavelength of the light. The intensity of the light determines the quantity of the electrons emitted. Potassium and cesium oxide will produce satisfactory emission when ordinary visible light is used. For most other metals, maximum emission is obtained when the wavelength of the light is in either the ultraviolet or infrared portions of the light spectrum.

A common application of photoelectric emission is found in the operation of the phototube (Fig. 3-7). In this tube, two metallic elec-

Fig. 3-7. Photoelectric emission occurring within a phototube.

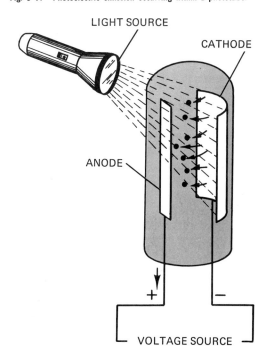

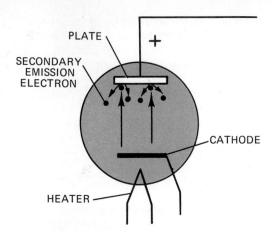

Fig. 3-8. Secondary emission taking place within an electron tube.

trodes—the anode (plate) and the cathode—are contained within a vacuum. When light strikes the cathode, the electrons emitted from it are attracted to the anode, which is positively charged. Since the magnitude of the current through the tube is dependent upon the intensity of light striking the cathode, such a tube can be used in various types of light-controlled circuits. Another important application of photoelectric emission is in the television camera or image-orthicon tube. This tube converts changes of light intensity into corresponding changes in the magnitude of current.

Secondary Emission. When electrons strike a surface such as the plate of an electron tube with sufficient velocity, they can dislodge other electrons from the surface (Fig. 3-8). While this "secondary emission," as it is called, is not commonly used as a source of electrons, it does occur during the operation of electron tubes. In the operation of most tubes it is undesirable and must be controlled by the use of a tube electrode known as a *suppressor grid*.

Field or Cold-cathode Emission. This type of emission occurs when the voltage existing between two surfaces produces an electrostatic field that is stronger than the attractive force of the surface barrier. As a result, electrons are emitted from the negatively charged surface and are attracted toward the positively charged surface (Fig. 3-9).

THE MULTIPLIER PHOTOTUBE

In the multiplier phototube, light is directed against the active surface of an electrode known as the *photocathode*. The electrons emitted from this cathode by the process of photoelectric emission are then electrostatically attracted to the first of a number of other electrodes called *dynodes* (Fig. 3-10). When normal operating voltage is applied to the first dynode, it emits electrons by the process of secondary emission. These electrons are then electrostatically attracted to the remaining consecutive dynodes, from which a larger and larger number of electrons are secondarily emitted. Following the last dynode, an anode "collects" the electrons and also serves as the output-signal electrode.

Because of the "multiplying effect" within the tube, the output current may be amplified to as much as several million times the current originally emitted from the photocathode. Hence, the multiplier phototube is very useful as a light detector or sensor in applications where the intensity of the light is extremely low. Such applications include the use of the multiplier tube in conjunction with light-control circuits, television camera tubes, and various kinds of photoelectric instruments.

Fig. 3-9. Field or cold-cathode emission.

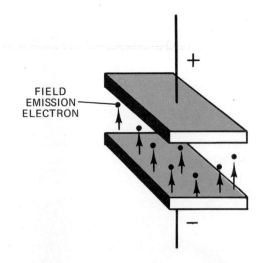

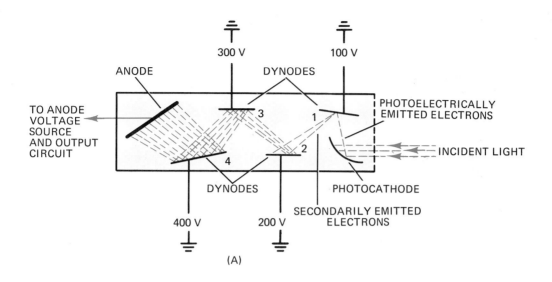

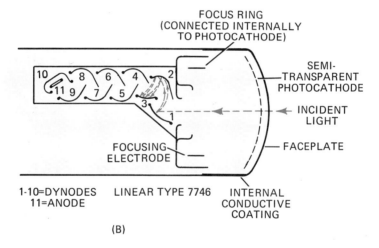

Fig. 3-10. Multiplier phototube: (A) basic operating action, (B) constructional features. (RCA)

For Review and Discussion

1. Describe the conduction of an electronic current through a metallic conductor.
2. What is meant by an electrical impulse?
3. Define direct current.
4. What is a pulsating direct current?
5. Define alternating current.
6. Draw a one-cycle sine wave of an alternating current and explain the conditions of the current it illustrates.
7. Define frequency and state the unit of frequency.
8. Name at least one type of circuit "load" that must be operated with direct current.
9. Define rectification.

10. What is meant by a positive temperature coefficient? A negative temperature coefficient?
11. What is meant by relative resistivity?
12. Describe the phenomenon of skin effect.
13. Describe the condition of superconductivity.
14. Define the breakdown voltage of an insulator.
15. Explain the process of thermionic emission.
16. What is photoelectric emission?
17. Explain the process of secondary emission.
18. What is meant by field or cold-cathode emission?
19. Describe the construction of a typical phototube and explain the operation of the device.
20. Describe the construction of a typical multiplier phototube and explain the basic operation of this device.

UNIT 4. Semiconductors

Semiconductors are a class of elements that are neither good conductors nor good insulators (Fig. 4-1). Because of their atomic structure, the resistance of such materials can be made to vary widely with the application of heat and light, or by the addition of small quantities of other materials. Because of these characteristics, semiconductor materials such as germanium (Ge) and silicon (Si) are widely used in the manufacture of devices such as transistors, diodes, photocells, and integrated circuits.

CRYSTALLINE STRUCTURE

Germanium and silicon are crystalline in structure. This means that the atoms in these materials are arranged in a definite pattern referred to as a *lattice*.

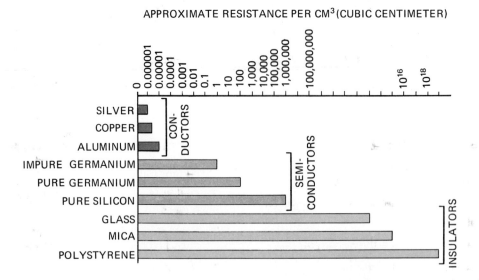

Fig. 4-1. Resistance of common semiconductors as compared with conductors and insulators.

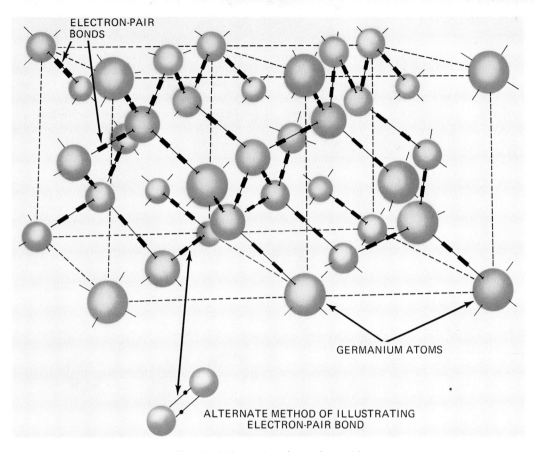

Fig. 4-2. Lattice structure of germanium crystal.

In the lattice structure of germanium, for example, each of the four valence electrons of each atom is shared with a valence electron of an adjacent atom. Each pair of shared electrons form what is known as a *covalent* or *electron-pair bond*. Thus, each germanium atom is bonded to four equidistant atoms by four electron-pair bonds that tend to maintain the geometric symmetry of the lattice (Fig. 4-2). Because the electrons are strongly bonded to the atoms that share them, the resistance of pure germanium is relatively high.

Since the atoms of silicon also have four valence electrons, silicon has a lattice structure and electrical characteristics that are similar to those of germanium. For this reason, the information presented in the following paragraphs, although related specifically to germanium, can also be applied to silicon. Both of these materials are actually insulators in their pure state. However, since they may easily become semiconductors under certain conditions, they are referred to as *intrinsic semiconductors*.

N-TYPE GERMANIUM

By applying heat, light, or a strong electrostatic field to pure germanium, enough energy can be imparted to the electrons to cause some to break their electron-pair bonds and become free. Nevertheless, since in semiconductor devices the flow of electrons must be carefully controlled, these methods of producing free electrons are not always suitable.

A much more effective method of producing free electrons within a semiconductor material is to add a material that acts as an impurity. This process is referred to as *doping*. Commonly used impurities, added to an extent usually no greater than one part impurity to 10 million parts of pure germanium, are arsenic, bismuth, phosphorus, and antimony. Because the atoms of these impurities contain five valence electrons, they are often referred to as *pentavalent* materials.

If, for example, arsenic (As) is added to pure germanium, four valence electrons from each arsenic atom will form electron-pair bonds with four adjacent germanium atoms. Under this condition, the fifth valence electron of each arsenic atom cannot enter into an electron-pair union with an electron from any adjacent germanium atom. As a result, it becomes an excess electron that is only slightly attracted by the nucleus of the parent (arsenic) atom (Fig. 4-3A). With this action repeated many, many times in a given mass of doped germanium, the excess or "free" electrons will form an electronic (negative charge) current through the material when a voltage is applied (Fig. 4-3B). This type of material is called *negative-type (n-type) germanium*.

Fig. 4-3. N-type germanium: (A) excess electron resulting from the addition of arsenic to a germanium crystal, (B) the flow of "free" electrons through n-type germanium.

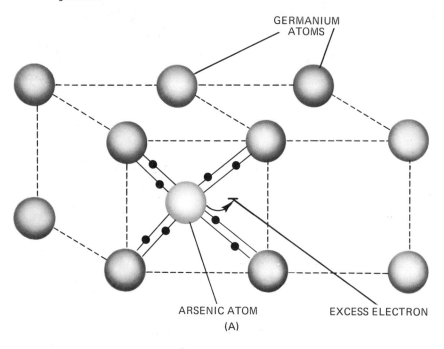

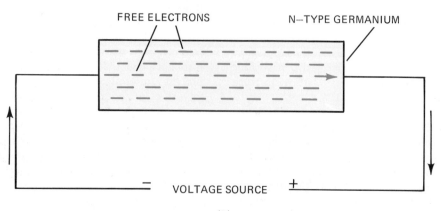

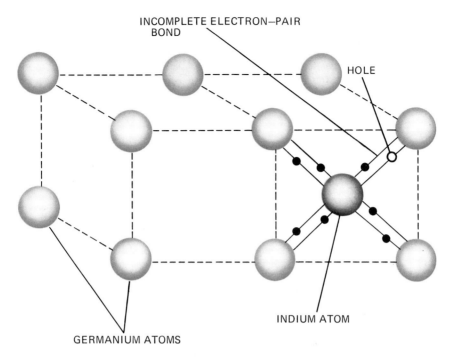

Fig. 4-4. Incomplete electron-pair bond formed by doping pure germanium with indium.

Since a pentavalent impurity, in effect, gives or "donates" free electrons to germanium, it is called a *donor impurity*. Whenever an "excess" electron leaves a donor impurity atom, the atom becomes a positive ion. However, since the positive charge of each donor ion is counteracted by the negative charge of the excess electron, the total (net) charge of a given mass of n-type germanium is zero.

P-TYPE GERMANIUM

The atoms of some impurity materials with which pure germanium (or silicon) is doped contain three valence electrons. These impurities are known as *trivalent* materials.

When a trivalent impurity such as indium (In) is added to pure germanium, the electrons of each indium atom form electron-pair bonds with three adjacent germanium atoms. In this case, one valence electron from an adjacent germanium atom cannot form a complete electron-pair bond with any indium atom since indium atoms have only three electrons to share.

As a result, one of the electron bonds between any indium atom and the four adjacent germanium atoms contains only one electron (Fig. 4-4). The space where an electron would be if the union between each indium atom and the four adjacent germanium atoms was complete is called a *hole*.

A hole represents the absence of an electron from a specific position within a given crystalline structure. It is considered to be a unit of positive charge. Since a very large number of holes exist within any given mass of indium-doped germanium, the material is now called *positive type (p-type) germanium*. In addition to indium, another trivalent impurity, gallium (Ga), is commonly used to produce p-type germanium.

It is possible for electrons within p-type germanium to acquire enough energy to break their individual electron-pair bonds. This action is illustrated in Fig. 4-5. Here, an electron from electron-pair bond 1 has left this bond to fill the hole in bond 2 (Fig. 4-5A). When this happens, two important changes take place. First, the indium atom that has accepted an electron be-

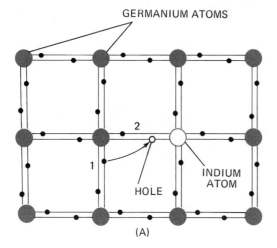

(A)

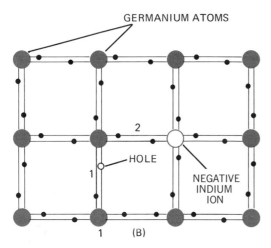

(B)

Fig. 4-5. "New" hole created within p-type germanium when an electron moves from one electron-pair bond to another.

trivalent material, accepts electrons, it is referred to as an *acceptor impurity*.

CONDUCTION OF CURRENT IN P-TYPE GERMANIUM

By applying a voltage across a mass of p-type germanium, it is possible to maintain a continuous transfer, or migration, of electrons and holes, thereby causing the germanium to conduct current. This action, in a very basic form, is shown in Fig. 4-6. Here, beginning at the right, an electron from electron-pair bond 2 is attracted by the positive terminal of the voltage source. As a result, this electron breaks its electron-pair bond and moves to fill hole 1. This creates hole 2. Now, an electron from electron-pair bond 3 breaks its bond and fills hole 2, thus creating hole 3.

With this action repeated throughout the germanium from right to left, there is, in effect, a continuous displacement of a hole through the material from right to left (positive to negative). Simultaneously, there is a movement of an electron from left to right (negative to positive).

When a hole reaches the negative end of the germanium, an electron from the negative terminal of the voltage source enters the germanium to fill the hole. At the same time, an electron from an electron-pair bond at the opposite (positive) end of the germanium breaks its bond and enters the positive terminal of the voltage

comes a negative ion (Fig. 4-5B). Second, the germanium atom (atom 1) that has lost the electron from bond 1 is left with only three valence electrons. Thus, it becomes a positive ion and a "new" hole is created in bond 1.

As a result of the actions just described, each negative indium ion produced within the germanium is counteracted by a corresponding positive germanium ion. This gives the entire mass of the p-type germanium a net charge of zero. Since, in p-type germanium, the impurity, whether it is indium, gallium, or some other

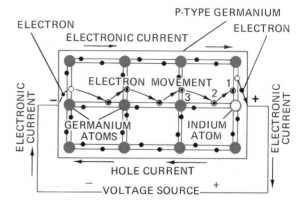

Fig. 4-6. Migration of electrons and holes through p-type germanium.

source. This action creates another hole at the positive end of the germanium, allowing the electron-hole displacement process to continue.

In actual practice, there are, of course, a great number of electrons and holes in action at any given instant, the exact number being dependent upon the nature of the germanium and the magnitude of the applied voltage. This produces a continuous flow of electrons (electronic current) in the external circuit consisting of the conductors that connect the germanium to the battery. However, it is very important to note that this electronic current is accompanied by a hole current moving through the germanium in the opposite (positive to negative) direction.

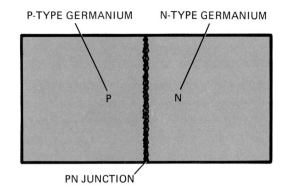

Fig. 4-7. A p-n junction.

PROPERTIES AND CHARACTERISTICS OF HOLES

In the study of semiconductor devices, it is most convenient to think of a hole as a specific particle possessing the mass of an electron and an equal but opposite (positive) charge. Holes in motion form an electrical current that, except for the charge polarity of the particles composing the current, is the same as an electronic current. As an example, holes are attracted and repelled by electrostatic and magnetic fields in accordance with the laws of electrical and magnetic attraction and repulsion.

In addition to the opposite charge that they possess, holes differ from electrons in two important respects. First, they are found only in semiconductor materials, since their existence depends upon the arrangement of electron-pair bonds in certain crystalline structures. Second, in the field of electronics, the electron is considered to be indestructible, whereas when a hole is filled by an electron, it no longer exists. This fact is supported by evidence that when a given mass of germanium contains an equal number of donor and acceptor impurity atoms, it does not exhibit any of the properties of either n-type or p-type germanium.

THE P-N JUNCTION

Most practical semiconductor devices are constructed by bringing into intimate contact p-type sections and n-type sections of a given semiconductor material. This is done by either fusing the sections or by adding an acceptor impurity and a donor impurity to different areas of a single body of material. The resulting point, or surface, of contact between any single p section and an n section is called a p-n junction (Fig. 4-7).

The Depletion Region. When the p-n junction is formed, free electrons within the n section diffuse (spread out) across the junction to combine with (fill) holes that are near the junction in the p section. At the same time, holes from the p section diffuse across the junction to combine with free electrons that are within the n section near the junction (Fig. 4-8A). As a result of these combinations, the atoms near the junction of each section become ionized. Since this action reduces the number of n-section electrons and p-section holes, the area where the process of ionization occurs is called a *depletion region*.

Potential Barrier. Because of the electron-hole combinations within the depletion region, atoms within the p section gain electrons and become negatively charged (negative ions), while atoms within the n section lose electrons and become positively charged (positive ions). This condition creates a small voltage across the junction that is referred to as a *potential barrier* (Fig. 4-8B). Because of the polarity of the potential barrier, there is no further significant diffusion of electrons from the n section into the p section or of holes from the p section into the n section.

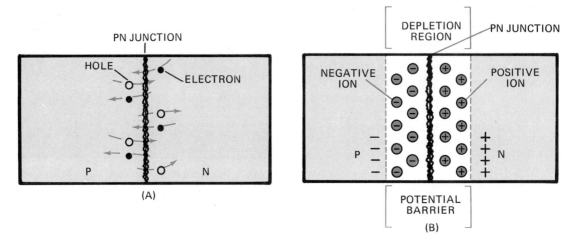

Fig. 4-8. Electron and hole activity at a p-n junction: (A) diffusion of electrons and holes through the junction, (B) creation of the potential barrier.

PHOTOCONDUCTIVITY

Photoconductivity is the process whereby the resistance of certain semiconductor materials such as silicon, germanium, lead sulfide, and cadmium sulfide decreases when exposed to rays of light. The energy imparted to these materials by light has the effect of weakening the bonds between each atom's nucleus and the electrons contained within its outer shell. As a result, there is an increase in the number of free electrons within the material and, therefore, a decrease in the amount of its resistance. This light-response characteristic is utilized in a device known as a *photoconductive cell*. Such cells are commonly employed in conjunction with relays, cameras, lighting circuits, and other circuits where it is necessary or desirable to control circuit operation by means of changes in light intensity.

The construction of a typical broad-area cadmium-sulfide photoconductive cell is shown in Fig. 4-9. This and other types of photoconductive cells are also often referred to as

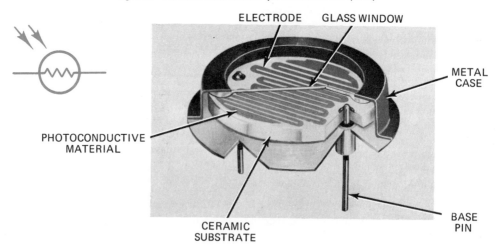

Fig. 4-9. Broad-area cadmium-sulfide photoconductive cell. (*RCA*)

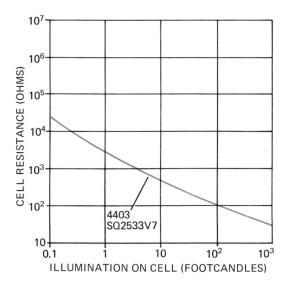

Fig. 4-10. DC cell resistance of 1-inch diameter broad-area photoconductive cells as a function of cell illumination. One footcandle is the direct illumination appearing upon a surface located one foot from the light source of an international candle. (RCA)

photoresistors or light-dependent resistors. A graph illustrating the relationship between the resistance of the cadmium-sulfide cell and the intensity of the illumination to which it is subjected is shown in Fig. 4-10.

THE PHOTOVOLTAIC CELL (PHOTODIODE)

A photovoltaic or solar cell is a semiconductor device that converts light energy into electrical energy. A common type of photovoltaic cell is made from silicon, usually in the form of a square or a rectangular wafer-like unit (Fig. 4-11).

The silicon photovoltaic cell is basically a p-n junction structure consisting of a p-type silicon base material in which a thin layer of n-type material is diffused (Fig. 4-12). A voltage (the barrier potential) exists across the junction because of the ionization of atoms within the depletion region. This is true of all p-n junctions. When light strikes the surface of the cell, the energy of the light is imparted to electrons and holes within the depletion region, causing

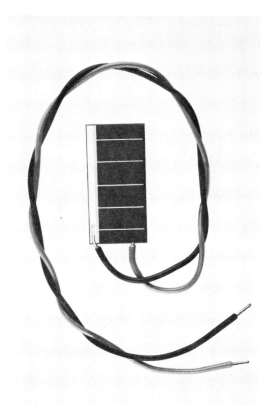

Fig. 4-11. Silicon photovoltaic cell. (RCA)

Fig. 4-12. Displacement of charges in a silicon photovoltaic cell.

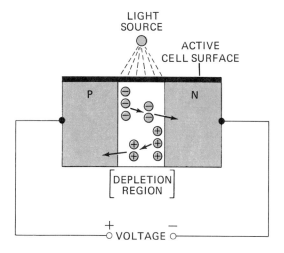

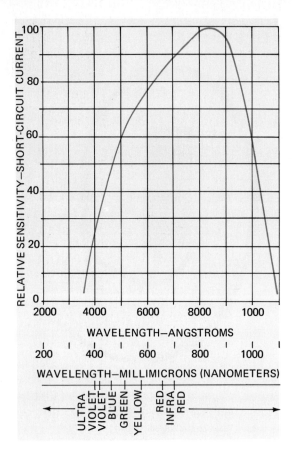

Fig. 4-13. Spectral sensitivity characteristic of a silicon photovoltaic cell (one angstrom is equivalent to one hundred-millionth of a centimeter). (*RCA*)

and greater power outputs are required, a number of cells can be connected in series or in series-parallel combinations.

The relative sensitivity of the cell, in terms of its short-circuit current-delivering capacity, is referred to as the *spectral response characteristic* and is a function of the wavelength of the light that strikes the active cell surface. This characteristic of a silicon cell under conditions of equal light intensities of different wavelengths is shown in Fig. 4-13.

Since chemical changes do not occur within a photovoltaic cell, it has an extremely long life under normal operating conditions. However, the low-voltage output and large surface areas that are necessary if the cell (or battery) is to have a practical current-delivering capacity limit its use to applications where space is not a critical factor.

An important and dramatic application of the silicon photovoltaic cell is its use as an energy source in space vehicles (Fig. 4-14). Such cells are also used in motion picture projectors, cameras, transistor radio receivers, data-processing equipment, alarm systems, and a wide variety of light-actuated, relay-controlled circuit systems.

them to be liberated. Since the electrons are under the influence of the negative "side" of the barrier potential (voltage), they are repelled into the n section of the structure. The positive "side" of the barrier potential repels holes in like manner into the p section. As a result of the displacement of charges, a voltage is developed across the p and n sections.

The direct voltage output of a typical silicon photovoltaic cell in bright sunlight ranges from approximately 0.5 to 0.8 volt per cell. Under this operating condition, the cell will deliver from 30 to 50 milliwatts (mW) of power per square inch of exposed surface area if it is connected to a low-resistance load. When higher voltage

For Review and Discussion

1. Define a semiconductor material in terms of its resistance characteristic.
2. Describe the lattice structure of pure germanium in its solid state.
3. State at least three ways electrons can be freed within a typical semiconductor.
4. What is meant by the process of "doping" a semiconductor?
5. Name at least four pentavalent impurity elements that act as "donors."
6. What is meant by n-type germanium or silicon?
7. Name at least two trivalent impurity elements that act as "acceptors."
8. Define a hole.

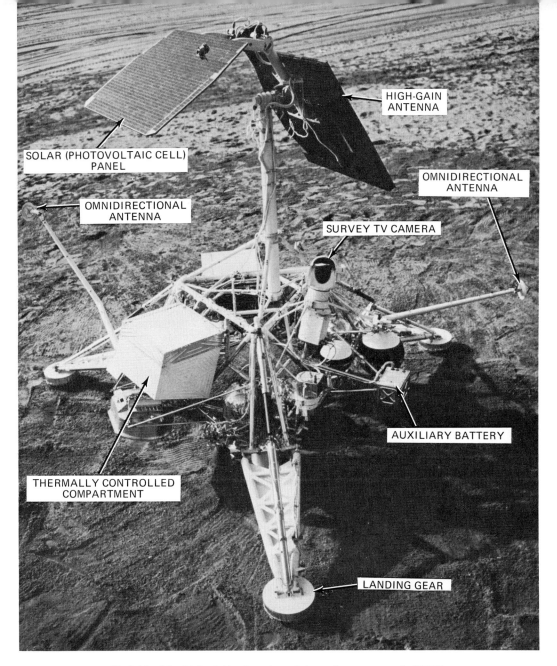

Fig. 4-14. Solar (photovoltaic) cell panel mounted upon Surveyor spacecraft. (NASA)

9. Explain how a hole is formed within a p-type semiconductor that has been doped with an acceptor impurity.
10. Describe the conduction of current in p-type germanium (or silicon).
11. Describe what is meant by the depletion region of a p-n junction.
12. Explain how a potential barrier is generated across a p-n junction.
13. What is meant by photoconductivity?
14. What is a photoconductive cell?
15. Explain the basic operation of a photovoltaic cell.

UNIT 5. Ohm's Law and Electrical Power

In a circuit where the only opposition to current is resistance, a definite relationship known as *Ohm's* law exists between the magnitude of the applied voltage E, the magnitude of the current I, and the resistance R. Since the current results from application of the voltage, it is directly proportional to the magnitude of the voltage (Fig. 5–1A). If the voltage applied to a circuit remains constant, the current will be reduced if the resistance is increased (Fig. 5–1B). The magnitude of the current, therefore, is inversely proportional to the amount of resistance.

OHM'S LAW FORMULAS

The relationships between the three basic quantities in any electrical circuit (voltage, current, and resistance) are conveniently expressed by stating them in terms of mathematical formulas. These formulas are

$$E = IR$$

$$I = \frac{E}{R}$$

$$R = \frac{E}{I}$$

where E = voltage, in volts
 I = current, in amperes
 R = resistance, in ohms

Problem 1. Compute the voltage that must be applied to a buzzer if its operating current is 0.65 ampere and its resistance is 9.2 ohms.

Solution. Since voltage is the unknown quantity, use the formula $E = IR$. Thus

$$E = 0.65 \times 9.2 = 5.98 \text{ volts}$$

Problem 2. The resistance of the coil in a motor is 18 ohms. Compute the magnitude of the current in the motor circuit when it is connected to a 6-volt battery.

Solution. Since current is the unknown quantity, use the formula $I = E/R$:

$$I = \frac{6}{18} = 0.333 \text{ ampere}$$

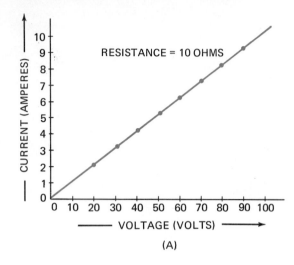

(A)

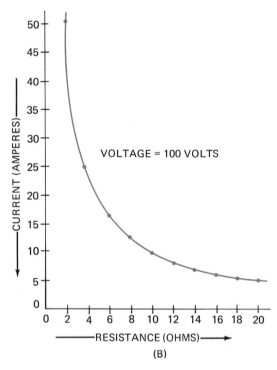

(B)

Fig. 5-1. Graphical illustrations of Ohm's law: (A) direct (linear) relationship between current and voltage, (B) inverse relationship between current and resistance.

Problem 3. The dial lamp in a 12-volt automobile radio has a current rating of 0.16 ampere. Compute the resistance of the lamp filament while the lamp is in operation.

Solution. Using the formula $R = E/I$,

$$R = \frac{12}{0.16} = 75 \text{ ohms}$$

ELECTRICAL POWER

Power is the rate at which work is performed, or the rate at which energy is expended. Work is often expressed in joules. In electrical terms, one joule of work is accomplished when a voltage of one volt causes one coulomb of electrons to pass through a circuit. When this amount of work is accomplished in one second, it is equal to one watt, the basic unit of power in the metric system. Thus, one watt is equal to one joule per second. However, since one coulomb per second is one ampere of current, one watt may also be defined as the amount of work that is accomplished when a voltage of one volt causes one ampere of current to pass through a circuit. This relationship between power P, voltage E, and current I is expressed by the following formula:

$$P = EI$$

where P = power, in watts
E = voltage, in volts
I = current, in amperes

As electrons collide within a conductor, their kinetic (motion) energy is transformed into heat energy. The amount of heat produced is directly proportional to the resistance of the conductor and to the square of the current flowing through it. This relationship between power, current, and resistance is expressed by the formula

$$P = I^2R$$

where P = power, in watts
I = current, in amperes
R = resistance, in ohms

A third power formula is derived from the basic power formula $P = EI$ as follows:

$$P = EI$$

but

$$I = \frac{E}{R} \quad \text{(Ohm's law)}$$

then

$$P = E \times \frac{E}{R}$$

or

$$P = \frac{E^2}{R}$$

Problem 4. A 1,100-watt heating appliance has a voltage rating of 120 volts. Compute the current that passes through the appliance circuit when it is operated at the rated voltage.

Solution. Since the known quantities in the circuit are power and voltage, start with the formula

$$P = EI$$

To solve for I, divide both sides of the formula by E:

$$I = \frac{P}{E}$$

Then

$$I = \frac{1{,}100}{120} = 9.17 \text{ amperes}$$

Problem 5. Compute the power dissipated by a circuit that has a resistance of 63 ohms, when 0.5 ampere of current passes through the circuit.

Solution. Since current and resistance are the known quantities, use the formula $P = I^2R$:

$$P = I^2R$$
$$= (0.5)^2 \times 63$$
$$= 0.25 \times 63 = 15.75 \text{ watts}$$

Problem 6. A 12-volt battery is used to operate a horn circuit that has a resistance of 3 ohms. Compute the power dissipated by the circuit when it is in operation.

Solution.

$$P = \frac{E^2}{R}$$
$$= \frac{12^2}{3}$$
$$= \frac{144}{3} = 48 \text{ watts}$$

Table 5-1. Derivation of the formulas given upon the Ohm's law and power formula memory-aid circle.

FORMULA	HOW DERIVED	FORMULA	HOW DERIVED
1. $E = IR$	Basic Ohm's law formula	7. $R = P/I^2$	$P = I^2R$ (Formula 10)
2. $E = \sqrt{PR}$	$P = E^2/R$ (Formula 11) By cross-multiplying, $E^2 = PR$ By taking the square root of both sides of this formula, $E = \sqrt{PR}$		By dividing both sides of this formula by I^2, $R = P/I^2$
		8. $R = E^2/P$	$P = E^2/R$ (Formula 11) By cross-multiplying, $PR = E^2$ By dividing both sides of this formula by P, $R = E^2/P$
3. $E = P/I$	$P = EI$ (Formula 12), basic power formula By dividing I into both sides of this formula, $E = P/I$	9. $R = E/I$	$E = IR$ (Formula 1) By dividing both sides of this formula by I, $R = E/I$
4. $I = E/R$	$E = IR$ (Formula 1) By dividing R into both sides of this formula, $I = E/R$	10. $P = I^2R$	$P = EI$ (Formula 12) and $E = IR$ (Formula 1) By substituting IR for E in Formula 12, $P = IRI$ or $P = I^2R$
5. $I = P/E$	$P = EI$ (Formula 12) By dividing E into both sides of this formula, $I = P/E$	11. $P = E^2/R$	$P = EI$ (Formula 12) and $I = E/R$ (Formula 4) By substituting E/R for I in Formula 12, $P = E \times E/R$ or $P = E^2/R$
6. $I = \sqrt{P/R}$	$P = I^2R$ (Formula 10) By dividing R into both sides of this formula, $I^2 = P/R$ By taking the square root of both sides of this formula, $I = \sqrt{P/R}$	12. $P = EI$	Basic power formula

LIMITATIONS OF OHM'S LAW AND THE POWER FORMULAS

The use of Ohm's law and the power formulas given in the preceding paragraphs is limited to those circuits in which ohmic resistance is the only significant opposition to current. This includes all direct-current circuits and those alternating-current circuits that do not contain a significant amount of inductance and/or capacitance.

The effects of inductance and capacitance in alternating-current circuits are discussed in later units. In these circuits, Ohm's law and the power formulas must be modified to include oppositions to current that are referred to as *inductive reactance* and *capacitive reactance*.

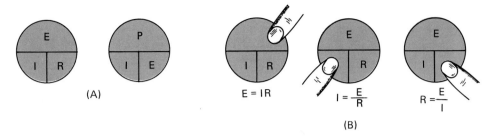

Fig. 5-2. Ohm's law and power circles and how they are used.

MEMORY AIDS

Since Ohm's law and the power formulas are so important in the solution of many electrical/electronic problems, they should be remembered and thoroughly understood. As an aid in remembering and using these formulas, it is often helpful to use circles divided and identified as shown in Fig. 5-2A. These aids are used by placing a finger over the quantity (E, I, R, or P) that is unknown and is to be computed. The uncovered quantities then indicate what the unknown quantity is equal to, as illustrated by the Ohm's law circles shown in Fig. 5-2B.

A memory-aid circle in which Ohm's law and the power formulas are combined is shown in Fig. 5-3. The mathematical processes used in deriving all but the basic formulas are indicated in Table 5-1. A thorough understanding of these processes will be most helpful not only in the use of Ohm's law and the power formulas but also in solving for unknown quantities that are stated in many other formulas.

OHM'S LAW CHART AND CALCULATOR

The Ohm's law chart provides a convenient method of graphically solving typical Ohm's law problems. An example of such a chart and the directions for using it are shown in Fig. 5-4. In addition to Ohm's law, this chart can be used to determine circuit power in watts. Another

Fig. 5-3. The Ohm's law and power formula memory-aid circle.

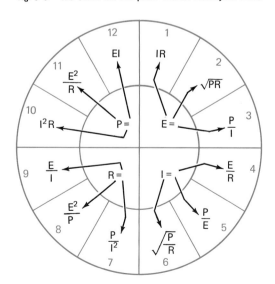

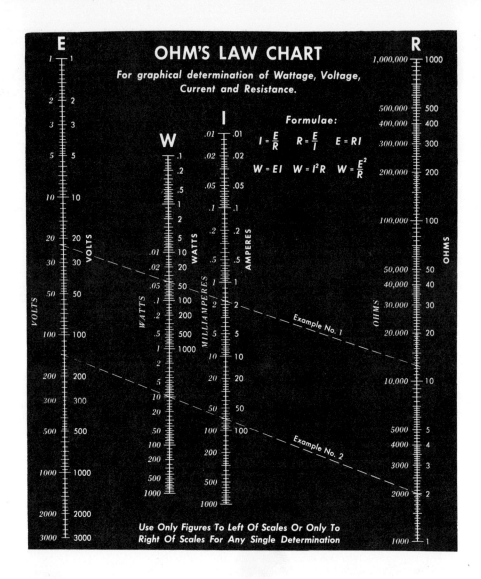

HOW TO USE THIS OHM'S LAW CHART

This alignment chart enables graphical solution of Ohm's Law problems. To use, place a ruler across any two known values on the chart; the points at which the ruler crosses the other scales will show the unknown values. The *italic* figures on the left of the scales cover one range of values and the figures on the right of the scales cover another range. For a given problem, all values must be read on the left set or right set of numbers only, as required.

Example No. 1: The current through a 12.5 ohm resistor is 1.8 amperes. What is the voltage across it? The wattage? Answer: Dotted line No. 1 through R = 12.5 and I = 1.8 shows E to be 22.5 volts and W to be 40.5 watts.

Example No. 2: What is the maximum permissible current through a 10 watt resistor of 2000 ohms? Answer: Dotted line No. 2 through W = 10 and R = 2000 shows I to be 70 milliamperes.

Fig. 5-4. The Ohm's law chart. (*Ohmite Manufacturing Co.*)

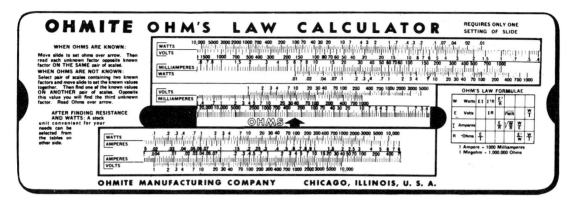

Fig. 5-5. The Ohm's law calculator. (*Ohmite Manufacturing Co.*)

device which provides a convenient method of solving typical Ohm's law and power problems is the Ohm's law calculator (Fig. 5-5). This and other similar calculators have the operating instructions printed on them.

For Review and Discussion

1. Name the three electrical quantities that are involved in any expression of Ohm's law. What letter symbols are used to represent these quantities?
2. State the three Ohm's law formulas.
3. Explain what is meant by the direct relationship between the voltage and the current in a circuit.
4. Explain what is meant by the inverse relationship between the current and the resistance in a circuit.
5. Define electric power.
6. Name and define the basic unit of power measurement.
7. Verbally express the relationship between the amount of heat produced within a conductor, the current through the conductor, and the resistance of the conductor.
8. State the three power formulas.
9. Show how the power formula $P = E^2R$ is derived from the basic power formula $P = EI$.
10. Describe the two general kinds of circuits to which the applications of the basic forms of Ohm's law and the power formulas are limited.

UNIT 6. Resistance and Resistive Devices

Although resistance is present in all but superconductive circuits, it may or may not perform a useful function. Because resistance can result in the waste of energy in the form of heat, it is not a desirable feature of conductors that are used to transmit electrical energy from one point to another. In many circuits, however, resistance is not only desirable but also necessary. Under these conditions, resistance is deliberately placed into circuits in the form of heating elements or devices known as *resistors*. The effect of resistance upon the operation of a circuit and the methods of computing the total resistance of circuits are discussed in detail in Unit 15.

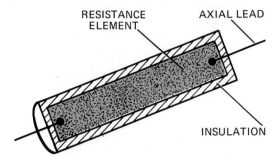

Fig. 6-1. Construction of the carbon (composition) resistor.

RESISTORS

Resistors are used to control the magnitude of current and voltage. Because of their versatility in performing these functions they are found in almost all electronic circuits.

Resistors are designed to have either a fixed resistance value or one that can be changed, and hence are generally classified as either "fixed" or variable.

Carbon (composition) Resistors. The carbon resistor (also called a composition resistor) is the most common type used in typical circuit applications. In this device the resistance element is composed of fine particles of carbon or graphite mixed into a ceramic-type core referred to as a "slug" (Fig. 6-1). The complete resistor consists of an insulating case, a core mixture, and leads, which protrude through the insulation. By varying the density of the carbon or graphite content within the core mixture, resistance values ranging from a fraction of an ohm to millions of ohms can be obtained. By varying the size of the resistor, its power rating can also be controlled (Fig. 6-2).

Because it is extremely difficult to precisely control the content of the mix used in the typical carbon resistor, such resistors are manufactured in accordance with established tolerance levels. Thus, the resistance of a given resistor is stated as being ± 5, ± 10, or ± 20 percent of the resistance value that is indicated in color on the resistor (see page 46).

Metal-glaze Resistors. In metal-glaze resistors, the resistance element consists of a film (layer) of glass and metal fused to a crystalline ceramic core, or substrate (Fig. 6-3). This type of construction makes possible approximately the same range of resistance values as those that are obtained with carbon resistors. However, the exact nature and thickness of the film can be easily controlled; therefore, these resistors have a resistance value very close to the rated value.

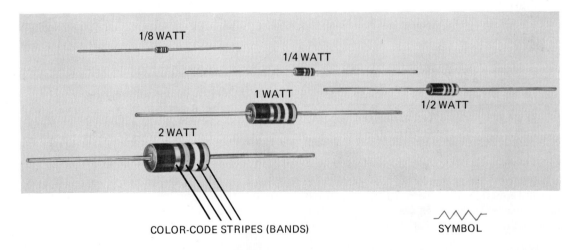

Fig. 6-2. Carbon resistors showing different wattage-rating sizes and color-code stripes. (*Ohmite Manufacturing Co.*)

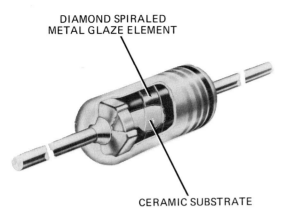

Fig. 6-3. Metal glaze resistor. (IRC, Inc.)

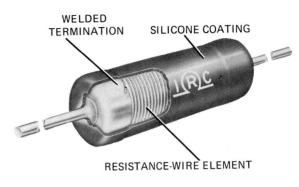

Fig. 6-5. Precision wire-wound resistor. (IRC, Inc.)

Resistors having this constructional feature are known as precision resistors. Another advantage of metal-glaze resistors, as compared with carbon resistors, is that their resistance remains considerably more stable under operating conditions subject to significant changes in temperature and humidity.

Glass-Tin Oxide Resistors. The glass-tin oxide resistor (Fig. 6-4) is manufactured by fusing tin oxide to the surface of a glass form (the substrate). This construction provides for excellent resistor stability under severe conditions of temperature and humidity.

Precision Wire-wound Resistors. The construction of a precision wire-wound resistor is shown in Fig. 6-5. Resistors of this type are manufactured in a wide range of resistance values, with tolerances extending as low as ±0.05 percent. Because of this and also because of relatively small size, they are widely used where accuracy and space factors are important circuit design considerations.

Power Resistors. The term "power resistors" is commonly applied to a wide variety of wire-wound units that are designed for heavy-duty applications. In the typical power resistor, the resistance wire, which is usually either a nickel-chromium or a copper-nickel alloy, is wound about a tubular ceramic form. The entire assembly is then covered with a heavy coating of vitreous enamel (Fig. 6-6).

Adjustable Resistors. The adjustable resistor is a form of power resistor that is designed so that any value of resistance within its maximum rating can be obtained. This is done by means of a slider arm that can be adjusted to contact

Fig. 6-4. Principal constructional features of a glass-tin oxide resistor. (Corning Electronics Division, Corning Glass Works)

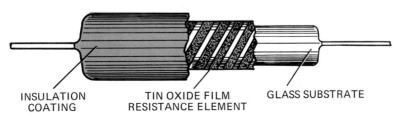

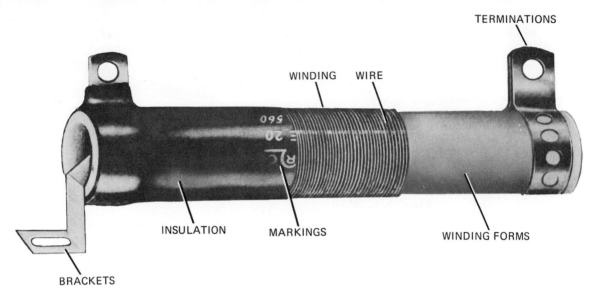

Fig. 6-6. Power resistor. *(IRC, Inc.)*

the resistance wire at any point between the end terminals (Fig. 6-7). Adjustable resistors are used in circuits where a variation in resistance is not often required and in circuits where a final change or adjustment must be made on the existing resistance.

Variable Resistors. Variable resistors are used when it is necessary to frequently vary the value of resistance in a circuit. In these devices, a sliding arm (the contact arm) is in contact with a carbon or wire-wound fixed resistance element that is most often circular in shape (Fig. 6-8). As the shaft to which the contact arm is connected is rotated, the resistance between the center (contact arm) terminal and each of the outside terminals is changed.

A variable resistor that is used to control the voltage applied to a circuit or to a portion of a circuit is known as a *potentiometer* or, simply, a control. When the resistor is used primarily to control the magnitude of current through the load in a circuit it is called a *rheostat*. Both of these devices are constructed in a similar man-

Fig. 6-8. Principal constructional features of a wire-wound variable resistor (rheostat). *(Ohmite Manufacturing Co.)*

Fig. 6-7. Adjustable resistor. *(Ohmite Manufacturing Co.)*

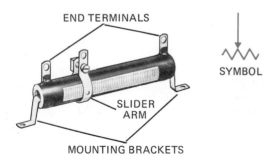

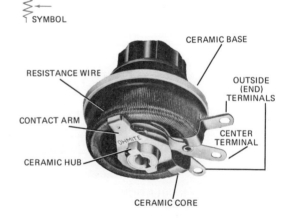

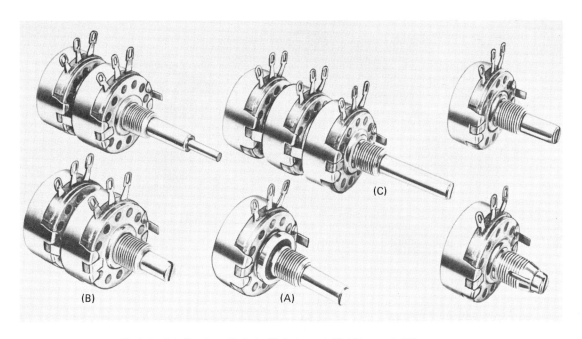

Fig. 6-9. Potentiometers: (A) single, (B) dual-ganged, (C) triple-ganged. (Allen-Bradley Co.)

Fig. 6-10. Potentiometer and rheostat resistance element tapers: (A) linear taper, (B) nonlinear taper.

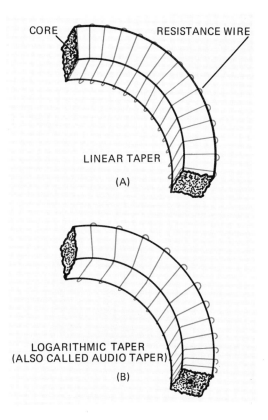

ner. However, since the rheostat is most often designed for heavy-duty applications, its resistance element is usually wire-wound. The resistance element in a potentiometer may be of either the wire-wound or the carbon type (Fig. 6-9).

Tolerance. Standard-use (replacement type) wire-wound power resistors are constructed with a tolerance of ±5 percent if their resistance is 1 ohm or greater. If the resistance is less than 1 ohm, the tolerance is ±10 percent. Adjustable and variable wire-wound resistors (rheostats) are constructed with a tolerance of ±10 percent unless otherwise specified.

Taper. Potentiometers and rheostats are constructed so that they have either a linear or a nonlinear (logarithmic) taper (Fig. 6-10). In the linear taper construction, the resistance change that occurs as the sliding arm is moved over a given distance between any two points of the fixed element is uniform.

In the nonlinear taper construction, the resistance change of the fixed element is not uniform, but varies from one end of the element to the other. Therefore, the resistance change varies as the sliding arm is moved over a given distance between any two points.

Both linear and nonlinear taper devices are used to control current and voltage. The most common application of the nonlinear taper potentiometer is its use as a volume control in audio amplifiers or in conjunction with loudspeaker system circuits. Use of the nonlinear taper control device is desirable in these circuits, since the human ear has an essentially logarithmic response. This means that the ear responds to different levels of sounds in direct proportion to the logarithms of the energies required to produce the sounds. For example, assume that the energy supplied to a loudspeaker is increased from a level of 10 to a level of 100 (the logarithm of 10 is 1, while the logarithm of 100 is 2). The ear would recognize this tenfold increase of energy as a sound only twice as loud as the sound originally produced by the loudspeaker.

The Trimmer Resistor. The trimmer resistor, also commonly referred to as a precision potentiometer, is a variable resistor used to make extremely small changes in the resistance of a circuit. The constructional features of one type of trimmer resistor are shown in Fig. 6-11.

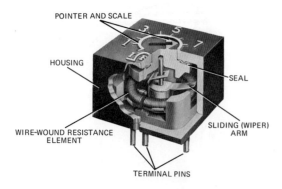

Fig. 6-11. Trimmer resistor. (IRC, Inc.)

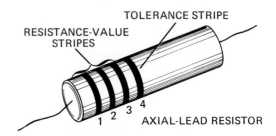

Fig. 6-12. Carbon resistor color code.

THE RESISTOR COLOR CODE

The resistance in ohms of all resistors except those of the carbon composition type is usually written upon the resistors. The resistance of fixed carbon resistors is most often indicated by means of a color code. This code is also used to show the tolerance limits of a resistor (Fig. 6-12).

In the resistor color-code system, the first three color stripes (bands) indicate the value of

Table 6-1. Color-numerical equivalent scheme used in conjunction with the carbon composition resistor color code.

FIRST THREE STRIPE COLOR-NUMBER CODE	
Black	= 0
Brown	= 1
Red	= 2
Orange	= 3
Yellow	= 4
Green	= 5
Blue	= 6
Violet	= 7
Gray	= 8
White	= 9
Gold	= Divide by 10
Silver	= Divide by 100
FOURTH STRIPE COLOR-TOLERANCE CODE	
Gold	= ±5%
Silver	= ±10%
None	= ±20%

the resistance. If there is a fourth stripe, its color indicates the tolerance rating of the resistor.

The color-number scheme used in conjunction with the color code is given in Table 6-1. To determine the resistance of any given resistor, the color code is used as follows:

1. The color of the first stripe indicates the first number of the resistance value.
2. The color of the second stripe indicates the second number of the resistance value.
3. The color of the third stripe indicates the multiplier. With the exception of the colors silver and gold, the color of this band indicates the number of zeros that are to be added following the first two significant figures of the resistance value. When silver is used as the third color band, the multiplier is 0.01. Gold indicates a multiplier of 0.1.

The color codes and respective resistance values of five resistors are shown in Fig. 6-13. A study of these resistors will be helpful in learning the application of the color code in determining the resistance and tolerance values of any carbon composition resistor.

THE COLOR CODER

The color coder is a device that provides a convenient means of determining the resistance of carbon composition resistors (Fig. 6-14). The device is used by first adjusting the three color "wheels" from left to right so that their colors correspond to the first three color stripes of a given resistor. The resistance value of the resistor is then indicated through the "windows" in the section marked "OHMS."

STANDARD RESISTANCE VALUES

Resistors of all types are manufactured in only a limited number of resistance values. These are often referred to as the standard resistance values. Some standard values are listed in Table 6-2. From this table it is seen that a carbon composition resistor having a resistance value of 26 ohms is not a stock item. However, in most

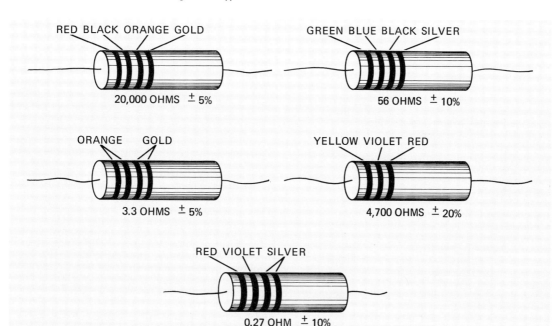

Fig. 6-13. Application of the carbon resistor color code.

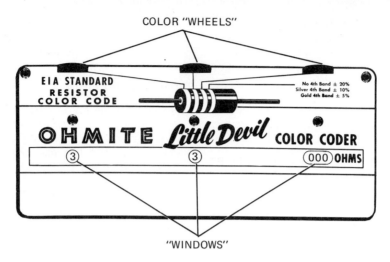

Fig. 6-14. Carbon resistor color coder. (*Ohmite Manufacturing Co.*)

practical circuits, a resistor of 24 or 27 ohms would be a satisfactory substitute for a resistance of 26 ohms. In other circuits where a precise value of resistance is required, the necessary resistor or resistors can be obtained from a manufacturer by special order.

Table 6-2. Typical standard or preferred values of resistors.

STANDARD (PREFERRED) RESISTANCE VALUES

Ohms	Ohms	Ohms
0.24	110	56K
0.36	180	100K
0.56	470	220K
1.1	560	270K
1.8	1000	470K
4.3	2200	1M
6.2	2400	1.6M
10	3600	2M
22	7500	4.7M
24	11K	5.6M
27	18K	10M
100	47K	22M

POWER RATING OF RESISTORS

In addition to resistance value, the operational characteristics of a resistor are also stated in terms of its power (or wattage) rating, expressed in watts. The power rating of a resistor is a measure of its ability to dissipate or "throw off" the heat produced by the current it conducts. The amount of this heat is directly proportional to the square of the current ($P = I^2R$). For this reason, even a relatively small increase in current can produce a damaging amount of heat if the power rating of the resistor used in a particular circuit is not sufficient.

Excessive heating causes a resistor to burn out or, in the case of a carbon resistor, it can cause the resistance to increase far beyond the normal tolerance limits. Therefore, a resistor that shows evidence of having been overheated (surface scorched) should be tested with an ohmmeter before it is put back into service.

Physical Size. Although the physical size of a resistor has no relationship to resistance value, it does provide an indication of the power rating. Since the current-carrying capacity of the conducting material within a resistor is dependent upon its cross-sectional area, a larger resistor of any given type will have a higher power (wattage) rating than a smaller one (see Fig. 6-2, page 42).

Safety Factor. In actual practice and under ordinary conditions, a power safety factor of at least 100 percent should be provided when a resistor is selected for use in a circuit. In other words, the resistor to be used should have a power rating approximately two times greater than the actual power dissipated by the resistor, as determined by the power formula. If a resistor is located within a tightly enclosed or otherwise unventilated space, a power rating of three or four times the computed value may be necessary.

Problem. The design of a given unenclosed circuit structure indicates that 0.15 ampere of current will pass through a 250-ohm resistor in the circuit. Compute the wattage rating of the resistor to be used if a safety factor of 100 percent is to be provided.

Solution.
$$P = I^2 R$$
$$= (0.15)^2 \times 250$$
$$= 0.0225 \times 250 = 5.625 \text{ watts}$$

Since 5.625 watts is the actual power dissipation of the resistor under the specified condition, a 10-watt resistor will provide the minimum safety factor. However, it would be more desirable to use a 12-watt resistor in the circuit. This will probably be a wire-wound power resistor because other types of resistors are seldom available with a power rating of this magnitude.

THE THERMISTOR

The *thermistor* (a combination of the words *thermal* and *resistor*) is a temperature-sensitive device with a high negative temperature coefficient (Fig. 6-15). While several different types

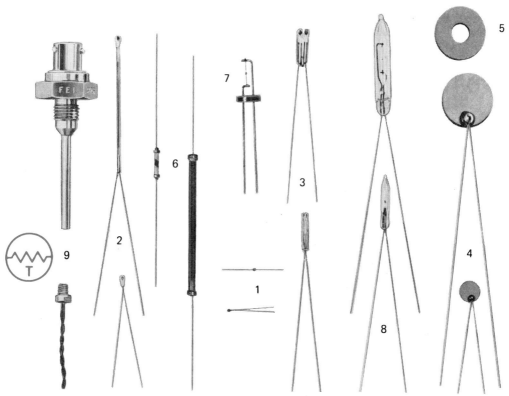

Fig. 6-15. Common types of thermistors. (*Fenwal Electronics, Inc.*)

1. BEADS (GLASS-COATED)
2. GLASS PROBES
3. INTERCHANGEABLE PROBES AND BEADS
4. DISKS
5. WASHERS
6. RODS
7. SPECIALLY MOUNTED BEADS
8. VACUUM AND GAS-FILLED PROBES
9. SPECIAL PROBE ASSEMBLIES

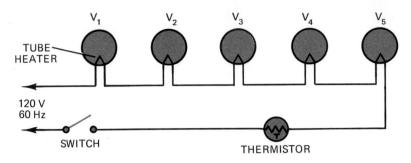

Fig. 6-16. Application of a thermistor in an electron tube heater circuit.

of thermistors have been developed, the resistance element in most of them is a mixture of metallic oxides, including oxides of nickel, manganese, cobalt, and iron.

An important application of the thermistor is its use as a current-control device in circuits where sudden surges of current can produce an overloaded or unstable operating condition. An example of this application is the use of a thermistor in an electron-tube heater circuit (Fig. 6-16). If the thermistor is not in the circuit, the relatively low resistance of the "cold" heaters will allow a surge of current to pass through them when the switch is first turned on. The magnitude of this current is often great enough to cause one of the heaters to burn out. With the thermistor in the circuit, the initial current is limited because of the high resistance of the thermistor when it is not heated. As current through the circuit continues, the heater resistance increases, while the resistance of the thermistor decreases as it becomes heated. Thus, the net effect of this action is to cause the circuit current to rise gradually until it has reached its normal value.

In addition to their use as current-control or temperature-compensation devices in electron-tube circuits, thermistors are used for the same purpose in transistor bias-stabilization circuits and in motor winding circuits.

Thermistors are also commonly used as heat sensors in temperature-measurement applications. In such circuits one or more thermistors are operated in conjunction with a resistance network or resistance bridge, which actuates some form of meter mechanism that serves as the final temperature-indicating device. A resistance-temperature curve illustrating the resistance response of a typical thermistor used as a heat sensor is shown in Fig. 6-17.

Fig. 6-17. Curve showing the resistance of a typical thermistor at various temperatures.

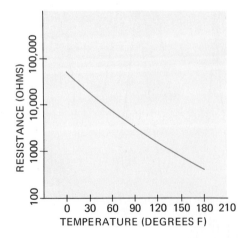

For Review and Discussion

1. State three uses of resistance devices.
2. What is meant by the tolerance of a resistor?
3. Define a precision resistor.
4. Describe the construction of a typical variable resistor.
5. Explain the difference between a linear and a nonlinear taper characteristic of a variable resistor.
6. For what purpose is a trimmer resistor used?

7. Describe the resistor color code.
8. What is meant by a standard resistance value?
9. Define the power rating of a resistor.
10. State the effects that the overheating of a carbon composition resistor can produce.
11. Explain a 100 percent safety factor as applied to the selection of a resistor.
12. What is a thermistor? Give at least one example of a thermistor application and explain the action of the thermistor in this application.

UNIT 7. Capacitance and Capacitors

Capacitance, or the ability to store an electric charge, is one of the most common characteristics of electronic circuits. Although capacitance is present in almost all operating circuits, it may or may not be useful. In circuits where it is desirable, the necessary capacitance is provided by a device called a *capacitor* (Fig. 7-1).

The capacitor is an extremely versatile device, and can be used to perform a number of different circuit functions. For example, the

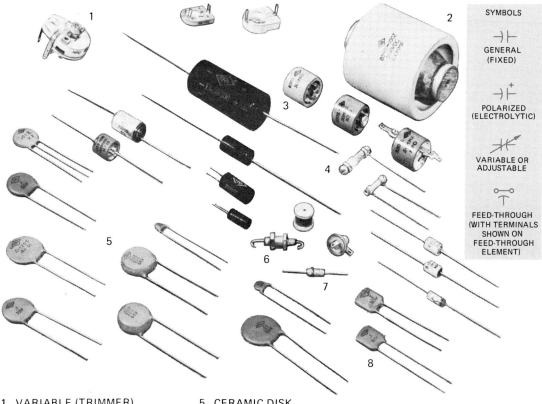

Fig. 7-1. **Common types of capacitors.** (Centralab Electronics Division of Globe-Union, Inc.)

1. VARIABLE (TRIMMER)
2. CERAMIC TRANSMITTING
3. ELECTROLYTIC
4. TRIMMER (TUBULAR)
5. CERAMIC DISK
6. CERAMIC FEED-THROUGH
7. CERAMIC TUBULAR (AXIAL LEAD)
8. CERAMIC FLAT PLATE

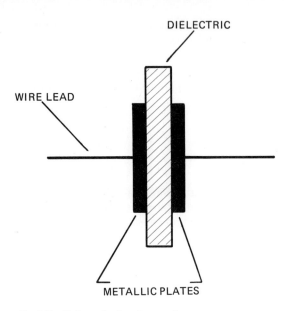

Fig. 7-2. Basic construction of a capacitor.

welding units. Capacitors are also commonly used to block direct current, smooth out variations of direct current, and couple or transfer energy from one point in a circuit to another. They are used in tuning circuits such as those found in radio receivers, in timing circuits, and in filter circuits, by means of which a current of a given frequency can be removed from a point in a circuit. The effects of capacitance upon the operation of circuits and the methods of computing the total capacitance of circuits are discussed in detail in Units 17 and 18.

CAPACITOR OPERATION

In its simplest form, a capacitor consists of two plates that are separated by an insulating material known as the *dielectric* (Fig. 7-2). In practical construction, the plates are made of metal and the dielectric may be any one of a number of materials including air, mica, glass, paper, ceramic, Mylar, oil, or metal oxide.

property of a capacitor that allows it to be charged at one time and discharged at a later time makes the capacitor of special value in the operation of photoflash, flasher signal, and

Basic Action. The basic charging action of a capacitor is illustrated in Fig. 7-3. With a battery connected to the plates of the capacitor, free (mobile) electrons are repelled to plate A, which

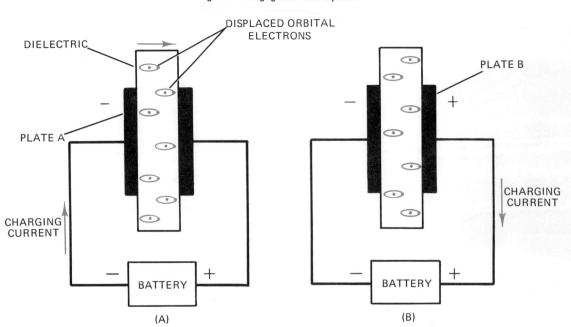

Fig. 7-3. Charging action of a capacitor.

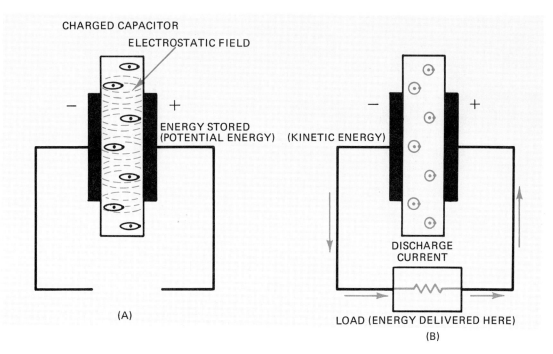

Fig. 7-4. Charge storage and discharge action of a capacitor.

is connected to the negative terminal of the battery. As a result, the negative polarity of this plate causes the electrons within the dielectric to be strained or distorted out of their normal orbital positions by the process of electrostatic induction (Fig. 7-3A). In turn, the polarity of the resulting charge across the dielectric causes electrons to be repelled from plate B. These electrons are then attracted to the positive terminal of the battery (Fig. 7-3B), causing plate B to become positively charged. It is important to note that during this process, free electrons were actually conducted *to* the capacitor plates, not *between* the plates themselves.

As the flow of electrons in the external circuit continues, the voltage developed across the plates of the capacitor increases. Since the polarity of each plate is identical to the polarity of the battery terminal to which it is connected, this voltage has the effect of opposing the applied (battery) voltage. For this reason, the voltage that is developed across the plates of a capacitor while it is charging is called a *counter electromotive force* (cemf), or *back voltage*. When the magnitude of the cemf becomes equal to the voltage of the battery, the flow of electrons in the external circuit ceases and the capacitor is said to be fully charged.

Charge Storage. When the battery is disconnected from the capacitor, excess electrons remain upon the negative plate, while the positive plate remains deficient in electrons. The energy that was supplied by the battery during the charging process is "stored" by the capacitor in the form of an electrostatic field within the dielectric (Fig. 7-4A).

Discharging. If the plates of a charged capacitor are connected with a conductor, the strain upon the electrons in the atomic orbits within the dielectric is released. In this condition, the energy of the electrostatic field that was stored in the dielectric is returned to the circuit, causing the excess electrons from the negative plate to flow to the positive plate (Fig. 7-4B). This flow of electrons, opposite in direction to the flow that occurred during the charging process, continues until the charges on both plates are neutralized (discharged).

In theory, a charged capacitor could retain its charge indefinitely. However, since no dielectric is a perfect insulator, excess electrons from the negative plate will, in time, "leak" through the dielectric, thus neutralizing the charges between the plates.

CAUTION

The voltage existing across the plates of a charged capacitor may cause serious electrical shock.

CAPACITIVE REACTANCE AND IMPEDANCE

Capacitors do not conduct in alternating-current circuits, but electrons do enter and leave the capacitor terminals. Since the direction of current flow is continually reversing, the capacitor tends to be constantly charging and discharging. In other words, the capacitor alternately takes energy from and delivers energy to the circuit.

Often, a capacitor will offer a significant opposition to the flow of alternating current. This opposition to current is called *capacitive reactance* and is represented by the symbol X_c. The combined opposition to current in an alternating-current circuit presented by both resistance and capacitive reactance is called *impedance*, which is represented by the letter symbol Z. The unit of measurement for both capacitive reactance and impedance is the ohm.

THE FARAD

The magnitude of the charge across the plates of a capacitor is expressed in terms of the difference in the number of electrons on its plates. When this difference is equal to one coulomb (6.28×10^{18} electrons) and the resulting voltage across the plates of a capacitor is one volt, the capacitor is said to have a capacitance of one farad (F). Stated in another way, a capacitor has a capacitance of one farad when a voltage that changes at the rate of one volt per second produces a current of one ampere in the capacitor circuit.

Although capacitors having a capacitance of one farad have been developed, common capacitors are rated in microfarads (μF) and picofarads (pF). One microfarad equals one-millionth of one farad and one picofarad equals one-millionth of one-millionth of one farad. It is important to note that the term picofarad has replaced the earlier term micromicrofarad ($\mu\mu$F). Thus, one picofarad equals one micromicrofarad.

The capacitance of a capacitor is also sometimes stated as the ratio of the charge upon its plates to the voltage that is applied to the plates. These relationships are expressed by the following formula:

$$C = \frac{Q}{E}$$

where C = capacitance, in farads
Q = charge, in coulombs
E = voltage, in volts

FACTORS AFFECTING CAPACITANCE

The ability to store a charge electrostatically is dependent upon the intensity of the electrostatic field that is generated between the plates of a capacitor. Hence, any condition that tends to increase the intensity of this field when a voltage is applied also increases the capacitance of the capacitor. The constructional features or factors that affect capacitance are the area of the plates, the distance between the plates, and the nature of the dielectric.

Plate Area. In general, a capacitor having a large plate area has a greater capacitance than one with a smaller plate area. The reason for this is that as the plate area is increased, a larger surface of the dielectric can contact the plates, thus increasing the area of the electrostatic field.

Distance Between Plates. A decrease in the distance between the plates of a capacitor produces a greater concentration of the electrostatic field between the plates. As a result, the capacitance is increased.

Nature of the Dielectric. The electrostatic field between the plates of a capacitor exists within the dielectric. The strength of this field is partially dependent upon the ability of the dielectric material to maintain the distorted orbital electron pattern of its atoms. Because different dielectric materials exhibit this property to a different extent, the kind of dielectric material used will affect the capacitance of a given capacitor.

Table 7-1. Dielectric constants of various materials.

Material	Constant
Air	1
Glass	7.5
Mica	5
Paper, Waxed	2.8
Polystyrene	2.6
Teflon	2.3

DIELECTRIC CONSTANT

The measure of the effect that a dielectric material has upon the capacitance of a capacitor, as compared with the capacitance when air is the dielectric, is known as the *dielectric constant K* of the material (Table 7-1). The dielectric constant table indicates, for example, that when mica is used as the dielectric in a given capacitor, the capacitance is from five to seven times greater than it would be if the dielectric was air.

TEMPERATURE COEFFICIENT

The temperature coefficient of a capacitor is a measure of the extent to which its capacitance varies with an increase in temperature. If its capacitance increases with an increase in temperature, the capacitor has a positive temperature coefficient. If the capacitance decreases with an increase in temperature, the coefficient is negative.

The temperature coefficient of a capacitor is usually stated in terms of millionths of a given capacitance change per Celsius degree. Therefore, a positive temperature coefficient of 450 indicates that the capacitance increases by 450/1,000,000 or 0.045 percent for every degree Celsius rise in temperature.

WORKING VOLTAGE

In addition to capacitance, a capacitor's operational characteristics are also stated in terms of the maximum direct voltage that can be applied to it without causing an excessive leakage of current through the dielectric. This characteristic is known as the *direct-current working voltage* (WVDC) *rating* of the capacitor.

The working voltage rating of a capacitor is dependent upon the type and thickness of its dielectric material. Thus, if two capacitors of the same type and of equal capacitance are compared, the one having the higher working voltage rating will be larger in physical size because a higher working voltage rating necessitates the use of a thicker dielectric.

Cost, like size, increases with an increase in the working voltage rating. If space and cost permit, a capacitor having a higher working voltage rating can always be substituted for one having a lower working voltage rating. Electrolytic capacitors, however, should be operated near their voltage rating to insure that their capacitance will be maintained.

Puncture. An excessive leakage current through a capacitor will often cause it to become overheated. If the excessive leakage continues for a period of time, the abnormal amount of heat produced will cause the dielectric to be punctured. This, in turn, usually results in a "short" between the plates of the capacitor. Since, in the majority of circuits a capacitor is connected across a comparatively high voltage, a shorted capacitor can cause an excessive current to pass through associated resistors, coils, and other components, thereby causing them to become overheated and often damaged.

Safety Factor. A capacitor used in a low-frequency circuit should have a working voltage rating of at least twice the actual alternating voltage to which it is connected. As the operating frequency increases, the temperature of the dielectric tends to increase significantly, thus lowering its dielectric strength. Therefore, when a capacitor is used in a high-frequency circuit (above one megahertz), the safety factor should be increased more than 100 percent.

FIXED CAPACITORS (NONELECTROLYTE)

Fixed capacitors have a definite value of capacitance that does not change under normal conditions. These capacitors are further identified by the specific dielectric material that is used in their construction. If the dielectric is formed by an electrochemical process, the capacitor is known as an electrolytic capacitor. The construction and characteristics of electrolytic capacitors are discussed in later paragraphs.

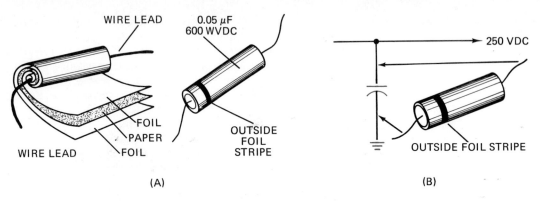

Fig. 7-5. Paper capacitor: (A) constructional features and outside foil stripe, (B) connection of the capacitor to a circuit.

The Paper Capacitor. The paper capacitor consists basically of plates made of aluminum or tin foil and a dielectric of paper that is impregnated (saturated) with an oil or a wax compound (Fig. 7-5A). These and other nonelectrolytic capacitors are nonpolarized devices. This means that they can be connected into a circuit without regard to voltage polarity.

The typical tubular paper capacitor usually has a color stripe located near one end. This indicates the lead of the capacitor that is attached to the outside foil, which can serve as a shield against stray electrostatic and magnetic fields. In practice, this lead is connected to the lower voltage or to the ground side of a circuit (Fig. 7-5B). If shielding is not a factor to be considered, the stripe can be disregarded.

Paper capacitors are manufactured in a capacitance range extending from approximately 0.001 to 1 microfarad. Although these capacitors provide a relatively high capacitance per unit volume, they have a higher temperature coefficient and a shorter service life than most other types of capacitors.

Metallized Capacitors. The metallized capacitor is a variation of the paper capacitor. Its plates are made by depositing a very thin layer of metal upon each side of a paper dielectric by means of an evaporation process. Since this plate material is thinner than metal foil, it is possible to construct metallized capacitors that are smaller in physical size than standard-type paper capacitors having equivalent capacitance values.

Film-type Capacitors. The construction of film-type capacitors is similar to that of paper capacitors but the dielectric is either Mylar, Teflon, or polystyrene film. The use of Mylar as the dielectric provides for a greater capacitance per unit volume, higher working voltage ratings, and smaller leakage current losses than are possible with paper capacitors.

The outstanding characteristic of the Teflon capacitor is its ability to operate satisfactorily under conditions of high temperature. The polystyrene capacitor is especially well suited for applications where a precise capacitance, an extremely long service life, and high reliability are important factors in circuit design. The polystyrene capacitor also has an exceptionally high resistance to leakage currents.

Mica Capacitors. In the mica capacitor, several strips of metal foil (aluminum, tin, or copper) are "sandwiched" between thin sheets of mica, which serves as the dielectric material. In the final assembly, alternate strips of foil are connected to form each plate (Fig. 7-6). The entire unit is then molded into a Bakelite case to keep out moisture.

Mica capacitors range from approximately 1 picofarad to 0.01 microfarad in capacitance. They have a stable temperature coefficient characteristic, a high working voltage rating, and a comparatively long service life. Because of their temperature stability, these capacitors are particularly useful for application in high-frequency circuits.

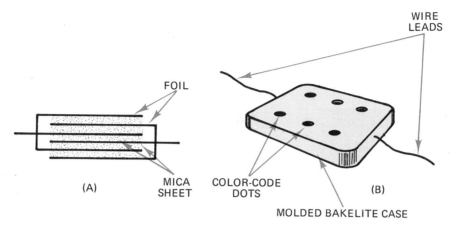

Fig. 7-6. Mica capacitor: (A) constructional features, (B) molded case and color-code dots.

In the silvered-mica capacitor, a silver compound is deposited upon one side of each mica sheet, thus forming the plate. This construction allows a reduction in physical size compared with a standard mica capacitor of equivalent capacitance.

Ceramic Capacitors. The plate material in a ceramic capacitor is a silver compound that is fired or deposited upon the surfaces of a ceramic form that serves as the dielectric. The form, made of titanium dioxide or a silicate compound, is either disk- or tubular-shaped. In the disk-type ceramic capacitor, the silver compound is located on each side of a disk-shaped form (Fig. 7-7).

In the tubular capacitor, the compound is deposited upon the inside and the outside surfaces of a ceramic tube. The completed assembly of both types of capacitors is then coated with a plastic material or some other high-quality insulator compound.

Since the ceramics used in the manufacture of these capacitors have an extremely high dielectric constant, it is possible to construct small units that have a relatively large capacitance value. The ceramics also have a high dielectric strength, thus providing for a high working voltage rating. Because of these advantages, ceramic capacitors having a capacitance range of from 1 picofarad to 0.1 microfarad are widely used in many types of circuits, including the circuits of space vehicle communications equipment, where both the working voltage rating and the physical space requirements are critical factors.

The Glass Capacitor. The glass-dielectric capacitor consists basically of aluminum plates hermetically sealed in a high-quality glass unit (Fig.

Fig. 7-7. The ceramic disk capacitor.

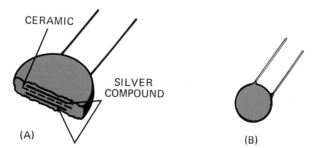

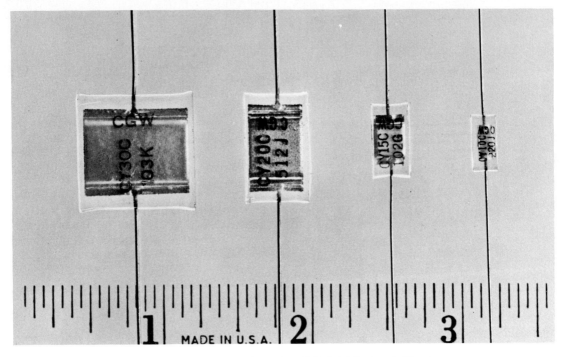

Fig. 7-8. Flat glass-dielectric capacitors (also available in tubular shapes). (*Corning Electronics Division, Corning Glass Works*)

7-8). This construction results in a capacitor that is extremely stable under the most severe operational and environmental conditions. Additional advantages of the glass capacitor are its relatively small size and light weight compared with other types of fixed capacitors having equivalent capacitance values.

ELECTROLYTIC CAPACITORS

Electrolytic capacitors are most commonly manufactured as aluminum "can" units or as tubular-shaped units encased in a paper-covered metallic container or a wax-filled cardboard tube (Fig. 7-9). While these capacitors have higher

Fig. 7-9. Electrolytic capacitors: (A) paper-covered tubular type, (B) "can" type.

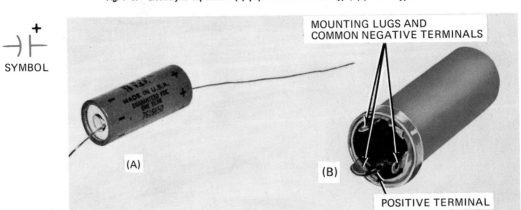

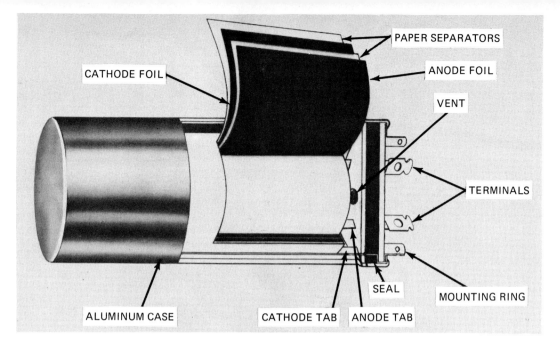

Fig. 7-10. Constructional features of a typical "can"-type aluminum electrolytic capacitor. (*P. R. Mallory and Co., Inc.*)

leakage current than other types of capacitors, they offer the advantages of lower initial cost and space savings, since their construction allows capacitances that usually range from approximately 1 microfarad to several thousand microfarads.

The Aluminum Electrolytic Capacitor. A typical aluminum electrolytic capacitor consists primarily of smooth or etched aluminum foils (the anode and cathode foils) and paper separators rolled into a tubular form that is impregnated with a suitable paste electrolyte and sealed within an aluminum can (Fig. 7-10). During the manufacturing process, the application of a direct "forming" voltage across the capacitor produces an electrochemical action that deposits a very thin coating of aluminum oxide upon the surface of the anode foil. This foil is the positive plate of the capacitor and the aluminum oxide is the dielectric. The exact thickness of the oxide determines the working voltage rating of the capacitor, which generally does not exceed 500 volts.

In addition to serving as the chemical agent, the electrolyte of the capacitor is also the negative plate. Since the electrolyte is in intimate contact with the dielectric, it provides for the high capacitance-to-volume ratio that is the outstanding characteristic of the electrolytic capacitor.

The cathode foil serves as the connecting medium between the electrolyte and the negative terminal, which may be either the aluminum can or an external terminal brought out at the insulator base of the can. The purpose of the paper separators is to reduce the possibility of shorts between the anode and cathode foils and to absorb the electrolyte, thus causing it to be uniformly distributed over all foil areas.

The typical electrolytic capacitor is a polarized device that must be operated under direct-voltage conditions (Fig. 7-11). If such a capacitor

Fig. 7-11. Electrolytic capacitors connected into a circuit.

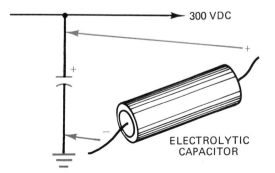

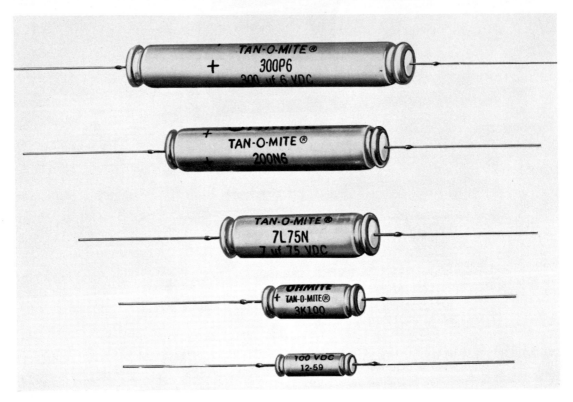

Fig. 7-12. Tantalum-foil electrolytic capacitors. (*Ohmite Manufacturing Co.*)

is connected across a voltage in the wrong way, the resulting electrolytic action will cause the dielectric to deteriorate quite rapidly. During this process, the capacitor becomes heated and often releases a quantity of gas.

Although the electrolytic capacitor operates satisfactorily under relatively severe temperature conditions, an abnormally high temperature sometimes causes the electrolyte to dry out. In this condition, its chemical activity is significantly reduced, with a corresponding decrease in capacitance.

The Tantalum Capacitor. The constructional features of the tantalum-foil electrolytic capacitor are similar to those of the aluminum electrolytic capacitor. In this capacitor, however, the metallic foils are made of tantalum and the electrolyte is an acid (Fig. 7-12).

The oxide coating that is formed upon one foil of the tantalum capacitor during the manufacturing process has almost twice the dielectric constant of aluminum oxide. For this reason, tantalum capacitors provide a considerably greater capacitance per unit volume as compared with the aluminum type. In addition, the tantalum capacitor is more rugged and possesses a greater degree of stability under conditions of varying temperatures. The primary disadvantage of the tantalum capacitor is that it normally has a much lower working voltage rating than an aluminum capacitor of equal capacitance.

The Nonpolarized Electrolytic Capacitor. The nonpolarized electrolytic capacitor is constructed by placing two anode foils (positive plates) within the same container. Such capacitors are used in motor-starting circuits, audio crossover networks, and in other applications where an electrolytic capacitor must be connected across an alternating voltage.

A single nonpolarized capacitor can be formed by connecting two electrolytic capacitors "back-to-back," as shown in Fig. 7-13. When this

Fig. 7-13. Connection of two electrolytic capacitors to form a single nonpolarized capacitor.

arrangement is used, the working voltage rating of each capacitor must be equal to or higher than the voltage of the circuit to which the capacitors are connected. The total capacitance (C_t) of the combination can be determined by using the following formula:

$$C_t = \frac{C_1 \times C_2}{C_1 + C_2}$$

Multisection Capacitors. Electrolytic capacitors are often constructed as multisection units, with each section acting as an individual capacitor. In the "can"-type capacitor, a terminal connected to the can serves as the common (negative) terminal for all the capacitors. The positive terminals are brought out through the insulator base of the container and are identified by the use of special symbols (Fig. 7-14A). The multisection paper-covered tubular capacitor is equipped with color-coded leads, which identify the common terminal and the positive terminals of the various sections (Fig. 7-14B).

VARIABLE CAPACITORS

The typical air-dielectric variable (or tuning) capacitor consists of a set of stationary (stator) plates and a set of movable (rotor) plates arranged so that they can mesh with each other without touching (Fig. 7-15A). The stator plates are mounted within the frame of the capacitor but are insulated from it. The rotor plates are mounted upon a movable shaft, which is also attached to the frame. Since the frame is in physical and electrical contact with the rotor plates, it serves as the terminal point for wire connections to the rotor plate assembly.

By rotating the shaft of the variable capacitor, the position of the rotor plates can be varied so that a smaller or a greater portion of their surface area is directly opposite the stator plates. As a result, the capacitance of the capacitor can be varied from a minimum (plates unmeshed) value to the maximum (plates fully meshed) value.

Solid-dielectric Variable Capacitors. Some types of variable capacitors have a solid dielectric material such as mica between their plates (Fig. 7-15B). This makes it possible to reduce the physical size of the units as compared with air-dielectric capacitors with the same capacitance range.

Fig. 7-14. Capacitance value and working voltage designations on multisection electrolytic capacitors: (A) "can" type, (B) paper-covered tubular type.

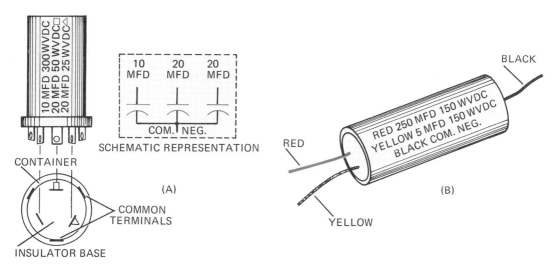

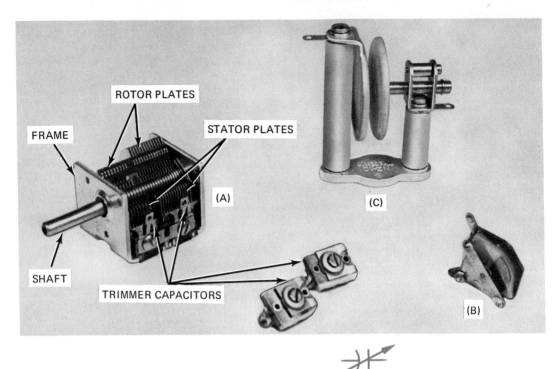

Fig. 7-15. Variable capacitors: (A) two-section receiver-type tuning capacitor, (B) solid-dielectric variable capacitor, (C) transmitter-type capacitor. (*Gail E. Henderson*)

Trimmer and Padder Capacitors. Mica and ceramic materials are also used as the dielectric in a type of small-capacitance-range variable capacitor known as a *trimmer*. Trimmer capacitors are often mounted upon the frame of a larger variable capacitor and connected in parallel with it (Fig. 7-15A) in order to make possible extremely small increases in the total capacitance of a circuit. When a small variable capacitor is connected in series with a larger capacitor for the purpose of adjusting the total capacitance of a circuit to a slightly lower value, it is referred to as a *padder*.

THE CAPACITOR COLOR CODE

A color code is used to indicate capacitance and the tolerance rating of most capacitors. On some, the code also indicates the temperature coefficient and working voltage rating. It is interesting to note, however, that there is a strong trend on the part of manufacturers to print the capacitance value and the working voltage rating directly on most common capacitors, thus eliminating the need for the color code.

Because so many different types and shapes of fixed capacitors are manufactured, it is difficult to prescribe a standard method of rating designation that will apply in all cases. However, there is an accepted code for the most commonly used types (Fig. 7-16). In these color codes, numerical values are represented by the same colors that are used with the resistor color code (refer to page 46). Unless otherwise specified, the capacitance value as indicated by the color codes is given in picofarads.

The class designation used with the molded-mica capacitor refers to a number of operational characteristics, including the leakage resistance and the temperature coefficient. When the specific class characteristics of any given mica capacitor are desired, these can be requested from the manufacturer.

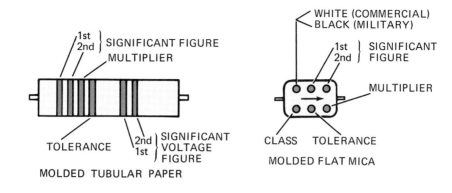
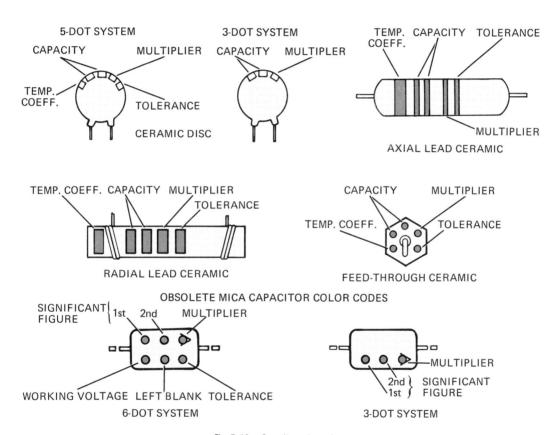

Fig. 7-16. Capacitor color codes.

For Review and Discussion

1. Define capacitance.
2. State at least four functions of capacitors.
3. Describe the construction of a capacitor.
4. Explain the charging and discharging actions of a capacitor.
5. Define capacitive reactance and give the letter symbol by which it is represented.
6. Define the farad unit of capacitance.

7. Name and define the two practical units of capacitance measurement.

8. State three factors governing the capacitance of any given capacitor.

9. Define dielectric constant.

10. What is meant by the temperature coefficient of a capacitor?

11. Define the working voltage rating of a capacitor.

12. Describe the basic construction of a typical paper capacitor and a ceramic capacitor.

13. Describe the basic construction of an aluminum electrolytic capacitor.

14. What important advantage is gained by utilizing the electrolytic-type capacitor construction?

15. Draw a diagram showing how two electrolytic capacitors can be connected to form a single nonpolarized electrolytic capacitor.

16. What is meant by a multisection capacitor?

17. Describe a typical air-dielectric variable capacitor.

18. State the purposes for which trimmer and padder capacitors are used.

UNIT 8. Safety

Safety is the responsibility of every individual. This involves both personal safety and a consideration for the safety of others. To meet this responsibility it is extremely important to realize the elements of danger that exist in the operation and servicing of electrical and electronic equipment. It is also just as important to know about the safety precautions to be observed and the practices to be followed in eliminating electrically dangerous conditions.

LOW VOLTAGES CAN BE DANGEROUS

The resistance of the human body to the passage of current varies from person to person. When the skin is dry, a person's bodily resistance ranges from approximately 100,000 to 600,000 ohms. When an abnormal amount of moisture exists at the point of contact between the body and electrical terminals, the bodily resistance may be reduced to 1,000 ohms or less.

Under this condition, even a relatively low voltage of 120 volts or less applied across portions of the body may result in electrocution. While we are inclined to think that electrocution is related to high voltages only, statistics show that the great majority of persons who have been electrocuted were handling or working with appliances or equipment operated at 120 volts.

THE NATURE OF ELECTRICAL SHOCK

As current passes through the body, it produces a physiological effect that causes a sharp nerve and muscular response. This is commonly known as an electrical shock. While the severity of the shock is determined primarily by the magnitude of the current and by its path through the body, other contributing factors include the nature and duration of the current, as well as the person's physical condition.

Magnitude of the Current. It is impossible to predict with any degree of accuracy exactly what magnitude of current will cause a specific sensation within the body. It is known that, in general, a current of approximately 0.001 ampere will produce a noticeable shock. At 0.01 or 0.02 ampere of current, the effect of the shock is often great enough to prevent voluntary control of the muscles. Thus a person may be unable to release the electrical terminals (or surfaces) with which he is in contact. This magnitude of current (0.01 to 0.02 ampere) also often causes permanent damage to body tissues and blood vessels. Currents of from 0.06 to 0.1 ampere passing through the chest region of the body for more than one second usually cause electrocution.

Path of the Current. The passage of current from finger to finger of one hand can result in a shock sensation that is usually harmless. Nevertheless, severe internal burns may result if the magnitude of the current is sufficient. However, when current passes from hand to hand or from hand to foot, it can cause death by disrupting the nervous system that controls the heart function or by paralyzing the muscles that operate the respiratory center.

Nature of the Current. Alternating currents having a frequency of from 50 to 160 hertz appear to produce the most dangerous reactions within the body. As the frequency increases above 20,000 hertz, the danger decreases significantly. The reason for this is that at the higher frequencies, a larger portion of current tends to pass over the surface of the body, instead of through it.

In addition to frequency, there is a marked relationship between the magnitude of the initial current surge through the body and the possibility of death occurring. When the current associated with voltages of 1,000 volts or more passes through the body, it causes an extremely violent contraction of the heart muscles. This condition, in turn, evidently decreases the danger of permanent damage to these muscles. As a result, those persons who are rendered unconscious by high-voltage currents such as those encountered in power transmission line work show a recovery rate that is substantially higher than in cases where shock has been produced by currents at much lower voltages.

SAFETY RULES AND PRACTICES

Safety measures can be applied in all phases of electrical and electronics work, including design, installation, maintenance, and servicing. By following accepted rules and practices, it is possible to greatly reduce or eliminate conditions that are potentially dangerous from the standpoint of both personal safety and the creation of fire hazards.

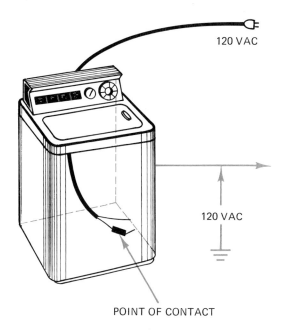

Fig. 8-1. An ungrounded equipment cabinet, which can cause a line voltage to exist between the cabinet and ground.

GROUNDING

Adequate precaution should always be taken to ensure that metal cases or cabinets of electrical equipment are well grounded. This practice is strongly recommended when using portable electric tools that are not equipped with reinforced or double insulation.

The importance of grounding is illustrated by Fig. 8-1. Here, an uninsulated portion of the ungrounded or "hot" side of the power line (cord) has come into contact with an appliance cabinet. As a result, the cabinet has become an extension of the shorted conductor. Since the line voltage exists between the cabinet and any grounded object (water pipe, gas pipe, etc.), this condition presents the danger of severe shock if bodily contact is made between the cabinet and ground.

Three-conductor Line Cords. Grounding is adequately provided for by the use of three-conductor line cords equipped with polarized three-prong plug caps. In these line cords, the green colored ground lead is connected to the equipment cabinet and to the round prong of

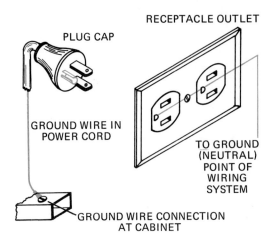

Fig. 8-2. Three-conductor plug cap and receptacle outlet.

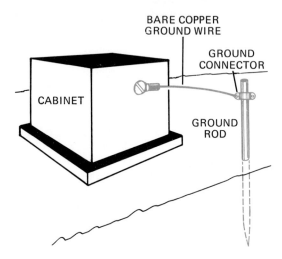

Fig. 8-4. Grounding an equipment cabinet with the use of a ground rod.

the plug cap. When the plug cap is inserted into its associated polarized outlet (receptacle), the equipment cabinet or case is automatically grounded to the wiring system ground (Fig. 8-2).

The Grounding Wire. If a three-conductor line cord is not equipped with a polarized plug cap, the green grounding wire is usually brought out of the cord as a separate lead (Fig. 8-3). This lead should always be properly grounded by connecting it to the wiring system ground, a water pipe, or a ground rod.

Fig. 8-3. Grounding lead extending from a two-conductor power cord.

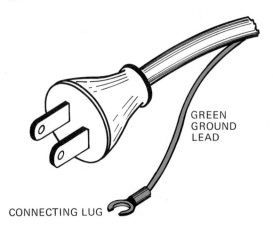

When the power line cord does not contain a grounding wire, a separate grounding wire should be added. This wire, which may be bare, is connected between a cabinet (or case) and the wiring system ground or a ground rod (Fig. 8-4). Grounding is particularly necessary in conjunction with the operation of laundry appliances, air conditioning condenser-compressor units, and other items of equipment that are located upon or near surfaces that may become damp.

CHECKING FOR VOLTAGE

The presence of a dangerously high voltage which may exist between an ungrounded metal appliance cabinet (or other form of housing) and ground can be detected by means of a voltmeter connected between the cabinet and a grounded conductor. This condition can also be conveniently detected with a neon lamp tester.

To use the neon lamp tester, one of its terminals is brought into contact with a bare surface of the cabinet and the other terminal is held between the thumb and forefinger of one hand (Fig. 8-5). If there is a line voltage between the cabinet and ground, the neon lamp of the tester will glow. Because of the resistance of the neon lamp and a resistor that is connected in series with it, this is a safe procedure. However, to be absolutely safe, the test should be performed

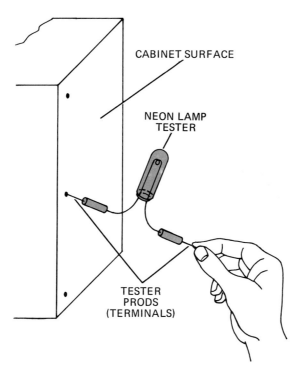

Fig. 8-5. Using a neon lamp tester.

while standing upon a dry, nonconducting surface, with no part of the body in contact with wires or other metallic surfaces. If the voltage check indicates the presence of a voltage, the equipment should be immediately unplugged and then serviced for correction of the shorted condition.

REINFORCED (DOUBLE) INSULATION

Reinforced insulation is now widely used in the construction of a number of portable electric tools. In such tools, additional insulation and the use of certain plastic parts provide protection from shock if the regular (functional) insulation breaks down or is damaged (Fig. 8-6). Reinforced insulation is now also commonly employed in the construction of many types of home appliances.

The cabinet (or case) of a reinforced insulation product should never be grounded. The principal reason for this is that an extra ground wire could increase the danger of shock in spite of the reinforced insulation. When using or installing any product with or without reinforced insulation, it is always good practice to become thoroughly familiar with the manufacturer's safety recommendations, as well as those specified by local operating and installation codes.

AC-DC CIRCUITRY

The so-called ac-dc or transformerless circuit system was formerly very popular and is still being used to some extent. In this circuit, one side (conductor) of the power line cord is connected directly to the chassis of products such as radio receivers, television sets, and record players. This feature creates a definite and dangerous shock hazard, since the conductor that is connected to the chassis may be the ungrounded or "hot" side of the power line. In this condition, the chassis becomes an extension of the ungrounded conductor; therefore, there is a line voltage between it and any grounded conductor or device.

When purchasing any electrical or electronic product, it is always good practice to check the circuitry to find out whether it is of the ac-dc type. This fact can be most readily ascertained

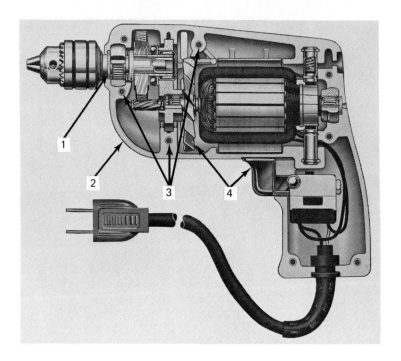

1. REINFORCE PHENOLIC IMPREGNATED SLEEVE INSULATES CHUCK SPINDLE AND PROTECTS THE OPERATOR IF THE DRILL BIT COMES INTO CONTACT WITH A "LIVE" POWER LINE.
2. REINFORCED FIBERGLASS HOUSING PROTECTS THE OPERATOR FROM ANY ELECTRICAL BREAKDOWN WITHIN THE DRILL.
3. THREADED SCREW INSERTS MOLDED INTO REINFORCED FIBERGLASS HOUSING TO INSULATE HOUSING SCREWS FROM ANY ELECTRICALLY CONDUCTIVE MATERIALS.
4. INSULATING MATERIALS USED IN CONJUNCTION WITH THE TRIGGER SWITCH AND FAN TO PROVIDE ADDITIONAL OPERATOR PROTECTION.

Fig. 8-6. Reinforced (double) insulated portable power drill. (*The Black and Decker Manufacturing Co.*)

by referring to the related schematic diagram (Fig. 8-7).

Checking for Voltage. When first placing an ac-dc circuit product into operation, always check for the presence of voltage between the chassis and ground. This can be done with a voltmeter or a neon lamp tester (see page 66). If there is a line voltage between the chassis and ground, reverse the line cord plug cap in its outlet receptacle. This will connect the grounded conductor of the power line circuit to the chassis, thereby eliminating the danger of shock.

Identifying Outlets and Plug Caps. It is a good practice to always identify the grounded and ungrounded sockets of outlets that are used in conjunction with ac-dc circuit products. The most convenient way of doing this is by using a neon lamp tester as previously described. The neon lamp of the tester will glow when its terminal is inserted into the socket of an outlet that is connected to the ungrounded side of the power line. The lamp will not glow when its terminal is inserted into the socket of an outlet that is connected to the ground side of the power line. If outlets have been wired in accordance with standard practice, a white insula-

VACUUM TUBE HEATERS IN SERIES

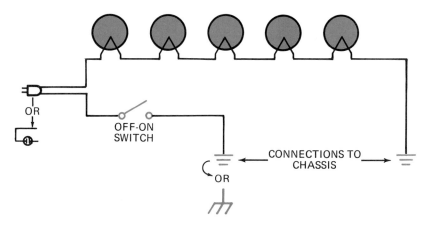

Fig. 8-7. Basic schematic diagram of an ac-dc radio receiver circuit.

tion wire will be connected to the grounded socket of each outlet. After the ungrounded and grounded sockets have been definitely identified, they can be marked by using paint or some other marking scheme (Fig. 8-8A).

To identify the prongs of a nonpolarized plug cap, first determine which prong is directly connected to the chassis of its associated equipment. This is most accurately done by making a continuity test with an ohmmeter. The prong that is connected to the chassis through the power line cord can then be marked by filing a small notch in its edge. The notched prong of the plug caps should then always be inserted into the grounded socket of an outlet (Fig. 8-8B).

CAPACITORS

The voltage across the terminals of a charged capacitor is approximately equal to the voltage of the charging source to which it was connected. A capacitor can "store" this charge for a relatively long period of time. The electrical energy stored in a capacitor can cause serious or even fatal electric shock. This energy can also cause considerable damage to some items of test equipment.

Discharging. In the interest of safety, all high-voltage, high-capacity capacitors in a circuit should be completely discharged before starting to work on the circuit. This can be done by bringing the bare ends of a short length of insulated wire into simultaneous contact with both terminals of each capacitor for a short period of time (Fig. 8-9A). In order to prevent a potentially dangerous spark, or discharge flash, when discharging high-voltage capacitors, a resistor of several thousand ohms with an adequate power rating can be connected in series with the shorting (jumper) lead (Fig. 8-9B). However, ample time for completed discharge must be allowed.

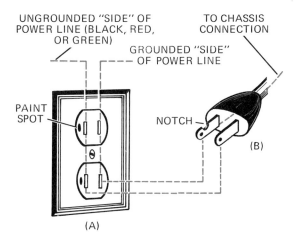

Fig. 8-8. Identifying the grounded and ungrounded connectors of outlet sockets and plug caps.

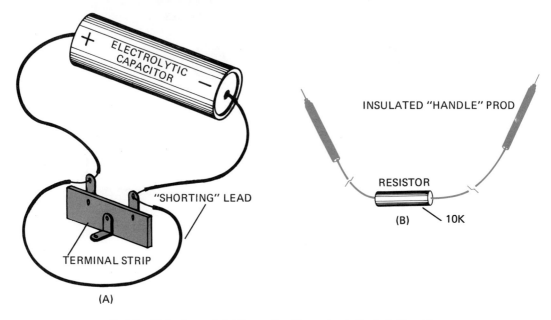

Fig. 8-9. Discharging an electrolytic capacitor: (A) use of a shorting lead, (B) resistor used in conjunction with a shorting lead.

The Bleeder Resistor. A bleeder resistor is one that is connected across the output of a rectifier power-supply circuit (Fig. 8-10). Such a resistor allows the filter capacitors (usually electrolytic) that are used in such a circuit to automatically discharge when the power switch is turned off. Thus, the danger of a capacitor-discharge shock is greatly reduced. Bleeder resistors may burn out, however, so one must not depend upon them unless it is definitely known that they are in good condition.

CATHODE-RAY TUBES

Extreme caution should always be observed when handling all types of cathode-ray tubes,

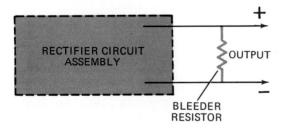

Fig. 8-10. The bleeder resistor.

including the common television picture tube. The high-vacuum space contained within the glass envelope of such tubes makes the normal atmospheric pressure on the outside surfaces of the glass envelope seem more intense, but only because there is no air within the glass envelope to help the glass withstand this pressure. If the glass breaks due to this pressure, glass particles move toward the center of the tube with great velocity. This action is referred to as an *implosion*.

To minimize the danger of flying glass from an implosion, safety goggles should always be worn when handling a cathode-ray tube. Never handle the tube by the "neck" alone, and avoid scratching its surfaces or otherwise handling it roughly. It is always good practice to "store" a cathode-ray tube in some kind of container, preferably its shipping carton. Always set the tube down on its "face" on a pad of cardboard, felt, or some other protective surface.

Before a cathode-ray tube is discarded or used for general demonstration purposes, it should be deactivated by filling the vacuum space. This can be done by first placing the tube within a covered container from which only the tube base projects. The plastic aligning key at

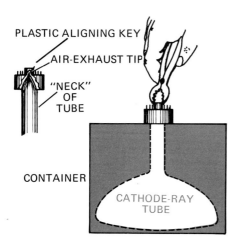

Fig. 8-11. Crushing the plastic aligning key and exhaust tip at the base of a television cathode-ray tube to deactivate the tube.

the center of the tube base should then be crushed with plier jaws (Fig. 8-11). This may or may not break the glass air-exhaust tip, which is located beneath the aligning key. If the tip

Fig. 8-12. Interlock-type switch and cheater cord.

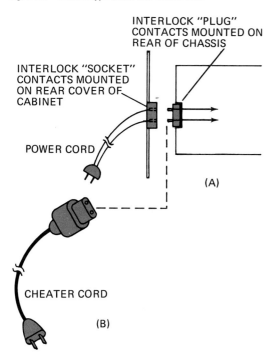

breaks, a hissing sound will be heard, indicating that air is entering the tube. If the exhaust tip does not break, it can be broken off by using the wire-cutting jaws of pliers. The tip can also be removed with the aid of a file.

INTERLOCK SWITCHES

An interlock switch is a type of connecting device that is used for the purpose of "breaking" the power circuit when a cover or lid is removed from an equipment enclosure (Fig. 8-12A). This device is an outstanding safety feature and should never be "wired" across or permanently removed.

A good-quality "cheater" cord should be used when it is necessary to energize equipment from which the interlock switch has been disconnected (Fig. 8-12B). Attempts to connect a power cord to the chassis plug contacts with clips or by twisting wire around these contacts often result in dangerous sparking and/or shorts because of improper contact.

CLEANING AIDS

Some of the chemicals and solutions used for cleaning switch, relay, and "wiper"-type contacts are highly flammable and should never be used near an open flame or a "live" circuit. Others, such as methyl alcohol and carbon tetrachloride, produce fumes, which can cause severe headaches, nausea, vomiting, and permanent liver damage. For these reasons, any cleaning aid that is used should be of a type that is specifically identified as being safe for use under the given condition.

GENERAL SAFETY RULES AND PRACTICES

1. Never completely trust a switch or any circuit control device. An electrical circuit is absolutely safe only after it has been "unplugged" and the capacitors discharged.

2. Replace defective cords and plug caps immediately.

3. Do not rely upon any type of insulation as a complete protection against electrical shock. Insulation must not replace common sense and caution.

4. Do not work on high-voltage equipment alone.

5. Whenever possible, work on an energized circuit with one hand only. Keep the other hand behind your back or in a pocket. This procedure will prevent current from passing through the chest region of your body.

6. It is always safe practice to use an isolation transformer in conjunction with the operation of ac-dc equipment. This procedure reduces the danger of shock and damage to instruments. For more information relating to the isolation transformer, refer to page 112.

7. Before purchasing any electrical or electronic device, check to see that it has been approved by the Underwriters Laboratories, Inc. This will ensure the maximum degree of safety under the conditions for which the product was designed to operate.

8. Use only carbon dioxide (CO_2) or dry chemical fire extinguishers for the control of fires involving electrical equipment that is connected to "live" power lines. If water, soda-acid, or foam-type extinguishers are used on such fires, the resulting liquid can provide dangerous conducting paths for current.

9. Avoid working on or otherwise handling an electrical-electronics device that is connected to a "live" power line while your hands are wet or damp from perspiration. Any amount of moisture on the surface of the skin greatly reduces its resistance to current, thereby increasing the danger of severe shock.

RESCUE OF SHOCK VICTIMS

In the case of an electrical accident, the victim may not be able to free himself from contact with the associated wires, terminals, and electrodes. Since time is of extreme importance, the first step to be taken is to promptly remove the victim from the contact. While doing this, the following precautions should be observed in order to prevent the rescuer himself from becoming an additional victim:

1. Do not touch the person with your bare hands until you are certain that the associated circuit has been broken or turned off.

2. If the circuit cannot be turned off, use a dry wooden stick, or other insulator material to free or, if necessary, to knock the victim from the contact. If none of these items are readily available, cover your hands thoroughly with dry clothing and, while standing upon a dry insulator material, push or pull the victim away. Be certain that your body does not come into contact with wires or terminals.

ARTIFICIAL RESPIRATION

After the victim has been removed from electrical contact and if he is not breathing normally, start artificial respiration immediately. Do not waste time. Seconds count.

The most effective procedure of administering artificial respiration is the mouth-to-mouth method, which is as follows:

1. Turn the victim on his back.

2. If necessary, clear the victim's mouth, nose, and throat so that the air passageway to his lungs is open. This can be done with a cloth or with the fingers.

3. Place the victim's head back as far as possible so that the front of the neck is stretched (Fig. 8-13A). This position of the head further clears the air passageway by causing the base of the tongue to move away from the back of the throat.

4. Open your mouth and place it firmly over the victim's open mouth. At the same time, pinch his nostrils shut (Fig. 8-13B).

5. Vigorously blow air into the victim's mouth. Then remove your mouth and turn your head to the side. If the air passageway is clear, you will hear the return rush of air. This indicates that air exchange is taking place. *Note:* If air exchange does not occur, turn the victim on his side and, with the palm of your hand, strike several sharp blows between his shoulder blades (Fig. 8-13C). Turn the victim on his back once

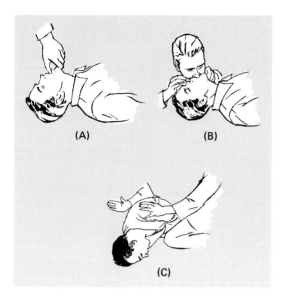

Fig. 8-13. Principal steps in the administration of mouth-to-mouth artificial respiration.

again and recheck the head and neck position.

6. If the victim is an adult, vigorously blow air into his mouth at the rate of approximately 12 breaths per minute. If the victim is a child, blow air less vigorously into his mouth at the rate of approximately 20 breaths per minute.

7. Continue artificial respiration until the victim is breathing normally or until there is, beyond any doubt, no hope of recovery.

8. After the victim has been revived, he should be kept warm and as quiet as possible. Do not feed him or give him liquids of any kind during the period immediately following.

9. Keep the victim under a doctor's care during the entire period of recovery. This is necessary since respiratory or other bodily disturbances may develop as the result of his experience.

For Review and Discussion

1. Explain why it is important to ground metallic cabinets and other forms of "housing" used in conjunction with nonreinforced insulation appliances and power tools.

2. Describe how the grounding function is provided for by the use of three-conductor plug caps and outlets (receptacles).

3. How is a neon lamp tester used in checking for voltage between an appliance cabinet and ground?

4. Describe the principal constructional features of a reinforced (double) insulated power tool.

5. What feature of a typical ac-dc circuit causes it to be a shock hazard?

6. Why is it important to identify the ungrounded and the grounded sockets of outlets that are used in conjunction with ac-dc circuit products?

7. Explain why it is important to discharge high-voltage electrolytic capacitors before working on a circuit in which such capacitors are present.

8. What is the function of a "bleeder" resistor?

9. State the safety procedures that should be followed when working with (or handling) cathode-ray tubes.

10. Define an interlock switch.

11. For what purpose is a "cheater" cord used?

12. State at least five general electricity-electronics safety rules and/or practices.

13. Give the procedures to be followed in removing a shock victim from an electrical contact.

14. State the principal procedures to be followed in administering mouth-to-mouth artificial respiration.

SECTION 2

MAGNETISM AND ELECTROMAGNETISM

The topic of magnetism cannot be separated from any thorough study of electricity and electronics because the source of magnetic energy is the electron. In this section, the nature of magnetism and electromagnetism is presented, giving special emphasis to the phenomenon of electromagnetic induction. This is the basic principle employed in the operation of a wide variety of products and devices, several of which are described in the following units. One such device is the generator. Because the generator commonly provides energy in the form of alternating voltage, the characteristics of this voltage and its associated current are also presented.

UNIT 9. The Nature of Magnetism

Magnetism is a property that is associated with an electric charge in motion. An electron, of course, is such a charge. As an electron moves about the nucleus of an atom in an orbital path, it also spins on its own axis (Fig. 9-1). This spinning motion, in effect, causes each electron to be a tiny magnet.

The polarity of the magnetic field that is generated by a spinning electron is determined by the direction of the spin. In most materials, the number of electrons spinning in one direction approximately equals the number of those spinning in the opposite direction. As a result, for all practical purposes, the materials are magnetically neutral and do not exhibit external magnetic characteristics. This is because the opposing magnetic effects tend to cancel one another.

When the magnetic effects of a large number of electrons within the atoms of a material are additive (added to each other), the material exhibits external magnetic characteristics. If these characteristics are retained over a long period of time, the material is said to become a *permanent magnet*.

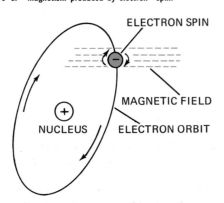

Fig. 9-1. Magnetism produced by electron "spin."

MAGNETIC DOMAINS

In certain materials such as iron, the atoms are arranged in groups that are called *domains*. The atoms within each particular domain are aligned so that a great majority of the electrons spin in the same direction. Therefore, the magnetic effects of the spins are combined, causing each domain to become an elementary magnet having opposite poles (Fig. 9-2A). These elementary magnets are often referred to as *dipoles*.

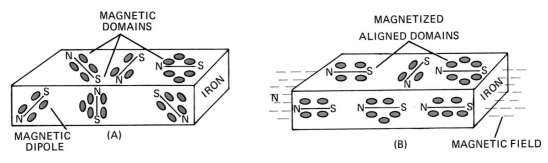

Fig. 9-2. Magnetic domain groups: (A) natural state, (B) aligned (magnetized) state.

In the unmagnetized state, the domain dipoles within a material are oriented in every possible direction. For this reason, the magnetic effects of the domains tend to be cancelled and the material does not exhibit any significant external magnetic characteristics.

When a mass of iron is placed under the influence of an external magnetic (magnetizing) field, many of the domains become aligned in the same direction (Fig. 9-2B). In this condition, the magnetic effects of the aligned domains are added to each other and the iron becomes magnetized. If the magnetizing field is strong enough, all of the domains within the iron will be aligned in the same direction. When this occurs, the iron has been magnetized to the maximum degree and is said to be *magnetically saturated*.

CLASSIFICATION OF MATERIALS

All materials are either magnetic or nonmagnetic. Classification is usually based on visible evidence that a material either exhibits or does not exhibit magnetic characteristics when it is placed near a relatively strong magnet. However, it is sometimes important to consider magnetic characteristics that are not readily apparent. For this reason, a more specific classification of materials is necessary.

Diamagnetic Materials. A *diamagnetic material* is one that is feebly repelled by a strong magnet and whose atoms do not form domain groups.

However, when a strong magnetizing field is applied, some of its atoms do become aligned to produce a very weak magnet whose polarity is such that it is repelled by a strong magnet. Common diamagnetic materials are antimony, copper, gold, mercury, silver, and zinc.

Paramagnetic Materials. A *paramagnetic material* is one that is feebly attracted by a strong magnet. Such a material contains some atoms that are naturally arranged in domains and others that are not. When an external magnetizing field is applied to such a material, many of the non-domain atoms will align. The polarity of a paramagnetic material is such that a strong magnet will attract it. Common paramagnetic materials include aluminum, chromium, manganese, and platinum.

FERROMAGNETIC MATERIALS

Ferromagnetic materials are materials that are strongly attracted by magnets. They are the most important magnetic materials insofar as the fields of electricity and electronics are concerned.

Ferromagnetic Elements. The ferromagnetic elements are cobalt, iron, and nickel. In these elements, a significant degree of domain alignment is readily accomplished by the application of an external magnetizing field. In addition, an electrostatic force (exchange interaction) that tends to maintain the domain alignment after it has been established is present within these ele-

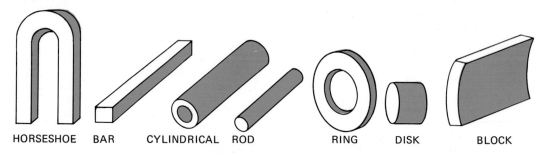

Fig. 9-3. Common physical shapes of Alnico magnets.

ments. If this force were not present, any amount of heat, even at ordinary room temperatures, would produce a sufficient amount of thermal agitation to cause a very rapid disalignment of the domain pattern.

Ferromagnetic Alloys. Many of the highly efficient permanent magnets that are found in industrial and commercial applications are made of ferromagnetic alloys. One common ferromagnetic alloy is Alnico (aluminum, nickel, cobalt, and iron). Permanent magnets, of many sizes and shapes, made of Alnico are used in a variety of devices including loudspeakers and motors, (Fig. 9-3). Other magnetic alloys include steel (iron and carbon), Vicalloy (vanadium, iron, and cobalt), Permalloy (nickel and iron), Lodex (iron and cobalt), and platinum-cobalt.

Ferrites. *Ferrites* are magnetic materials that are ceramic-like in structure. They consist primarily of metallic elements and metallic oxides. After it is mixed, the ferrite material is pressed into the desired shape and "baked" at a high temperature.

There are two basic kinds of ferrites: "soft" and "hard." Soft ferrites are made from nickel-zinc or manganese-zinc base materials to which copper oxide, magnesium oxide, and cobalt have been added. Soft ferrites are used primarily as core materials in various kinds of inductor (coil) assemblies (Fig. 9-4).

The hard ferrites most often consist of a barium compound as the base material, with lead oxide, copper oxide, and bismuth oxide added. These ferrites are used primarily in the manufacture of permanent magnets such as those found in loudspeakers.

THE MAGNETIC FIELD

A *magnetic field* exists in the region surrounding a magnet. This field consists of lines of flux moving from the north pole to the south pole within the magnet (Fig. 9-5). The entire group of flux lines associated with any particular magnetic field is referred to as the *flux field* and is represented by the Greek letter *phi* (ϕ).

Flux Density. The *metric* system unit of magnetic flux is the *maxwell,* which equals one flux line. *Flux density,* represented by the letter symbol B, is the number of flux lines or maxwells per unit area when the area is in a plane perpendicular to the direction of the flux lines. The unit of flux density is the *gauss*. One gauss indicates the presence of one flux line (one maxwell) in an area one centimeter square.

Magnetomotive Force. The *magnetomotive force* (mmf) of a magnet is that force which tends to produce a magnetic field. The metric system unit of magnetomotive force is the *gilbert,* which is represented by the symbol F. One gilbert is equal to that magnitude of force required to establish a flux density of one maxwell in a flux path or magnetic circuit having a reluctance of one unit (refer to page 79).

Magnetizing Force. *Magnetizing force,* represented by the letter symbol H, is the total force that tends to establish flux lines in a magnetic field. The metric system unit of magnetizing force is the *oersted* (Oe). One oersted is equivalent to a magnetomotive force of one gilbert when this force is exerted across a magnetic circuit that is one centimeter in length.

Fig. 9-4. Ferrite cores. *(Allen-Bradley Co.)*

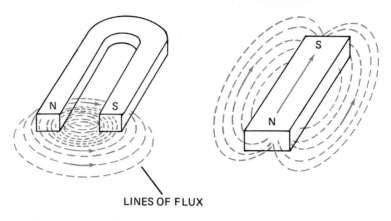

Fig. 9-5. Magnetic fields surrounding horseshoe- and bar-shaped permanent magnets.

MAGNETIC POLES

The poles of a permanent magnet are determined by the direction of magnetization or domain alignment. The magnetic poles of bar-, cylindrical-, and disc-shaped magnets are shown in Fig. 9-6. The poles have been named in accordance with the direction in which a bar magnet will "point" when it is freely suspended. The end of a magnet that points toward the north is the north-seeking or north pole and the end that points toward the south is the south-seeking or south pole.

Law of Attraction and Repulsion. The law of *magnetic attraction and repulsion* states that unlike magnetic poles will attract each other, while like poles will repel each other. This is the underlying behavior for the operation of many devices, including motors, solenoid-plunger assemblies, and most electrical meters. In each of these devices, two interesting and very important conversions of energy occur. First, the energy of electrons in motion (*electric energy*) is converted into magnetic or *magnetostatic energy* in the form of a magnetic field. Second, the force of attraction (or repulsion) exerted by the magnetic field results in the motion of a mass, thus producing a conversion of magnetostatic energy into *mechanical energy* (Fig. 9-7).

Coulomb's Law. *Coulomb's law* for magnetism states that the force existing between two magnetic poles is directly proportional to the flux density at the poles and inversely proportional to the square of the distance between the poles. The latter statement of this law means, for example, that if the distance between two unlike poles is doubled, the force of attraction between the poles is reduced to one-fourth the original force. Similarly, if the distance between two like poles is reduced by one-half, the force of repul-

Fig. 9-6. Common directions of magnetization used with bar-, cylindrical-, and disk-shaped permanent magnets.

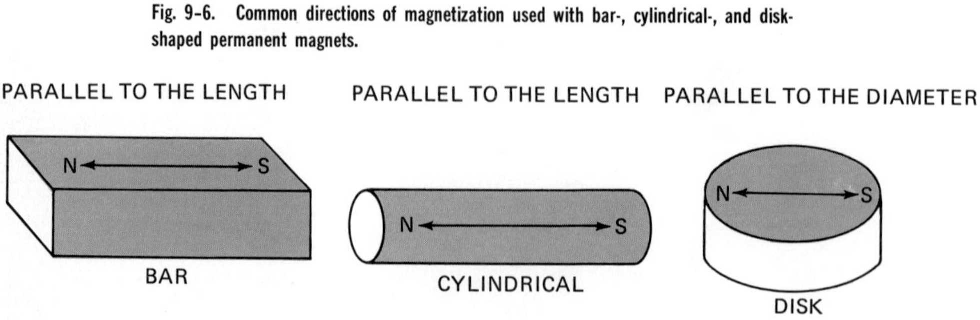

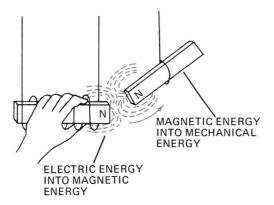

Fig. 9-7. Energy conversions occurring when a magnet is repelled by another magnet and caused to move.

sion between the poles is increased by four times the original force.

PERMEABILITY

Permeability is defined as a measure of the flux density produced within a material as the result of the application of a given magnetizing force. More specifically, permeability is the ratio of the flux density B to the magnetizing force H. In accordance with these definitions, permeability can also be thought of as a measure of how efficiently a material will conduct lines of flux. The relative permeability of a material is its permeability as compared with that of a vacuum. A vacuum space is considered to have a permeability of one.

The permeability of a material is not constant but varies with the flux density within the material. As the flux density increases, an increase of the magnetizing force does not produce a proportional increase of flux density. Hence, permeability decreases with an increase of flux density. This decrease continues until the material has become saturated. Under this condition, the permeability of the material is, for all practical purposes, the same as that of air.

When a material having a high permeability is placed within a magnetic field that exists in an air space, the flux lines are concentrated within the material. This produces a condition that is referred to as a *magnetic field distortion* (Fig. 9-8).

RELUCTANCE

Since iron, for example, conducts flux lines more readily than does air, it can be assumed that air presents a greater opposition to flux lines. This opposition is called *reluctance* and is inversely proportional to permeability. Reluctance can be compared to resistance that opposes the flow of electrons through a conductor. As in the case of permeability, the reluctance of different materials is stated as a comparison with the reluctance of air or of a vacuum.

The reluctance of a given material is directly proportional to its length and inversely proportional to its cross-sectional area. Expressed as a formula,

$$\text{Reluctance} = \frac{L}{\mu A}$$

where L = length, in inches or in centimeters
μ = permeability of the material
A = cross-sectional area, in square inches or in square centimeters

MAGNETIC INDUCTION

Magnetic induction is the process whereby a magnetic field produces a domain alignment in a magnetic material that is placed within the field. As a result of this alignment, the material

Fig. 9-8. Example of magnetic field distortion.

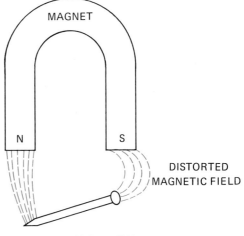

NAIL EXHIBITS GREATER PERMEABILITY THAN AIR

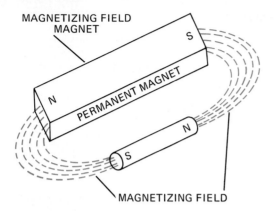

Fig. 9-9. Polarity of a magnetic material produced by magnetic induction.

becomes magnetized. The magnetism that is produced by magnetic induction is referred to as *induced magnetism*.

Polarity of Induced Magnetism. The polarity of a pole of a magnet that is produced by magnetic induction is always opposite to the polarity of an adjacent pole of the magnetizing field magnet (Fig. 9-9). It is because of this effect that a force of attraction exists between a magnet and a magnetic material that is placed within the field of the magnet.

Residual Magnetism. The magnetism that a magnetic material exhibits after it has been removed from the influence of the induction (magnetizing) field is known as *residual magnetism*. If the residual magnetism is retained by a material for a relatively long period of time, the material is said to possess a high retentivity.

MAGNETIC STABILITY

One of the most important fundamental laws relating to permanent magnets is that no energy is required to maintain a magnetic field. For this reason, permanent magnets made of high-efficiency materials will, for all practical purposes, produce a constant flux density under all but the most adverse conditions. The most common of these adverse conditions are high temperatures and exposure to external magnetic fields.

The degree of magnetization that a permanent magnet retains after it has been removed from the influence of heat is referred to as *remanence*. Because of the increased agitation of molecular particles produced by heat, the remanence of a permanent magnet normally decreases with an increase of temperature. This reduction of remanence continues until the temperature of a given material reaches what is known as its *Curie temperature* (Table 9-1). At this temperature, the ferromagnetic properties of a material cease to exist; therefore the remanence becomes zero.

THE MAGNETIC CIRCUIT

The primary purpose of any magnet is to store magnetic energy and/or to convert this energy into another form of energy. This is most commonly achieved in one of two ways: by storing magnetic energy within a magnetic material, or by setting up a magnetic field within a space called an *air gap* (Fig. 9-10).

The typical magnetic circuit (utilizing a permanent magnet) may be divided into two parts: the permanent magnet, which is the source of the magnetomotive force, and the structures or devices that form the path through which lines of flux pass as they move from one pole of the magnet to the other.

The design of any magnetic circuit is complicated because no materials act as magnetic

Table 9-1. Curie temperatures of common magnetic materials. (*General Electric Co.*)

MATERIAL	CURIE TEMPERATURE ±10° C
Alnico 1	780
Alnico 2	815
Alnico 3	760
Remalloy	900
Cunico	860
Cunife	450
Vicalloy	855
Platinum Cobalt	490
Barium Ferrite	450

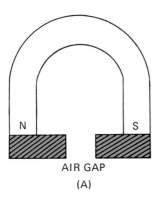

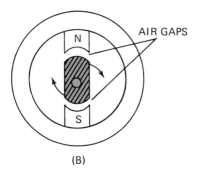

Fig. 9-10. Magnetic air gaps: (A) fixed or stationary air gap, (B) variable air gap.

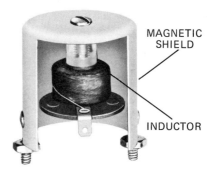

Fig. 9-11. Cutaway view of a magnetic shield used to protect an inductor from the influence of external magnetic fields. (*Bud Radio, Inc.*)

insulators. Because of this, it is often difficult to design a magnetic circuit that will result in the maximum application of flux lines to those parts of the circuit where they will be most useful. When this condition does not exist to a satisfactory degree, the circuit is said to exhibit a high flux leakage.

MAGNETIC SHIELDING

The property of permeability is widely used to protect or shield certain components and instruments from the effects of magnetic fields. In many instruments and other circuits, the effects of "stray" magnetic fields can cause significant errors in measurement or other forms of interference.

A basic form of magnetic shield is shown in Fig. 9-11. In this type of shielding, a cover or "can" made of a high-permeability ferromagnetic material is placed over a circuit component. As a result, any external magnetic field that may be present in the vicinity of the component is "bypassed" (Fig. 9-12).

MAGNETOSTRICTION

Magnetostriction is the term applied to the effect of a magnetic field that causes certain ferromagnetic materials to expand or to contract. Nickel, for example, is reduced in size when exposed to a relatively strong magnetic field. Iron and certain iron-nickel alloys expand when subjected to a magnetic field. The property of magnetostriction, although not commonly utilized, is applied in the operation of some communications equipment frequency-selective circuits.

Fig. 9-12. Principle of magnetic shielding.

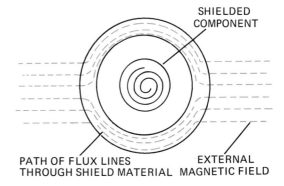

For Review and Discussion

1. Describe the source of magnetism in terms of electron "spins."
2. What is a magnetic domain?
3. Explain the process of magnetization in terms of magnetic domain alignment.
4. Define magnetic saturation.
5. Name three ferromagnetic elements.
6. Name at least three ferromagnetic alloys.
7. Define a ferrite.
8. For what purpose are "soft" ferrites commonly used?
9. Describe the magnetic field surrounding a bar magnet.
10. Name and define the basic unit of magnetic flux density.
11. State the law of magnetic attraction and repulsion.
12. Express Coulomb's law for magnetism.
13. Define magnetic permeability and reluctance.
14. Describe the process of magnetic induction.
15. What is meant by residual magnetism?
16. Define magnetic remanence.
17. What are the basic components of a magnetic circuit?
18. Describe a magnetic shield and give the function of such a shield.

UNIT 10. Electromagnetism

As we have just seen, electron "spins" generate what may be referred to as *atomic magnetism*. A magnetic field is also generated whenever free electrons move through a conductor. This extremely important relationship between electricity and magnetism is known as the *magnetic effect of current*.

The magnetic field accompanying the flow of electrons consists of flux lines generated in concentric circles in a plane perpendicular to the direction of the current (Fig. 10-1A). Movement of the flux lines about the conductor is in either a clockwise or a counterclockwise direction, depending upon the direction of current flow (Fig. 10-1B).

AIDING AND OPPOSING MAGNETIC FIELDS

The direction of the flux lines constituting the magnetic field about a conductor gives the field a definite magnetic polarity. As a result, the field either aids or opposes the magnetic field that is generated around an adjacent conductor. Each

Fig. 10-1. The magnetic effect of current: (A) magnetic field surrounding a current-carrying conductor, (B) change in the direction of flux lines with change in direction of current.

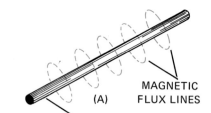

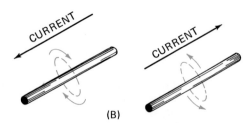

of these conditions is illustrated by Fig. 10-2. Fig. 10-2A shows current moving through two adjacent conductors in the same direction. This produces magnetic fields that move in opposite directions in the space between the conductors. Since these are, in effect, magnetic fields having opposite polarity, they will attract or aid each other. The total magnetic field existing between the two conductors is equal to the sum of their individual magnetic fields.

Fig. 10-2B shows current moving through the conductors in opposite directions. This causes the magnetic fields between the conductors to move in the same direction (to be of like polarity). In this condition the fields repel each other.

The effects produced by aiding and opposing magnetic fields may or may not be useful. In the electromagnet, for example, the aiding magnetic fields between the windings of a coil produce a desired strong, total magnetic field.

Fig. 10-2. Aiding and opposing magnetic fields surrounding adjacent current-carrying conductors.

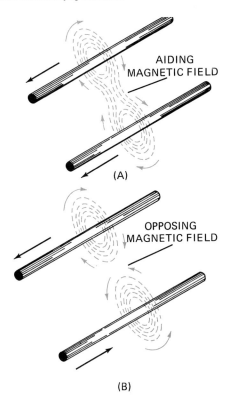

Fig. 10-3. Basic form of noninductive winding.

In other types of coils and in wire-wound resistors where the aiding interaction between windings must be kept to a minimum, the coils are often wound as shown in Fig. 10-3 in order to eliminate the cumulative magnetic effect. In this noninductive winding scheme, the current in adjacent conductors moves in opposite directions. Therefore, the magnetic fields oppose each other and interaction between the fields surrounding the conductors is minimized.

In long-distance telephone lines, the interaction of the magnetic fields between parallel conductors often produces a condition known as *crosstalk*. Crosstalk is a form of distortion whereby the information transmitted over one circuit interferes with the information transmitted by the conductors of an adjacent circuit.

To reduce crosstalk, conductors associated with the same open-wire circuit are often transposed. Transposition causes the magnetic fields surrounding the conductors to oppose each other, and as a result, the interaction between the fields is substantially reduced. In cable lines, transposition is achieved by twisting the conductors associated with the same circuits.

THE MOVING MAGNETIC FIELD

The flux density of a magnetic field at any instant of time is directly proportional to the magnitude of the current that produced it. Suppose, for example, that the magnitude of a direct current through a conductor steadily increases from zero to a given maximum. As the current increases, the flux density of the associated magnetic field also increases. This action results in an expanding magnetic field, which may be thought of as moving outward from the conductor (Fig. 10-4A). When the current increase is sudden, the magnetic field expands or moves rapidly. When the current increase is gradual, the field expands much more slowly. If, after reaching its maximum magnitude, the current remains constant, the flux density also remains constant. In this condition, the field is said to be *fixed* or *stationary*.

When the magnitude of the current decreases, the flux density of the magnetic field also decreases. This again produces a movement of the field, but now the movement is in the opposite direction or toward the conductor (Fig. 10-4B). This action is commonly referred to as a *collapsing magnetic field*.

When an alternating current passes through a conductor, the associated magnetic field constantly expands and collapses about the conductor as it "keeps in step" with the changing magnitude of current. Since an alternating current also changes in direction, the direction of the flux lines around the conductor will change accordingly.

THE ELECTROMAGNET

While the magnetic field about a single conductor is important in certain circuits, it is generally much too weak to be of any practical value. To produce a somewhat stronger magnetic field, a conductor is formed into a loop. When this is done, the flux lines along all points of the con-

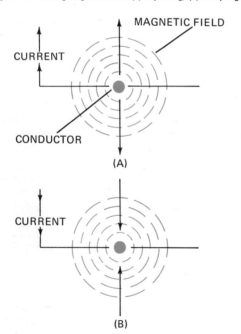

Fig. 10-4. Moving magnetic fields: (A) expanding, (B) collapsing.

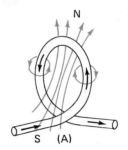

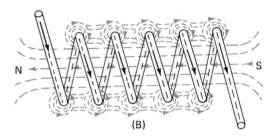

Fig. 10-5. Simplified magnetic action of the electromagnet.

ductor pass through the inside of the loop in the same direction. As a result, the ends of the loop assume a definite magnetic polarity (Fig. 10-5A).

The Solenoid. A stronger electromagnet is produced when a coil consisting of several loops or turns of wire is formed. The increased strength results because the magnetic field around every turn of the coil aids the magnetic field of every other turn (Fig. 10-5B).

If the solenoid is operated (energized) with direct current, fixed magnetic poles are produced at its ends. These poles and the magnetic field existing between them have properties that are identical to those of a bar-type permanent magnet. If the solenoid is energized with alternating current, its magnetic polarity reverses with each change in the direction of the current.

Adding an Iron Core. Powerful magnets are made by winding a coil of insulated wire around a core form made of a low-retentivity ferromagnetic material such as soft iron (Fig. 10-6). Because of the high permeability of such a core, the flux density is greatly increased. This,

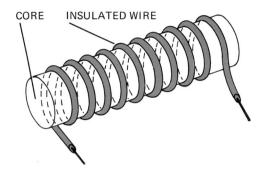

Fig. 10-6. Iron-core electromagnet.

of course, produces a much stronger electromagnet than would be the case if the core was not used.

Most iron-core electromagnet coils are wound with magnet wire. This wire is usually coated with a special kind of varnish or plastic that provides the necessary insulation.

The Left-hand Rule. The magnetic polarity of a solenoid coil or an electromagnet can be determined by using the left-hand rule, which states that when the fingers of the left hand are placed around the coil in the direction of the current, the extended thumb points toward the north pole as shown in Fig. 10-7.

MAGNETOMOTIVE FORCE

The magnetomotive force of an electromagnet is determined primarily by four factors:

1. The magnitude of the current with which it is energized
2. The number of turns or loops of wire used in winding the coil
3. The physical size of the core
4. The material from which the core is made

Ampere-turns. The magnetomotive force of either an air-core or an iron-core electromagnet is generally expressed in terms of *ampere-turns*. The number of ampere-turns is equal to the product that is obtained when the energizing current is multiplied by the number of coil turns.

Expressed as a formula,

$$\text{Ampere-turns} = IN$$

where I = current, in amperes
N = number of turns

Problem. Compute the ampere-turns rating of an electromagnet wound with 600 turns of wire when it is energized with 2 amperes of current.

Solution.

$$\text{Ampere-turns} = 2 \times 600 = 1{,}200$$

If the energizing current was increased, the number of turns required to construct the 1,200 ampere-turn electromagnet could be reduced. Similarly, the same ampere-turn rating could be obtained by decreasing the energizing current and increasing the number of turns.

Conversion to Gilberts. The magnetomotive force of an electromagnet (in ampere-turns) can be converted to magnetomotive force in gilberts by using the following formula:

$$F = 0.4\pi\, IN$$

where F = magnetomotive force, in gilberts
π = 3.1416
I = current, in amperes
N = number of turns

If ampere-turns (IN) in the above formula is equal to one, the magnetomotive force is equal to 0.4 × 3.1416, or 1.257. Thus, one ampere-turn is equivalent to 1.257 gilberts.

Fig. 10-7. Applying the left-hand rule for determining the polarity of an electromagnet.

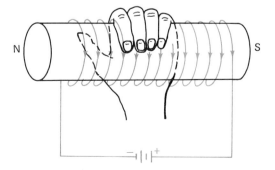

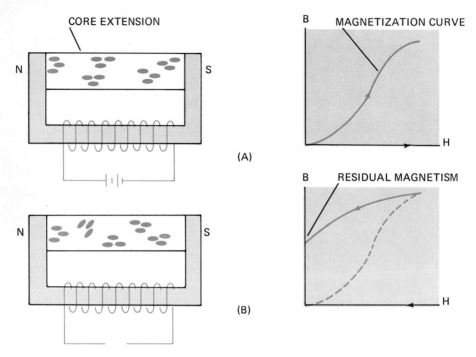

Fig. 10-8. Circuits and graphical curves illustrating the processes of magnetization and magnetic hysteresis.

MAGNETIC HYSTERESIS

When an iron-core electromagnet is energized, many of the domains within the core become aligned as it assumes a definite magnetic polarity. This condition can be illustrated by the graph of Fig. 10-8A, with the magnetizing force H produced by the current forming the horizontal axis and the flux density B forming the vertical axis.

When the current is stopped, some of the domains do not return to their preenergized (demagnetized) state. As a result, a certain amount of residual magnetism remains in the core (Fig. 10-8B). This tendency of the domains to lag behind the magnetizing force is known as *hysteresis*.

Hysteresis is caused by the resistance to movement that the domain dipoles exhibit as a result of the internal friction between them and the molecular particles within the core. By reversing the flow of current through the coil of an electromagnet, hysteresis can be overcome, thereby reducing the residual magnetism of its core to zero. This is because a reverse in current direction causes a reverse or opposite-direction magnetizing force $(-H)$ to be applied to the core. The extent of the magnetizing force necessary to produce a complete demagnetization of a core is referred to as the *coercive force* of the core material.

Hysteresis Loss. When an electromagnet is energized with alternating current, the magnetizing force produced by the current is also alternating in nature. This causes the hysteresis effect to occur during a part of each current cycle. As a result, a portion of the magnetizing force must be utilized to overcome hysteresis. This portion of the magnetizing force does not, therefore, enter directly into magnetization activity. Instead, it is used to overcome the frictional resistance of the domain dipoles to movement. In overcoming this resistance, the magnetizing force produces heat within the core. Since this heat represents a loss of energy, it is referred to as *hysteresis loss* or *hysteresis power loss*.

Hysteresis Vs. Frequency. The effect of hysteresis within a material is dependent not only upon the nature of the material but also upon the frequency of the magnetizing force to which it is subjected. At comparatively low frequencies, such as 60 hertz, hysteresis losses are usually considered to be insignificant. However, as the frequency of the energizing current within the coil increases, the extent of the hysteresis loss can be considerable. In order to reduce this loss, cores are often made of a special low-hysteresis loss material such as a ferrite.

THE HYSTERESIS LOOP

The pattern of magnetization or induction within a given material can be illustrated by graphically plotting the magnitudes of flux density and the magnetizing force that is applied to the material. The pattern produced by the use of an alternating magnetizing force is known as a *hysteresis loop* or a *B-H curve* (Fig. 10-9). The magnetizing force–flux density response that is shown by this loop is described in the following paragraphs.

When the demagnetized material is subjected to a magnetizing force that gradually increases from zero to H_{max}, the flux density within the material increases from zero to B_{max}, the saturation point. If the magnetizing force is then gradually reduced to zero, the flux density decreases from B_{max} to B_r on the vertical axis. This point on the curve represents residual magnetism.

If the direction of the magnetizing force is now reversed and the magnitude increased, the flux density is further reduced and becomes zero when the magnetizing force is of a magnitude represented by H_c on the horizontal axis. The distance on the horizontal axis from 0 to H_c corresponds to the coercive force of the material. A further increase in the reverse magnetizing force causes the flux to reverse its direction. This condition is indicated by the portion of the vertical axis that extends from 0 to $-B$.

As the reverse magnetizing force continues to increase to the maximum magnitude $(-H)$, the flux density once again reaches the saturation point at $-B_{max}$. If the magnetization force is now reversed and increased from $-H_{max}$ to H_{max}, the flux density will again increase to B_{max} as represented by the $-B_{max}$, $-B_r$, B_{max} portion of the curve. This completes the hysteresis loop.

Width of the Loop. The width of a hysteresis loop, or distance across the loop along the hori-

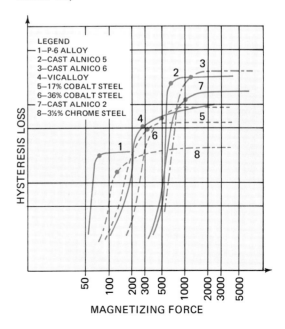

Fig. 10-10. Graphical representation of the relationship between the relative hysteresis losses and the magnetizing force applied to a number of common magnetic materials. (*General Electric Co.*)

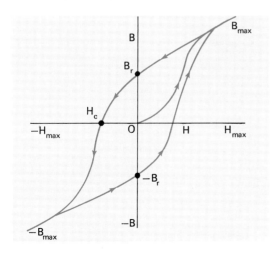

Fig. 10-9. Hysteresis loop.

zontal axis, provides an accurate indication of the hysteresis loss that can be expected from the use of a given material. If the loop is narrow, as in the case of soft iron, the effect of hysteresis within the material will be small. Silicon steels are often used for narrow hysteresis loops. On the other hand, Alnico and other high-retentivity materials have wide hysteresis loops. This indicates that these materials are excellent for making permanent magnets but are undesirable for use as cores in alternating-current devices.

Hysteresis Loss Vs. Magnetizing Force. The relative hysteresis losses occurring within a number of commonly used permanent-magnet alloys under conditions of a varying magnetizing force are shown by the curves of Fig. 10-10. These curves show that the greatest loss is developed within Alnico 6 as the magnetizing force is increased to the maximum.

ADVANTAGES OF ELECTROMAGNETS

The use of electromagnets instead of permanent magnets provides several advantages that are very important in the application of magnetic devices to the areas of electricity and electronics. The principal advantages are: the flux density can be controlled and easily varied by varying the magnitude of the current, the polarity can be reversed by simply changing the direction of the current, and the magnetomotive force is not affected by conditions such as aging, heat, or severe mechanical vibration. Each of these factors tends to reduce the magnetomotive force of a permanent magnet.

MAGNETIZING AND DEMAGNETIZING

To transform a high-retentivity ferromagnetic object into a useful permanent magnet, the object must be brought under the influence of an

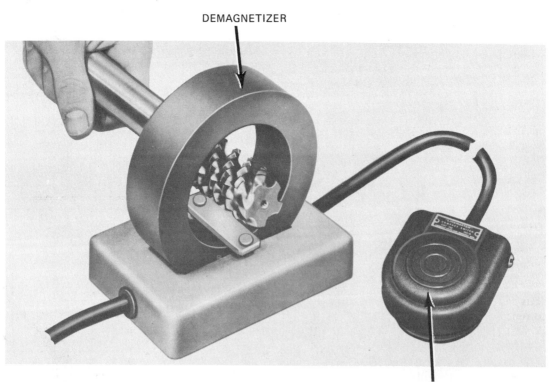

Fig. 10-11. Using a demagnetizer with a metal-cutting tool. (*The Martindale Electric Co.*)

intense magnetizing field. Such a field can be generated by a permanent-magnet magnetizer unit energized with direct current. The object to be magnetized is placed between or across the poles (end pieces) of the magnetizer, which is then energized. As a result, many of the magnetic domains within the object are aligned in accordance with the magnetic polarity of the poles.

It is often necessary to "remove" the magnetism from objects such as watches and metal-cutting tools. This is accomplished by placing the object to be demagnetized within the field of an electromagnet that is energized with alternating current, and then slowly removing the object from the field (Fig. 10-11). This action causes the alternating magnetizing force of the magnetic field to seriously distort the domain alignment, thereby significantly reducing the residual magnetism within the object. The process of *demagnetization* is also referred to as "degaussing."

For Review and Discussion

1. Define the magnetic effect of current, and describe the magnetic field generated about a current-carrying conductor.

2. Draw appropriate diagrams illustrating the principles of the aiding and opposing magnetic fields generated about adjacent current-carrying conductors.

3. What is meant by a noninductive winding?

4. Explain how a moving magnetic field is generated about a current-carrying conductor.

5. Explain the electromagnetic effect that results when a current-carrying wire is formed into a series of turns or loops.

6. What electromagnetic advantage is gained by inserting an iron core into a solenoid coil? Give the reason for this condition.

7. State the left-hand rule for determining the polarity of a solenoid coil or an electromagnet.

8. State four factors that govern the magnetomotive force of an electromagnet.

9. Define the ampere-turns unit of magnetomotive force.

10. Define magnetic hysteresis.

11. What is meant by a hysteresis loss?

12. Express the relationship between hysteresis loss and the frequency of the current with which a given electromagnet is energized.

13. Draw a sample hysteresis loop and explain the magnetic conditions that this loop represents.

14. What is the significance of the width of a hysteresis loop?

15. State three advantages that are gained by using electromagnets instead of permanent magnets in certain applications.

16. Define degaussing.

UNIT 11. Alternating Voltage and Current

Thus far, our study of alternating voltage and current has been limited to general definitions. While this was sufficient for purposes of introduction, it does not satisfy the need for a further examination and a more detailed study of these quantities. This is particularly true since alternating voltage and current create circuit conditions that must be understood if the study of circuit behavior and analysis is to proceed in a logical manner.

ELECTROMAGNETIC INDUCTION

As indicated in Unit 10, the movement of electrons within a conductor is always accompanied by the generation of a magnetic field. In addition to this magnetic effect of current, there is another extremely important relationship existing between a magnetic field and the motion of

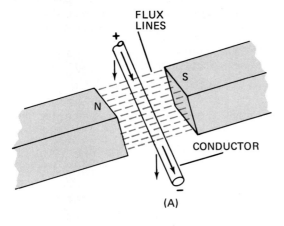

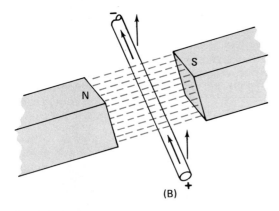

Fig. 11-1. Producing a voltage across the ends of a conductor by means of electromagnetic induction.

electrons. It is possible to cause a movement of electrons within a conductor when there is relative motion between the conductor and magnetic flux lines. The process by which this is accomplished is called *electromagnetic induction*.

Basic Action. If a conductor is moved downward between the poles of a magnet, the action as it cuts across flux lines forces electrons to move through the conductor in the direction shown in Fig. 11-1A. This displacement of electrons causes a voltage to be established between the ends of the conductor (induced voltage). If the conductor is now moved upward (the opposite direction), the direction of the electron movement reverses, thus causing a reversal in the polarity of the induced voltage (Fig. 11-1B). The voltage-producing effect of electromagnetic induction also occurs when the magnetic field is moved while the conductor is held stationary and when both the magnetic field and the conductor are in motion.

The effect of generating a voltage by means of a relative motion between a conductor and magnetic flux lines is the basic principle behind the operation of a generator (a machine in which mechanical energy is changed into electrical energy).

Magnitude of the Induced Voltage. The magnitude of the induced voltage produced by electromagnetic induction at any instant of time is directly related to the number of flux lines that are being cut. The number of flux lines cut is dependent upon:

1. The flux density of the magnetic field through which the conductor is moving. The greater the flux density, the greater the magnitude of the induced voltage.

2. The velocity of the conductor motion. As the velocity is increased, the magnitude of the induced voltage increases.

3. The angle at which the conductor cuts across the flux lines.

The magnitude of the induced voltage is greatest when the conductor cuts across flux

Fig. 11-2. Conductor cutting across flux lines: (A) perpendicular to the direction of the flux lines, (B) parallel with the flux lines.

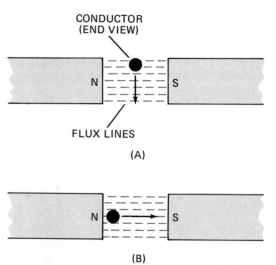

lines in a direction that is perpendicular to the direction of the flux lines (Fig. 11-2A). The magnitude of the induced voltage is at a minimum when the conductor is moving in a direction that is parallel to the direction of the flux lines (Fig. 11-2B).

GENERATION OF THE SINE WAVEFORM

The basic construction of a simple two-pole single-coil generator with rotating armature and alternating voltage is shown in Fig. 11-3. This type of generator is not nearly as common as the type with a rotating field. It does, however, provide a convenient means of studying the generation of the sine waveform, which is essentially the same in all types of generators.

The Voltage Waveform. The relationship between the position of the armature (coil) of the generator and the magnitude and polarity of the voltage induced within the coil during one complete cycle is illustrated in Fig. 11-4. Here, the heavy portions of the coil represent the "sides" or conductors of the coil that effectively cut across flux lines. The circles represent the position of the conductors as the coil revolves through 360 degrees, and the sine waveform represents the changing magnitude of the induced voltage. The magnitude of the voltage is plotted with reference to a horizontal base line that indicates the angular position of the coil.

In this generator, each side of the coil can be thought of as a single conductor connected in series with the other. The total voltage appearing across the ends of the coil at any instant is then, in effect, the sum of the voltages induced across the conductors at that instant. This voltage is referred to as the *instantaneous voltage* and is represented by the letter symbol e.

As indicated by the sine waveform, the output voltage of the generator is an alternating voltage that constantly changes in magnitude and periodically reverses its polarity.

The magnitude of the induced voltage changes because of the change in the number of flux lines that the coil conductors cut across as they come into positions that are at different angles to these lines. Voltage polarity reverses because the conductors begin to cut across the flux lines in a different direction when they reach the 0- and 180-degree positions of their revolutions. An alternating voltage that exhibits these cyclical characteristics is known as a *sinusoidal voltage*.

The Current Waveform. When a sinusoidal voltage is applied to a circuit whose only effective opposition to current is resistance, there is, at any instant, a direct relationship between the magnitudes of the voltage and the current. Because of this, there is no difference between their starting points during any cycle. Thus, the voltage and the current "rise" and "fall" simultaneously and reach their respective maximum

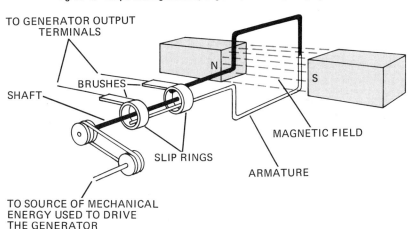

Fig. 11-3. Simple rotating-armature, single-coil, alternating-voltage generator.

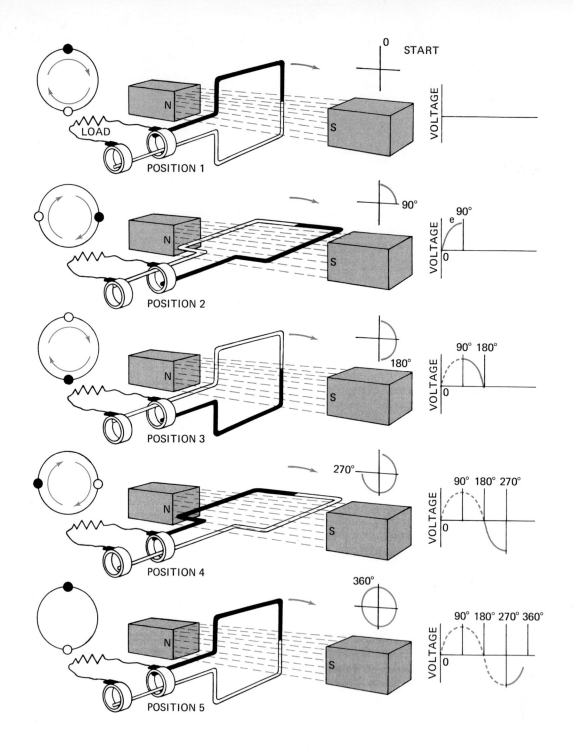

Fig. 11-4. Generation of the output-voltage waveform by a single-coil, alternating-voltage generator.

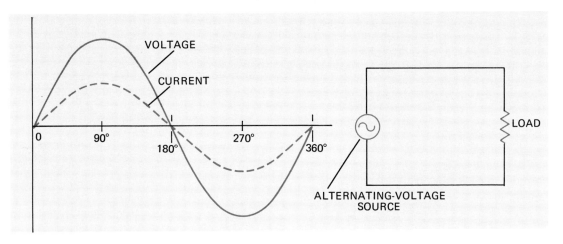

Fig. 11-5. "In-phase" relationship between a sinusoidal voltage and current.

and minimum magnitudes at exactly the same time. As a result, the waveform that represents the response of the current in the circuit will also be sinusoidal. The voltage and the current are said to be *in phase* with each other (Fig. 11-5).

THE MATHEMATICAL BASIS OF THE SINE WAVEFORM

Thus far, the sine waveform has been discussed only in terms of the result that is obtained when a conductor is moved within a magnetic field with a steady, circular motion. At this time, it is desirable to analyze the development of the waveform from a mathematical viewpoint. This will provide the basis for establishing further relationships that are important in the study of alternating voltage and current.

The rotation of the armature or of the field coil in a generator is an action that is periodic in nature, or one that repeats itself in the same order and at regular intervals of time. The result of this action, the generation of voltage, can then be analyzed by means of a trigonometric function that expresses the relationship between the position of the armature and the magnitude of the voltage at any instant.

The Sine Curve. The position of one conductor of a simple single-coil two-pole generator in relation to the magnetic field can be illustrated as shown in Fig. 11-6A. Here, the end view of

Fig. 11-6. The relationship between the position of one conductor of a single-coil, alternating-voltage generator and the magnitude of the instantaneous voltage generated across it.

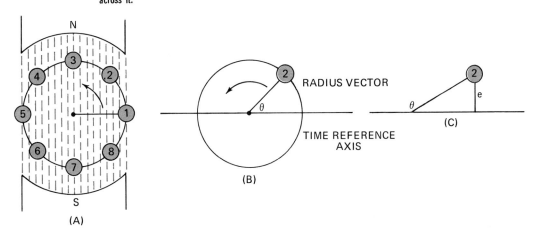

the conductor that is rotating in a counterclockwise direction is shown at eight different positions as it moves along a circular path. At each of these positions, the instantaneous value of the voltage generated across the conductor is proportional to the sin (sine) of the angle through which the conductor has rotated from the starting point of 0 degrees. This angle, represented by the Greek letter theta (θ), is the angle that exists between the horizontal time reference axis and the radius vector drawn to a given position (Fig. 11-6B). With the conductor at position 2, for example, the instantaneous voltage e can then be represented by a line drawn perpendicular to the time reference axis and extending from position 2 to the axis (Fig. 11-6C). If the radius vector is assigned a value of 1, the magnitude of the instantaneous voltage generated across the conductor at any position can be expressed by the following formula:

$$e = \sin \theta$$

When the above formula is graphed by plotting the values of e that correspond to a number of different angles of conductor rotation, the result is a *sine curve* or a *sinusoid* (Fig. 11-7). Therefore, the sine curve is a mathematical indication of the effects that are observed when the phenomenon of electromagnetic induction is applied to generator operation.

Relationship Between e and Maximum Voltage. As shown by the sine curve, the voltage generated across a conductor is at a maximum magnitude when its angle of rotation is 90 degrees. Therefore, at this angle, the instantaneous voltage is equal to the maximum voltage, or

$$e = E_{max}$$

$$\frac{e}{E_{max}} = 1$$

Then, since the sine of a 90-degree angle is 1,

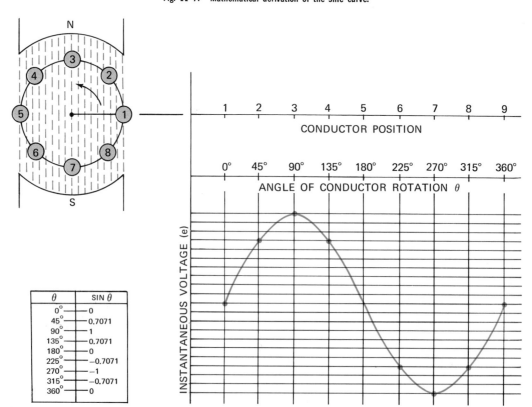

Fig. 11-7. Mathematical derivation of the sine curve.

$$\sin\theta = \frac{e}{E_{max}}$$

and, by cross-multiplying,

$$e = E_{max} \sin\theta$$

The Voltage Vector Diagram. When the maximum magnitude of the generated voltage is known, the relationship between this voltage, the instantaneous voltage, and the angle of conductor rotation can be conveniently illustrated by means of a vector diagram. On such a diagram, the rotating conductor is represented by a line or vector that represents the maximum magnitude E_{max} of the generated voltage at a given angle of rotation. With this vector and the time reference axis as the sides of the angle, the magnitude of the instantaneous voltage e can then be represented as the vertical component of the vector diagram (Fig. 11-8).

A single conductor is, in effect, one unit of any number of conductors that may be formed into the shape of a coil such as the armature coil of a practical generator. Therefore, a voltage vector diagram can also be used to represent the relationships that exist between the voltages and the angle of rotation as indicated by any sine waveform.

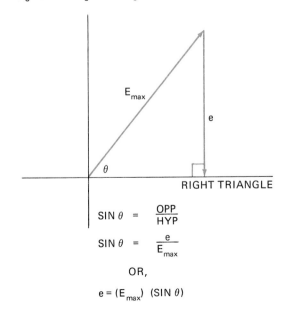

Fig. 11-8. Voltage vector diagram.

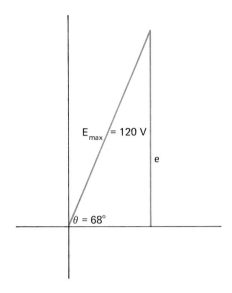

Fig. 11-9. Relationship between voltage and angle of rotation.

Problem 1. The maximum magnitude of a sinusoidal voltage is 120 volts. Compute the instantaneous magnitude of the voltage at the 68-degree point of its cycle.

Solution. The vector diagram that represents the conditions of this problem is shown in Fig. 11-9. A reference to a table of natural trigonometric functions will show that the sine of a 68-degree angle is 0.9272. Since

$$e = E_{max} \sin\theta$$

then

$$e = 120 \times 0.9272 = 111.3 \text{ volts}$$

PEAK, AVERAGE, AND EFFECTIVE VALUES

The instantaneous magnitude of a sinusoidal voltage or current changes constantly during a cycle. For this reason, it is not practical to use this indication of magnitude when different voltages or currents are compared with each other. To make this comparison, the magnitude of sinusoidal voltages and currents is expressed in terms indicating a specific quantity that remains constant for any given voltage or current. These *terms*, commonly referred to as *values*, are the *peak*, the *average*, and the *effective values*.

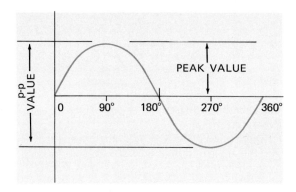

Fig. 11-10. Peak and peak-to-peak values of a sinusoidal voltage.

Peak Value. The peak value of a sinusoidal voltage is its maximum magnitude, which occurs at the 90- and the 270-degree points of the cycle as represented by the sine waveform. This value may apply to either the positive or the negative "peak" since they are equal in magnitude. In some circuit applications, it is desirable to express the magnitude of a given voltage in terms of its peak-to-peak (p-p) value. This is equal to twice the value of a single peak (Fig. 11-10).

Average Value. The average value of one complete cycle of a sinusoidal voltage or current is zero since the positive alternation (one half cycle) of its sine waveform is equal to the negative alternation. Therefore, when the term "average" is used in conjunction with such a waveform, it is applied to either the positive or the negative alternation.

Fig. 11-11. Instantaneous magnitudes of a sinusoidal voltage used in computing the average value of the voltage.

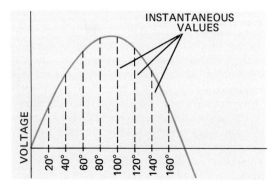

The average value can be determined quite accurately by finding the mean (average) of a number of instantaneous magnitudes that are separated by an equal number of degrees on the sine waveform time reference axis (Fig. 11-11). By using this and other methods of computation, the mathematical relationships between the average value and the peak or maximum value have been found to be as follows:

$$E_{av} = 0.637 E_{max}$$
$$I_{av} = 0.637 I_{max}$$
$$E_{max} = 1.57 E_{av}$$
$$I_{max} = 1.57 I_{av}$$

Effective (RMS) Value. When a direct current passes through a resistor, the energy of moving electrons is converted into heat energy at a constant rate that is proportional to the square of the current multiplied by the resistance. When an alternating current passes through a resistor, the rate of heat production is not constant since the instantaneous magnitude of the current is constantly changing. However, it can be determined that a certain amount of alternating current will produce the same heating effect as does direct current when both pass through equal resistors for equal periods of time. This amount or magnitude of the alternating current is known as the *effective value* of the current.

The effective value of one complete cycle of a sinusoidal voltage or current is computed by squaring a large number of instantaneous values that occur during the cycle, finding the average of these squared values, and extracting the square root of this average. As a result of this mathematical process, the effective value is more commonly known as the root-mean-square (rms) value.

The rms value is the most common method of specifying the magnitudes of alternating voltage and current. It is also the magnitude that is indicated by typical voltmeters and ammeters. Therefore, unless otherwise stated, a given magnitude of alternating voltage or current can always be assumed to be the rms value.

It has been determined that the rms value of a voltage or a current is very nearly equal to the instantaneous magnitude occurring at the 45-degree point of their respective cycles. Since this is 70.7 percent of the peak (maximum) value, these relationships can be stated by the following formulas:

$$E = 0.707 E_{max}$$
$$I = 0.707 I_{max}$$
$$E_{max} = 1.414 E$$
$$I_{max} = 1.414 I$$

where E, I = rms values
$$E_{p\text{-}p} = 2.828 E_{rms}$$

Problem 2. Compute the rms value of an alternating current that has a maximum magnitude of five amperes.

Solution.

$$I = 0.707 I_{max}$$
$$= 0.707 \times 5 = 3.535 \text{ amperes}$$

Problem 3. Compute the peak, the peak-to-peak, and the average voltage values of the typical 120-volt home-appliance circuit.

Solution.

$$E_{max} = 1.414 E$$
$$= 1.414 \times 120 = 169.68 \text{ volts}$$
$$E_{p\text{-}p} = 2 \times 169.68 = 339.36 \text{ volts}$$
$$E_{av} = 0.637 E_{max}$$
$$= 0.637 \times 169.68 = 108.09 \text{ volts}$$

THREE-PHASE VOLTAGE AND CURRENT

The voltage induced within a generator winding that is, in effect, a single coil is known as a single-phase voltage. When this voltage is applied to a circuit, it results in a single-phase current. In large, power-station generators and

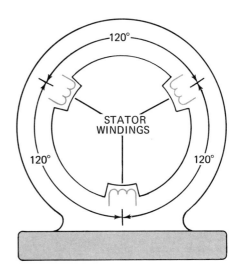

Fig. 11-12. Position of stator winding coils in a three-phase generator.

in automobile alternators, which are usually of the rotating-field type, the stator windings consist of three separate coils equally spaced around the circularly positioned frame pole pieces (Fig. 11-12). As a result, the coils are 120 degrees apart both mechanically and electrically.

The operation of such a generator produces three different voltages, which are said to be 120 degrees out of phase with each other. This means that each of the voltages attains a peak value at a point that is 120 degrees from the points at which the two other voltages reach their peak values during any given cycle (Fig. 11-13). Since the current waveforms that result

Fig. 11-13. Three-phase voltage waveforms.

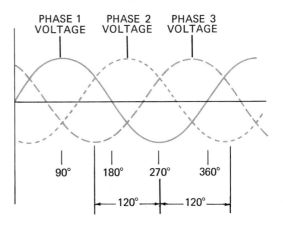

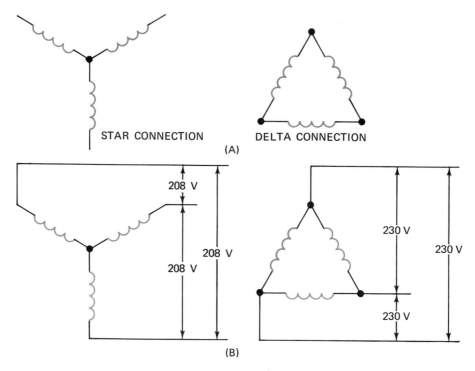

Fig. 11-14. Three-phase connections: (A) star and delta connections, (B) voltage relationships between typical star- and delta-connected three-phase distribution lines.

when a three-phase voltage is applied to a circuit are similar to the voltage waveforms, the current is referred to as a three-phase current.

When three-phase current is used, a more constant supply of power is delivered to a given load. This is because during any particular voltage-current cycle, there is less fluctuation of the current delivered to the load than there is in a system using single-phase current. Because of the reduced fluctuations, three-phase electricity is very often used for the operation of heavy-duty devices, including motors, welders, and heating appliances.

The stator coils of a three-phase generator are connected together in one of two ways: the star or Y connection or the delta (Δ) connection (Fig. 11-14). In both of these connections, three-phase energy is transmitted over a three-wire distribution system.

When the star connection is used, each pair of line wires is connected in series with two stator coils. Under this condition, the voltage present across any two lines is equal to $\sqrt{3}$, or 1.73 times the voltage generated across any single coil. The most common line voltage (voltage across any two leads of the three-phase generator) produced by typical, star-connected systems utilizing three-phase distribution transformers is 208 volts.

A fourth line wire (the neutral) is commonly added to the star-connected system. This arrangement makes it possible, for example, to obtain three-phase voltage at 208 volts and three additional single-phase lines, each with a voltage of 120 volts (Fig. 11-15). Because of this feature and other factors, the star connection is rapidly replacing the delta connection in modern electrical distribution systems.

In the delta-connected system, the output voltage across any two of the three lines is equal to the voltage generated across any given coil. The most common line voltage produced by a typical delta-connected distribution system is 230 volts.

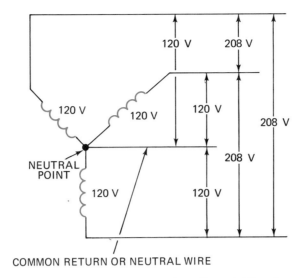

Fig. 11-15. Single-phase 120-volt lines obtained from a Y-connected three-phase system.

NONSINUSOIDAL WAVEFORMS

The sinusoidal-waveform voltage generated by the conventional rotating-field or rotating-armature generator is, for all practical purposes, "pure" in nature. However, a voltage that exhibits an absolutely pure sine waveform is very seldom generated.

Certain transistor and electron tube circuits known as *oscillators* produce what is essentially a pure sinusoidal voltage known as the *fundamental* and a number of component sinusoidal voltages known as *harmonics*. The harmonic of a given fundamental frequency is a frequency that is an exact, whole multiple of the fundamental. For example, the second harmonic of a fundamental frequency of 1,000 kilohertz is 2,000 kilohertz, the third harmonic is 3,000 kilohertz, and the fourth harmonic is 4,000 kilohertz.

The result of the combination of a fundamental sine waveform and its component harmonics is a distorted nonsinusoidal waveform called the *composite*. Such waveforms are often used in special circuit applications and are produced by clipping, clamping, or limiting circuits that reshape a given sinusoidal waveform into the particular waveform desired. The most commonly used nonsinusoidal waveforms are peaked, square, and sawtooth waveforms.

The Peaked Waveform. The composition of the peaked waveform is shown in Fig. 11-16. Here, the fundamental waveform *A* is associated with its third harmonic, waveform *B*. The magnitude of the composite waveform *C* is at any point equal to the sum of waveforms *A* and *B*. As a result, the composite waveform is peaked. In practical circuits where this type of waveform is desired, the peaked effect is usually made more pronounced by the introduction of additional harmonics.

Fig. 11-16. Composition of peaked waveform.

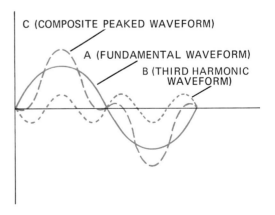

99

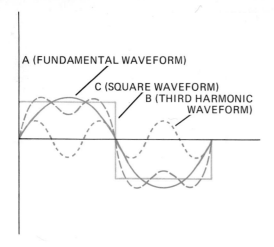

Fig. 11-17. Composition of square waveform.

The Square Waveform. The square waveform is also the result of a combination of the voltages of the fundamental frequency and its third harmonic (Fig. 11-17). In this case, however, unlike that of the peaked waveform, the first alternation of harmonic B is of the same polarity as the first alternation of the fundamental waveform A. Thus, the composite, waveform C, tends to be square.

The Sawtooth Waveform. When the fundamental frequency waveform and its second harmonic are combined and their first alternations are of like polarity, the resulting composite waveform tends to be sawtooth in shape (Fig. 11-18A). To "sharpen" the sawtooth, it would be necessary to introduce certain other harmonics, which would then produce the waveform shown in Fig. 11-18B.

For Review and Discussion

1. Explain the phenomenon of electromagnetic induction.
2. What factors govern the magnitude of an induced voltage?
3. Explain how a sine waveform voltage is generated by a simple two-pole, single-coil, rotating-armature generator.
4. Define a sinusoidal voltage or current.
5. What is meant by the in-phase relationship between voltage and current?
6. By using appropriate vector diagrams, illustrate the generation of a sine waveform voltage by plotting the values of the instantaneous voltage generated across a rotating conductor at the 0-, 90-, 180-, 270-, and 360-degree positions of its rotation.
7. At which angles of rotation is the instantaneous voltage generated across a rotating conductor equal to the maximum voltage? What is this value of the voltage called?

Fig. 11-18. Composition of sawtooth waveform.

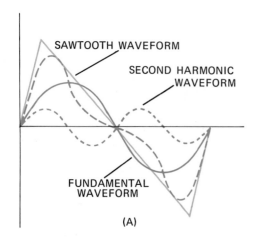

(A)

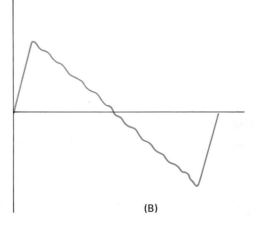

(B)

8. Draw a sine wave and use it to illustrate the peak-to-peak value of a sinusoidal voltage or current.

9. State the formula that expresses the relationship between the maximum value of a sinusoidal voltage and the average value of the voltage.

10. Define the effective (rms) value of a sinusoidal current.

11. Explain how the rms value of a sinusoidal voltage or current is computed.

12. State the formula that expresses the relationship between the rms value of a sinusoidal voltage and the maximum value of the voltage.

13. Describe the stator winding construction of a basic, rotating-field type, three-phase generator.

14. Describe and illustrate the waveforms of a three-phase voltage or current.

15. Draw diagrams representing the basic star and the Delta three-phase generator stator winding connections.

16. Draw a diagram showing the dual-voltage (208- and 120-volt) conditions between the various lines of a star-connected three-phase power distribution system.

17. What advantage is gained by the use of three-phase power as compared with single-phase power?

18. Define a harmonic frequency.

19. What is the third harmonic of a fundamental frequency of 930 kilohertz?

20. Describe the basic composition of a non-sinusoidal waveform.

21. Draw a peaked waveform and describe its composition.

22. Draw waveforms that represent square and sawtooth waveforms.

UNIT 12. Self-inductance

The phenomenon of self-inductance is present to a significant degree in any alternating-current circuit containing a coil-like device. Although self-inductance may not be useful in some circuits, it is essential for the operation of many others. Self-inductance, provided by coils or inductors, is used to limit current, and to provide the self-inductance necessary for the operation of oscillator, tuned, and filter circuits. The effect of self-inductance upon the operation of circuits is discussed in detail in Units 16 and 18.

CHARACTERISTICS OF SELF-INDUCTANCE

When the current through a conductor varies in magnitude, the flux density of the magnetic field generated about the conductor also varies in direct proportion. This change in the flux density, regardless of whether the magnetic field is expanding or collapsing about the conductor, is equivalent to a magnetic field in motion. Therefore, the magnetic field is capable of inducing a voltage within the conductor itself. The property whereby a conductor induces a voltage within itself by means of its own magnetic field is called *self-inductance* and is represented by the letter symbol L.

Basic Action. The basic action of self-inductance can be illustrated by the circuit shown in Fig. 12–1A. When the circuit's switch is closed (from position 1 to position 2), current instantly begins to move through the coil. This, in turn, generates an expanding magnetic field about the coil while the current is increasing to its maximum magnitude. As a result, a voltage is induced across the coil. According to *Lenz' law*, this induced voltage will always be of a polarity that will cause it to oppose the voltage applied by the battery. The induced voltage is therefore a *counter-electromotive force* (cemf) that, in effect, opposes the increase of current.

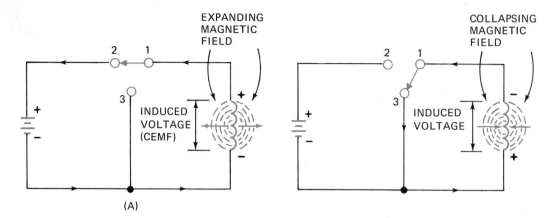

Fig. 12-1. Basic action of self-inductance.

When the current in the circuit reaches its maximum magnitude, it remains constant for as long as the circuit is operated in this manner. The magnetic field about the coil is no longer in motion (expanding), and therefore the cemf ceases to exist. In this condition, the current is limited only by the magnitude of the battery voltage and the ohmic resistance of the circuit.

If the switch is now closed from position 1 to position 3, the energy that was stored in the magnetic field about the coil will be returned to the circuit as the field collapses (Fig. 12-1B). This action once again causes a voltage to be induced across the coil. The polarity of the induced voltage is now reversed, or in a direction that tends to sustain the current in the coil circuit even though the battery voltage has been removed from the circuit. As a result, the induced voltage again, and in accordance with Lenz' law, tends to oppose a change (decrease) of current in the circuit. Because of this effect, the current does not instantly become zero. Self-inductance can then be defined as the circuit property that opposes any change in current.

Inductive Reactance and Impedance. It should now be apparent that self-inductance will generate a cemf in an alternating-current circuit at all times, since self-inductance tends to oppose any change (increase or decrease) in the magnitude of current. This opposition to current is called *inductive reactance* and is represented by the letter symbol X_L. The unit of inductive reactance is the ohm. The total opposition to current in an alternating-current circuit presented by any combination of resistance, capacitive reactance, and inductive reactance is called *impedance* and is represented by the letter symbol Z. The unit of impedance is also the ohm.

THE HENRY

The unit of self-inductance is the henry (H). A coil has an inductance of one henry if a voltage of one volt is induced across it when the current through the coil is changing at the average rate of one ampere per second (Fig. 12-2). Expressed as a formula,

$$L = \frac{E_{ind}}{\Delta I / \Delta t}$$

where L = inductance, in henrys
E_{ind} = induced voltage, in volts
$\frac{\Delta I}{\Delta t}$ = rate of current change in amperes per second

We can solve for E_{ind} in the self-induction formula by using the process of cross-multiplication. Thus

$$E_{ind} = L \frac{\Delta I}{\Delta t}$$

Note: The symbol Δ is the Greek letter delta that is commonly used to designate a "variation in" or a "change of."

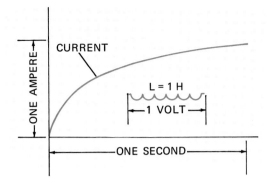

Fig. 12-2. Graphical representation of the henry unit of self-inductance.

An examination of this formula will show that the induced voltage is inversely proportional to the time during which a given variation of current takes place. In other words, as t decreases with any given variation of current, the magnitude of the induced voltage increases. This is because a more rapid variation of current increases the velocity of the magnetic field motion as it either expands or collapses. This results in a corresponding increase of induced voltage.

Problem. Compute the self-inductance of a coil through which a current that changes from 0.1 to 1.3 amperes per second causes a voltage of 8 volts to be induced across it.

Solution.

$$L = \frac{E_{ind}}{\Delta I / \Delta t}$$

$$= \frac{8}{1.3 - 0.1/1}$$

$$= \frac{8}{1.2} = 6.67 \text{ henrys}$$

FACTORS AFFECTING INDUCTANCE

A given coil has a fixed inductance that is determined by a number of constructional factors. Among the most important of these factors are:

1. The number of turns of wire wound upon the coil. The inductance is proportional to the square of the number of turns.

2. The cross-sectional area of the core. The inductance increases as the cross-sectional area increases.

3. The length of the coil. The inductance is inversely proportional to the length of the coil.

4. The permeability of the core material. The inductance increases as the permeability increases.

The approximate inductance of a solenoid having a coil that is at least ten times longer than the diameter can be determined by using the following formula:

$$L = \frac{1.26 N^2 \mu A}{\ell 10^8}$$

where L = inductance, in henrys
N = number of coil turns
μ = permeability of core material
A = core area, in square centimeters
ℓ = core length, in centimeters

INDUCTORS

The effects of self-inductance exhibited by both iron-core and air-core coils called *inductors* or *choke coils* are utilized in a wide variety of tuning and filter circuits. Iron-core inductors are generally used in circuits where the frequencies are in the audio range (below 20 kilohertz). When frequencies higher than this are present, air-core and ferrite-core inductors are used. Several types of inductors are shown in Fig. 12-3.

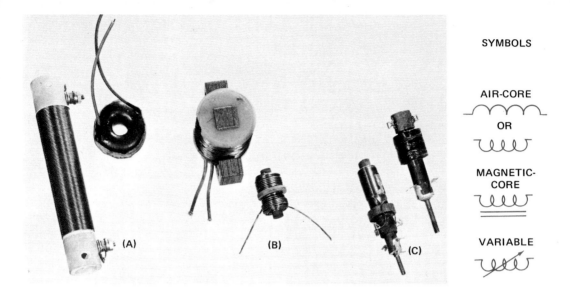

Fig. 12-3. Common types of inductors: (A) air-core, (B) fixed-ferrite-core, (C) variable-ferrite-core. (Gail E. Henderson)

Variable Inductors. Inductors are constructed to be either fixed or variable. In the variable inductor, which is usually used in high-frequency circuits only, the position of the core within its coil can be adjusted. This is most often accomplished by threading the core so that its position can be controlled (moved further into or out of the coil) with the aid of a screwdriver (Fig. 12-4).

When the core is moved into the coil the permeability of the coil's magnetic circuit is increased, thus increasing the inductance. When the core is unscrewed, the permeability of the magnetic circuit decreases and the inductance decreases. This process of changing the inductance of an inductor by adjusting the position of the core is commonly referred to as either *slug* or *permeability* tuning.

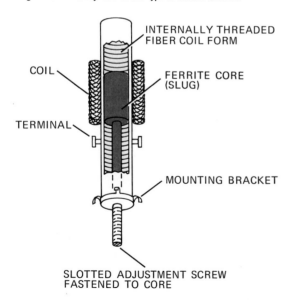

Fig. 12-4. Cutaway view of one type of variable inductor.

Q of an Inductor. The Q or *quality factor* of an inductor is the ratio of its inductive reactance in any given circuit to its ohmic resistance when both of these quantities are stated in ohms. Therefore

$$Q = \frac{X_L}{R}$$

This factor provides an indication of how efficiently a coil will serve as an inductor under a specific circuit condition. As the Q increases, the efficiency of the inductor also increases.

Rating. Iron-core inductors, known as *filter* and *smoothing chokes*, are rated in terms of inductance, ohmic resistance, and the safe current carrying capacity of the coil wire. High-frequency air or ferrite-core inductors, commonly referred to as *radio-frequency (r-f) chokes*, are usually rated in terms of inductance, Q, and the frequency at which they are designed to operate.

For Review and Discussion

1. Explain the process of self-inductance.
2. Does self-inductance occur in a direct-current circuit after the current has reached the maximum (steady-state) magnitude? Give the reason for your answer.
3. Explain why self-inductance is constantly present in an alternating-current circuit that contains one or more inductors.
4. Define inductive reactance and give the letter symbol by which it is represented.
5. Name and define the basic unit of self-inductance.
6. State four factors governing the self-inductance of an inductor or coil.
7. Describe the basic construction of one common type of variable inductor.
8. What is meant by the Q of an inductor?
9. Name three rating characteristics commonly used in conjunction with an iron-core inductor such as a filter or smoothing choke.
10. Name three rating characteristics commonly used in conjunction with an r-f choke.

UNIT 13. Transformers

The phenomenon of electromagnetic induction is also exhibited in the process of *mutual induction*. The principal application of this process is found in the operation of the transformer, a device that utilizes a moving magnetic field to transfer electrical energy from one circuit to another.

CHARACTERISTICS OF MUTUAL INDUCTION

Mutual induction occurs when the magnetic field existing within the area surrounding one conductor cuts across another conductor and induces a voltage within it. While this condition exists even if the conductors are straight wires placed near each other, a much greater effect is produced by forming the conductors into coils.

When one of the coils is energized with an alternating current, a moving magnetic field cuts across the turns of the second coil (Fig. 13-1). The two coils are then said to be *linked* or *inductively coupled* together by magnetic flux lines. Since the turns of any coil are, in effect, connected in series with each other, the magnitude of the voltage induced across the second coil will be equal to the sum of the voltages induced in its individual turns.

Coefficient of Coupling. The portion of the total flux lines that establishes an effective magnetic link between two coils is referred to as the *coefficient of coupling*. By placing the coils closer together, by placing them parallel to each

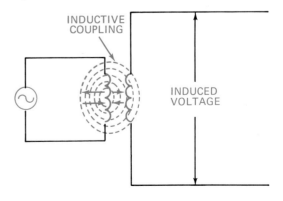

Fig. 13-1. Mutual induction.

105

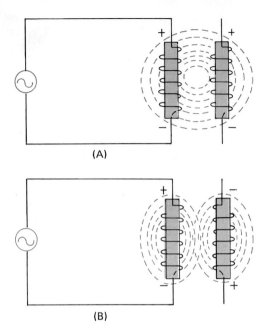

Fig. 13-2. Polarity of voltage generated by mutual induction: (A) coils wound in the same direction, (B) coils wound in opposite directions.

other, or by winding the coils around a common ferromagnetic core the coefficient of coupling can be increased.

Magnitude of the Induced Voltage. The magnitude of the voltage induced across a coil by means of mutual induction depends primarily upon three factors:

1. The number of flux lines that cut across the coil—the greater this number, the greater the magnitude of the induced voltage.
2. The frequency at which the flux lines cut across the coil. As the frequency is increased, the magnitude of the induced voltage increases.
3. The number of turns. Under any given condition, the magnitude of the induced voltage will increase as more turns of wire are wound around the coil.

Polarity of the Induced Voltage. If two inductively coupled coils are wound in the same direction (clockwise or counterclockwise), the polarity of the voltages across these coils will be identical at any instant (Fig. 13-2A). In this case, the voltages are said to be *in phase* with each other. When the coils are wound in opposite directions, the polarity of the voltage across the coil in the energized circuit is, at any instant, opposite the polarity of the induced voltage (Fig. 13-2B). Now, the voltages are said to be 180 degrees *out of phase* with each other.

THE TRANSFORMER

A transformer consists basically of two coils electrically insulated from each other and wound upon a common core. As in the case of inductors, the core is either air or a ferromagnetic material, depending upon the frequency at which the transformer is designed to operate.

In the transformer, magnetic coupling is used to transfer electrical energy from one coil to another. The coil that is energized by connecting it to a source of energy is called the *primary winding*. The coil to which electrical energy is transferred and that delivers this energy to a circuit load is called the *secondary winding* (Fig. 13-3).

Fig. 13-3. Principal parts of a basic transformer.

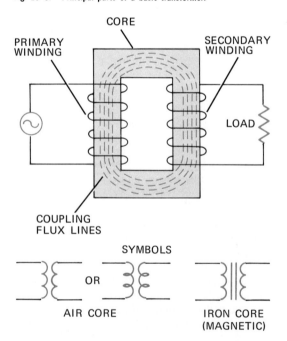

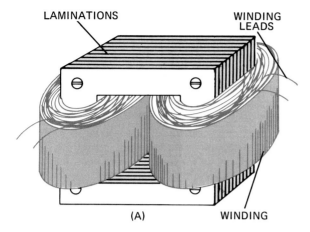

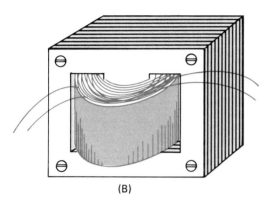

Fig. 13-4. Laminated transformer cores: (A) shell core, (B) closed core.

Core Construction. In most transformers (and inductors) designed to operate at audio frequencies, the core is of a type that provides a continuous ferromagnetic path through which flux lines can pass. This type of core makes it possible to obtain a high degree of transformer efficiency because of the resulting high coefficient of coupling and high permeability.

Although many different materials are available for the construction of power and audio transformer cores, an alloy consisting of pure iron and from 1 to 5 percent silicon is most commonly used. This material has the desired permeability characteristic and is very efficient in minimizing hysteresis losses within the core. Sheets of this alloy are first coated with a naturally formed insulating oxide scale or with an insulating varnish, and then sandwiched together to form laminated cores (Fig. 13-4).

The cores of high-frequency transformers (and inductors) are most often made of ferrite and usually do not form a closed magnetic circuit. Instead, the cores are simply shaped into the form of rods or bars (Fig. 13-5). Since ferrite cores are extremely permeable and because high-frequency currents provide for a high degree of mutual induction (because of the velocity of their magnetic fields), the "open" core is quite efficient.

Eddy Currents. As a magnetic field expands and collapses about the windings of an iron-core transformer, its flux lines cut across both the turns of the windings and the core. As a result, voltages are induced within the core itself. These voltages, in turn, establish currents called *eddy*

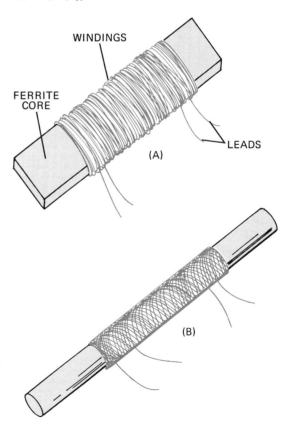

Fig. 13-5. Radio receiver antenna transformers: (A) flat type, (B) rod (stick) type.

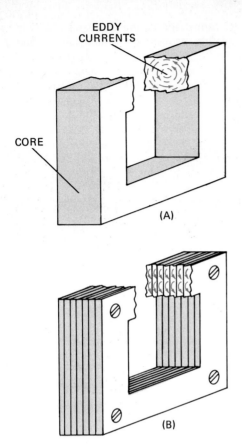

Fig. 13-6. Eddy currents in transformer cores.

currents that move through the core in circular paths (Fig. 13-6A).

Since eddy currents create heat within the core and do not aid in the process of mutual induction, they result in a waste of energy that is referred to as an *eddy-current loss*. In low-frequency iron-core transformers and inductors, laminated cores are used to reduce eddy current losses. Because the sheets or laminations of such cores are insulated from each other, the cross-sectional resistance of the cores is high. Thus, the magnitude of eddy currents is reduced (Fig. 13-6B). In high-frequency transformers and inductors, eddy-current losses are reduced to a minimum by the extremely high resistance of the ferrite core material.

Copper Losses. In addition to the energy losses resulting from hysteresis and eddy currents, a third energy loss occurs during the operation of a transformer. This loss, known as *copper loss*, results from the ohmic resistance of the windings and is equivalent to I^2R as stated by the power formula $P = I^2R$.

One of the factors influencing copper loss is resistance. Because resistance increases with temperature, copper loss also increases with an increase of temperature. For this reason, large electrical power distribution system transformers that operate under high-current conditions are often cooled by means of forced air, water, or oil circulation (Fig. 13-7). In addition to reducing copper losses, the cooling of such transformers helps to prevent high temperature damage to winding insulation materials.

OPERATING CHARACTERISTICS OF TRANSFORMERS

A transformer, in addition to transferring electrical energy from one circuit to another, can also be designed to "deliver" this energy under different conditions of voltage and current. This is accomplished by constructing the transformer so that a definite ratio exists between the number of primary- and the number of secondary-winding turns.

Voltage Ratio. If an iron-core transformer is assumed to be operating under an ideal (perfect) condition of 100 percent or unity coupling, the ratio of the primary voltage to the secondary (induced) voltage is equal to the ratio existing between the number of primary and secondary winding turns. This relationship is expressed by the following formula:

$$\frac{E_p}{E_s} = \frac{N_p}{N_s}$$

where E_p = primary voltage, in volts
E_s = secondary voltage, in volts
N_p = number of primary winding turns
N_s = number of secondary winding turns

This formula shows that the voltage ratio can be varied by changing the ratio between the primary and secondary winding turns. If the primary winding has more turns than the secondary, the voltage is reduced or "stepped down." If the primary winding has fewer turns than the secondary, the voltage is increased or "stepped up" (Fig. 13-8).

HEAT EXCHANGERS

Fig. 13-7. Power transformers with forced oil cooling through heat exchangers mounted on the sides. (Pumps at the top of each heat exchanger force hot oil through the heat exchanger, which is cooled by means of integrally mounted fans.) (*Allis-Chalmers Manufacturing Company*)

Fig. 13-8. Simplified diagrams of step-down and step-up transformers: (A) step-down transformer, (B) step-up transformer.

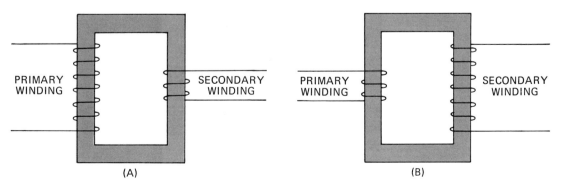

109

Needless to say, a coefficient of coupling of 100 percent is never achieved in the operation of a typical transformer. This is because magnetic circuits, like electrical circuits, are never perfect. However, modern power transformers often have a coefficient of coupling of 98 percent or more.

Problem 1. A 60-hertz, iron-core transformer that is operated from a 120-volt line has a primary winding consisting of 500 turns and a secondary winding of 100 turns. Compute the secondary winding voltage.

Solution.

$$\frac{E_p}{E_s} = \frac{N_p}{N_s}$$

$$\frac{120}{E_s} = \frac{500}{100}$$

$$500 \times E_s = 120 \times 100$$

$$500 E_s = 12{,}000$$

$$E_s = \frac{12{,}000}{500} = 24 \text{ volts}$$

Power Transfer. Although a transformer can "step up" or "step down" voltage, it cannot, under any condition, transfer more power into the secondary winding than exists in the primary winding circuit. If core and copper power losses are neglected, the power in the primary circuit will equal the power that is available to a load in the secondary circuit, or

$$E_p I_p = E_s I_s$$

where I_p = primary winding current, in amperes
I_s = secondary winding current, in amperes

This formula shows that if the voltage is stepped up, the current must be reduced or stepped down if the EI products in the windings are to be equal. Similarly, if the voltage is stepped down, the current available to the secondary load will be greater than the current in the primary winding.

Problem 2. When the primary winding of a certain transformer is operated at 120 volts, the current in the winding is 2 amperes. Compute the maximum current that will be available to the secondary winding load if the voltage is stepped up to 600 volts.

Solution.

Since $E_p I_p = E_s I_s$

then

$$I_s = \frac{E_p I_p}{E_s}$$

$$= \frac{120 \times 2}{600}$$

$$= \frac{240}{600} = 0.4 \text{ ampere}$$

Efficiency. The efficiency of a transformer is a measure of the amount of power delivered by its secondary winding when a given amount of power is applied to the primary winding. The efficiency of a transformer is also an indication of the extent of the core and copper losses occurring within the transformer. Expressed as a formula,

$$\text{Efficiency (in percent)} = \frac{P_s}{P_p} \times 100$$

where P_p = power delivered to primary winding, in watts
P_s = power available to load, in watts

Problem 3. A transformer whose primary operates at 100 watts delivers 95 watts of power to a resistance wire heating element. Compute the efficiency of this transformer.

Solution.

$$\text{Efficiency} = \frac{P_s}{P_p} \times 100$$

$$= \frac{95}{100} \times 100$$

$$= 0.95 \times 100 = 95 \text{ percent}$$

Operation without Load. When a transformer is operated without a load connected to its secondary winding, its operation is similar to that of a single iron-core coil. In this condition, the voltage applied to the primary winding is opposed by the cemf that is induced across the winding because of self-inductance. Since the applied voltage is only slightly higher than the cemf, the impedance of the coil limits the current to the comparatively small value necessary to magnetize the core. This current is referred to as the *energizing* or *excitation current* of the transformer.

Operation with Load. When the secondary winding of a transformer is connected to a load, the current in this winding produces a magnetic field that opposes the magnetic field of the primary winding. The effect of this action is to decrease the flux density of the primary magnetic field and also to cause a reduction of the cemf that is induced across the primary winding. As a result, the voltage applied to the primary winding has a greater influence upon the primary current, and the current increases. This, in turn, causes the primary field flux density to increase, thus increasing the induced voltage, the current, and the magnetic field flux density of the secondary winding. Thus, the increase in the flux density of the primary field is counteracted and the flux density within the core remains approximately the same as that which existed under the no-load condition.

Operation with Direct Current. If the primary winding of a transformer is energized with "pure" direct current, a moving magnetic field will not be generated about the winding. This will prevent mutual induction and self-induction within the winding itself. As a result, since the current is limited only by the relatively low ohmic resistance of the winding, an excessive current will pass through the primary winding, causing the transformer to become overheated.

When a transformer is operated in a direct-current circuit, a breaker or vibrator device must be used to produce a variation in the magnitude of the current. A common application of such a device is found in the typical automotive ignition system. Here, breaker (contact) points located within the distributor are connected between the battery and the primary winding of the ignition coil, which is an iron-core, step-up transformer (Fig. 13-9). The points act as a switch that "breaks up" the battery current into a pulsating direct current. Pulsating direct current, in turn, generates the moving magnetic field needed to operate the ignition coil.

It is important to note that, when the primary winding of a transformer is energized with

Fig. 13-9. Principal components of typical battery-breaker point automobile ignition system. (*Delco-Remy, Division of General Motors Corp.*)

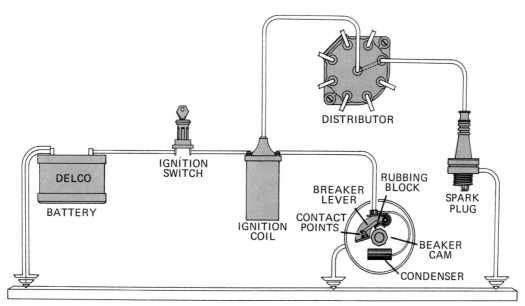

pulsating or some other form of varying direct current, an alternating voltage is induced across the secondary winding. This is because the "on" and "off" conditions of the current generate an expanding-collapsing magnetic field that has the effect of reversing the polarity of the induced voltage.

THE POWER TRANSFORMER

The power transformer is used primarily to couple electrical energy from a power-supply line to a circuit system or to one or more components of the system. In one type of power transformer, there are three separate secondary windings, each designed for a different voltage and current output (Fig. 13-10A). These include a high-voltage winding that is usually center-tapped and two low-voltage windings that may or may not be center-tapped. All of the windings, including the primary, are identified by means of a color code (Fig. 13-10B). Power transformers used in conjunction with solid-state circuits are commonly referred to as *rectifier transformers*.

Rating. The rating of a power transformer is usually given in terms of the maximum voltage and current-delivering capacity of its secondary windings. For example, the specifications pertaining to a particular transformer that is to be operated from a 60-hertz, 120-volt power line may read as follows: 600 V. CT @ 90 ma, 6.3 V. @ 3 amp, 5 V. @ 2 amp. The total power or wattage rating of this transformer can be determined by adding the power (*EI*) ratings of the individual secondary windings. Thus, its total power rating is equal to (600 × 0.09) + (6.3 × 3) + (5 × 2) = 82.9 watts.

Efficiency. The efficiency of power transformers having a power rating of from 50 to 400 watts ranges from approximately 82 to 94 percent. Transformers that have higher power ratings are generally more efficient because of the larger wire that is used in their windings and the larger cross-sectional area of their cores.

THE AUTOTRANSFORMER

The autotransformer is a special type of power transformer. It consists of a single, continuous winding that is tapped to provide the necessary step-up or step-down function. When it is used as a step-up transformer, the entire primary winding is a part of the secondary winding. When used as a step-down transformer, the entire secondary winding is part of the primary winding (Fig. 13-11).

The autotransformer is a compact, highly efficient unit that provides good voltage regulation. This means that the voltage across the secondary winding remains rather constant, even though the secondary current may vary over a wide range. The chief disadvantage of the autotransformer is that it does not electrically isolate the primary power line from the secondary load. Under certain conditions, this situation can introduce the danger of electrical shock.

Fig. 13-10. Power transformer: (A) external view, (B) winding color code.

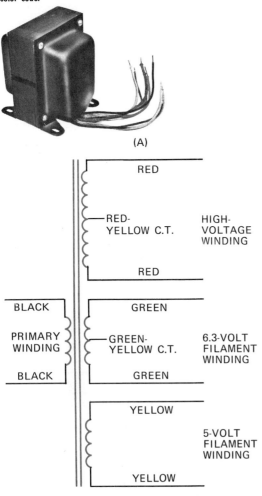

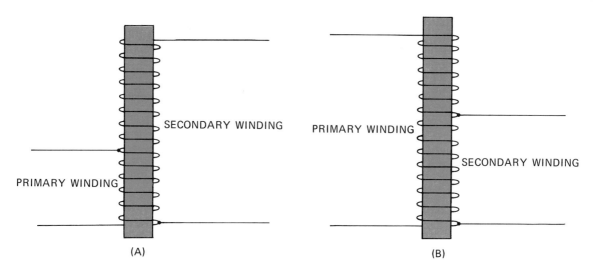

Fig. 13-11. Diagrams of basic autotransformer: (A) step-up transformer, (B) step-down transformer.

THE ISOLATION TRANSFORMER

An isolation transformer has a 1:1 turns ratio. Therefore, it does not step voltage up or down. Its primary function is to serve as a safety device by isolating the grounded conductor of a power line from a chassis or from any portion of a circuit load (Fig. 13-12). The use of an isolation transformer does not, however, reduce the danger of shock if contact is made across the secondary winding of the transformer itself.

IMPEDANCE-MATCHING TRANSFORMERS

A maximum amount of power is transferred from a source to a given load when the impedance (or resistance) of the load is equal to the internal impedance of the source of power. This condition can be illustrated by means of a simple battery resistor circuit in which the impedance of the battery is, for all practical purposes, equal to its internal resistance R_B (Fig. 13-13). In this circuit, the resistance of the load, R_L, is varied from 4 ohms to 0.25 ohm (as shown by the accompanying table), while the internal resistance of the battery remains constant. Notice that, as the load resistance approaches the internal resistance of the battery, the power delivered to the load increases toward the maximum, which occurs when R_L equals (matches) R_B. When the load resistance is either less or more than the internal resistance of the battery, the power delivered to the load decreases.

The Transformer as an Impedance-Matching Device. The transformer is a very useful impedance-matching device. By constructing its winding so that it has a definite turns ratio, the transformer can be made to perform any impedance-matching function. The turns ratio automatically establishes the proper relationship between the primary and secondary winding impedances. This relationship is expressed by the following formula:

$$\frac{N_p}{N_s} = \sqrt{\frac{Z_p}{Z_s}}$$

where N_p = number of primary winding turns
N_s = number of secondary winding turns
Z_p = impedance of primary winding, in ohms
Z_s = impedance of secondary winding, in ohms

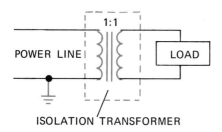

Fig. 13-12. Application of the isolation transformer.

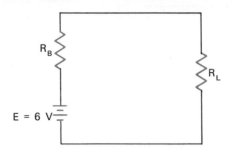

E (volts)	R_B (ohms)	R_L (ohms)	Total circuit resistance (R_t) $R_t = R_B + R_L$	I (amperes) $I = E/R_t$	Power delivered to load (watts) $P = I^2 R_L$
6	1	4	5	1.2	5.76
6	1	3	4	1.5	6.75
6	1	2	3	2	8
6*	1	1	2	3	9
6	1	0.5	1.5	4	8
6	1	0.33	1.33	4.51	6.71
6	1	0.25	1.25	4.8	5.76

*Maximum power transfer occurs when the load resistance equals the source resistance.

Fig. 13-13. Circuit and tabular data showing that the maximum amount of power is transferred from any source to a load when the resistance or the impedance of the load is equal to the internal impedance of the source of energy.

Problem 4. A certain energy source that has an impedance of 14,400 ohms is to be used in conjunction with a load that has an impedance of 400 ohms. Compute the turns ratio of the transformer that should be used to provide the proper impedance match between the source and the load in order to allow maximum power transfer to the load.

Fig. 13-14. Circuit illustrating Problem 4.

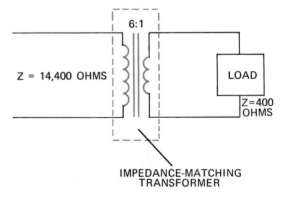

Solution.

$$\frac{N_p}{N_s} = \sqrt{\frac{Z_p}{Z_s}}$$

$$= \sqrt{\frac{14,400}{400}}$$

$$= \sqrt{36} = 6$$

Therefore, the transformer used in this circuit would be a step-down transformer having a primary-to-secondary turns ratio of 6:1 (Fig. 13-14).

The Output Transformer. The output transformer is often used in radio and television amplifier circuits to connect the last (power stage) transistor or electron tube of the audio amplifier to the loudspeaker. The purpose of this transformer is to match the relatively high impedance of the transistor or the tube output circuit to the low impedance of the loudspeaker voice coil (Fig. 13-15).

Output transformers are usually rated in terms of their primary and secondary winding impedances, and the maximum power in watts that the secondary winding can deliver to a load. When selecting an output transformer for use in a given circuit, four factors should be considered:

1. The primary winding impedance should be approximately the same as the impedance of the circuit stage to which it is to be connected.

2. The impedance of the secondary winding should be approximately the same as the impedance of the loudspeaker voice coil to which it is to be connected.

3. The power (wattage) rating of the transformer should be at least equal to the power rating of the loudspeaker.

4. The frequency response of the transformer should provide the expected range of sounds.

The Microphone Transformer. The microphone transformer is commonly used to match the impedance of a given microphone cartridge to either a high- or a low-impedance line (Fig. 13-16). This makes it possible to use one microphone with both high- and low-impedance amplifier-input circuits.

When a microphone is to be located a relatively long distance from its associated amplifier, the low-impedance operation provides the greatest degree of efficiency. When the distance between the microphone and the amplifier is comparatively short, the microphone is normally operated as a high-impedance device.

The Interstage Transformer. The interstage transformer is used to inductively couple the various transistor or electron-tube amplifier stages in many different kinds of circuit systems, including radio and television receivers. In a typical receiver, one type of interstage transformer known as an intermediate-frequency (i-f) transformer is designed to operate at one specific frequency. Such a transformer often consists

Fig. 13-16. Microphone transformer.

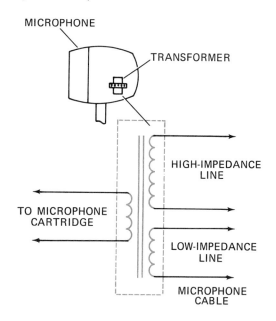

Fig. 13-15. Output transformer used to match the output impedance of a transistor amplifier stage to a loudspeaker.

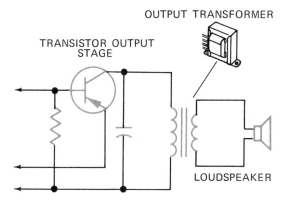

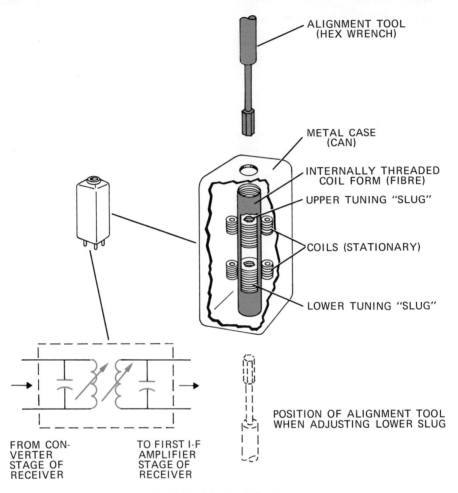

Fig. 13-17. Interstage i-f transformer.

of two windings wound upon a common tubular form housing adjustable ferrite cores or slugs (Fig. 13-17). This makes it possible to adjust the inductance of each of the winding circuits so that it will correctly match the impedance of the circuit stage to which it is connected. The slugs are adjusted to the proper position with a plastic or fibre tool called an *alignment tool*.

The intermediate-frequency transformer is most often operated in conjunction with capacitors that are connected across each of its windings (usually located within the transformer case). These capacitors act in conjunction with the windings to provide the correct frequency response.

For Review and Discussion

1. Explain the process of mutual induction.
2. Define coefficient of coupling.
3. In what three ways can the coefficient of coupling between two coils be increased?
4. State four factors that influence the magnitude of the induced voltage generated by means of mutual induction.
5. Describe the construction of a basic transformer and define its principal components.
6. What advantage is gained by the use of a laminated transformer core?

7. Define eddy currents and explain why these currents result in a transformer power loss.

8. What is meant by a transformer copper loss?

9. How is the voltage ratio that is to exist between the primary and the secondary windings of a transformer determined?

10. Explain the power transfer characteristic of a transformer that is expressed by the formula $E_p I_p = E_s I_s$.

11. Define the efficiency of a transformer.

12. Why is it impossible to operate a transformer with a "pure" direct current?

13. Draw a diagram illustrating the windings of a typical multisecondary winding power transformer and show the color code scheme that is used to identify the various windings.

14. Describe the construction of a typical autotransformer.

15. Explain the safety advantage that is gained by the use of an isolation transformer.

16. Under what resistance (or impedance) condition is the maximum amount of power transferred from a source of power to any given load?

17. For what purpose is an impedance-matching transformer used?

18. State the function of a typical output transformer.

UNIT 14. Miscellaneous Magnetic Devices

In addition to its application to inductors, generators, transformers, and motors, the phenomenon of magnetism has been widely used in many other electrical and electronic devices. In this unit, the constructional features, theory of operation, and practical uses of a number of these devices are discussed.

THE MAGNETIC PHONO CARTRIDGE (PICKUP)

The magnetic phono cartridge performs the same function as crystal and ceramic cartridges. While there are several different types of magnetic cartridges available, all of them operate in accordance with the principle of electromagnetic induction.

In one common type of magnetic cartridge, the needle is mechanically coupled to a soft-iron armature located within the air gap of a permanent magnet (Fig. 14-1). As the needle moves in accordance with the variations in the groove of a record, the armature is set into motion. As a result, the permeability of the armature material causes the reluctance of the air gap to vary. This, in turn, varies the flux density of the magnetic circuit and produces what is, in effect, a moving magnetic field. The flux lines of this field then cut across the stationary coil, thereby inducing a voltage across the coil. Since the intensity of this voltage is proportional to the intensity of needle motion, it will therefore be an electrical representation of the physical variations in the groove of the record.

In the moving-magnet-type cartridge, the needle is coupled to a small, permanent magnet.

Fig. 14-1. Basic constructional features of a variable-reluctance-type magnetic phono cartridge.

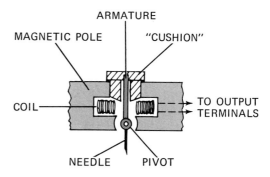

117

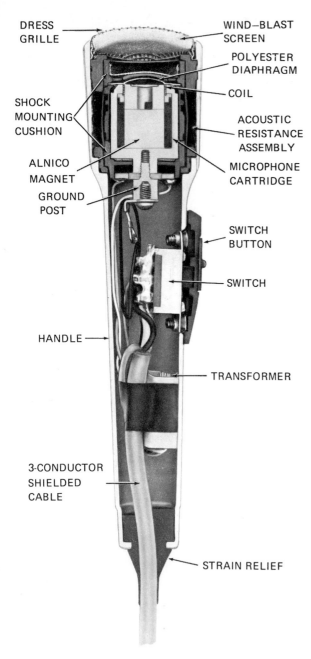

Fig. 14-2. Dynamic microphone. (*Sonotone Corp.*)

As the needle moves, the moving field of the magnet cuts across stationary coils, inducing a voltage within them.

Magnetic phono cartridges are rated in terms of two related characteristics: *frequency response* and *compliance*. Frequency response is the range of audio frequencies to which a magnetic phono cartridge will respond. In a typical high-fidelity cartridge, the range may extend from 20 hertz to 20 kilohertz. This means that the cartridge is constructed in a manner that will permit its moving parts to vibrate at any frequency within this range without producing a distorted signal voltage output.

Compliance is a measure of the ability of the needle to move freely within the cartridge assembly. The compliance of a cartridge is indicated in centimeter per dyne units as, for example, 25×10^{-6} cm/dyne. This means that the needle will move 25×10^{-6} centimeters when the force applied to it is equal to one dyne. A high cartridge compliance is a desirable characteristic because it increases the frequency response of the cartridge and also reduces needle wear.

THE DYNAMIC MICROPHONE

In the dynamic microphone, a diaphragm is connected to a coil that is wrapped around a lightweight form and is free to move over one pole of a permanent magnet "yoke" (Fig. 14-2). Sound waves striking the diaphragm cause the coil to move. As the coil moves, it cuts across the magnetic field of the permanent magnet and a voltage is induced across it. The magnitude and frequency of this voltage vary in proportion to the intensity of the sound waves. Therefore, the voltage is an electrical representation of the sound waves.

Dynamic and other types of microphones are rated in terms of their impedance, frequency response, directional characteristics, and power output. Many microphones are equipped with a switching arrangement that allows them to be operated at both high and low impedances. Often their frequency response will extend to 15 kilohertz or more.

Directional Characteristic. The directional characteristic of a microphone indicates how well it will respond to sound waves originating from

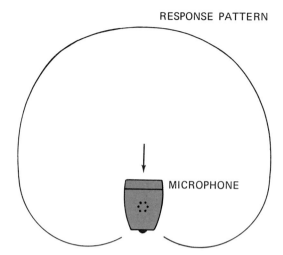

Fig. 14-3. Cardioid response pattern of a microphone.

Since a microphone has no electrical power input, its power output is compared at a zero decibel reference level, which is usually considered to be 0.006 watt or 6 milliwatts. With the reference level known, the power output of a microphone can be determined in decibels by substituting the reference level for P_1 in the preceding formula and the output power for P_2.

Problem. Compute the power output of a microphone that is rated at -50 dB, or 50 decibels below the zero-decibel reference level.

Solution.

$$dB = 10 \log \frac{P_o}{6}$$

where $P_o =$ power output of the microphone, in milliwatts. Therefore

$$-50 = 10 \log \frac{P_o}{6}$$

$$\frac{-50}{10} = \log \frac{P_o}{6}$$

$$-5 = \log P_o - \log 6$$

$$\log P_o = \log 6 - 5$$

$$= 0.7782 - 5$$

$$P_o = \text{antilog} - 5.7782$$

$$= 0.00006 \text{ milliwatt} = 0.06 \text{ microwatt}$$

By following a similar mathematical process, it is found that the power output of a microphone that is rated at -40 decibels is 0.6 microwatt. Thus, a microphone that is rated at -50 decibels is less sensitive to a given intensity of sound than one that is rated at -40 decibels.

a given direction. Dynamic microphones are often designed to have a directional characteristic that is described as being either *cardioid* or *omnidirectional*.

The cardioid (heart-shaped) response pattern is shown in Fig. 14-3. With this directional characteristic, a microphone will respond most effectively to sound originating within the cardioid area.

The term "omni" is a combining form that means "all." Thus, an omnidirectional microphone is one that will respond equally well to sounds originating from any direction.

Power Output. The power output of a microphone indicates its sensitivity, or how effectively it will convert a given intensity of sound into electrical energy. This characteristic is usually stated in terms of the *decibel* unit, which is a measure of the ratio between two different amounts of power. Expressed as a formula:

$$\text{Decibels (dB)} = 10 \log \frac{P_2}{P_1}$$

where $P_2 =$ output power, in watts
$P_1 =$ input power, in watts

Note: P_1 and P_2 can also be given in terms of milliwatts (mW) or microwatts (μW)

THE LOUDSPEAKER

The principal constructional features of a permanent-magnet (PM) type loudspeaker are shown in Fig. 14-4. In this loudspeaker, the voice coil is designed to move freely over (but not touch) the pole piece that is formed by one end of a disk (or cylindrical) Alnico or ceramic magnet.

In practical application, the voice coil is connected to the secondary winding of its associated audio-circuit output transformer by means of terminals mounted upon the loudspeaker frame. The alternating-voltage audio signal de-

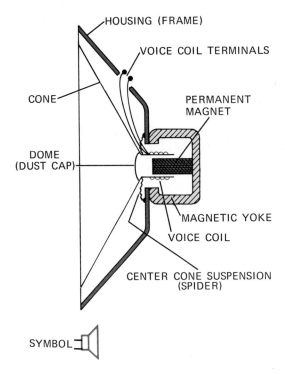

Fig. 14-4. Constructional features of a typical permanent-magnet type loudspeaker.

sponse or the range of frequencies that are reproduced without significant distortion. The frequency-response range varies from approximately 50 hertz to 8 kilohertz in the low-cost loudspeaker and from approximately 20 hertz to 20 kilohertz in the more expensive loudspeakers that are designed for high-fidelity applications.

Use as a Microphone. Since a loudspeaker contains a coil that can be moved within a magnetic field, it can be operated to function as a microphone. This is often done in intercommunications systems where the loudspeaker is switched to serve as the sound-producing unit when the control switch is on "receive," and as the microphone when the switch is on "send."

The Electrodynamic Loudspeaker. Permanent-magnet loudspeakers have replaced the electrodynamic (EM) type of loudspeaker, which was formerly very popular. In the electrodynamic loudspeaker, the magnetic field that interacts with the magnetic field of the voice coil is produced by energizing a field coil with direct current. This, of course, requires the expenditure of power, a condition that is eliminated by the use of a permanent magnet.

veloped across the secondary winding is applied to the voice coil, thus energizing it to produce a comparatively weak electromagnet. The electromagnetic field is then alternately attracted and repelled by the field of the permanent magnet, causing a motor action that moves the cone back and forth in a horizontal plane. The intensity of this movement varies in proportion to the magnitude and frequency of the signal voltage. Therefore, the sound waves that are produced by the moving cone will also vary in intensity and will be proportional to the signal that was originally produced by a device such as a microphone or a record-player cartridge.

Rating. A loudspeaker is rated in terms of the size of its cone, the impedance of its voice coil, the power capacity of its voice coil, its frequency response, and the weight of its magnet.

If similar loudspeakers have magnets of the same type but differing in weight, the loudspeaker with the heavier magnet will generally be more efficient with respect to frequency re-

THE HEADSET

In the typical magnetic headset, a soft-iron diaphragm is suspended a short distance from the cores of two coils. The cores are in contact with the poles of a permanent magnet (Fig. 14-5). When a varying audio signal voltage is applied to the coils, they become energized to form electromagnets with a flux density that varies in proportion to the applied voltage. These electromagnets either aid or oppose the field of the permanent magnet, depending upon their polarity at any instant of time. As a result, the diaphragm is attracted toward the poles of the permanent magnet with a varying intensity, thereby producing sound waves.

Headsets are usually rated in terms of their impedance and frequency response. Economy-type units most often have a frequency response ranging from approximately 50 hertz to 5 kilohertz, while more expensive headsets may have a frequency response that extends from approximately 20 hertz to 20 kilohertz.

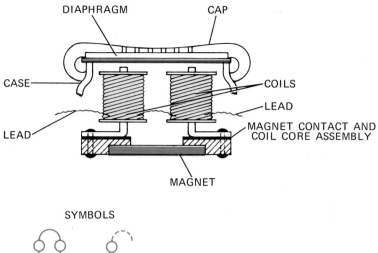

Fig. 14-5. Basic constructional features of a magnetic headset.

TAPE RECORDING

The process of tape recording involves the magnetization of an iron-oxide film that has been placed upon a thin-tape base made of a plastic compound such as cellulose, acetate, polyester, or Mylar.

To Record. The tape is magnetized by bringing it under the influence of an electromagnetic field. This field is produced by a coil located in a device known as the *recording head* (Fig. 14-6). When an audio signal voltage is applied to the recording head, it generates a magnetic field around the coil. Since the strength of the field is proportional to the voltage applied, the iron-oxide (dull) surface of the tape is magnetized in a pattern that represents the variations of the voltage (Fig. 14-7) as it passes over the recording head. In this way, it is possible to magnetically "store" information.

Fig. 14-6. Relative positions of the recording head and tape used in the process of magnetic tape recording. (*Ampex Corp.*)

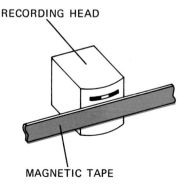

Fig. 14-7. Example of a magnetic pattern impressed upon single-track magnetic tape. (*Ampex Corp.*)

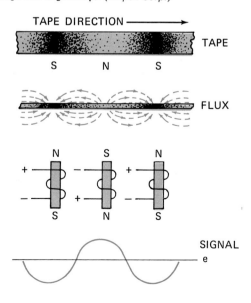

121

Playback. After the tape has been rewound and before it is to be "played back," a switch on the recorder is turned to the "playback" position. This connects another coil, which is very often located within the recording head unit, to the input circuit of the recorder amplifier. As the tape passes over this coil in the same direction it followed during the recording process, its magnetic field induces a voltage across the coil. The magnitude and frequency of this voltage are proportional to the variations of the magnetic patterns on the tape. Since these variations are, in turn, proportional to the magnitude and the frequency of the recording voltage, the resulting sound produced by the loudspeaker (or headset) is very nearly the same as the sound that was originally recorded.

Erasing. The magnetic patterns impressed upon a tape during the recording process can be removed by the process of *erasing*. This is accomplished by passing the tape over still another coil, which may or may not be located within the recording head assembly.

In the more expensive tape recorders, the erase coil is energized with an alternating current at a supersonic (above the audio range) frequency. A bias oscillator circuit, located within the recorder, supplies the high-frequency current that is required. The resulting magnetic field demagnetizes the tape, thus preparing it for rerecording. In the less expensive models, erasure is accomplished by simply passing the tape over a permanent magnet or by energizing the coil with direct current. While this method of erasing does effectively "remove" voice or music from the tape, it sometimes leaves a magnetic pattern that results in a hissing noise when the tape is played back.

For Review and Discussion

1. Explain the operation of a magnetic phono cartridge.
2. What is meant by the frequency response and the compliance characteristics of a phono cartridge?
3. Describe the principal constructional features of the dynamic microphone and explain the operation of this device.
4. For what purpose is a transformer often used in conjunction with a dynamic microphone?
5. Define the directional characteristic of a microphone.
6. Draw a diagram illustrating a cardioid microphone response pattern.
7. Define an omnidirectional microphone-response characteristic.
8. Name and define the basic unit by means of which the power output of a microphone is indicated.
9. Describe the principal constructional features of a permanent-magnet loudspeaker and explain its operation.
10. Name four rating characteristics commonly given in conjunction with a permanent-magnet loudspeaker.
11. Describe the basic processes of tape recording, playback, and erasing.

SECTION 3
CIRCUIT ANALYSIS

The study of electricity and electronics is basically the study of the effects produced in circuit systems and components as a result of the flow of electrons. These effects are, in turn, strongly influenced by resistance, inductance, and capacitance. The first unit of this section presents the resistive circuits: series, parallel, and series-parallel circuits, and their associated characteristics. In the following units, special emphasis is given to the effects of inductance and capacitance upon the relationship between voltage and current, the total opposition to current, and power dissipation. A knowledge of these subjects is essential to a thorough understanding of the operation and functions of different kinds of circuit systems.

UNIT 15. Resistive Circuits

A resistive circuit is one in which the only effective opposition to current is ohmic resistance. All of the energy that is supplied to it in the form of a flow of electrons is converted into heat energy. The heat energy, which may or may not serve a useful purpose, is not returned to the circuit and is, therefore, referred to as a *power loss*.

The behavior of a resistive circuit, as defined by Ohm's law, indicates that a simple linear (direct) relationship exists between voltage and current. Therefore, the voltage and current in a purely resistive circuit are "in phase" at all times. Thus, as mentioned in Unit 11, these quantities will attain their respective maximum and minimum magnitudes at exactly the same time regardless of whether the voltage applied to the circuit is a direct voltage or an alternating voltage.

THE SERIES CIRCUIT

In a series circuit, there is only one path through which electrons can flow. This path consists of the conductor and the components that are connected "into" it in a tandem or "one-after-another" pattern (Fig. 15-1A). If the circuit is broken or "open" at any point, it becomes inoperative since there is no longer a continuous path through which electrons can move (Fig. 15-1B).

Fig. 15-1. The series circuit: (A) basic construction of the circuit, (B) an "open," which in any part of a series circuit disrupts the operation of the entire circuit.

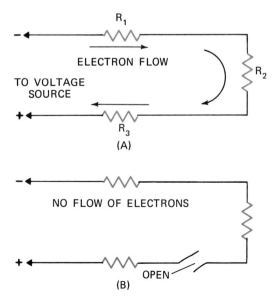

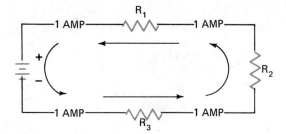

Fig. 15-2. Series circuit, showing that equal amounts of current pass through all parts of the circuit.

Current. Electrons are not consumed as they flow through a circuit. There is as much current moving from any point in a series circuit as there is moving to that point. Therefore, the same magnitude of current passes through all of the components in a series circuit (Fig. 15-2).

Resistance. When resistive components are connected in series, the current in the circuit must overcome the resistance of each component as it passes through the complete circuit. The total resistance to current in the circuit is then, in effect, equal to the sum of the various resistances in the circuit, or

$$R_t = R_1 + R_2 + R_3 \cdots$$

where R_t = total resistance
R_1, R_2, R_3 = resistances in the circuit

Problem 1. A 27-ohm, a 150-ohm, and a 2,700-ohm resistor are connected in series (Fig. 15-3). Compute the total resistance of this combination.

Solution.

$$R_t = R_1 + R_2 + R_3$$
$$= 27 + 150 + 2{,}700 = 2{,}877 \text{ ohms}$$

Voltage. Since the magnitude of the current flowing through each component of a series circuit is the same, the circuit voltage must be distributed among the components of the circuit in direct proportion to their resistance in order for the current to be maintained.

The voltage that exists across any given component in a series circuit is known as a *voltage drop* because it reduces the voltage that is available for application across the remaining components in the circuit. Since, according to Ohm's law, the magnitude of this voltage E is equal to the current times the resistance (IR), it is also commonly referred to as an *IR drop*.

The sum of the voltage drops in any series circuit is always equal to the voltage that is applied to the circuit. This relationship, known as *Kirchhoff's law*, is expressed by the following formula:

$$E_t = E_{R_1} + E_{R_2} + E_{R_3} \cdots$$

where E_t = total (applied) voltage
$E_{R_1}, E_{R_2}, E_{R_3}$ = voltage drops across circuit components

Problem 2. A 90-volt battery is connected in series with a 20-ohm, a 100-ohm, and a 180-ohm resistor (Fig. 15-4). Compute the voltage drop across each of these resistors.

Solution. To solve this problem, it is first necessary to determine the magnitude of the current in the circuit:

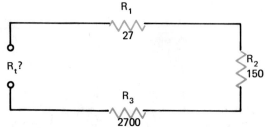

Fig. 15-3. Series circuit, showing a 27-ohm, a 150-ohm, and a 2,700-ohm resistor in series.

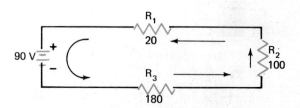

Fig. 15-4. Series circuit voltage drop.

$$I = \frac{E}{R_t} \quad \text{(Ohm's law)}$$

$$= \frac{90}{20 + 100 + 180}$$

$$= \frac{90}{300} = 0.3 \text{ ampere}$$

then

$$E_{R_1} = IR_1 = 0.3 \times 20 = 6 \text{ volts}$$
$$E_{R_2} = IR_2 = 0.3 \times 100 = 30 \text{ volts}$$
$$E_{R_3} = IR_3 = 0.3 \times 180 = 54 \text{ volts}$$

Polarity of a Voltage Drop. A voltage drop across a resistor indicates that one end of the resistor is more negative than the other end. This condition must exist if current is to pass through the resistor. The polarity of the voltage drop is determined by the direction of the current flow. Since current moves from negative to positive, the end of a resistor from which current moves is always negative and the end toward which current moves is always positive (Fig. 15-5).

Disadvantages of the Series Circuit. The operation of loads in a series-circuit arrangement presents several problems or disadvantages that must be considered when using such a circuit. Since the same amount of current is present in all parts of the circuit, all of the loads must be designed to operate at this current value. This limits the kinds of loads that can be connected into a series circuit.

The use of a series circuit also limits the number of loads that can be operated at a given applied voltage. If more than the proper number of loads are connected into the circuit, the additional resistance will cause the current to decrease to below the normal value at which the loads operate. The current can, of course, be increased by increasing the voltage, but this is often impractical.

Another disadvantage of the series circuit is that any break or open in a conductor or load

Fig. 15-5. Polarity of the voltage drop developed across a resistor relative to the direction of the current through the resistor.

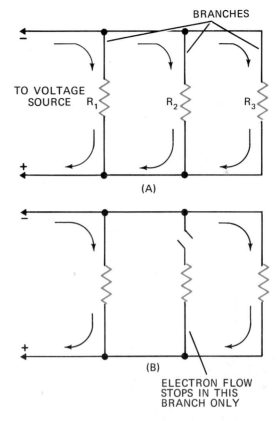

Fig. 15-6. The parallel circuit: (A) basic construction of the circuit, (B) the loads of a parallel circuit, which operate independently of each other.

causes the entire circuit to become inoperative. This factor often increases the problems encountered in maintenance and servicing.

THE PARALLEL CIRCUIT

The loads (or the components) of a parallel circuit are connected across the power line to form branches (Fig. 15-6A). The loads operate independently of each other, and therefore a break in any one branch does not prevent the line voltage from being applied to the remaining branches (Fig. 15-6B). Parallel connections of loads and other devices are also commonly called *shunts*.

Voltage. The voltage applied to a parallel circuit is the same across all branches of the circuit

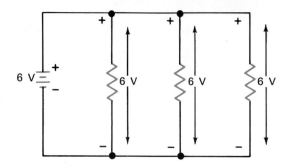

Fig. 15-7. Equal voltage existing across all branches of a parallel circuit.

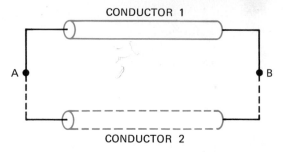

Fig. 15-9. The addition of conductor 2 in parallel with conductor 1 reduces the total resistance between points A and B.

(Fig. 15-7). As a result, all of the loads must be designed to operate at the same voltage.

Current. The total current is divided among the various branches of the circuit. Since the voltage applied across all branches is of the same magnitude, the current through each branch is inversely proportional to its resistance. The total current I_t is then equal to the sum of the individual branch currents (Fig. 15-8).

Resistance. The total resistance of a parallel circuit decreases as more branches are added. The basic reason for this often confusing situation is illustrated by Fig. 15-9. When conductor 1, having a given length and cross-sectional area, is connected to A and B, it presents a certain amount of resistance between these points. When conductor 2 (which is the same size as conductor 1) is connected in parallel with conductor 1, the resistance presented by both conductors is the same as the resistance that would result if a single conductor of the same length but with twice the cross-sectional area were connected between A and B. Because of this, the addition of conductor 2 reduces the resistance of the circuit to one-half the original value. The addition of more conductors in parallel with conductors 1 and 2 would further decrease the total resistance between A and B, since each of them would provide an additional path through which current can pass as it moves from point A to point B.

The total resistance R_t of a parallel circuit is always less than the resistance of any of its branches and is therefore less than the value of the lowest resistance in the circuit. The total resistance of any parallel circuit can be determined by using one of the following formulas.

If there are two resistances of unequal value in the circuit (Fig. 15-10A),

$$R_t = \frac{R_1 R_2}{R_1 + R_2} \qquad (1)$$

If the resistances (two or more) are equal in value (Fig. 15-10B),

$$R_t = \frac{\text{Value of one resistance}}{\text{No. of resistances}} \qquad (2)$$

If there are three or more resistances of unequal value in the circuit (Fig. 15-10C),

$$R_t = \frac{1}{1/R_1 + 1/R_2 + 1/R_3 \cdots} \qquad (3)$$

Note: Formula (3) can be used to find the total resistance of any parallel circuit. It is more convenient, however, to use Formulas (1) and (2) when circuit conditions permit.

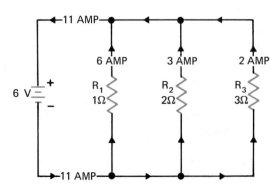

Fig. 15-8. The total current in a parallel circuit is equal to the sum of the individual branch currents.

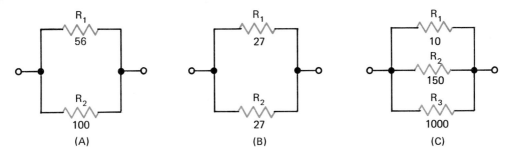

Fig. 15-10. Resistor circuits illustrating parallel resistance formulas.

The manner in which Formula (3) is derived presents an interesting study in the application of Ohm's law to the solution of a parallel circuit problem such as the one illustrated in Fig. 15-11. In this circuit,

$$I_t = \frac{E}{R_t}$$

and

$$I_t = \frac{E}{R_1} + \frac{E}{R_2} + \frac{E}{R_3}$$

Therefore

$$\frac{E}{R_t} = \frac{E}{R_1} + \frac{E}{R_2} + \frac{E}{R_3}$$

and, by dividing both sides of this formula by E,

$$\frac{1}{R_t} = \frac{1}{R_1} + \frac{1}{R_2} + \frac{1}{R_3} = G$$

where G = conductance.

Fig. 15-11. Circuit used in explaining the derivation of the general parallel resistance formula.

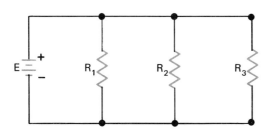

This form of the parallel resistance formula is known as the *conductance formula*. It indicates that the total conductance G of a parallel circuit is equal to the sum of the various conductances of the circuit. Conductance is the reciprocal of resistance, or $G = 1/R$. Therefore, since the reciprocal of any quantity is equal to 1 divided by that quantity, the conductance formula can be revised to indicate an expression of resistance relationships. This revision results in Formula (3).

Problem 3. A 10- and a 56-ohm resistor are connected in parallel (Fig. 15-12). Compute the total resistance of the combination.

Solution. Using Formula (1),

$$R_t = \frac{10 \times 56}{10 + 56}$$

$$R_t = \frac{560}{66} = 8.485 \text{ ohms}$$

Fig. 15-12. Circuit illustrating Formula 1.

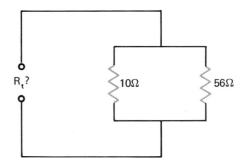

127

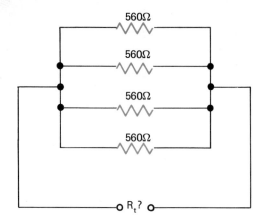

Fig. 15-13. Circuit illustrating Formula 2.

Problem 4. Four 560-ohm resistors are connected in parallel (Fig. 15-13). Compute the total resistance of this combination.

Solution. Using Formula (2),

$$R_t = \frac{560}{4} = 140 \text{ ohms}$$

Problem 5. Three resistors with resistance values of 4, 5, and 20 ohms respectively are connected in parallel to a 6-volt battery (Fig. 15-14).

Disregarding the internal resistance of the battery and the resistance of the conductors, compute (a) the total resistance of the circuit, (b) the current in each branch of the circuit, and (c) the total circuit current.

Fig. 15-14. Circuit illustrating Formula 3.

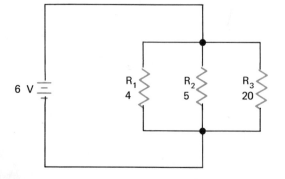

Solution.

a. Using Formula (3),

$$R_t = \frac{1}{1/4 + 1/5 + 1/20}$$

$$= \frac{1}{0.25 + 0.20 + 0.05} = \frac{1}{0.50}$$

$$= 2 \text{ ohms}$$

b.

$$I_{R_1} = \frac{E}{R_1} = \frac{6}{4} = 1.5 \text{ amperes}$$

$$I_{R_2} = \frac{E}{R_2} = \frac{6}{5} = 1.2 \text{ amperes}$$

$$I_{R_3} = \frac{E}{R_3} = \frac{6}{20} = 0.3 \text{ ampere}$$

c.

$$I_t = \frac{E}{R_t} = \frac{6}{2} = 3 \text{ amperes}$$

or

$$I_t = I_{R_1} + I_{R_2} + I_{R_3}$$
$$= 1.5 + 1.2 + 0.3 = 3 \text{ amperes}$$

SERIES-PARALLEL CIRCUITS

The series-parallel circuit combines one or more series and parallel circuits. Such circuits make it possible to combine the different voltage characteristic of a series circuit with the different current characteristic of a parallel circuit within a single network. This condition is particularly advantageous when it is necessary to operate loads that have different voltage and current requirements from the same source of energy.

The ability to analyze a series-parallel circuit in terms of its voltage, current, and resistance characteristics is very important, since many practical circuits are arranged in this manner. While there are no specific procedures that can be applied to the solution of all series-parallel problems, a first step usually involves the identification of the various parts of the circuit as essentially series and/or parallel. To do this it is often convenient to rearrange a given circuit diagram so that its series and parallel components are clearly illustrated. Ohm's law and the resistance formulas can then be applied to specific portions of the circuit to further simplify it and to complete the analysis.

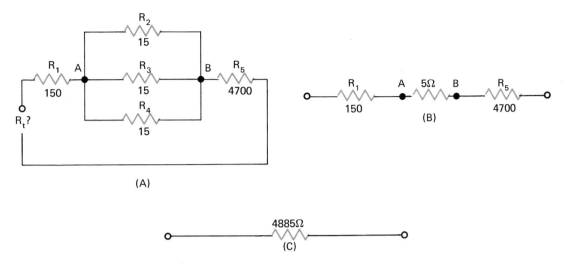

Fig. 15-15. Circuit illustrating Problem 6.

Problem 6. Compute the total resistance of the circuit shown in Fig. 15-15A.

Solution. By studying the diagram, this can immediately be recognized as a series-parallel circuit. To simplify the circuit, it is desirable to first determine the resistance of the parallel network (between A and B), which consists of resistors R_2, R_3, and R_4:

$$R_{A \text{ to } B} = \frac{15}{3} = 5 \text{ ohms}$$

The circuit can now be arranged as an equivalent—but simpler—combination of resistors, as shown in Fig. 15-15B. Here, a single 5-ohm resistor is used to represent the resistance between points A and B. The result is a series circuit. Therefore

$$R_t = R_1 + 5 + R_5$$
$$= 150 + 5 + 4{,}700 = 4{,}855 \text{ ohms}$$

The circuit can now be represented in its simplest, but equivalent, form insofar as the total resistance is concerned. This would be a single resistor of 4,885 ohms (Fig. 15-15C).

Problem 7. Compute the total resistance and total current of the circuit shown in Fig. 15-16A.

Solution. When analyzing such a circuit it is often convenient to rearrange the network into an electrically equivalent circuit that is illustrated in a more conventional manner (Fig. 15-16B). This makes it possible to more clearly determine the relationships between the various resistors in the circuit.

To further simplify the circuit, it is desirable to first determine the resistance of the parallel network consisting of R_3 and R_4:

$$R_{R_3,R_4} = \frac{6 \times 4}{6 + 4} = \frac{24}{10} = 2.4 \text{ ohms}$$

The circuit can now be illustrated as shown in Fig. 15-16C. The total resistance of the upper branch of this circuit consisting of R_2, R_{R_3,R_4}, and R_5 in series is

$$R_{\text{upper branch}} = 8 + 2.4 + 10 = 20.4 \text{ ohms}$$

The equivalent circuit can then be reduced to a simple parallel circuit (Fig. 15-16D). Therefore

$$R_t = \frac{20.4 \times 24}{20.4 + 24}$$

$$= \frac{489.6}{44.4} = 11.03 \text{ ohms}$$

and

$$I_t = \frac{12}{11.03} = 1.09 \text{ amperes}$$

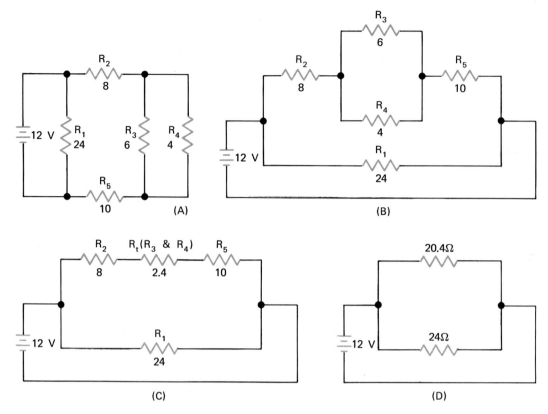

Fig. 15-16. Circuit illustrating Problem 7.

Problem 8. Compute the total resistance and the total current of the circuit shown in Fig. 15-17. Also, compute the current through each resistor and the voltage drop across each resistor.

Solution. The steps followed in simplifying this circuit into equivalent circuits for the purpose of computing the total resistance and total current are shown in Fig. 15-18. The current through each resistor and the voltage drop across each resistor are most conveniently computed by applying Ohm's law to individual parts of the circuit illustrated by step 1 as follows:

1. $I_{R_1,R_2} = E/(R_1 + R_2) = 90/(10 + 20) = 90/30 = 3$ amperes
2. $E_{R_1} = IR_1 = 3 \times 10 = 30$ volts
3. $E_{R_2} = IR_2 = 3 \times 20 = 60$ volts
4. $I_{R_3} = E/(R_3 + R_{R_4,R_5,R_6}) = 90/(18 + 2) = 90/20 = 4.5$ amperes
5. $E_{R_3} = I_{R_3} R_3 = 4.5 \times 18 = 81$ volts
6. $E_{R_4,R_5,R_6} = 90 - 81 = 9$ volts
7. $I_{R_4} = E_{R_4}/R_4 = 9/4 = 2.25$ amperes
8. $I_{R_5} = E_{R_5}/R_5 = 9/5 = 1.8$ amperes
9. $I_{R_6} = E_{R_6}/R_6 = 9/20 = 0.45$ ampere

Fig. 15-17. Circuit illustrating Problem 8.

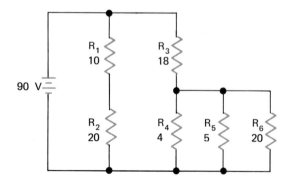

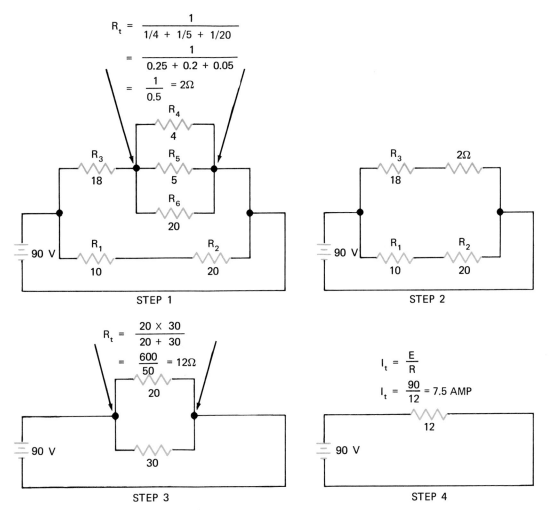

Fig. 15-18. Circuit illustrating solution to Problem 8.

POWER

The total power delivered to a resistive series or parallel circuit is equal to the sum of the power dissipated by each resistance of the circuit. The power dissipation of each resistance is, in turn, determined by using the appropriate power formula.

The power dissipated in a sinusoidal alternating-current resistive circuit is most often stated in terms of effective power, which is computed by using rms values of voltage and current. The power dissipated in such a circuit can also be stated in terms of instantaneous power, which is equal to the product of the voltage and current existing at any instant during a cycle. Thus

$$p = ei \qquad p = i^2 R \qquad p = \frac{e^2}{R}$$

where p = instantaneous power
e = instantaneous voltage
i = instantaneous current

The magnitude of instantaneous power during one complete cycle of voltage and the resulting in-phase current in a sinusoidal resistive circuit are represented by the power curve

131

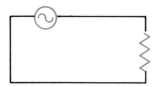

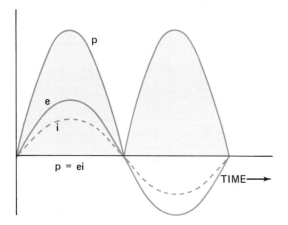

Fig. 15-19. Sinusoidal resistive circuit power curve.

Solution.
$$P = \frac{E^2}{R_t}$$
$$R_t = R_1 + \frac{R_2}{2}$$
$$= 6 + \frac{4}{2} = 6 + 2 = 8 \text{ ohms}$$

Therefore
$$P = \frac{12^2}{8} = \frac{144}{8} = 18 \text{ watts}$$

SOLVING STATEMENT-TYPE CIRCUIT PROBLEMS

The general procedures useful when analyzing circuits that are represented by diagrams have been discussed in the preceding paragraphs. In many cases, however, the information relating to a given circuit problem is not presented in the form of a diagram, but rather, in the form of a verbal or a written statement. This is particularly true when circuits are being designed or modified to satisfy a certain operating condition.

To solve a statement-type problem, it is first necessary to correctly interpret the statement of the problem. The problem should then be represented by a circuit sketch indicating the known conditions and values. As the final step, the circuit is analyzed by using the appropriate laws and mathematical formulas in the conventional manner.

shown in Fig. 15-19. On this curve, the magnitude of *p* at any instant is equal to the product of *e* and *i*. Since *e* and *i* are both either above or below the time reference axis at any given instant, their product results in two positive power "loops" during each cycle. These loops represent the rate at which energy is being delivered to a circuit.

Problem 9. Compute the power dissipated in the series-parallel circuit shown in Fig. 15-20.

Problem 10. It is necessary to operate a miniature lamp that operates at 6 volts and draws 0.15 ampere from a 12-volt battery. Determine the resistance value and the power rating of the voltage-dropping resistor that must be con-

Fig. 15-20. Circuit illustrating Problem 9.

Fig. 15-21. Circuit illustrating the solution to Problem 10.

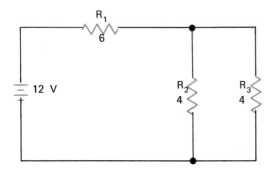

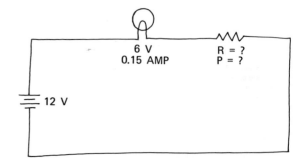

nected in series with the lamp to permit this operation.

Solution. The sketch of the circuit upon which the conditions of the problem are given is shown in Fig. 15-21. The voltage drop that is to occur across the resistor is equal to 12 − 6, or 6 volts. Therefore

$$R = \frac{E}{I} = \frac{6}{0.15} = 40 \text{ ohms}$$

$$P_R = \frac{E^2}{R} = \frac{36}{40} = 0.9 \text{ watt}$$

The use of a 2-watt resistor would provide slightly more than the desirable 100 percent safety factor.

Fig. 15-22. The parallel resistor chart. (Ohmite Manufacturing Co.)

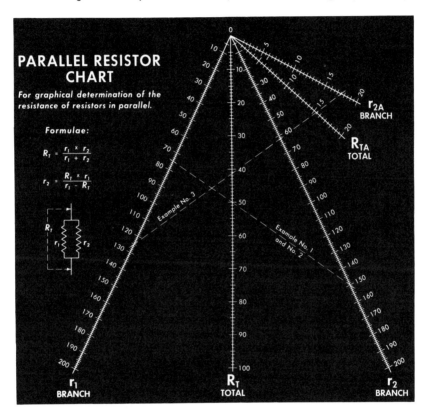

HOW TO USE THIS PARALLEL RESISTOR CHART

This alignment chart enables graphical solution of problems involving resistances connected in parallel. The values of the parallel resistors r_1 and r_2 and of the total effective resistance R_T must be read on the scales marked with the corresponding letters. To use, place a ruler across the two known values; the point at which the ruler crosses the third scale will show the unknown value. Pairs of resistances that will produce a given parallel resistance can be obtained by rotating a ruler around the desired value on scale R_T. The range of the chart can be increased by multiplying the values on *all* the scales by 10, 100, 1000, etc., as required. Scales r_{2A} and R_{TA} are used with scale r_1 when the values of r_1 and r_2 differ greatly.

EXAMPLE NO. 1: What is the total resistance of a 75-ohm resistor and a 150-ohm resistor connected in parallel? Answer: From dotted line No. 1, R_T is 50 ohms.

EXAMPLE NO. 2: What resistance in parallel with 750 ohms will give a combined value of 500 ohms? Answer: From dotted line No. 1, r_2 is 1500 ohms.

EXAMPLE NO. 3: What is the combined resistance of 1750 ohms and 12,500 ohms? Answer: Scales r_1 and r_{2A} are used and from dotted line No. 3, R_{TA} is 1535 ohms.

EXAMPLE NO. 4: What is the combined resistance of 400, 600 and 800 ohm resistors in parallel? Answer: First find R_T for 400 ohms and 600 ohms. Then set the 240 ohms thus found as a new r_1 and the final answer is found to be 185 ohms.

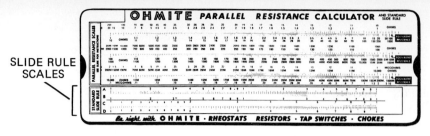

Fig. 15-23. The parallel resistance calculator. (*Ohmite Manufacturing Co.*)

ACCURACY OF COMPUTATIONS

Although mathematics is considered to be an exact science, it is important to note that in the majority of practical circuits only approximations of values or magnitudes are usually obtained. For example, a meter reading is an approximation and not a precise indication of an electrical quantity; a slide rule answer is an approximation of a given number; and a resistor may have a slightly different resistance than is indicated by its color-code markings. Therefore, while we must constantly strive to express the results of a mathematical analysis as accurately as possible, it is true that in most practical situations a precise indication of any value or magnitude is not required.

ROUNDING OFF NUMBERS

The most common method of stating a numerical value in terms of a reasonable approximation is to round off numbers. For example, suppose that it is necessary to compute the quotient that results when 10 is divided by 3. The quotient, 3.33333333 · · · , could be continued indefinitely. However, it is usually rounded off to two or three significant figures beyond the decimal point. The quotient of 10/3 could then be expressed as simply 3.3 or 3.33. The significant figures are the numbers 1 through 9 and any zeros that are between these numbers or that may have been retained in rounding off the numbers.

If the last figure to be dropped when rounding off a number is less than 5, the last figure that is retained in an answer is left unchanged. If the last figure to be dropped is 5 or greater than 5, the last figure retained is increased by 1. Therefore, the number 5.63427 rounded off to three significant figures is 5.63 and the number 49.67889 rounded off to five significant figures is 49.679.

When rounding off numbers, it must be noted that zero cannot be counted as a significant figure if it appears immediately following a decimal point. Such zeros must be retained, and the count of significant figures must begin at the first significant figure beyond them. Thus, the number 0.0039267 rounded off to three significant figures is 0.00393.

PARALLEL RESISTOR CHART AND CALCULATOR

The parallel resistor chart provides a convenient method of graphically solving parallel resistance problems when not more than two resistors are connected into a circuit. An example of such a chart and the directions for its use are shown in Fig. 15-22.

The parallel resistance calculator is a "sliderule"-type device that also provides a convenient method of solving two-branch resistance circuit problems (Fig. 15-23). In addition to the parallel resistance feature, the calculator is equipped with the basic standard sliderule scales.

For Review and Discussion

1. Define a resistive circuit.
2. What is the phase relationship between the voltage and the current in a resistive circuit?
3. Define a series circuit and state the current, the resistance, and the voltage characteristics of the circuit.
4. Draw a diagram that illustrates the polarity of the voltage drop across a resistor in relation to the direction of current through the resistor.
5. What are the two principal "disadvantages" or limitations of the series circuit?
6. Define a parallel circuit and state the current, the resistance, and the voltage characteristics of the circuit.

7. Explain why the total resistance of a parallel circuit decreases as branch loads are added to it.
8. Define conductance.
9. State the principal procedures to be followed in the mathematical analysis of a typical series-parallel circuit.
10. What is the relationship between the total power delivered to a series or to a parallel circuit and the power dissipated by each resistance in the circuit?
11. Define effective power.
12. Describe the process of "rounding off" numbers.

UNIT 16. Inductive Circuits

An inductive circuit is one in which the property of self-inductance is present to a significant degree. As a result, such a circuit has definite characteristics that affect the phase relationship between voltage and current, the opposition to current, and the power dissipation. A knowledge of these characteristics is essential to understanding the functions of inductors and the effects that are produced in a circuit when inductance is combined with resistance and capacitance.

As current in an inductive circuit increases, the magnetic field about the inductor in the circuit also increases. While these increases are taking place, the energy supplied to the circuit is stored within the magnetic field. When the current decreases, the magnetic field collapses and energy is returned to the circuit (Fig. 16-1).

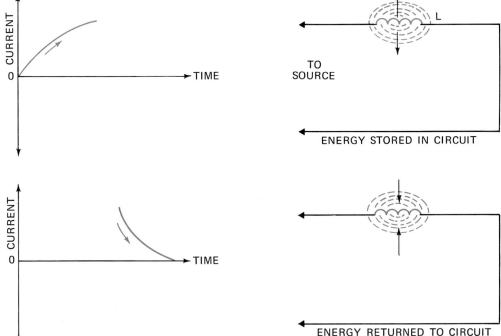

Fig. 16-1. Energy conditions in a purely inductive circuit.

If the circuit is purely inductive (without resistance), no energy will be expended in it. However, because all circuits contain resistance, some of the electrical energy supplied to an inductive circuit is converted into heat or some other form of energy.

EFFECT OF INDUCTANCE IN A DC CIRCUIT

Since the counter electromotive force (cemf) induced across an inductive component always opposes the applied voltage, the current in an inductive circuit lags behind the applied voltage whenever there is a change in the magnitude of the current. Thus, the current in a direct-current inductive circuit is not in phase with the applied voltage during the period of time when the current is changing. The manner in which the current in such a circuit responds to a change in the magnitude of the applied voltage is referred to as the *transient response* of the circuit.

The current responses in a typical series *RL* circuit are illustrated in Fig. 16-2. Responses that occur when voltage is applied or removed are not instantaneous. The straight portion of the curve between the current-increasing and current-decreasing responses indicates the steady state of the circuit. During this time, no cemf is induced across the coil. Therefore, the current and the voltage are in phase because the current is limited only by the resistance of the circuit.

Time Constant. The time required for the current in an inductive circuit to rise to its maximum magnitude I_{max} after the operating voltage has been applied depends upon both the inductance and the resistance of the circuit. With a fixed amount of resistance in the circuit, the time period will be increased if the inductance is increased, because the greater the inductance, the greater the magnitude of the cemf that is present to oppose the change in current. If the inductance of the circuit remains constant and

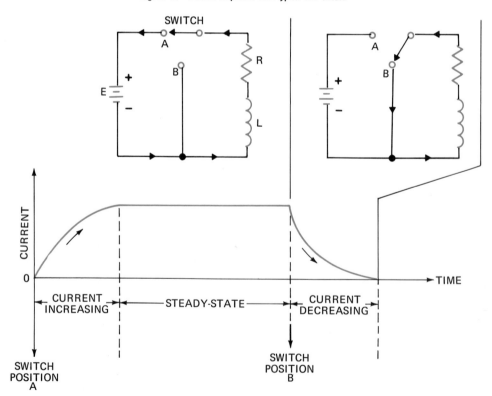

Fig. 16-2. Current responses in a typical *RL* circuit.

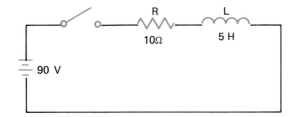

Fig. 16-3. Circuit illustrating Problem 1.

the resistance is increased, the period of time required for attaining maximum current is decreased. This is because the increase in resistance reduces the current flow in the circuit, and as a result a smaller magnitude of cemf is generated to oppose the change in current.

The time required for the current in a series RL circuit to increase to 63.2 percent of its maximum (steady-state) magnitude is known as the L/R time constant of the circuit. Expressed as a formula,

$$t = \frac{L}{R}$$

where t = time constant, in seconds
 L = inductance, in henrys
 R = resistance, in ohms

The time constant also governs the rate at which a current decrease occurs after the source of the applied voltage is removed from a circuit (if a "discharge" path is provided by the remaining components of the circuit). Thus, if the current in a given circuit increases to 63.2 percent of its maximum in 0.25 second, it will also decrease by 63.2 percent of its maximum in 0.25 second.

Problem 1. Compute the time constant of the circuit shown in Fig. 16-3. Also compute the time that is required for the current in this circuit to reach its maximum magnitude.

Solution.

$$t = \frac{L}{R} = \frac{5}{10} = 0.5 \text{ second}$$

The transient response characteristic of the circuit and the method of determining the magnitude of the current at the end of each time constant period are shown in Fig. 16-4. This curve shows that, for all practical purposes, the current reaches a maximum magnitude after five time-constant periods (5 × 0.5 second, or 2.5 seconds).

The curve also shows that, as the magnitude of the current increases, the rate of current change per unit of time becomes slower and slower. As a result, the curve "flattens out." This indicates that the current is gradually reaching a point where its magnitude will be determined only by the battery voltage and the resistance of the circuit.

Voltage Characteristics. During the time that the current in a series RL circuit is increasing, there

Fig. 16-4. Circuit illustrating the solution to Problem 1.

TIME CONSTANT PERIOD	CURRENT
1	0.632 × 9 = 5.678 amp
2	5.678 + 0.632 (9 − 5.678) = 7.778 amp
3	7.778 + 0.632 (9 − 7.778) = 8.55 amp
4	8.55 + 0.632 (9 − 8.55) = 8.834 amp
5	8.834 + 0.632 (9 − 8.834) = 8.939 amp

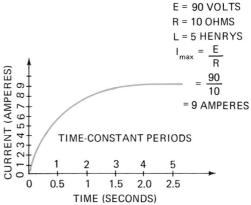

$E = 90$ VOLTS
$R = 10$ OHMS
$L = 5$ HENRYS
$I_{max} = \frac{E}{R} = \frac{90}{10} = 9$ AMPERES

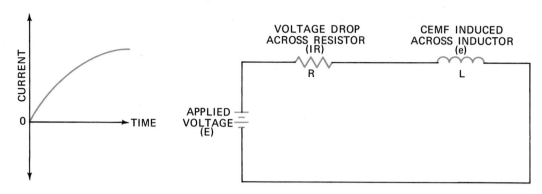

Fig. 16-5. Voltages present in a series *RL* circuit.

are three different voltages present in the circuit: the applied voltage *E*, the *IR* voltage drop across the resistance, and the cemf (e) generated across the inductive component or components (Fig. 16-5).

Since the polarity of the cemf opposes the applied voltage, the cemf can be considered a voltage drop. Therefore, according to Kirchhoff's law,

$$E = IR + e$$

The magnitude of the cemf in a series *RL* circuit depends upon the inductance of the circuit and the instantaneous rate of current change (the rate at which the current is changing at any given instant). This rate of change is represented by the symbol *di/dt*, and

$$\frac{di}{dt} = \frac{E}{L} - \frac{IR}{L}$$

where *E* = applied voltage, in volts
 L = inductance, in henrys
 I = current, in amperes
 R = resistance, in ohms

Expressed as a formula,

$$\text{cemf} = L\frac{di}{dt}$$

or, from Kirchhoff's law,

$$\text{cemf} = E - IR$$

High Induced Voltages. When the current in an inductive circuit is increasing, the magnitude of the cemf induced across any coil in the circuit cannot become greater than the applied voltage. However, when the switch in such a circuit is opened, the resistance of the circuit suddenly approaches infinity. As a result, the time constant *L/R* becomes extremely small, and the current decreases toward zero very rapidly. This in turn causes a voltage many times greater than the applied voltage to be induced across the coil. A spark can often be observed between a switch and other contacts when an inductive circuit is suddenly opened.

EFFECTS OF INDUCTANCE IN AC CIRCUITS

In a sinusoidal alternating-current *RL* circuit, the current is constantly changing in magnitude and periodically changing in direction. This condition causes the induced voltage (cemf) to be present at nearly all times, thereby introducing the effects of inductive reactance and phase shift into the circuit.

Inductive Reactance. The opposition offered by a cemf to an applied voltage tends to limit the flow of electrons in an alternating-current circuit. It is important to note that the nature of this opposition does not result from electronic collisions within the inductor itself, but from the reaction of the induced cemf against the applied voltage. It is for this reason that the opposition is called *inductive reactance*, X_L.

Because of the direct relationship between inductive reactance and the cemf generated across an inductor, any condition that produces a greater cemf also causes a greater inductive reactance. Thus, the inductive reactance of an inductive component is determined by its in-

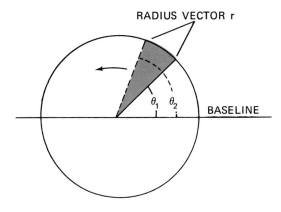

Fig. 16-6. Angular velocity.

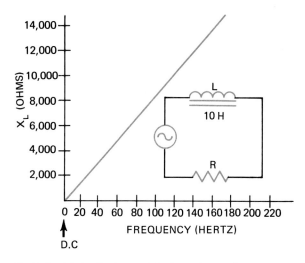

Fig. 16-7. Relationships between inductive reactance and frequency in a series *RL* circuit having a fixed inductance.

ductance and by the rate at which circuit current changes.

Angular Velocity. In a sinusoidal alternating-current circuit, the rate of current change is governed by the *angular velocity* of the applied voltage. Angular velocity is the rate at which an angle is formed between a base reference line and a rotating line (radius vector) that represents one section of the rotating coil in a generator (Fig. 16-6).

During each cycle, such a radius vector will rotate through 360 degrees, or 2π radians (a radian equals approximately 57.3 degrees, and is the unit of angular measurement in what is commonly referred to as the natural system of angular measurement). Since one cycle of sinusoidal alternating voltage is developed during 2π radians of coil rotation, the angular velocity of a sinusoidal waveform is the product of 2π and the frequency, or

$$\omega = 2\pi f$$

where ω (omega) = angular velocity
$\pi = 3.14$
f = frequency, in hertz

The inductive reactance of a circuit can then be expressed as the product of its inductance and angular velocity, or

$$X_L = 2\pi f L$$

where X_L = inductive reactance, in ohms
f = frequency, in hertz
L = inductance, in henrys

A graphical representation of the relationship between inductive reactance and frequency in a series *RL* circuit having a fixed inductance is shown in Fig. 16-7. As indicated by the inductive reactance formula, there is a direct (linear) relationship between inductive reactance and frequency.

Problem 2. The ballast coil used in a 120-volt, 60-hertz fluorescent lamp circuit (Fig. 16-8) has an inductance of 4.2 henrys. Compute the inductive reactance of the coil when the circuit is in operation.

Fig. 16-8. Circuit illustrating Problem 2.

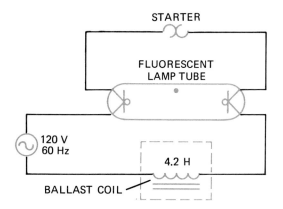

Solution.

$$X_L = 2\pi f L$$
$$= 6.28 \times 60 \times 4.2 = 1{,}582.5 \text{ ohms}$$

Problem 3. The inductive reactance of a motor winding operated at a frequency of 60 hertz is 560 ohms. Compute the inductance of this winding.

Solution. Since

$$X_L = 2\pi f L$$

then

$$L = \frac{X_L}{2\pi f}$$

$$= \frac{560}{6.28 \times 60}$$

$$= \frac{560}{376.8} = 1.48 \text{ henrys}$$

Phase Shift. In addition to limiting the current in an alternating-current circuit, inductive reactance has the effect of opposing any change in the magnitude of the current. Since the current in such a circuit is constantly changing in magnitude, the opposition to current change presented by inductive reactance is always present. As a result, it produces a continuous phase shift or a difference in the time between the starting points and between the minimum and maximum magnitude points of the applied voltage and the current.

To understand the phase shift occurring between a sinusoidal voltage and current in an inductive circuit, it is first necessary to recall that the cemf induced across an inductive component has a polarity that is always opposite to the polarity of the applied voltage. The applied voltage and the cemf are, therefore, said to be 180 degrees out of phase. This condition is represented by the waveforms of these voltages in a purely inductive circuit (Fig. 16-9).

At this time, it is also necessary to recall that the magnitude of the cemf is directly proportional to the rate of current change in a circuit. Thus, the rate of current change must be greatest when both the cemf and the applied voltage are at their maximum, or at the 90- and 270-degree points of their respective waveforms. At this same time, the current is passing through zero (Fig. 16-10). Therefore, the current in a purely inductive circuit lags the applied voltage by 90 degrees. In this condition, the current is 90 degrees out of phase with the applied voltage.

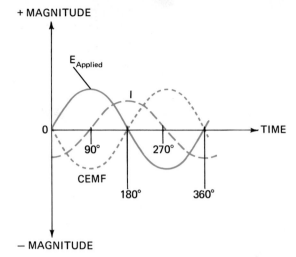

Fig. 16-10. Waveforms showing the phase relationships between the applied voltage, the cemf, and the current in a purely inductive circuit.

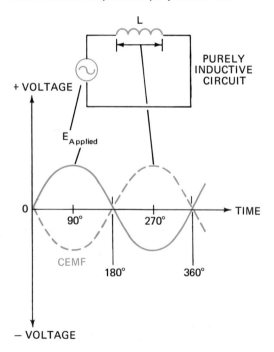

Fig. 16-9. Waveforms showing the 180-degree out-of-phase relationship between the applied voltage and the cemf developed across the inductive component of a purely inductive circuit.

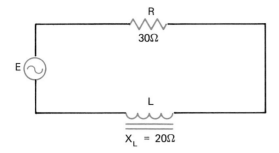

Fig. 16-11. Circuit illustrating Problem 4.

Phase Angle. Phase angle, θ, is a measure of the time difference between the applied voltage and the current in a sinusoidal alternating-current circuit. In practical circuits, which of course contain both inductance and resistance, the phase angle is a function of the ratio of inductive reactance to resistance and can be determined by means of a vector diagram.

Problem 4. Compute the phase angle between the applied voltage E and the current in the series RL circuit shown in Fig. 16-11.

Solution. To solve this problem, the values of resistance and inductive reactance are represented by vectors. The vectors are used to simultaneously represent magnitude and phase angle. Since resistance does not affect the phase relationship between voltage and current, and since inductive reactance causes a 90-degree phase shift between voltage and current, the vectors are drawn at an angle of 90 degrees to each other. Their length is proportional to the magnitudes that they represent (Fig. 16-12A).

Next, draw a rectangle with resistance (30 ohms) and inductive reactance (20 ohms) as the sides and then draw the diagonal of this rectangle (Fig. 16-12B). The angle θ that is formed by the diagonal and the resistance vector is the phase angle:

$$\tan \theta = \frac{X_L}{R}$$

$$= \frac{20}{30} = 0.666$$

and, from a table of trigonometric functions,

$$\theta = 33.7 \text{ degrees}$$

A study of the vector diagram used in solving the preceding problem shows that, if the inductive reactance of a circuit is increased while the resistance of the circuit remains constant, the ratio of inductive reactance to resistance also increases. This, in turn, causes the phase angle to become larger. Since inductive reactance is directly proportional to both inductance and frequency, the angle by which the current lags the voltage in a series RL circuit will increase if either the inductance or the frequency is increased without a corresponding increase of resistance.

IMPEDANCE

The properties that oppose current in an inductive circuit are resistance and inductive react-

Fig. 16-12. Vector diagram illustrating the solution to Problem 4.

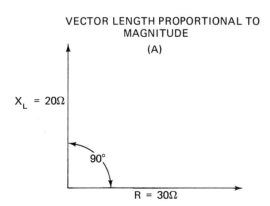

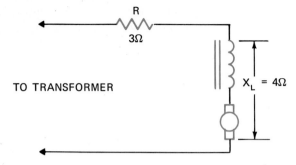

Fig. 16-13. Circuit illustrating Problem 5.

ance. Because inductive reactance tends to produce a 90-degree phase shift between the applied voltage and the current, its opposition acts upon the current at a right angle to the opposition presented by resistance. For this reason, impedance (the total opposition to current) cannot be computed as the simple arithmetic sum of resistance and inductive reactance. Instead, the impedance is computed by means of vector addition, which involves the use of the Pythagorean theorem. Expressed as a formula,

$$Z^2 = R^2 + X_L^2$$

where Z = impedance, in ohms
 R = resistance, in ohms
 X_L = inductive reactance, in ohms

Problem 5. A small universal-type motor is connected to the secondary winding of a power transformer and is operated with a speed-control

Fig. 16-14. Vector diagram illustrating the solution to Problem 5.

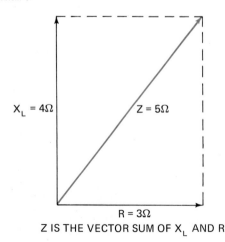

Z IS THE VECTOR SUM OF X_L AND R

resistance of 3 ohms. Compute the impedance of the circuit if the inductive reactance of the motor windings is 4 ohms (Fig. 16-13).

Solution. Draw the vectors representing the inductive reactance and resistance of the circuit. Complete the rectangle, and draw its diagonal (Fig. 16-14). The length of the diagonal will then represent the impedance, which is the vector sum of R and X_L:

$$Z^2 = R^2 + X_L^2$$
$$Z = \sqrt{R^2 + X_L^2}$$
$$= \sqrt{3^2 + 4^2}$$
$$= \sqrt{25} = 5 \text{ ohms}$$

Problem 6. The coil of a 120-volt, 60-hertz relay has a resistance of 3 ohms and an inductance of 2 henrys. To operate the relay, it is necessary to connect a 300-ohm voltage-dropping resistor in series with the line (Fig. 16-15). Compute the impedance of the circuit and the phase angle by which the current lags the voltage when the coil is energized.

Solution.

$$R = 3 + 300 = 303 \text{ ohms}$$
$$X_L = 2\pi f L$$
$$X_L = 6.28 \times 60 \times 2 = 753.6 \text{ ohms}$$
$$Z = \sqrt{R^2 + X_L^2}$$
$$= \sqrt{303^2 + 753.6^2}$$
$$= \sqrt{660,177} = 813.1 \text{ ohms}$$
$$\tan \theta = \frac{X_L}{R}$$
$$= \frac{753.6}{303} = 2.4877$$

Fig. 16-15. Circuit illustrating Problem 6.

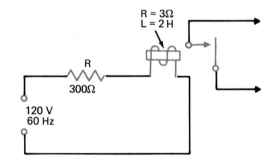

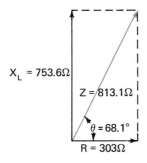

Fig. 16-16. Vector diagram illustrating the solution to Problem 6.

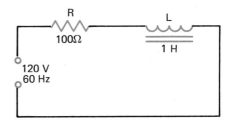

Fig. 16-17. Circuit illustrating Problem 7.

and
$$\theta = 68.1 \text{ degrees}$$
(see Fig. 16-16).

If the impedance of the circuit is known, the phase angle can also be determined from the ratio of resistance to impedance:

$$\cos \theta = \frac{R}{Z}$$
$$= \frac{303}{813.1} = 0.3739$$

and
$$\theta = 68.1 \text{ degrees}$$

OHM'S LAW FOR INDUCTIVE CIRCUITS

The forms of Ohm's law that have been discussed previously can be used to determine the relationships between voltage, current, and resistance in a circuit where the only opposition to current is resistance or when the effects of inductance are insignificant. However, in an alternating-current circuit, inductive reactance may have a considerable effect upon the magnitude of the current. For this reason, Ohm's law must be modified when it is to be applied to the analysis of alternating-current inductive circuits:

1. In circuits where the resistance is insignificant,

$$E = IX_L$$
$$I = \frac{E}{X_L}$$
$$X_L = \frac{E}{I}$$

where E = voltage, in volts
I = current, in amperes
X_L = inductive reactance, in ohms

2. In circuits where the value of resistance is significant,

$$E = IZ$$
$$I = \frac{E}{Z}$$
$$Z = \frac{E}{I}$$

where E = voltage, in volts
I = current, in amperes
Z = impedance, in ohms

Problem 7. An *RL* circuit that is operated from a 120-volt, 60-hertz power line consists of a 100-ohm resistor in series with a coil that has an inductance of 1 henry (Fig. 16-17). Compute the magnitude of the current in this circuit.

Solution.

$$X_L = 2\pi f L$$
$$= 6.28 \times 60 \times 1 = 376.8 \text{ ohms}$$
$$Z = \sqrt{R^2 + X_L^2}$$
$$= \sqrt{100^2 + 376.8^2}$$
$$= \sqrt{151{,}977} = 389.84 \text{ ohms}$$

and
$$I = \frac{120}{389.84} = 0.3077 \text{ ampere}$$

VOLTAGE RELATIONSHIPS IN SERIES CIRCUITS

The instantaneous value of the voltage applied to a series *RL* circuit is equal to the sum of the voltage drops occurring in the circuit at the same instant. However, when rms values of voltages are considered, it is necessary to illustrate the

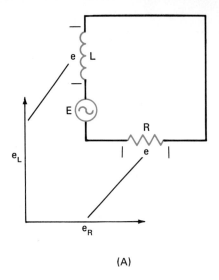
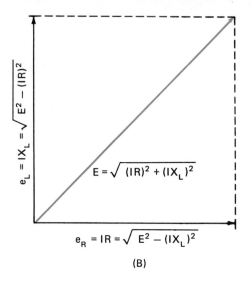

Fig. 16-18. Vectorial representations of the voltage relationships in series *RL* circuits.

relationships between the various voltages in the circuit by means of a vector diagram.

The voltage drop e_R across any resistive component in a series *RL* circuit is in phase with the applied voltage *E*, while the voltage drop across any inductive component e_L is 90 degrees out of phase with the applied voltage. As a result, the vectors representing the voltage drops on a vector diagram are drawn at right angles or 90 degrees apart (Fig. 16-18A).

As in a resistive circuit, the same magnitude of current is present in all parts of a series *RL* circuit. Therefore, the voltage drop across any resistive component in the circuit is equal to *IR*. The voltage drop across any inductive component is then equal to the vector sum of *IR* and IX_L (Fig. 16-18B).

PARALLEL *RL* CIRCUITS

In some practical electrical and electronic circuits, resistors and inductive components are operated in parallel combinations (Fig. 16-19). The characteristics of the individual components in such circuits are similar to those that have been previously discussed. However, when the components are connected in parallel, their effect upon the behavior of the total circuit is quite different from the effect produced in a series circuit. Basically, the method of analyzing parallel resistive circuits can also be applied to the parallel *RL* circuit. This is particularly true insofar as the impedance of the circuit is concerned.

Voltage. The voltages across the resistive and the inductive branches of a parallel *RL* circuit are equal to each other and to the applied voltage. As a result, all of these voltages are in phase with each other.

Current. The current I_L in any inductive branch of a parallel *RL* circuit lags the applied voltage, while the current through any resistive branch I_R is in phase with the applied voltage *E*. Assuming a two-branch circuit and neglecting the resistance of the coil in the inductive branch, the current through the coil will lag the applied voltage by 90 degrees. Therefore, the current in the resistive branch leads the current in the

Fig. 16-19. A parallel *RL* combination.

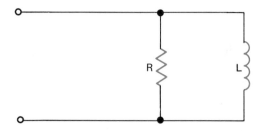

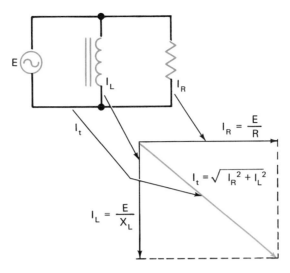

Fig. 16-20. Current relationships in a parallel RL circuit.

inductive branch by 90 degrees. The total or line current in the circuit I_t is then equal to the vector sum of the branch currents (Fig. 16-20). Notice that in this vector diagram the coil current represented by the I_L vector is drawn below or in a negative direction from the horizontal vector that represents the current through the resistor, I_R. This is done to indicate vectorially that the current through the coil lags the current through the resistor by 90 degrees.

Since the total circuit current I_t is the vector sum of I_R and I_L, and is represented by the hypotenuse of a right triangle, I_t is always greater in magnitude than the current in either branch. It follows, then, that the impedance of a parallel RL circuit must be less than the opposition presented to current by any one branch. This relationship is, of course, similar to the relationship existing between the total resistance and the branch circuit resistances of a parallel resistive circuit.

Phase Angle. The phase angle of a parallel RL circuit is the angle between the applied voltage and the total (line) current. Since the applied voltage and the voltage across the resistive branch of the circuit are in phase, the phase angle can also be represented vectorially by the angle between the resistive branch current vector and the total current vector (Fig. 16-21).

Problem 8. A 200-ohm resistor and a coil having an inductance of 0.5 henry are connected in parallel to a 120-volt, 60-hertz power line (Fig. 16-22). Compute the total current and the phase angle of this circuit if the ohmic resistance of the coil is neglected.

Fig. 16-21. Phase angle of a parallel RL circuit.

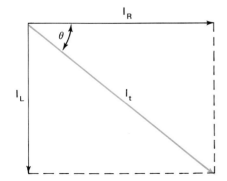

Fig. 16-22. Circuit illustrating Problem 8.

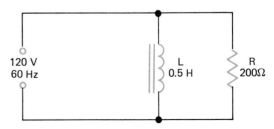

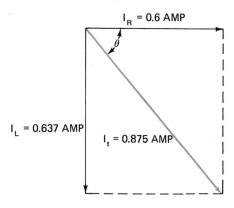

Fig. 16-23. Vector diagram illustrating the solution to Problem 8.

Solution.

$$I_R = \frac{E}{R}$$
$$= \frac{120}{200} = 0.6 \text{ ampere}$$
$$X_L = 2\pi fL$$
$$= 6.28 \times 60 \times 0.5 = 188.4 \text{ ohms}$$
$$I_L = \frac{E}{X_L}$$
$$= \frac{120}{188.4} = 0.637 \text{ ampere}$$
$$I_t = \sqrt{I_R^2 + I_L^2}$$
$$= \sqrt{0.6^2 + 0.637^2}$$
$$= \sqrt{0.766} = 0.875 \text{ ampere}$$
$$\tan \theta = \frac{I_L}{I_R}$$
$$= \frac{0.637}{0.6} = 1.0617$$

(see Fig. 16-23) and

$$\theta = 47.9 \text{ degrees}$$

Impedance. If the total current and the applied voltage of a parallel RL circuit are known,

$$Z = \frac{E}{I_t}$$

If the total current in a two-branch parallel RL circuit is unknown,

$$Z = \frac{RX_L}{\sqrt{R^2 + X_L^2}}$$

where $Z =$ impedance, in ohms
$R =$ resistance of resistive branch, in ohms
$X_L =$ inductive reactance of inductive branch, in ohms

INDUCTORS IN SERIES

When inductors or other forms of coils are connected in series and widely spaced so that there is no significant interaction (coupling) of their magnetic fields, the total inductance L_t of the combination is equal to the sum of the individual inductances, or

$$L_t = L_1 + L_2 + L_3 + \cdots$$

If the coils are wound in the same direction and located near each other, there may be a significant aiding interaction among their magnetic fields (Fig. 16-24A). In this case, the voltages induced across the coils by the process of

Fig. 16-24. Inductors in series: (A) coils wound in same direction, (B) coils wound in opposite directions.

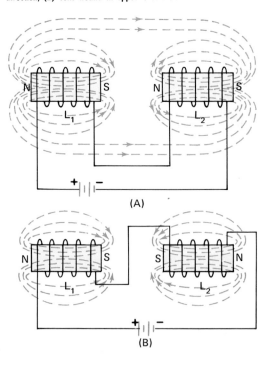

mutual induction M will be in phase, or additive. As a result, the total inductance will be greater than the sum of the individual inductances, as expressed by the following formula:

$$L_t = L_1 + L_2 + L_3 + \cdots + 2M$$

where M = mutual induction, in henrys. Note that

$$M = K\sqrt{L_1 L_2 L_3 \cdots}$$

where K = coefficient of coupling, expressed as a decimal factor

L_1, L_2, L_3 = inductances of individual coils, in henrys

When coils in series are wound in opposite directions and closely coupled, there may be a significant opposing interaction of their magnetic fields (Fig. 16-24B). In this case, the voltages induced across the coils by mutual induction will be out of phase and will, therefore, tend to oppose each other. This condition will result in a total inductance that is less than the sum of the individual inductances, or

$$L_t = L_1 + L_2 + L_3 + \cdots - 2M$$

INDUCTORS IN PARALLEL

When inductors are connected in parallel, they are usually shielded from each other, or spaced so that there is an insignificant degree of magnetic coupling between them. Under this condition, the total inductance of a particular combination is determined by formulas that are similar to those used for finding the total resistance of resistors in parallel:

1. If there are two inductances of unequal value in the circuit,

$$L_t = \frac{L_1 L_2}{L_1 + L_2}$$

2. If the inductances are equal in value,

$$L_t = \frac{\text{Value of one inductance}}{\text{No. of inductances}}$$

3. If there are three or more inductances of unequal value in the circuit,

$$L_t = \frac{1}{1/L_1 + 1/L_2 + 1/L_3}$$

POWER IN INDUCTIVE CIRCUITS

In a purely inductive (no resistance) circuit having sinusoidal waveform voltage and current characteristics, an equal amount of power is alternately delivered to the circuit load and returned to the source. This condition is illustrated by Fig. 16-25. Here, the waveforms of voltage e and current i represent instantaneous values of applied voltage and current. The current, of course, lags the applied voltage by 90 degrees.

Since power p equals voltage E times current I, the instantaneous power in this circuit is equal to ei. When e and i are of like sign (both either above or below the horizontal time reference axis), the product of ei is positive. This product represents positive power p that is delivered to the circuit load. When e and i differ in sign (one is above the horizontal axis and the other is below it), the product of ei is negative. This product represents negative power that is returned to the source of energy. During any given cycle, the amount of positive power equals the negative power. Therefore, the amount of power dissipated by a purely inductive circuit is zero.

Apparent Power. In a practical inductive circuit where the properties of resistance and induct-

Fig. 16-25. Power curve of a sinusoidal, purely inductive circuit.

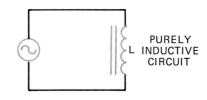

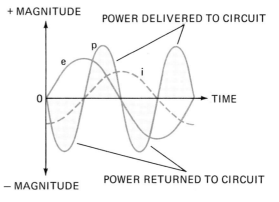

ance are both present, only the resistive components of the circuit dissipate power. However, the source of energy (such as a generator) must also supply the energy that is returned to it by the inductive components of the circuit. In terms of power, the generator output to the circuit is then equal to the product of the effective or rms values of voltage and current. This power is referred to as *apparent power* and is often expressed in terms of the voltampere or VA (volts × amperes) unit.

True Power. *True power* is that portion of the apparent (VA) power delivered to an *RL* circuit which is actually dissipated by the resistive components of the circuit and the resistance of the circuit conductors. True power P is measured in watts. In a sinusoidal circuit,

$$P = \cos\theta\, EI$$

and

$$P = I^2 R$$

Problem 9. A series *RL* circuit operated at 120 volts has a resistance of 200 ohms and an impedance of 300 ohms. Compute the values of apparent power and true power in this circuit.

Solution.

$$I = \frac{E}{Z}$$

$$= \frac{120}{300} = 0.4 \text{ ampere}$$

Apparent power $= EI$

$$= 120 \times 0.4 = 48 \text{ voltamperes}$$

$$\cos\theta = \frac{R}{Z}$$

$$= \frac{200}{300} = 0.6666$$

True power $= \cos\theta\, EI$

$$= 0.6666 \times 120 \times 0.4$$

$$= 32 \text{ watts}$$

or

True power $= I^2 R$

$$= 0.4^2 \times 200$$

$$= 0.16 \times 200 = 32 \text{ watts}$$

POWER FACTOR

The relative amounts of apparent power and true power are important considerations in power-distribution circuit design and efficiency. From the viewpoint of the public utility company that provides electrical power, the generator and the power-distribution system used to deliver electrical power to a customer must be designed to supply a given apparent power to any given circuit. If the apparent power is significantly greater than the true power, the power lines and other components of the distribution system must be considerably larger than necessary in order to satisfy the voltampere (apparent power) requirements of a circuit. This factor lowers the efficiency of the system and increases the cost of construction and maintenance. These conditions, in turn, increase the unit cost of electrical energy because they result in a greater investment that must be made by the power company.

The most economical and efficient operation of an inductive circuit is obtained when the amount of apparent power delivered to the circuit is as near as possible to the amount of true power dissipated in the circuit. The ratio of the true power to the apparent power in a circuit is known as the *power factor* PF of the circuit. Thus

$$PF = \frac{\text{True power}}{\text{Apparent power}}$$

Series Circuits. The power factor of a series *RL* circuit is most commonly determined by reference to the cosine of the phase angle θ that exists between the applied voltage and the current. The cosine of the phase angle provides an indication of the extent to which the power delivered to a circuit is actually dissipated by the resistive components of the circuit. With the phase angle known,

$$PF = \cos\theta$$

and

$$\cos\theta = \frac{R}{Z}$$

Problem 10. A series *RL* circuit has a resistance of 30 ohms and an inductive reactance of 40 ohms. Compute the power factor of this circuit.

Solution.

$$Z = \sqrt{R^2 + X_L^2}$$
$$= \sqrt{30^2 + 40^2}$$
$$= \sqrt{2{,}500} = 50 \text{ ohms}$$
$$PF = \cos \theta$$
$$\cos \theta = \frac{R}{Z} = \frac{30}{50} = 0.6$$

Therefore

$$PF = 0.6 = 60 \text{ percent}$$

Parallel Circuits. The power factor of a two-branch parallel RL circuit is also equal to the cosine of the phase angle between the applied voltage and the current. Since the phase angle in such a circuit is a function of the ratio of the current through the resistive branch I_R to the total current I_t,

$$PF = \cos \theta$$

and

$$\cos \theta = \frac{I_R}{I_t}$$

Correction of Power Factor. A study of the power factor ratio (true power/apparent power) shows that, in a circuit having a low power factor, the apparent power is considerably greater than the true power dissipated in the circuit. Thus, such a circuit is inefficient because it produces a relatively small amount of work in comparison with the amount of energy that must be supplied to it. In order to increase circuit efficiency, it is desirable to design circuits that have a high power factor or a power factor that approaches unity (1) or 100 percent.

The basic design of circuits that contain large iron-core coils or windings usually results in a low power factor. To remedy this condition, it is necessary to raise or "correct" the power factor. This is commonly accomplished by the use of capacitors in conjunction with coils or windings. The effect of capacitance decreases the phase angle in an inductive circuit; therefore, it also increases the power factor.

For Review and Discussion

1. Define a resistance inductance (RL) circuit.
2. Explain the energy characteristics of a purely inductive circuit.
3. What is meant by the transient response of a circuit?
4. Describe the effect of inductance in a direct-current circuit.
5. Define the L/R time constant of a series inductive circuit.
6. In accordance with the time-constant formula, what effect does an increase of inductance have upon the time-constant period? What effect does an increase of resistance have upon the time-constant period?
7. Draw a diagram illustrating the three different voltages that are present in a series RL circuit at the time that the current through the circuit is increasing.
8. What factors determine the magnitude of the cemf induced in a series RL circuit?
9. Explain the inductive reactance opposition to current in a sinusoidal alternating-current circuit.
10. Define angular velocity.
11. Give the inductive reactance formula and state the relationships between inductive reactance, frequency, and inductance that are indicated by this formula.
12. Draw a graph illustrating the relationship that exists between inductive reactance and frequency in a series RL circuit having a fixed inductance.
13. Describe the phase shift that occurs between the voltage and the current in a sinusoidal RL circuit.
14. Define phase angle and give the letter symbol that represents it.
15. Draw a vector diagram illustrating the relationships between the resistance, inductive reactance, and phase angle of a series RL circuit.

16. Define the impedance of an *RL* circuit and state the formula that indicates the relationship between the impedance, resistance, and inductance of such a circuit.

17. Explain Ohm's law for inductive circuits.

18. State the voltage and the current characteristics of a parallel *RL* circuit.

19. What is the effect of connecting inductors in series? In parallel?

20. Define apparent power and true power in an *RL* circuit.

21. What is meant by the power factor of an *RL* circuit?

22. Explain why it is desirable to operate an *RL* circuit under a high power factor condition.

UNIT 17. Capacitive Circuits

A capacitive circuit is one in which the property of capacitance is present to a significant degree. As in the case of inductance, capacitance produces definite characteristics in a circuit that affect the phase relationship between voltage and current, the total opposition to current, and the power dissipation. However, as we shall learn, the effects of capacitance are opposite to those of inductance.

ENERGY IN A CAPACITIVE CIRCUIT

While a capacitor is charging, energy from the charging source (generator or battery) is expended in displacing electrons from one plate of the capacitor to the other. As a result of this expenditure of energy, a voltage is developed across these plates. When the capacitor becomes fully charged, the energy supplied by the charging source is stored in the form of an electrostatic field within the dielectric (Fig. 17-1).

As the capacitor discharges and the voltage across its plates decreases, the electrostatic field becomes weaker. The energy that was stored in the electrostatic field is then returned to the charging source. Thus, there is no expenditure of energy in a purely capacitive circuit. However, since all practical capacitive circuits contain resistance, some of the energy that is supplied to a capacitive circuit is converted into heat energy.

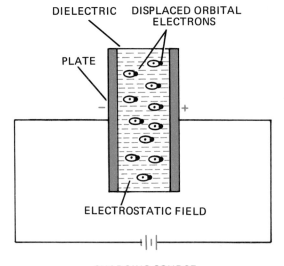

Fig. 17-1. Energy stored within the electrostatic field of a capacitor.

EFFECT OF CAPACITANCE IN A DC CIRCUIT

The transient response of a series *RC* circuit or the manner in which the current responds to a change in applied voltage resulting from the capacitor charging process can be described by referring to the circuit shown in Fig. 17-2A. At the instant the switch is closed from position 1 to position 2, electrons are transferred from one plate of the capacitor to the other. This, of

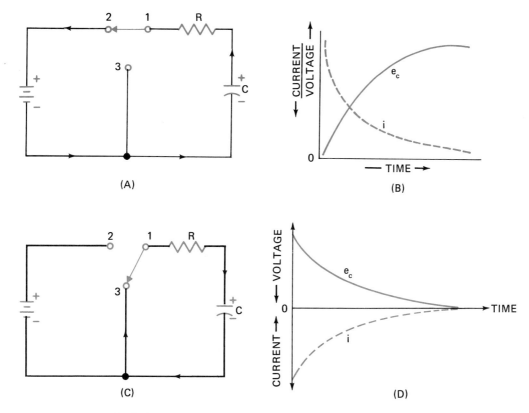

Fig. 17-2. Illustrations used to explain the effect of capacitance in a dc circuit.

course, causes a charging current i to flow in the circuit. At this same instant, the voltage across the capacitor e_c is at a minimum, since the voltage drop IR across R is at a maximum (Fig. 17-2B). Therefore, the initial current in the circuit at the instant the capacitor begins to charge is very nearly equal to E/R.

As the current flow in the circuit continues, the capacitor gradually becomes charged and a voltage begins to develop across it. Because the capacitor voltage is a cemf that opposes the applied voltage, the effective voltage applied to the circuit gradually decreases; as a result, the circuit current also decreases. When the capacitor is fully charged, the voltage across it is equal to the battery voltage and the magnitude of the current is equal to zero.

If the switch is closed from position 1 to position 3, the capacitor immediately begins to discharge through the resistor (Fig. 17-2C). Since the voltage across the capacitor is at a maximum, the initial discharge current is also at a maximum (Fig. 17-2D). As the capacitor continues to discharge, the current gradually decreases until both the circuit current and the voltage across the capacitor become zero.

Time Constant. As indicated in the preceding paragraphs, a certain period of time must elapse before a capacitor in an RC circuit becomes fully charged. The period of time required for the charging process to be completed is dependent upon the capacitance and the resistance of the circuit. With a given amount of resistance in the circuit, the charging time will be increased if the capacitance is increased. This is because a greater number of electrons must be transferred from plate to plate before the voltage developed across the capacitor is equal to the applied voltage.

If the capacitance of a circuit remains constant and the resistance is increased, the transfer

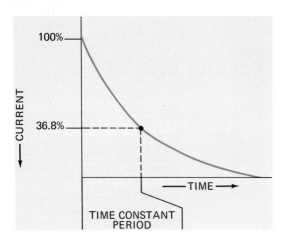

Fig. 17-3. Time constant of an *RC* circuit.

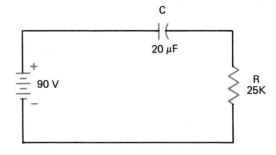

Fig. 17-4. Circuit illustrating Problem 1.

of electrons from plate to plate occurs at a slower rate. As a result, the time required for the voltage developed across the capacitor to become equal to the applied voltage is once again increased.

The time required for a capacitor in a series *RC* circuit to become charged to 63.2 percent of its full charge is known as the *RC time constant* of the circuit. The time constant may also be defined as the time required for the voltage across the capacitor to attain a magnitude that is equal to 63.2 percent of the applied (source) voltage. Stated in terms of the circuit current, the time constant of an *RC* circuit is the time that elapses before the current in the circuit decays (decreases) to $100 - 63.2$ or 36.8 percent of the initial maximum magnitude (Fig. 17-3). Expressed as a formula,

$$t = RC$$

where t = time, in seconds
C = capacitance, in microfarads
R = resistance, in megohms

Problem 1. A series circuit consists of a 90-volt battery, a 20-microfarad capacitor and a 25-kilohm resistor (Fig. 17-4). Compute the time constant of this circuit and the time required for the capacitor to become fully charged following the application of the battery voltage.

Solution.

$$t = RC$$

$$= 0.025 \times 20 = 0.5 \text{ second}$$

where t = time constant.

After one time-constant period (0.5 second), the voltage across the capacitor will be

$$E_c = 90 \times 0.632 = 56.88 \text{ volts}$$

In the next time-constant period, the capacitor voltage will be equal to 56.88 volts plus 63.2 percent of the additional voltage that must be developed across the capacitor before it is fully charged. Therefore, after two time-constant periods, the capacitor voltage is

$$E_c = 56.88 + 0.632(90 - 56.88)$$
$$= 77.81 \text{ volts}$$

By following a similar mathematical procedure, it can be determined that the voltage across the capacitor increases to a maximum magnitude of nearly 90 volts after five time-constant periods or after 2.5 seconds (Fig. 17-5). This is, for all practical purposes, the time that is required for the capacitor to become fully charged. After this period of time, the circuit is said to have reached its steady-state (no current) condition.

The circuit will, of course, remain in a steady-state condition for as long as it is connected to the battery. If the battery is removed from the circuit and the capacitor is allowed to discharge through the resistor, its rate of discharge will also be determined by the time-constant period. As in the case of the charging process, a total of five time-constant periods will elapse before the capacitor becomes completely discharged and the circuit once again returns to a steady-state condition.

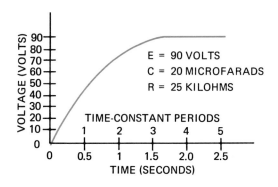

Fig. 17-5. Graph illustrating the solution to Problem 1.

Although the time constants of different RC circuits may vary widely, the transient response of all circuits is always equal to five time-constant periods. This factor is very commonly used for the operation of certain filter, oscillator, and timing circuits that are discussed in later units.

Voltage Characteristics. During the time that the capacitor in a direct-current series RC circuit is charging, there are three different voltages present in the circuit: the applied voltage E, the IR voltage drop across any resistive component, and the voltage developed across the capacitor, E_c.

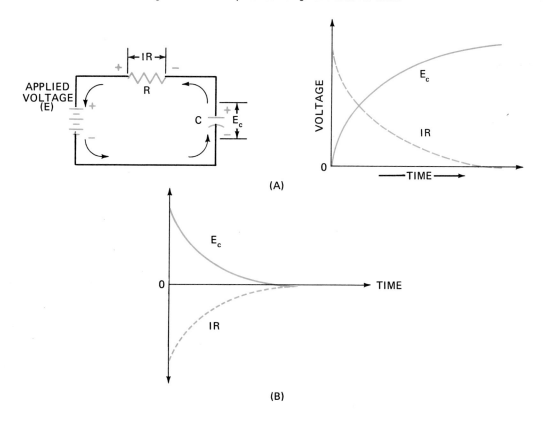

Fig. 17-6. Relationships between voltages in a series RC circuit.

Since the voltage across the capacitor opposes the applied voltage, it is, in effect, a cemf and can be considered as a voltage drop. Thus, at any instant of time, the sum of the voltage drops in the circuit must equal the applied voltage or

$$E = IR + E_c$$

The relationship between the voltages in an RC circuit is illustrated by Fig. 17-6A. At time zero, the current in the circuit is at a maximum and the IR drop across the resistor is also at a maximum. As the capacitor charges its voltage increases, and the effect of this cemf causes a corresponding decrease in the circuit current and the IR drop across the resistor. When the capacitor has become fully charged, no further current flows in the circuit. In this condition, the IR drop across the resistor is zero and the capacitor voltage must therefore be equal to the applied voltage.

When the applied voltage is removed from the circuit and the capacitor is allowed to discharge through the resistor, only two voltages, IR and E_c, are present in the circuit. As E_c decreases, the current through the resistor also decreases and, as a result, there is a corresponding decrease of the IR drop. The relationship between these voltages is illustrated by Fig. 17-6B.

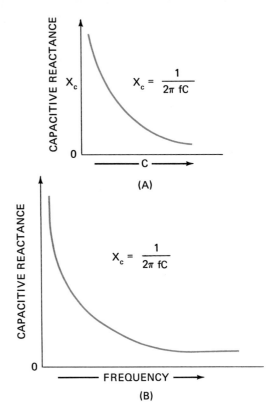

Fig. 17-7. Relationships between capacitive reactance, capacitance, and frequency in an ac capacitive circuit.

EFFECTS OF CAPACITANCE IN AC CIRCUITS

The effects of capacitance in a direct-current circuit are present only during the times that there are changes in the voltage applied to the circuit. These changes occur when the voltage is first applied to the circuit and when the applied voltage is removed from the circuit. In a sinusoidal alternating-current circuit, however, the applied voltage is constantly changing in magnitude and periodically changing in polarity. Because of this condition, the effect of capacitance, which tends to oppose changing voltage, is present at all times. Thus, the properties of capacitive reactance and phase shift are constantly introduced into the circuit.

Capacitive Reactance. While a capacitor is charging, the cemf developed across it opposes the voltage that is applied to the capacitor circuit. Since this action, in effect, causes a decrease in the magnitude of current in the circuit, it tends to oppose the current. As in the case of inductive reactance, this opposition does not result from resistance but from the reaction of the cemf against the applied voltage. For this reason, the opposition is called *capacitive reactance*, X_c.

If the capacitance of a circuit is increased, a greater charging current must be present before a given cemf is developed across any given capacitor. Therefore, at any given frequency, there is an indirect or inverse relationship between capacitive reactance and capacitance (Fig. 17-7A). Also, the charge and discharge currents in a capacitive circuit will increase if the rate of change of the applied voltage is increased. Since, in an alternating-current circuit, this rate of change is governed by the angular velocity of the applied voltage, there is also an

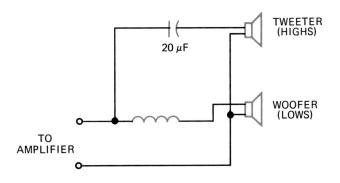

CAPACITOR BLOCKS LOW FREQUENCIES FROM TWEETER, WHILE INDUCTOR BLOCKS HIGH FREQUENCIES FROM WOOFER

Fig. 17-8. Circuit illustrating Problem 2.

inverse relationship between capacitive reactance and angular velocity (ω or $2\pi f$). These inverse relationships of capacitive reactance to capacitance and to angular velocity are expressed by the following formula:

$$X_c = \frac{1}{2\pi f C}$$

where X_c = capacitive reactance, in ohms
π = 3.14
f = frequency, in hertz
C = capacitance, in farads

A graphical representation of the relationship between capacitive reactance and frequency in a capacitive circuit is shown in Fig. 17-7B.

Problem 2. A crossover network used in a high-fidelity loudspeaker system is shown in Fig. 17-8. Compute the capacitive reactance of the capacitor at a frequency of 10 kilohertz.

Solution.

$$X_c = \frac{1}{2\pi f C}$$

$$= \frac{1}{6.28 \times 10{,}000 \times 0.00002}$$

$$= \frac{1}{1.256} = 0.796 \text{ ohm}$$

Phase Shift. In addition to limiting the current in an alternating-current circuit, capacitive reactance also has the effect of opposing any change in the magnitude of the applied voltage. Since the applied voltage is constantly changing in magnitude, this opposition is present in the circuit at all times. As a result, it produces a continuous phase shift or a difference in the time between the starting points and between the minimum and maximum magnitude points of the applied voltage and the current.

As indicated previously, the cemf, or voltage developed across a capacitor while it is charging, opposes the applied voltage. In a purely capacitive circuit, the magnitude of the cemf is equal to the applied voltage at any instant. Thus, these two voltages are 180 degrees out of phase with each other, as shown by the sinusoidal waveforms of Fig. 17-9A.

A study of these waveforms shows that both the applied voltage and the cemf are at a maximum at the 90- and 270-degree points of their cycles. At these same points (times), the current in the circuit is at a minimum because the rate of change of the applied voltage is least, as indicated by the comparatively gradual slope of the waveform (Fig. 17-9B). Since the current in the circuit is also sinusoidal in nature, it will be at a maximum when both the applied voltage and the cemf are at a minimum or at the 0-, 180-, and 360-degree points of their waveforms. As a result, the applied voltage passes through zero 90 degrees after the current. Thus, in a purely capacitive circuit, current leads the applied voltage by 90 degrees. It is important to note that

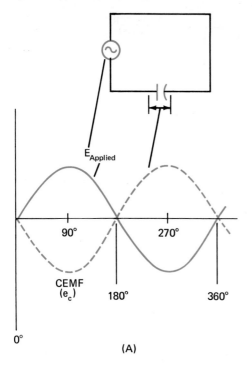

Fig. 17-9. Waveforms showing voltage relationships and current-voltage relationship in a purely capacitive circuit.

this condition is opposite to the phase shift that takes place in a purely inductive circuit. The inductive circuit current lags the voltage by 90 degrees.

Phase Angle. The phase shift of 90 degrees between the applied voltage and the current in a purely capacitive circuit is referred to as the *phase angle* θ. In practical RC circuits, the phase angle by which the current leads the applied voltage is always less than 90 degrees. The exact phase angle in such a circuit is a function of the ratio of the capacitive reactance to resistance, or

$$\tan \theta = \frac{X_c}{R}$$

Problem 3. A series RC circuit has a resistance of 50 ohms and a capacitive reactance of 60 ohms (Fig. 17-10). Compute the phase angle of this circuit and draw the vector diagram that represents the resistance and capacitive reactance conditions of the circuit.

Solution.

$$\tan \theta = \frac{X_c}{R}$$

$$= \frac{60}{50} = 1.200$$

and

$$\theta = 50.2 \text{ degrees}$$

Fig. 17-10. Circuit illustrating Problem 3.

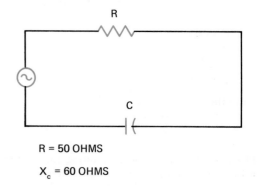

R = 50 OHMS

X_c = 60 OHMS

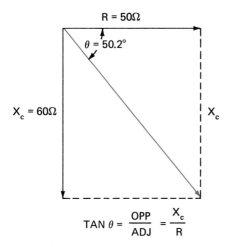

Fig. 17-11. Vector diagram illustrating the solution to Problem 3.

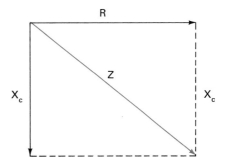

Fig. 17-12. Impedance vector diagram of a series *RC* circuit.

Since resistance alone does not affect the phase relationship between current and voltage, and since capacitive reactance alone causes a 90-degree phase shift between current and voltage, the vectors representing the resistance and the capacitive reactance of the circuit are drawn perpendicular to each other or at an angle of 90 degrees. Since capacitive reactance is an opposition to current that tends to cause the current to lead the applied voltage by 90 degrees, its vector is drawn below the reference or *R* vector (Fig. 17-11).

A study of this vector diagram shows that the phase angle between current and voltage becomes larger if there is an increase of capacitive reactance, while the resistance of the circuit remains constant. Capacitive reactance is inversely proportional to capacitance and to frequency. Therefore, the phase angle in an *RC* circuit will increase if either the capacitance or the operating frequency of the circuit is decreased without a corresponding increase of resistance.

IMPEDANCE

As in the case of a series *RL* circuit, the impedance or total opposition to current in a series *RC* circuit is a vector sum. Thus, the impedance is equal to the hypotenuse of a right angle having R and X_c as its sides (Fig. 17-12). From this vector diagram it is apparent that

$$Z^2 = R^2 + X_c^2$$

and

$$Z = \sqrt{R^2 + X_c^2}$$

where Z = impedance, in ohms
  R = resistance, in ohms
  X_c = capacitive reactance, in ohms

Problem 4. Compute the impedance and the phase angle of a series *RC* circuit that has a capacitive reactance of 30 ohms and a resistance of 40 ohms (Fig. 17-13).

Solution.

$$\begin{aligned}Z &= \sqrt{R^2 + X_c^2} \\ &= \sqrt{30^2 + 40^2} \\ &= \sqrt{2{,}500} = 50 \text{ ohms}\end{aligned}$$

$$\tan \theta = \frac{X_c}{R}$$

$$= \frac{30}{40} = 0.7500$$

Fig. 17-13. Circuit illustrating Problem 4.

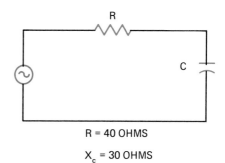

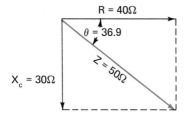

Fig. 17-14. Vector diagram illustrating the solution to Problem 4.

and

$$\theta = 36.9 \text{ degrees}$$

(see Fig. 17-14).

With the impedance of the circuit known, the phase angle can also be determined from the ratio of the resistance to impedance:

$$\cos \theta = \frac{R}{Z}$$

$$= \frac{40}{50} = 0.8000$$

and

$$\theta = 36.9 \text{ degrees}$$

OHM'S LAW FOR CAPACITIVE CIRCUITS

In alternating-current circuits where the effect of capacitance is significant, capacitive reactance may have a considerable effect upon the magnitude of the current. Therefore, when Ohm's law is applied to the analysis of capacitive circuits, it must be modified to include the effects of both capacitive reactance and resistance. This is done by substituting either X_c or Z for R as the total opposition to current.

1. In circuits where the resistance is insignificant,

$$E = IX_c$$

$$I = \frac{E}{X_c}$$

$$X_c = \frac{E}{I}$$

where E = voltage, in volts
I = current, in amperes
X_c = capacitive reactance, in ohms

2. In circuits where the value of resistance is significant,

$$E = IZ$$

$$I = \frac{E}{Z}$$

$$Z = \frac{E}{I}$$

where E = voltage, in volts
I = current, in amperes
Z = impedance, in ohms

Problem 5. A series RC circuit consisting of a 200-ohm resistor and a 0.001-microfarad capacitor is connected to a signal generator with an output voltage of 0.75 volt at 455 kilohertz (Fig. 17-15). Compute the magnitude of current in this circuit.

Solution.

$$X_c = \frac{1}{2\pi fC}$$

$$= \frac{1}{6.28 \times 455{,}000 \times 0.000000001}$$

$$= \frac{1}{0.00286} = 349 \text{ ohms}$$

$$Z = \sqrt{R^2 + X_c^2}$$

$$= \sqrt{200^2 + 349^2}$$

$$= \sqrt{161{,}801} = 402.4 \text{ ohms}$$

$$I = \frac{E}{Z}$$

$$= \frac{0.75}{402.4} = 0.00186 \text{ ampere}$$

Fig. 17-15. Circuit illustrating Problem 5.

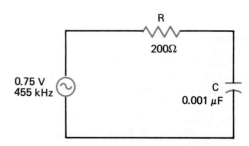

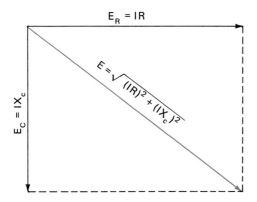

Fig. 17-16. Vectorial representation of the voltage relationships in series *RC* circuits.

VOLTAGE RELATIONSHIPS IN SERIES CIRCUITS

The resistance and the capacitive reactance of an alternating-current series *RC* circuit can, as previously mentioned, be represented by vectors that are drawn at an angle of 90 degrees. Since the current in such a circuit is a common factor, the voltage drop E_R across a resistive component in the circuit is equal to IR and the voltage across a capacitor in the circuit (E_c) is equal to IX_c. These voltages are 90 degrees out of phase with each other and can be represented by vectors IR and IX_c. The voltage E applied to the circuit will then be equal to the vector sum of IR and IX_c, or

$$E = \sqrt{(IR)^2 + (IX_c)^2}$$

(see Fig. 17-16).

PARALLEL *RC* CIRCUITS

When capacitors and resistors are connected in parallel, the voltage and current characteristics of the individual components are similar to those previously discussed. However, the effect of this combination upon the behavior of the total circuit is somewhat different from the effect produced in a series circuit.

Voltage. The voltages across the resistive and capacitive branches are equal to each other and to the applied voltage. Consequently, all these voltages are also in phase with each other.

Current. The current in any capacitive branch of a parallel *RC* circuit leads the applied voltage, while the current through any resistive branch is in phase with this voltage. If, for example, the capacitive branch of a two-branch circuit has a negligible amount of resistance, the current in this branch will lead the applied voltage by 90 degrees. Therefore, the current in the capacitive branch leads the current in the resistive branch by 90 degrees.

The total or line current of the circuit is equal to the vector sum of the branch currents as indicated by the vector diagram shown in Fig. 17-17. In this diagram, the capacitive branch current (I_c) vector is drawn above or in a positive direction from the horizontal vector (I_R) that represents the current in the resistive branch. This is to indicate that the current in the capacitive branch leads the current in the resistive branch by 90 degrees.

Phase Angle. The phase angle of a parallel *RC* circuit is the angle by which the line current leads the applied voltage. Because the applied voltage in such a circuit is in phase with the

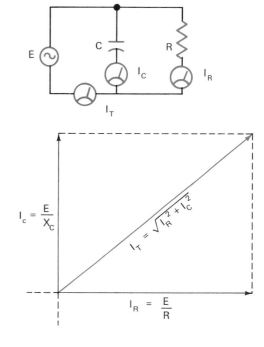

Fig. 17-17. Current relationships in a parallel *RC* circuit.

voltage across the resistive branch, the phase angle can be represented vectorially by the angle between the resistive branch current vector and the total current (I_t) vector (Fig. 17-18). Expressed as a formula,

$$\tan \theta = \frac{I_c}{I_R}$$

Impedance. If both the total current and the applied voltage of a parallel RC circuit are known,

$$Z = \frac{E}{I_t}$$

If the total current in a two-branch parallel RC circuit is unknown,

$$Z = \frac{RX_c}{\sqrt{R^2 + X_c^2}}$$

where Z = impedance, in ohms
 R = resistance of resistive branch, in ohms
 X_c = capacitive reactance of capacitive branch, in ohms

CAPACITORS IN SERIES

When capacitors are connected in series, the effect is the same as that produced when the distance between the plates of a given capacitor is increased. Because of this, the total capacitance of a series combination of capacitors is always less than the capacitance of any capacitor in the circuit. This condition, you will recall, is

Fig. 17-18. Phase angle of a parallel RC circuit.

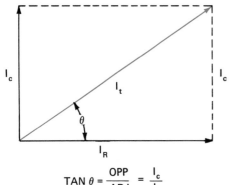

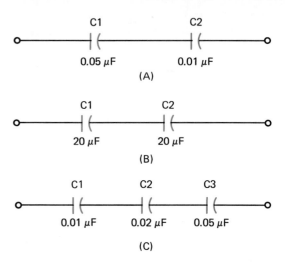

Fig. 17-19. Illustrations used in conjunction with the capacitors-in-series formulas.

similar to the relationship between total resistance and individual resistances when resistors are connected in parallel. The following formulas, which are similar to the parallel resistor formulas, are used to determine the total capacitance of capacitors in series.

If there are two capacitors having unequal capacitances in the circuit (Fig. 17-19A),

$$C_t = \frac{C_1 C_2}{C_1 + C_2} \quad (1)$$

If the capacitors (two or more) have equal capacitances (Fig. 17-19B),

$$C_t = \frac{\text{Capacitance of one}}{\text{No. of capacitors}} \quad (2)$$

If there are three or more capacitors having unequal capacitances in the circuit (Fig. 17-19C),

$$C_t = \frac{1}{1/C_1 + 1/C_2 + 1/C_3 + \cdots} \quad (3)$$

One reason for connecting capacitors in series is that it provides a means of using two or more capacitors in a circuit where the applied voltage is greater than the voltage rating of any one of the capacitors. For example, if two 0.02-

Note: Formula (3) can be used to find the total capacitance of any series capacitor combination.

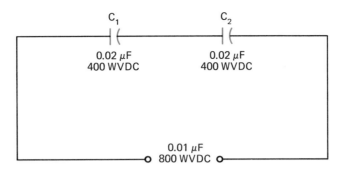

Fig. 17-20. When capacitors are connected in series, the working voltage rating of the combination is greater than the working voltage rating of any individual capacitor.

microfarad capacitors, each having a working voltage rating of 400 volts, are connected in series, the total capacitance of the combination is reduced to 0.01 microfarad, but the working voltage rating of the capacitors is increased to 800 volts (Fig. 17-20).

The sum of the voltage drops across two or more capacitors in series is equal to the voltage applied to the combination. The voltage drop across any given capacitor of the combination is inversely proportional to its capacitance value. This factor must be considered when a series connection of capacitors is used for any purpose.

For example, assume that C_1 and C_2, having capacitance values of 0.01 and 0.05 microfarad respectively, are connected across a 600-volt line. Since the capacitance ratio of C_1 to C_2 is 1:5, the voltage drop across C_1 will be equal to 500 volts, or 5 times the voltage drop across C_2 (Fig. 17-21). Thus, C_1 should have a minimum

Fig. 17-21. When capacitors are connected in series, the voltage drop across any given capacitor is inversely proportional to its capacitance.

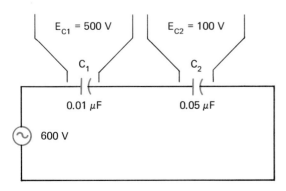

working voltage rating of 1,000 volts, and C_2 should have a minimum working voltage rating of 200 volts if the desirable 100 percent safety factor is to be provided.

CAPACITORS IN PARALLEL

When capacitors are connected in parallel, the effect is the same as that produced when the plate area of a given capacitor is increased. Because of this, a parallel connection of capacitors can be used to increase the total capacitance of a circuit. In this regard, it is important to note that, when electrolytic capacitors are connected in parallel, they must be connected across the line in like polarity (Fig. 17-22). The formula that is used to determine the total capacitance of capacitors in parallel is similar to that used to find the total resistance of resistors in series. Thus

$$C_t = C_1 + C_2 + C_3 + \cdots$$

The working voltage rating of a parallel combination of capacitors is limited by the capacitor with the lowest rating. Therefore, if two

Fig. 17-22. A "like polarity" connection of capacitors in parallel.

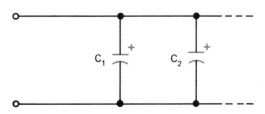

capacitors having working voltage ratings of 200 and 600 volts are connected in parallel, the maximum voltage that should be applied across the combination is 100 volts. This magnitude of applied voltage is, of course, based on the assumption that a 100 percent safety factor is to be provided.

POWER IN CAPACITIVE CIRCUITS

As in a purely inductive circuit, the power dissipated in a purely capacitive sinusoidal circuit is zero. This condition is illustrated by the power curve shown in Fig. 17-23. Here, the products of *e* and *i* result in positive and negative "loops" of power that are equal to each other during the course of any given cycle. It is important, however, to note that there is a significant difference in the power curve of a purely inductive circuit (Fig. 16-25) and the power curve of a purely capacitive circuit. This difference is that the positive power loops in the capacitive circuit occur 90 degrees ahead of the positive power loops in the inductive circuit. The reason for this is that in the capacitive circuit the current leads the applied voltage by 90 degrees, while in an inductive circuit the current lags the voltage by 90 degrees.

The true (effective) power dissipated in a sinusoidal *RC* circuit, as in the *RL* circuit, is determined by the following formula:

$$P = \cos \theta \, EI$$

The ratio of true power to apparent power is the power factor of the circuit and is equal to the cosine of the phase angle between the current and the applied voltage. Since the current leads the voltage in an *RC* circuit, such a circuit is said to have a leading power factor, while an inductive circuit has a lagging power factor.

DISTRIBUTED CAPACITANCE

Stray or random capacitance produced by the capacitor-like effect of closely spaced conductors and the metallic structural elements of certain components is present throughout an alternating-current circuit. This capacitance is referred to as "distributed" capacitance, as distinguished from the "lumped" capacitance that is introduced into a circuit by the use of capacitors (Fig. 17-24).

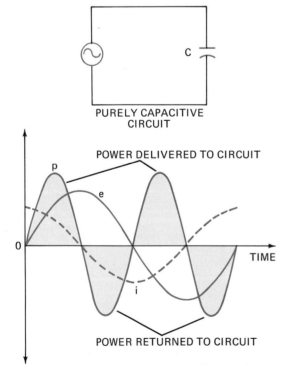

Fig. 17-23. Power curve of a sinusoidal, purely capacitive circuit.

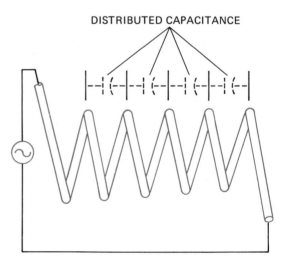

Fig. 17-24. The distributed capacitance existing between the turns of a coil.

In low-frequency circuits, the effects of distributed capacitance are generally insignificant and can be disregarded. However, in high-frequency circuits, the effects of distributed capacitance must be kept to a minimum. This is accomplished by design considerations involving factors such as the location of conductors and the position and physical construction of components.

For Review and Discussion

1. Define a capacitive circuit.
2. Explain the energy characteristics of a capacitive circuit.
3. Describe the effect of capacitance in a direct-current circuit.
4. Define the RC time constant of a capacitive circuit.
5. What effect does an increase of capacitance have upon the RC time-constant period? What effect does an increase of resistance have upon the time-constant period?
6. What factor limits the magnitude of the voltage that is developed across a capacitor in any capacitive circuit?
7. Name and define the three voltages that are present in a series RC circuit when the voltage applied to the circuit is changing in magnitude.
8. Explain the capacitive reactance opposition to current in a sinusoidal alternating-current circuit.
9. Give the basic capacitive reactance formula and state the relationships among capacitive reactance, frequency, and capacitance that are indicated by this formula.
10. Draw a graph illustrating the relationship between capacitive reactance and frequency in a capacitive circuit having a negligible amount of resistance.
11. Describe the phase shift occurring between the voltage and the current in a sinusoidal RC circuit.
12. Draw a vector diagram illustrating the relationships among the resistance, the capacitive reactance, and the phase angle of a series RC circuit.
13. Define the impedance of an RC circuit and state the formula that indicates the relationships among the impedance, resistance, and capacitance of such a circuit.
14. Explain Ohm's law for capacitive circuits.
15. Describe the voltage relationships in a series RC circuit.
16. State the voltage and the current characteristics of a parallel RC circuit.
17. State the general formula that can be used to compute the total capacitance of any combination of capacitors connected in series.
18. What is the relationship between the total working voltage rating of a series combination of capacitors and the working voltage ratings of the individual capacitors of the combination?
19. State the formula that can be used to compute the total capacitance of any combination of capacitors connected in parallel.
20. What factor limits the working voltage rating of a combination of capacitors connected in parallel?
21. Explain why the power dissipated by a purely capacitive circuit is zero.
22. Define distributed capacitance.

UNIT 18. Resonance and Tuned Circuits

In many circuits, inductors and capacitors are connected in series or in parallel. Such circuits are often referred to as *RLC* or combination circuits. One of the most important characteristics of an *RLC* circuit is that it can be made to respond most effectively to a single given frequency. When operated in this condition, the circuit is said to be in *resonance* or *resonant* to the operating frequency.

A series or a parallel *RLC* circuit that is operated at or near resonance exhibits specific properties of impedance that allow it to respond selectively to certain frequencies while rejecting others. When an *RLC* circuit is operated for the purpose of providing such frequency selectivity it is called a *tuned circuit*. Tuned circuits are found in a wide variety of electronic circuit systems and, in addition to frequency selectivity (tuning), are very commonly used for impedance matching and for achieving the process of oscillation, which generates an alternating voltage.

RESONANCE

As indicated by their respective formulas, inductive reactance is directly proportional to frequency, while capacitive reactance is inversely proportional to frequency. Therefore, in either a series or a parallel *RLC* circuit, there is one frequency at which the two reactances are equal. When this occurs, the circuit is said to be resonant, or in a state of resonance, and the frequency at which this condition is achieved is called the *resonant frequency* f_r (Fig. 18-1).

The formula for determining the resonant frequency of a given circuit is derived from the mathematical law stating that quantities that are equal to the same quantity are equal to each other. Thus, since in a resonant circuit

$$X_L = 2\pi f L \quad \text{and} \quad X_c = \frac{1}{2\pi f C}$$

and at resonance

$$X_L = X_c$$

then at resonance

$$2\pi f L = \frac{1}{2\pi f C}$$

and, by cross multiplication,

$$(2\pi f L)(2\pi f C) = 1$$
$$4\pi^2 f^2 LC = 1$$

Solving for f,

$$f^2 = \frac{1}{4\pi^2 LC}$$

Therefore

$$f_r = \frac{1}{2\pi \sqrt{LC}}$$

where f_r = resonant frequency, in hertz
L = inductance, in henrys
C = capacitance, in farads

and

$$f_r = \frac{1,000,000}{2\pi \sqrt{LC}}$$

where f_r = resonant frequency, in kilohertz
L = inductance, in microhenrys
C = capacitance, in picofarads

LC Product. An examination of the resonant frequency formula shows that for any product of L and C there is only one resonant frequency. Thus, various combinations of L and C may be

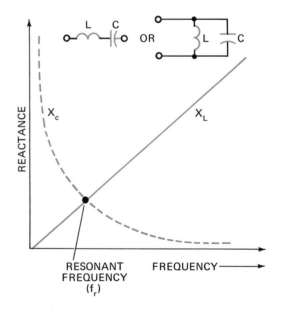

Fig. 18-1. Graphical representation of resonance in an *RLC* circuit.

Table 18-1. Example of various combinations of L and C required to achieve resonance in an RLC circuit at a given frequency.

RESONANT FREQUENCY (kHz)	L (µH)	C (pF)	LC PRODUCT
600	1,000	0.072	72
600	500	0.144	72
600	100	0.72	72
600	20	3.60	72
600	1	72	72
600	0.2	360	72
600	0.05	1,440	72

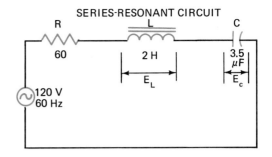

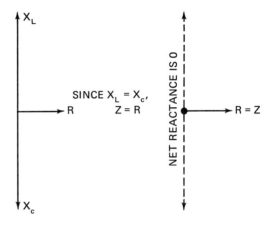

Fig. 18-2. Comparison of the voltages developed across the components of a series-resonant RLC circuit.

used in a circuit to achieve resonance if their product remains the same (Table 18-1).

The formula also shows that the resonant frequency is inversely proportional to the LC product. For this reason, the values of inductance and capacitance necessary to establish the condition of resonance in high-frequency circuits are very small.

Comparison of Voltages. A comparison of the voltages across the inductor, the capacitor, and the applied voltage in a series resonant circuit shows an interesting feature that must be considered in the design of any series RLC circuit. For example, refer to the circuit shown in Fig. 18-2. Under these conditions, the circuit is resonant at a frequency of 60 hertz. Therefore, the current in the circuit is

$$I = \frac{E}{R}$$
$$= \frac{120}{60} = 2 \text{ amperes}$$

and

$$X_L = 2\pi f L$$
$$= 6.28 \times 60 \times 2 = 753.6 \text{ ohms}$$
$$X_c = X_L = 753.6 \text{ ohms}$$
$$E_L = I X_L$$
$$= 2 \times 753.6 = 1{,}507.2 \text{ volts}$$
$$E_c = E_L = 1{,}507.2 \text{ volts}$$

Since, at resonance, the high reactive voltages E_L and E_c cancel each other, they have no effect upon the load, but these voltages do actually exist across the coil and the capacitor. As a result, the insulation of the coil conductors and the dielectric of the capacitor must be designed to withstand a voltage that is approximately 13 times greater than the voltage applied to the circuit.

THE SERIES-TUNED CIRCUIT

Since $X_L = X_c$ at the resonant frequency of a series RLC circuit, the two reactances completely cancel each other. Because of this, the impedance of the circuit at the resonant frequency is at a minimum and is equal to the resistance of the circuit. Thus, at resonance, a series RLC circuit behaves as a simple resistive circuit in which the current is limited only by the resistance. This

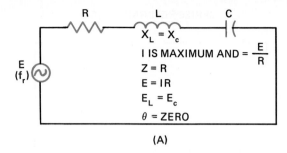

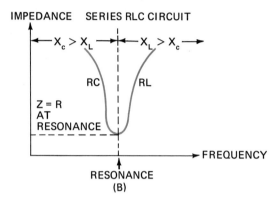

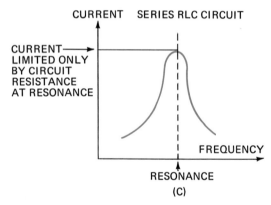

Fig. 18-3. Characteristics of a series-tuned (*RLC*) circuit at resonance.

Selectivity. At frequencies below the resonant frequency, the impedance of a series *RLC* circuit consists of resistance and capacitive reactance, while at frequencies above the resonant frequency the impedance consists of resistance and inductive reactance (Fig. 18-3B). Therefore, at frequencies other than the resonant frequency, the impedance of the circuit is greater than the resistance. The property of a circuit that allows it to respond to the resonant frequency with a minimum amount of impedance, is referred to as *selectivity*. The effect of selectivity upon the current response characteristic of the circuit at resonance is graphically illustrated by Fig. 18-3C.

Identification. In many tuned *RLC* circuits, the inductive component is the secondary winding of a transformer across which a capacitor is connected (Fig. 18-4). Notice that the voltage applied to this circuit is the voltage induced across the secondary winding, which is in series with the capacitor. Thus, while the circuit has the structural characteristics of a parallel circuit, it is, in fact, a series circuit because of the series voltage input. The ability to recognize a series *RLC* circuit in terms of its voltage input is im-

Fig. 18-4. A parallel-like tuned circuit and its series-tuned circuit equivalent.

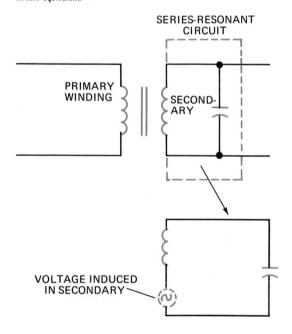

condition produces a circuit response that provides for maximum current and the development of maximum and equal voltages across the inductive and the capacitive components of the circuit. However, since these voltages are 180 degrees out of phase with each other, they cancel each other. Hence, the applied voltage is equal to *IR* and the phase angle of the circuit is zero (Fig. 18-3A).

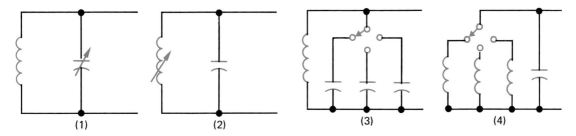

Fig. 18-5. Four common methods of tuning a circuit.

portant, since both series- and parallel-tuned circuits are often present in the same circuit system.

Tuning. The process of adjusting the components of a circuit so that it will respond to one frequency or to a selected band of frequencies is referred to as *tuning, aligning,* or *peaking*. In practical applications such as the tuning circuit of a radio receiver, tuning must be accomplished over a relatively wide band of frequencies. This is most commonly done in one of four ways: by changing the value of capacitance with a variable capacitor, by changing the value of inductance with a variable inductor, by switching different fixed capacitors into a given circuit, or by switching different fixed inductors into a given circuit (Fig. 18-5).

Effect of Resistance. The resistance of a series-tuned circuit has no effect upon the frequency at which the circuit resonates. However, since resistance does limit the current at resonance, it has a direct effect upon the selectivity of the circuit or the manner in which the circuit responds to the resonant frequency. For this reason, the inductors used in frequency-selective circuits should have a high Q (see page 104).

In a circuit with a low resistance, the current increases sharply toward and decreases sharply from its maximum magnitude as the circuit is tuned to and away from the resonant frequency. This condition, of course, provides good selectivity. With an increase of resistance, there is both a decrease in current and a much more gradual rate of current change (Fig. 18-6). This results in a lower degree of selectivity because the current change is more gradual as the circuit is tuned over a band of frequencies.

Q of a Circuit. The degree to which a series-tuned circuit is selective is proportional to the ratio of its inductive reactance to its resistance. This ratio X_L/R is known as the Q of the circuit and is dependent primarily upon the physical properties of the inductor used in the circuit.

Problem 1. The series "tuning" circuit of a broadcast-band radio receiver consists of a 100-microhenry fixed inductor and a variable capacitor with a capacitance range of from 10

Fig. 18-6. Effect of resistance upon the frequency selectivity of a tuned circuit.

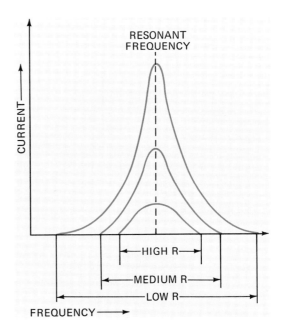

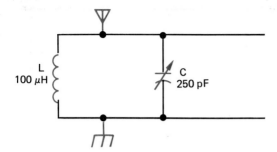

Fig. 18-7. Circuit illustrating Problem 1.

to 365 picofarads (Fig. 18-7). Compute the resonant frequency of this circuit when the capacitor is adjusted to 250 picofarads.

Solution.

$$f_r = \frac{1{,}000{,}000}{2\pi\sqrt{LC}}$$

$$= \frac{1{,}000{,}000}{2\pi\sqrt{100 \times 250}}$$

$$= \frac{1{,}000{,}000}{2\pi\sqrt{25{,}000}}$$

$$= \frac{1{,}000{,}000}{6.28 \times 158}$$

$$= \frac{1{,}000{,}000}{992} = 1{,}008 \text{ kilohertz}$$

BANDPASS

From a strictly theoretical standpoint, resonance in a series-tuned circuit occurs at only one specific frequency. However, in the operation of practical circuits, it is important to realize that a tuned circuit will respond not only to the resonant frequency but also to frequencies that are near it. The range of frequencies that a tuned circuit will accept with approximately the same degree of efficiency with which it accepts the resonant frequency is called the *bandpass* of the circuit.

The bandpass of a series-tuned circuit is defined by design standards as the total range of frequencies (above and below the resonant frequency) that the circuit will accept with a current response of not less than 70.7 percent of the response I_{max} at the resonant frequency

(Fig. 18-8). For any frequency within these bandpass limits (range), the impedance of the circuit is less than 1.414 times the impedance or resistance of the circuit at the resonant frequency. When a circuit is tuned to frequencies outside the limits of its bandpass, the impedance rises to a level that results in a significant decrease of current in the circuit.

Relationship of Bandpass to Frequency. In general, series-tuned circuits that resonate at frequencies from approximately 50 to 500 kilohertz can be designed to have rather high selectivity. When higher frequencies are involved, it becomes more difficult to design circuits that are highly selective. At frequencies above 500 kilohertz, the Q of a series-tuned circuit tends to decrease. This results because of skin effect, increased inductor core, and other component losses. Therefore, at the higher frequencies, the tuning of a circuit will become less selective or broader. Expressed as a formula,

$$\text{Bandpass} = \frac{f_r}{Q}$$

where $Q = X_L/R$ at resonance, and bandpass is expressed in hertz.

Problem 2. The inductor of a series-tuned circuit that has a resonant frequency of 500

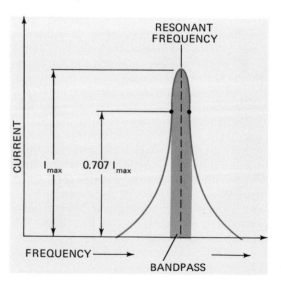

Fig. 18-8. Bandpass of a series-tuned circuit.

kilohertz has a Q of 80. Compute the bandpass of this circuit if the resistance of the inductor wire is the only effective resistance in the circuit.

Solution.

$$\text{Bandpass} = \frac{f_r}{Q}$$

$$= \frac{500{,}000}{80}$$

$$= 6{,}250 \text{ hertz} = 6.25 \text{ kilohertz}$$

Thus, the circuit will effectively accept or pass all frequencies between 6.25/2 kilohertz below and 6.25/2 kilohertz above the resonant frequency. This is equivalent to a band of frequencies that extends from 500 − 3.125 to 500 + 3.125 kilohertz; that is to say, from 496.875 to 503.125 kilohertz.

THE PARALLEL-TUNED CIRCUIT

A parallel-tuned circuit is one in which the applied voltage appears across, or in parallel with, the circuit components (Fig. 18-9). Notice that in such a circuit, unlike the series-tuned circuit, the applied voltage is effectively introduced across (in parallel with) the components of the circuit.

THE PURE *LC* CIRCUIT

From a practical standpoint, the pure *LC* (no resistance) parallel-tuned circuit does not exist. However, it is useful to imagine such a circuit for the purpose of explaining the impedance characteristic of all parallel-tuned circuits.

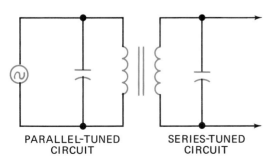

Fig. 18-9. A parallel-tuned circuit used in conjunction with a series-tuned circuit.

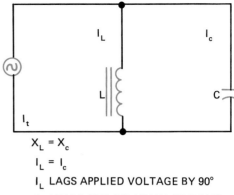

$X_L = X_c$

$I_L = I_c$

I_L LAGS APPLIED VOLTAGE BY 90°

I_c LEADS APPLIED VOLTAGE BY 90°

I_t (LINE CURRENT) = ZERO

Z APPROACHES INFINITY

Fig. 18-10. Characteristics of a purely *LC* parallel-tuned circuit at resonance.

In the parallel-tuned circuit, as in the series-tuned circuit, resonance occurs when the circuit is tuned to a frequency that has the effect of causing the inductive reactance to equal the capacitive reactance. Therefore, the branch currents are also equal because the applied voltage is common to both branches.

The current in the capacitive branch of a pure *LC* parallel-tuned circuit leads the applied voltage by 90 degrees, while the current in the inductive branch lags the voltage by 90 degrees. Since these currents are equal at the resonant frequency, their vector sum is zero and the line (total) current is also zero. Under this condition, the impedance of the circuit at the resonant frequency must be infinite in value (Fig. 18-10).

THE PRACTICAL PARALLEL-TUNED CIRCUIT

In a practical parallel-tuned circuit there is, of course, some resistance, most of which is due to the ohmic resistance of the inductor wire. While this resistance does not alter the resonant frequency of the circuit, it does have the effect of changing the phase relationship between the applied voltage and the current in the inductive branch. The resistance causes the current in this branch to lag the voltage by somewhat less than 90 degrees. As a result, the branch currents do not cancel each other completely and a line

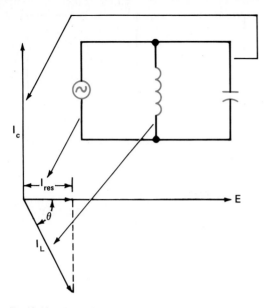

Fig. 18-11. Vector diagram showing the relationships between the currents in a parallel-tuned circuit.

current is present. The line current is in phase with the applied voltage and equal in magnitude to the resistive component of the current that is present in the inductive branch (Fig. 18-11).

Impedance. The presence of line current in a parallel-tuned circuit indicates that resistance has the effect of decreasing the impedance of the circuit at resonance. While this is true, the impedance of the circuit continues to be maximum at the resonant frequency and decreases at frequencies below and above the resonant frequency (Fig. 18-12).

The maximum impedance at the resonant frequency that is characteristic of a parallel-tuned circuit is very commonly used for the purpose of selectively matching the output impedance of a transistor or an electron tube amplifier stage to its associated load, as shown in Fig. 18-13. In this circuit, the primary winding of the intermediate-frequency interstage transformer and C_1 operate as a parallel-tuned circuit to provide the correct impedance match for the collector circuit of the transistor. The secondary winding of the transformer and C_2 act as a series-tuned circuit. As a result, the maximum amount of energy is coupled from the transistor to the following amplifier stage. The impedance characteristic of a parallel-tuned circuit is also widely used in band-reject filter circuits, which filter out currents of one band of frequencies from a circuit while allowing currents of other frequencies to pass through the circuit.

The impedance of a parallel resonant circuit can be determined by the following formula:

$$Z = \frac{X_L^2}{R}$$

but since

$$Q = \frac{X_L}{R}$$

then also

$$Z = QX_L$$

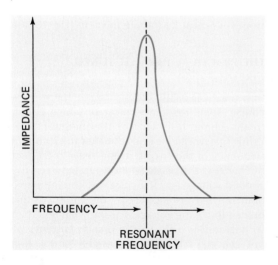

Fig. 18-12. Relationship of impedance and frequency in a parallel-tuned circuit.

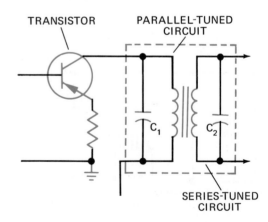

Fig. 18-13. A parallel-tuned circuit used for the purpose of impedance matching.

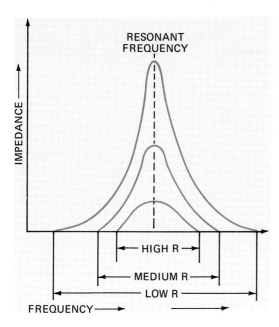

Fig. 18-14. Effect of resistance upon the impedance of a parallel-tuned circuit.

Selectivity. In addition to decreasing the impedance of a parallel-tuned circuit, an increase in resistance has the effect of causing the impedance to vary less "sharply" as the circuit is tuned over a band of frequencies immediately below and above the resonant frequency (Fig. 18-14). Because of this, the number of different frequencies that the circuit will effectively reject is increased and therefore the circuit becomes less selective.

Conditions Off Resonance. At frequencies below resonance, the reactance of the inductive branch of a parallel-tuned circuit decreases, while the reactance of the capacitive branch increases. Thus, the circuit is inductive in nature and the line current is equal to the vector sum of the reactive $(I_L - I_c)$ current and the resistive component of the current in the inductive branch.

At frequencies above resonance, the reverse conditions exist. Now the inductive reactance exceeds the capacitive reactance and the circuit is capacitive in nature. As the frequencies increase above or below resonance to a greater extent, the difference between the reactive currents increases, thereby causing the line current to increase proportionately.

Bandpass. The bandpass of a parallel-tuned circuit is defined as the total number of different frequencies (above and below the resonant frequency) that the circuit will tend to reject with an impedance of not less than 70.7 percent of the impedance Z_{max} at resonance (Fig. 18-15). The bandpass is determined by the same formula that is used to determine the bandpass of a series-tuned circuit.

Problem 3. Compute the bandpass of a parallel-tuned circuit that has a resonant frequency of 600 kilohertz and that consists of a 25-ohm, 2-millihenry inductor and a 36-picofarad capacitor.

Solution.

$$X_L = 2\pi fL$$
$$= 6.28 \times 600{,}000 \times 0.002$$
$$= 7{,}536 \text{ ohms}$$

$$Q = \frac{X_L}{R}$$
$$= \frac{7{,}536}{25} = 301.4$$

$$\text{Bandpass} = \frac{f_r}{Q}$$
$$= \frac{600{,}000}{301.4} = 1{,}990.7 \text{ hertz}$$
$$= 1.9907 \text{ kilohertz}$$

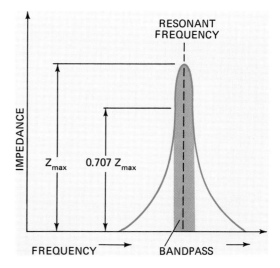

Fig. 18-15. Bandpass of a parallel-tuned circuit.

Thus, the circuit will effectively reject all frequencies between 1.9907/2 kilohertz below and 1.9907/2 kilohertz above the resonant frequency. This is equivalent to a band of frequencies that extends from 600 − 0.9954 to 600 + 0.9954 kilohertz, that is to say, from 599.0047 to 600.9954 kilohertz.

OSCILLATORY RESPONSE

The preceding discussion of the parallel-tuned circuit has been limited to the impedance and the line current characteristics, which are of greatest importance when such a circuit is used to perform the frequency-reject function. However, when the circuit is used to inductively couple signal energy from one circuit to another, or when it is used in an oscillator circuit system, the oscillatory response of the branch circuit currents is of prime importance.

Basic Oscillatory Action. *Oscillatory action*, or *oscillation*, is the process whereby an interchange of energy takes place between the inductor and the capacitor of a tuned circuit. This interchange of energy is brought about by shock-exciting an *LC* circuit by means of a voltage applied to the circuit from an external source. This excitation produces a circulating current within the circuit, which then acts as a series circuit. The circulating or oscillating current is sinusoidal in nature and has a frequency

Fig. 18-16. Basic oscillatory action of a parallel-tuned circuit.

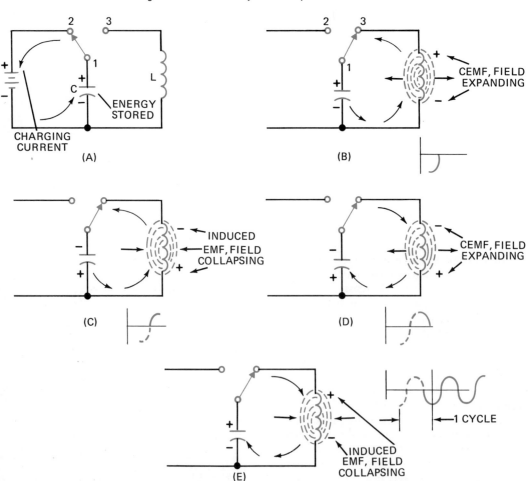

that is determined by the resonant conditions of the circuit.

Because of oscillatory response, the energy that was originally "fed" into an LC circuit by the source of excitation can be thought of as being temporarily "stored" in the circuit during the time that oscillation continues. For this reason, a parallel-tuned circuit is commonly referred to as a "tank" circuit.

The basic oscillatory action of an LC circuit is illustrated by Fig. 18-16. When the switch of the circuit is closed from position 1 to position 2, the capacitor charges to the voltage of the battery (Fig. 18-16A). If the switch is now closed from position 1 to position 3, the capacitor acts as a source of energy and begins to discharge through the inductor (coil) that serves as the circuit load (Fig. 18-16B). As the capacitor discharge current passes through the inductor, the energy that was stored within the dielectric of the capacitor is converted into energy in the form of a magnetic field that expands about the inductor.

After the capacitor has become completely discharged, the magnetic field about the inductor begins to collapse. This action induces a voltage across the inductor, which opposes the change of current (decrease of capacitor discharge current) and causes current to move through the circuit in the same direction. As a result, the capacitor once again becomes charged, but with the opposite polarity (Fig. 18-16C). The capacitor now begins to discharge again, causing current to pass through the circuit in the opposite direction. This reverses the polarity of the cemf across the inductor (Fig. 18-16D). When the capacitor is completely discharged, the induced voltage acts to produce a current in the same direction. Therefore, the capacitor becomes charged to its original polarity and the capacitor discharge-charge cycle is ready to be repeated (Fig. 18-16E).

Effect of Resistance. As described in the preceding paragraphs, the oscillatory response of a parallel-tuned circuit results in oscillation, or the "back and forth" movement of current in the tank circuit. The circulating current would continue indefinitely under no-resistance circuit conditions. However, the ability of a practical circuit to sustain oscillation is limited by the resistance of the circuit. Because of the resist-

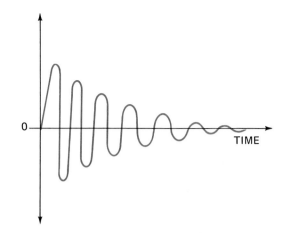

Fig. 18-17. A "damped" sine wave.

ance, each successive cycle of the circulating current decreases in magnitude to produce what is referred to as a *damped sine wave* (Fig. 18-17).

In order to replace the energy that is lost as a result of the resistance of a tank circuit and to maintain continuous oscillation within it, the circuit must be supplied with energy from an external source. This energy is most often in the form of pulses that are applied to the circuit from the output of a transistor or an electron tube. The manner in which the energy pulses are applied to produce continuous oscillation in various types of oscillator circuits is discussed in Unit 24.

FILTER CIRCUITS AND WAVETRAPS

A filter or wavetrap circuit is one that is designed to pass or to reject a certain band of frequencies. Although the specific design of such circuits may vary, all of them employ two characteristics of tuned circuits for their operation: the variation of inductive reactance and capacitive reactance with frequency, and the variation of impedance in series and in parallel circuits.

The Low-pass Filter. A basic low-pass filter circuit arrangement is shown in Fig. 18-18. In this circuit, currents of high frequencies meet significant opposition because of the inductive reactance of inductor L. At the same time, these currents are effectively bypassed from the output circuit because of the short-circuiting effect of

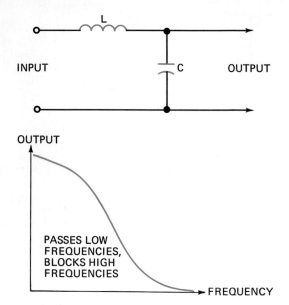

Fig. 18-18. A low-pass filter arrangement.

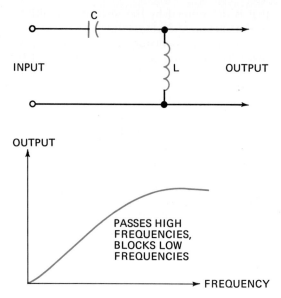

Fig. 18-19. A high-pass filter.

the low capacitive reactance presented by capacitor C. Conversely, at low operating frequencies, the inductive reactance of L is relatively small, while the capacitive reactance of C is significant. Therefore, this filter circuit will effectively pass (from input to output) only a band of frequencies that is lower than those frequencies the circuit is designed to suppress or reject.

The High-pass Filter. In the high-pass filter circuit, the position of L and C is reversed (Fig. 18-19). Now, low-frequency currents meet significant opposition because of the capacitive reactance presented by C and are bypassed from the output because of the low inductive reactance of L. Conversely, currents at high frequencies are opposed by an insignificant capacitive reactance, while the inductive reactance of L at these frequencies is high. As a result, this circuit will pass only a band of frequencies that is above those frequencies the circuit is designed to suppress.

The Bandpass Filter. The basic bandpass filter circuit consists of a series LC combination (Fig. 18-20). Since the impedance of such a circuit is minimum at the resonant frequency and increases at all other frequencies, this circuit will effectively pass only a band of frequencies with lower and upper limits that are fairly near the resonant frequency to which the circuit is tuned.

Fig. 18-20. A series LC combination used as a bandpass filter.

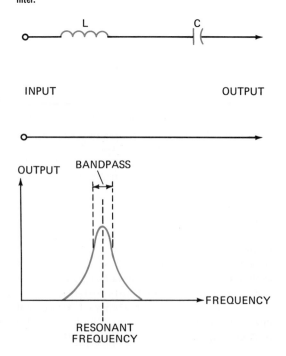

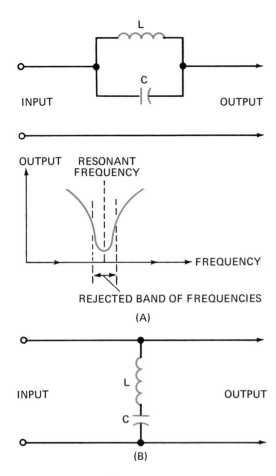

Fig. 18-21. Band-reject filters.

The Band-reject (wavetrap) Filter. In one type of band-reject filter circuit, a parallel-tuned LC combination is connected in series with one side of the line (Fig. 18-21A). Since the impedance of such a circuit is maximum at resonance and decreases at all other frequencies, this circuit will effectively reject from the output circuit only the band of frequencies with lower and upper limits that are relatively near the resonant frequency.

Another type of band-reject filter employs a series LC combination that is connected in parallel with the line (Fig. 18-21B). This circuit effectively bypasses across the line a band of frequencies with upper and lower limits near the resonant frequency. Hence, this band of frequencies is rejected from the output portion of the circuit.

For Review and Discussion

1. Define resonance as related to an *RLC* circuit.
2. Write the basic resonant frequency formula and state the relationship between the resonant frequency and the *LC* product that is indicated by this formula.
3. State the relationship between the reactive voltages in a resonant series *RLC* circuit.
4. What is meant by a tuned circuit?
5. What is the purpose for which tuned circuits are used?
6. Describe the reactance and the resistance characteristics of a series-tuned circuit at resonance.
7. Describe the reactance and the resistance characteristics of a series-tuned circuit at frequencies below and above the resonant frequency.
8. What operational characteristic identifies a series-tuned circuit?
9. Name four methods by means of which circuit tuning is commonly accomplished.
10. What is meant by the bandpass of a series-tuned circuit?
11. Define a parallel-tuned circuit.
12. Describe the impedance characteristic of a parallel-tuned circuit at resonance.
13. Give two important applications or uses of the parallel-tuned circuit.
14. Define the bandpass of a parallel-tuned circuit.
15. Describe the oscillatory response of a "tank" circuit.
16. Draw a schematic diagram illustrating a basic form of low-pass filter circuit, and state the purpose for which such a circuit is most commonly used.
17. Draw a schematic diagram illustrating a basic form of high-pass filter circuit.
18. Define a band-reject (wavetrap) filter circuit and draw a schematic diagram illustrating one form of such a circuit.

SECTION 4

FUNCTIONS OF DEVICES AND CIRCUITS

This section presents the devices and the basic circuit systems that are commonly employed to achieve the major electronic functions of rectification, detection, amplification, and oscillation. The material is arranged so that each device is first described in terms of its structural characteristics, theory of operation, and ratings. The device is then integrated into a circuit system by means of which it is able to perform a specific function. This information will be most useful to an understanding of the complete receiving, transmitting, and control circuit systems that are discussed in the following sections.

UNIT 19. Rectifiers and Rectifier Circuits

The process of *rectification,* or the *conversion* of alternating voltage and current into direct voltage and current, is an important and widely used circuit function. Rectification makes it possible to operate all types of equipment from alternating-current power lines by converting this energy to the specific direct-current voltage and current required. Rectification in the form of detecting, clipping, clamping, and limiting circuit functions is also widely used in communications systems circuits and in various special circuit applications.

The most commonly used rectifier devices are semiconductor (silicon) diodes. In this unit, the constructional and operational features of these devices are discussed, together with their application in complete rectifier circuits. Other devices discussed here include the crystal diode, the zener diode, and the silicon controlled rectifier.

A knowledge of semiconductor structure and theory is essential to understand the operation of semiconductor diodes. Therefore, it is recommended that Unit 4, "Semiconductors," be thoroughly reviewed before studying the following pages.

THE SEMICONDUCTOR JUNCTION DIODE

The semiconductor junction diode, as the name implies, is a p-n junction device (Fig. 19-1). Like all rectifying devices, the semiconductor junction diode readily conducts current when the applied voltage is of a given polarity, but is highly resistive to current flow when the polarity of the applied voltage is reversed. These conducting and nonconducting states of the diode are referred to as conditions of *forward bias* and *reverse bias.*

Forward Bias. For the diode to conduct, the barrier potential across the p-n junction, which may range from 0.1 to 0.3 volt in the typical diode, must be overcome by an externally applied bias voltage. To do this the negative terminal of the external voltage source is connected to the n (cathode) section of the diode and the positive terminal of the voltage source to the p (anode) section of the diode (Fig. 19-2A). The diode is now said to be *forward-biased.* In this

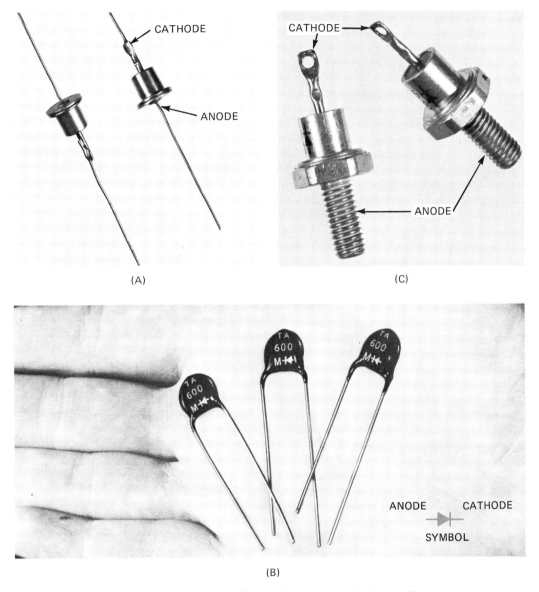

Fig. 19-1. Silicon rectifier diodes: (A) "top hat" or flange type, (B) disk type, (C) stud-mounted type.

condition, free electrons from within the n section of the diode are repelled by the negative terminal of the battery, and drift toward the junction. At the same time, holes from within the p section are repelled by the positive terminal of the battery and they also move toward the junction (Fig. 19-2B). Because of the energy imparted to these electrons and holes by the bias voltage, they are able to penetrate the depletion region and combine.

As the result of each combination of a free electron and a hole within the depletion region, an electron from the negative terminal of the external battery enters the n section and drifts

177

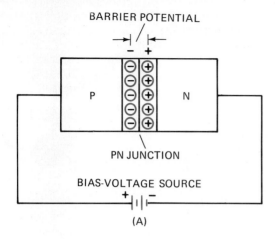

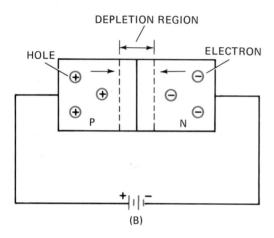

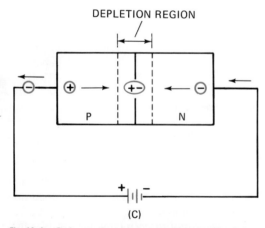

Fig. 19-2. Basic operation of the semiconductor junction diode: (A) forward-bias condition, (B) holes and electrons migrating toward the p-n junction, (C) combination of holes and electrons within the depletion region.

toward the junction (Fig. 19-2C). Simultaneously, an electron from an electron-pair bond within the p section near the positive terminal of the battery breaks its bond and enters the positive terminal of the battery. This creates a new hole within the p section, which then drifts toward the junction. Thus, electrons and holes within the depletion region continue to combine and will do so for as long as the diode is forward-biased.

The combination of electrons and holes within the diode while it is forward-biased

Fig. 19-3. Reverse-bias conditions of the semiconductor junction diode.

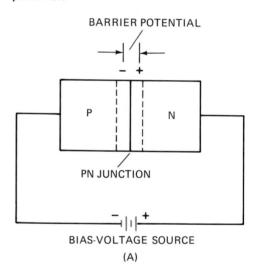

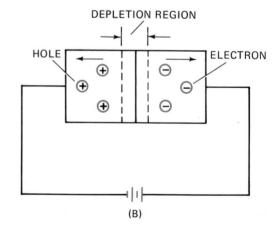

produces a continuous electron current in the external (diode-to-battery) circuit in the forward-biased direction, or from the negative terminal of the battery to the positive terminal. During this time, the current within the diode itself consists of holes in the p section and electrons in the n section.

Reverse Bias. When the connections from the battery to the diode are reversed, the diode is said to be *reverse-biased* (Fig. 19-3A). In this condition, the battery voltage tends to aid the voltage across the potential barrier, thereby increasing the effect of the barrier voltage. Now, electrons within the n section are attracted toward the positive terminal of the battery, while holes within the p section are attracted toward the negative terminal (Fig. 19-3B). Because of this action, the amount of combination of electrons and holes within the depletion region and the amount of current in the external circuit are insignificant.

Reverse Current. Ideally, a semiconductor diode that is reverse-biased within its rated limit should not conduct any current. However, a very small current is always present in the diode circuit, even when it is reverse-biased. This is called a *reverse* or *leakage current* and is illustrated by the voltage-current curve of Fig. 19-4.

The reverse current is due to current carriers present in the semiconductor crystal as a result of thermal agitation. The only current carriers that one would expect to find in the p region of the diode would be holes. However, electrons will also be present because absorption of heat energy will cause some of them to escape from their electron-pair bonds. These *uncontrolled* or *minority carriers,* as they are called, enable the diode to conduct very slightly when reverse-biased. As the temperature of the diode increases, the reverse current also increases, since more minority carriers are produced. For this reason, the temperature of the diodes found in many rectifier circuits is controlled by mounting the diodes upon a metallic surface that serves as a heat sink (Fig. 19-5). The purpose of a heat sink is to conduct heat from the diode and to dissipate this heat, thereby preventing excessive temperature from damaging the diode.

The Diode as a Rectifier. When an alternating voltage is applied to a diode in a rectifier circuit, its polarity changes with each half-cycle or alternation. This causes the diode to be alternately forward-biased and reverse-biased. Thus, the diode conducts only during that half-cycle when the cathode becomes negative with respect to the anode. As a result, a pulsating direct current

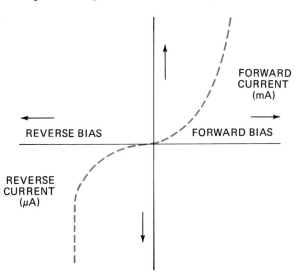

Fig. 19-4. Voltage-current characteristic of the junction diode.

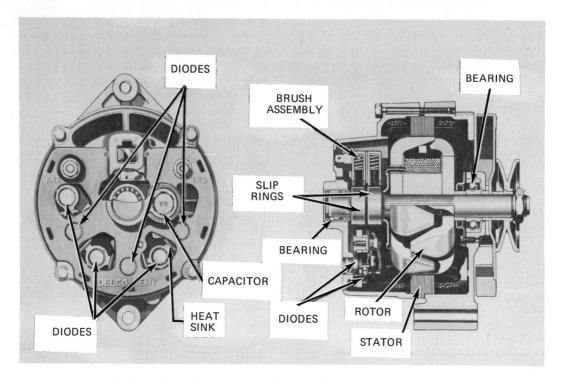

Fig. 19-5. The end frame of an automobile alternator (generator) upon which diodes are mounted serves as a heat sink for the diodes. (*Delco-Remy Division of General Motors Corp.*)

passes through the rectifier circuit load (Fig. 19-6). This forms what is referred to as a *half-wave rectifier circuit*.

Rating. Silicon power diodes, or diodes that are used for rectification in power supply circuits, are rated in terms of the maximum rms input or supply voltage, the maximum average output (direct) current, and the peak-inverse voltage (PIV). The *peak-inverse voltage* rating corresponds to the reverse-bias voltage previously discussed. It is therefore the maximum reverse-bias voltage that can safely be applied across the diode. For this reason, the PIV rating of a diode is often referred to as the *peak-reverse voltage* (PRV).

Avalanche Breakdown. If the peak-inverse voltage applied to a diode is increased beyond the rated limit, the leakage electrons passing through the p-n junction gain energy in the form of increased acceleration. At a given inverse voltage (*breakdown* or *avalanche voltage*), these electrons collide with electrons within the crystal lattice structure with sufficient force to dislodge them from their respective electron-pair bonds. This action is referred to as an *avalanche breakdown*. Because of the breakdown, the current through the diode rises rapidly, heating it to a temperature that causes permanent damage to the entire crystalline structure of the semiconductor material.

Cathode Marking. The fact that the cathode of a semiconductor diode (indicated by a straight line in the symbol) is sometimes identified by a + sign marked upon the diode is very often a cause for some misunderstanding. Since a diode does not generate a voltage, the + sign cannot be an indication of the polarity of a

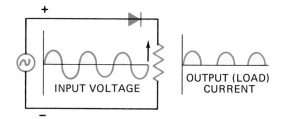

Fig. 19-6. Operation of the diode in a half-wave rectifier circuit.

voltage developed by a diode. Instead, it identifies the electrode (the cathode) to be connected to the end of a load that is to become positive with respect to the other end because of the direction of the current through the load. This direction is from the negative to the positive terminal of the applied voltage and is opposite to the direction toward which the arrowhead in the symbol is pointing.

THE FILTER CIRCUIT

The half-wave rectifier circuit shown in Fig. 19-6 produces an output voltage that varies or pulsates at the applied alternating-voltage rate. As a result, the current through any load connected to the circuit also varies at the same rate. These variations are often referred to as ac *ripple* or as the ac *component* of the output voltage and current. While such variations in voltage and current do not affect the operation of some circuit loads, they do produce an objectional 60-hertz hum in the loudspeaker output of receivers and amplifiers.

In order to minimize the 60-hertz hum, a smoothing filter circuit is added to the rectifier circuit (Fig. 19-7A). When the varying output voltage of the rectifier is applied to the capacitors, they become charged during the time that a ripple-voltage peak exists across the output of the rectifier. Since the time constant of the filter-load circuit does not permit the capacitors to discharge immediately, the capacitors retain their charge. Then, when the rectifier output voltage begins to decrease from a peak magnitude, the capacitors discharge, thus delivering energy to the load (Fig. 19-7B). Because of this action, the voltage across the output of the filter circuit is relatively smooth. Resistor R_1 in the rectifier circuit acts as a "fuse" to protect the circuit in case of a short circuit.

Fig. 19-7. Rectifier filter circuit: (A) diagram of the circuit, (B) basic action of the filter capacitors in the circuit.

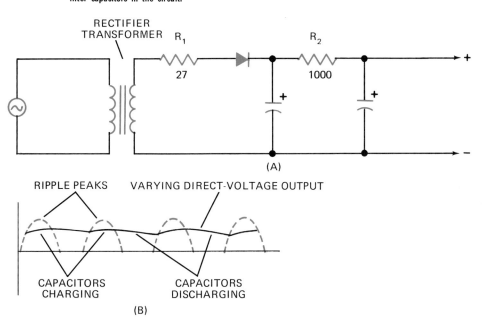

The electrolytic capacitors used in rectifier filter circuits usually have a capacitance rating of from 20 to 100 microfarads, depending upon the design of a particular circuit. To prevent excessive leakage and "breakdown" puncture, any capacitor used in a filter circuit must have a direct-current working voltage (WVDC) rating that is greater than the peak voltage appearing at the rectifier output.

Charge Storage Function. The use of a capacitor in a smoothing filter circuit represents an application of the charge storage function of a capacitor. This characteristic, which is utilized in a number of different kinds of circuits, allows a capacitor to be charged at one time and discharged at another.

Rectifier Transformer. Although the diode in a half-wave rectifier circuit can be connected directly to an input-voltage "power" line, the circuit is usually operated in conjunction with a rectifier transformer. The use of the transformer provides two important advantages. First, the transformer can be designed as either a step-up or a step-down unit to furnish any desired voltage to the rectifier. Second, the transformer in such a circuit also serves as an isolation transformer, thereby providing a definite safety feature as explained in Unit 13.

Additional Filter Circuits. The capacitor-resistor filter combination previously described provides an adequate amount of smoothing action for some circuit-load applications. However, it does not provide the quality of smooth, unfluctuating direct-voltage output necessary for high-fidelity receiver and amplifier operation. To meet this need, various types of networks utilizing the filtering characteristics of both capacitors and choke coils (inductors) are used in conjunction with both half-wave and full-wave rectifier circuits (Fig. 19-8).

FULL-WAVE RECTIFIER CIRCUITS

A full-wave diode rectifier circuit is highly efficient since the total input waveform is included in its output. In the circuit shown in Fig. 19-9, a voltage of the correct diode forward-bias polarity is alternately applied to a diode (− to cathode and + to anode) during each half-cycle of the alternating voltage appearing across the center-tapped secondary winding of the rectifier transformer. This produces a 120-hertz ripple "peak" across the output of the diodes. Because of this higher ripple frequency, as compared with that of the half-wave rectifier circuit, the output of the full-wave rectifier circuit is more easily filtered to provide a smooth output voltage and current.

Fig. 19-8. Basic filter circuits: (A) Pi (π) section filter with filter choke coil, (B) capacitance-input L-section filter, (C) choke-input L-section filter.

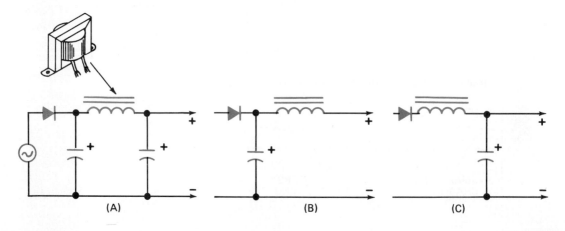

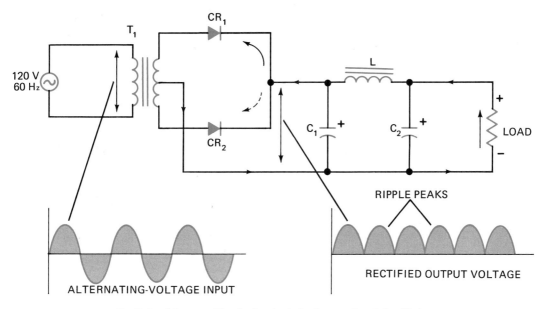

Fig. 19-9. Full-wave rectifier circuit and output voltage waveform before filtering.

The Center-tapped Rectifier Unit. Instead of using two separate diodes in the rectifier circuit, a single rectifier unit is often used (Fig. 19-10). This unit, which consists of two diodes with a common cathode lead or a common anode lead, is connected into the circuit as shown in Fig. 19-11.

The Bridge Circuit. A typical bridge rectifier circuit is one in which either four separate diodes or a single bridge rectifier unit may be used

Fig. 19-10. Center-tapped silicon rectifiers. (*P. R. Mallory and Co., Inc.*)

Fig. 19-11. Negative and positive tap center-tapped full-wave rectifier circuits.

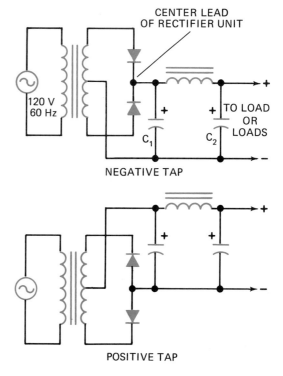

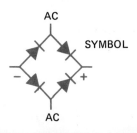

Fig. 19-12. Full-wave silicon bridge rectifier units. (P. R. Mallory and Co., Inc.)

(Fig. 19-12). A schematic diagram illustrating the rectifying action of the bridge circuit is shown in Fig. 19-13. Since this circuit does not require a center-tapped transformer secondary winding, it can be operated with or without a rectifier transformer.

The principal advantage of the bridge circuit is that when it is used with a given rectifier transformer, the voltage output is approximately twice that obtained by using the two-diode, full-wave circuit with the same transformer. The reason for this is that in the bridge circuit, the total secondary voltage is available for rectification, while in the two-diode circuit, the voltage applied to each diode (from the center tap to each end of the secondary winding) is equal to only one-half the secondary voltage.

THE VOLTAGE-DOUBLER CIRCUIT

The only way that direct output voltage can be significantly increased in the rectifier circuits

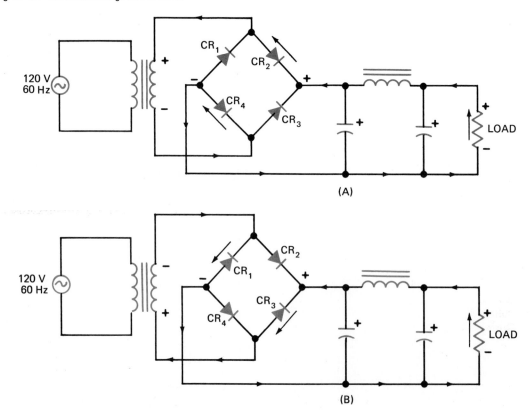

Fig. 19-13. The full-wave bridge rectifier circuit.

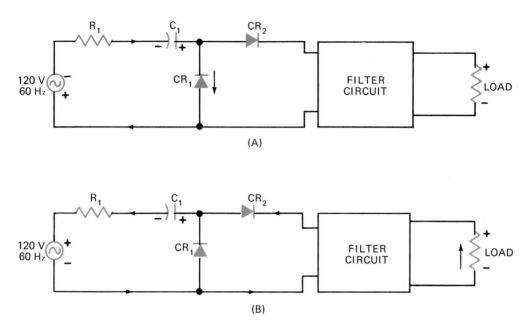

Fig. 19-14. The half-wave voltage-doubler rectifier circuit.

discussed thus far is to use a step-up transformer that will apply the desired alternating voltage to the rectifier or rectifiers. However, the use of a transformer increases the cost of circuit construction and results in a larger and heavier circuit assembly.

The voltage-doubler circuit is used to increase the output voltage of a rectifier circuit without using a transformer. Because of this feature, the voltage-doubler circuit, which has an output voltage of approximately 2.5 times the rms value of the input (power line) voltage, is commonly used in radio and television receivers.

The Half-wave Doubler Circuit. The schematic diagram of a basic, half-wave voltage-doubler rectifier circuit is shown in Fig. 19-14. When the polarity of the input voltage to this circuit is the same as that shown in Fig. 19-14A, capacitor C_1 charges through CR_1 to the peak value of the line voltage in the polarity shown.

During the next half-cycle of the input voltage, the polarity of the line voltage reverses (Fig. 19-14B). Under this condition, the charge across C_1 is retained, but because of the reversal in the polarity of the line voltage, CR_1 no longer conducts. However, CR_2 now conducts, and as a result the voltage across capacitor C_1 is effectively applied in series with the line voltage. Since voltages in series are additive, the voltage applied to the filter circuit at this time is approximately twice the line voltage. This, of course, produces a direct voltage output to the load that is approximately double, or twice the magnitude of the alternating input voltage, during each cycle of the input voltage. The function of R_1 in the circuit is to limit the initial charging (surge) current of capacitor C_1 and to act as a "fuse" in case of a short circuit.

The Full-wave Doubler Circuit. A full-wave, voltage-doubler rectifier circuit is shown in Fig. 19-15. The polarity of the input voltage causes diode CR_1 to conduct current, thus charging capacitor C_1 (Fig. 19-15A). During the next half-cycle of the input voltage, diode CR_2 conducts, causing capacitor C_2 to become charged (Fig. 19-15B). Therefore, during any given cycle of the input voltage, the two capacitors become charged. Since the capacitors are connected in series, the voltages across them are additive, and as a result the total direct output voltage applied to the load is approximately twice the alternating input voltage.

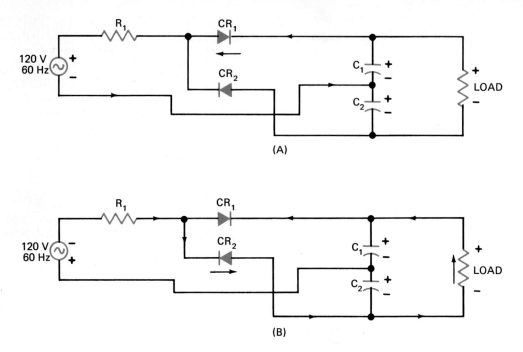

Fig. 19-15. Full-wave voltage-doubler rectifier circuit.

VOLTAGE TRIPLER AND QUADRUPLER CIRCUITS

It is possible, from a theoretical standpoint, to multiply or increase the alternating-voltage input to a rectifier circuit to any magnitude of direct output voltage by using the proper combination of rectifiers and capacitors. However, when cost and efficiency are considered, the practical limit of voltage-multiplier circuits is the voltage quadrupler, which produces an output voltage that is approximately four times the input voltage.

VOLTAGE REGULATION

The voltage output of a rectifier circuit tends to decrease when a load is connected to it, as a result of the internal resistance of the circuit. Since the resistance produces an *IR* drop across certain components of the circuit, the effective voltage that is available for application to a given load decreases in proportion to the current delivered to the load. If a rectifier circuit does not have the proper power output capacity necessary for the operation of its associated load, the internal resistance of the circuit will cause the output voltage to drop considerably.

Variations in the output voltage of a rectifier circuit are also caused by fluctuations in the magnitude of the alternating input voltage. Thus, it is often necessary to equip a rectifier circuit with a voltage-regulation device that will maintain the output voltage at a constant magnitude.

Most electronic circuits will operate quite satisfactorily with a certain amount of variation in the supply voltage. However, in some circuits, even a slight deviation from the normal supply voltage will cause undue loss of operating efficiency. Hence, a voltage-regulation device is inserted into the associated power supply circuit, either between the rectifier and the load or between the power line and the rectifier.

The Bleeder Resistor. The simplest type of voltage regulator consists of a resistor that is connected across the output terminals of the rectifier circuit (Fig. 19-16). This resistor is referred to as a *bleeder resistor*. In practical circuits, it has a resistance that will allow approximately 10 percent of the full load current to pass through it. By providing a fixed load for the rectifier, the

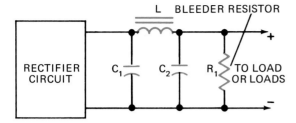

Fig. 19-16. The rectifier circuit bleeder resistor.

bleeder resistor prevents the filter capacitors from charging up to the peak voltage output of the rectifier. It also prevents the load current from ever reaching zero. Therefore, a more constant voltage is available to the load.

In addition to improving the voltage regulation of a rectifier circuit, the bleeder resistor provides a discharge path for the filter capacitors after the power switch has been turned off. This is an extremely important safety feature, particularly in circuits where, without the resistor, the capacitors could remain charged to a dangerously high voltage for a considerable period of time.

The Zener Diode. The zener or voltage reference diode is a semiconductor device which, unlike ordinary diodes, is capable of conducting a relatively high reverse current without being damaged (Fig. 19-17). In practical operation as a voltage-regulation device, the diode is connected across the output of a rectifier circuit and is used in conjunction with a regulating resistor (Fig. 19-18).

Under conditions of normal output voltage, the reverse bias applied to the diode causes a breakdown of electron-pair (covalent) bonds within the diode material. As a result, the diode conducts a given amount of reverse current. When the output voltage begins to decrease, the breakdown effect is less pronounced and the internal impedance of the diode increases. Now, the diode conducts less current and the voltage drop across the resistor decreases, thereby holding the voltage that is applied to the load constant.

When the output voltage of the rectifier circuit increases, the voltage across the diode also increases, producing a greater breakdown effect within the diode and causing it to conduct a greater reverse current. Since this action increases the voltage drop across the resistor, it has the effect of keeping the voltage applied to the load constant.

Zener diodes are usually employed in comparatively low-voltage circuits and are designed to operate effectively over a given range of current at a fixed voltage. When the voltage to be regulated exceeds the voltage rating of a particular diode, two or more diodes may be connected in series.

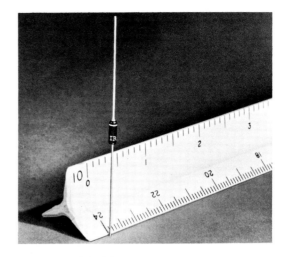

SYMBOL

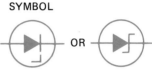

Fig. 19-17. Zener (voltage reference) diode. (*International Rectifier*)

Fig. 19-18. Zener diode voltage regulation.

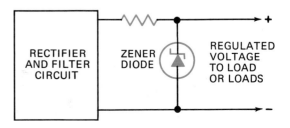

Percentage of Voltage Regulation. The degree of voltage regulation occurring, in any given rectifier circuit is usually stated in terms of the change between the full-load output voltage as compared with the no-load voltage. Expressed as a formula,

$$\text{Percentage of voltage regulation} = \frac{E_{NL} - E_{FL}}{E_{FL}} \times 100$$

where E_{NL} = no-load voltage
E_{FL} = full-load voltage

THE VOLTAGE DIVIDER

A voltage divider consists basically of a load resistor that is either a single-unit tapped resistor or a combination of individual resistors connected in series (Fig. 19-19). By utilizing the voltage drops across the resistors, the voltage divider makes it possible to obtain two or more different fixed voltages from a single power supply circuit.

The no-load voltages between the common B− (ground, or G) terminal and the various other terminals of a voltage divider are dependent upon the total current and the resistance of each resistor. Thus, the magnitude of each of these voltages can be determined by using Ohm's law as indicated in Fig. 19-20.

Voltages under Load Conditions. If the rectifier circuit with which the voltage divider shown in Fig. 19-20A is used is well regulated, the full output voltage of 440 volts between terminals 1 and B− will remain fairly constant under all normal load conditions. However, it is important to note that when a load is connected across B− and any intermediate terminal, the voltage between these terminals will change. This factor must, of course, be considered in the design of any voltage-divider circuit.

As an example, assume that a load R_4 with a resistance of 40,000 ohms is connected between terminals 3 and B− (Fig. 19-20B). Since this load resistance is in parallel with R_3, the total resistance between 3 and B− is now

$$R_{\text{B− to 3}} = \frac{40{,}000}{2} = 20{,}000 \text{ ohms}$$

As a result, the total resistance between terminals 1 and B− is now

$$R_{\text{B− to 1}} = 20{,}000 + 60{,}000 + 120{,}000 = 200{,}000 \text{ ohms}$$

and the total current is increased to 440/200,000, or 2.2 milliamperes.

Therefore, the voltages between B− and the intermediate terminals are reduced to

$$E_{\text{B− to 3}} = 0.0022 \times 20{,}000 = 44 \text{ volts}$$
$$E_{\text{B− to 2}} = 0.0022 \times 80{,}000 = 176 \text{ volts}$$

Negative Voltage. In addition to providing voltages that are of a positive polarity with respect to ground, many voltage-divider circuits are designed to provide a voltage that is negative with respect to ground. This is particularly true in some transmitter circuit systems where a comparatively high negative bias voltage must be applied to the control grids of certain electron tubes.

To obtain a negative voltage from a voltage-divider network, one of the intermediate terminals of the voltage-divider resistor is grounded (connected to chassis—see Fig. 19-21). Since current passes through this voltage divider from the B− terminal to terminal 1, current passes through R_3 from B− to C. Thus, the polarity of terminal B− is negative with respect to C by an amount that is equal to the voltage drop across R_3, or 60 volts.

THREE-PHASE RECTIFICATION

The basic half-wave and full-wave circuits used to rectify a three-phase alternating voltage are shown in Fig. 19-22. Since the ripple frequency of the output voltage in such circuits is relatively high, the circuits may or may not be operated in conjunction with associated filter sections. The full-wave circuit is used in the automobile alternator to rectify the three-phase output of the alternator to the direct voltage (and current) necessary to operate the automotive electrical system (see Fig. 19-5).

Fig. 19-19. Tapped (voltage-divider) wire-wound resistor. (IRC, Inc.)

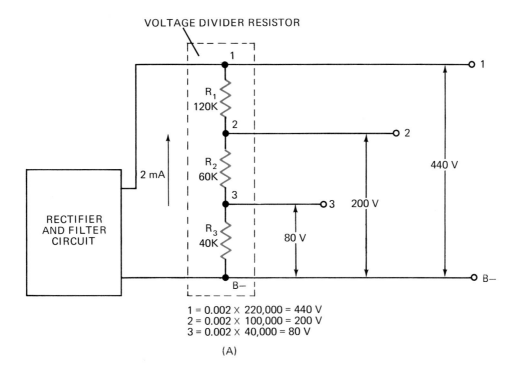

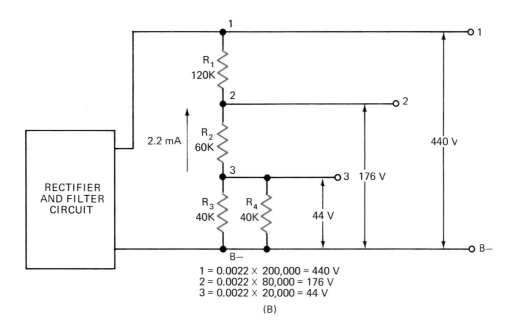

Fig. 19-20. The voltage-divider circuit.

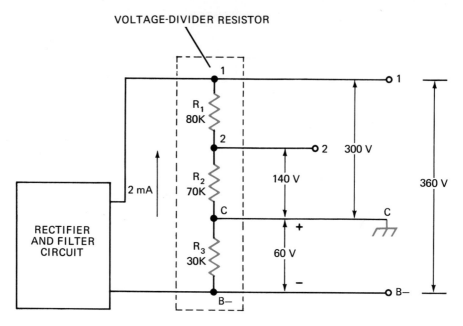

Fig. 19-21. Obtaining a negative voltage from a voltage-divider network.

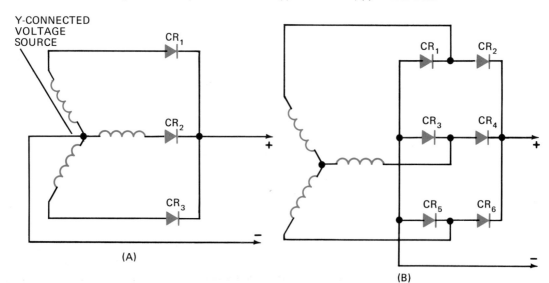

Fig. 19-22. Three-phase rectifier circuits: (A) half-wave circuit, (B) full-wave circuit.

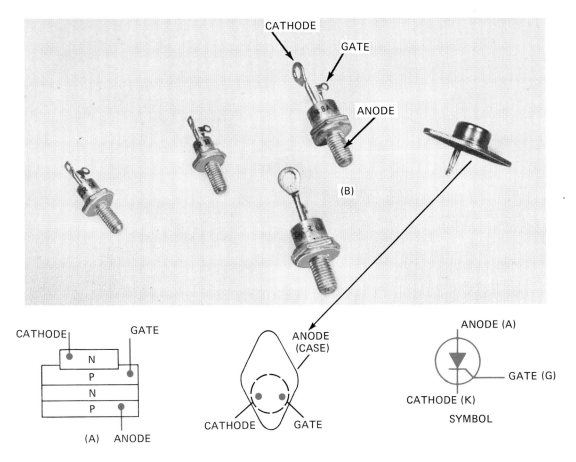

Fig. 19-23. Silicon controlled rectifiers: (A) basic construction, (B) common types. (Gail E. Henderson)

THE SILICON CONTROLLED RECTIFIER

The typical silicon controlled rectifier (SCR) is a four-layer n-p-n-p or p-n-p-n device. It is equipped with three electrodes, which are identified as the cathode K, the anode A, and the gate G that serves as the control electrode (Fig. 19-23). As in the case of the semiconductor diode, the silicon controlled rectifier will conduct current only in the forward direction or when it is forward-biased. However, this rectifier will not conduct a significant current in the forward direction until the voltage applied across the cathode and anode is greater in magnitude than a given minimum voltage that is referred to as the *forward breakover voltage*.

The special characteristic of the silicon controlled rectifier is that it can be made to conduct current at a voltage much less than the forward breakover voltage if a positive voltage, the control signal, is applied to the gate. This feature of the rectifier makes it a very useful switching and current-regulating device, which is employed in many different kinds of power-control circuits.

A basic silicon controlled rectifier circuit with the gate "open" (no control voltage applied to the gate) is shown in Fig. 19-24A. In this circuit, assume that the rectifier has a forward breakover voltage that is higher than the peak alternating voltage applied across the anode and cathode through the load. Therefore, the rectifier

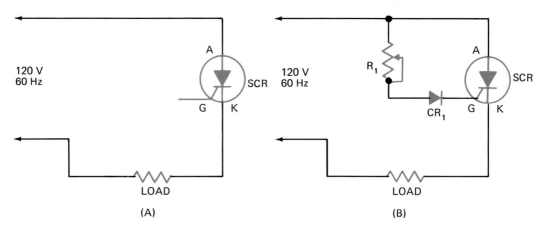

Fig. 19-24. Silicon controlled rectifier circuits: (A) gate-open condition, (B) control voltage applied to gate.

is in the "off" condition and will not conduct current through the load.

To provide a gate-control voltage, a diode CR_1 is connected into the circuit as shown in Fig. 19-24B. Because of the rectifying action of CR_1, a positive voltage is applied to the gate. The magnitude of this voltage can, in turn, be varied by rheostat R_1.

R_1 can be adjusted to provide a given voltage to the gate at the beginning of the positive half-cycle of the input (line) voltage that is applied to the anode. This "triggers" the rectifier, which will conduct current for the remainder of the half-cycle. However, R_1 can also be adjusted to control the conduction of forward current through the rectifier (and the load) so that the rectifier will conduct during any portion of the positive half-cycle of the input voltage. Hence, the power delivered to the load can be controlled by the adjustment of R_1. Such circuit modifications are commonly employed in light-dimmer circuits and in controlling the speed of dc and universal-type motors.

The silicon controlled rectifier is a member of a group of semiconductor devices that perform a current-control function similar to that of the thyratron tube. These devices are classified by the generic term *thyristor*, which, in general, applies to any device that is "triggered" into conductivity by the application of a voltage pulse with a given polarity and magnitude.

THE CRYSTAL DIODE

The germanium "crystal" diode is a rectifier device used primarily to produce a rectifying action known as *detection*. The process of detection or *demodulation* results in the separation of an audio signal from a modulated carrier wave that is used in radio, television, and other forms of "wireless" transmitting systems.

A very popular type of crystal diode consists of a phosphor-bronze wire (the catwhisker) in

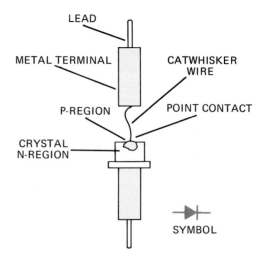

Fig. 19-25. Constructional features of a typical germanium crystal diode.

contact with a wafer section of n-type germanium. During the final fabrication (construction) process, a relatively high surge of direct current is passed through the diode from the wire to the germanium. This causes a small region of p-type germanium to be formed around the wire contact point as shown in Fig. 19-25. This fabrication process is known as *hole injection*. The result is a p-n junction. The electronic characteristic of this junction enables the diode to conduct current easily from the n-type germanium (the cathode) to the wire, but presents a high resistance to current flowing in the opposite direction.

THE SELENIUM RECTIFIER

A selenium rectifier cell consists basically of an aluminum base plate that is coated with a layer of selenium by means of a high-temperature evaporation process (Fig. 19-26). For most circuit applications, several cells are then "stacked" together in series, parallel, or series-parallel combinations to form a rectifier unit (Fig. 19-27).

Current will readily pass through the selenium rectifier from the base plate to the selenium (cathode to anode). However, the rectifier will not conduct a significant current in the

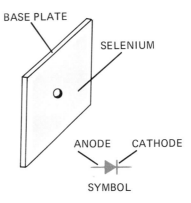

Fig. 19-26. Single selenium rectifier cell.

Fig. 19-27. Selenium rectifiers. (*Gail E. Henderson*)

opposite (anode to cathode) direction. These features make it possible to use the device in rectifier circuits that are similar to the silicon-diode rectifier circuits discussed previously.

Selenium rectifiers are rated in terms of the maximum rms input voltage and the maximum direct current that they will continuously deliver to a circuit load. If either of these operational ratings is exceeded, a selenium rectifier may be permanently damaged.

For Review and Discussion

1. Explain the forward-biased and the reverse-biased conditions of a semiconductor junction diode.
2. What is meant by a diode reverse current?
3. For what purpose is a heat sink sometimes used with a silicon diode?
4. Explain the operation of the diode half-wave rectifier circuit and state the difference between the half-wave circuit and the full-wave circuit.
5. Define the peak inverse voltage rating of a silicon diode.
6. What is meant by avalanche breakdown?
7. For what purpose is a filter circuit used in a rectifier circuit?
8. Describe the action of the electrolytic capacitors in a filter circuit.
9. What advantages are gained by the use of a rectifier transformer?
10. Explain the operation of the basic full-wave diode rectifier circuit shown in Fig. 19-9.
11. Draw a schematic diagram of a diode bridge rectifier circuit and explain its operation.
12. Draw a schematic diagram of a basic half-wave voltage-doubler circuit and explain its operation.
13. What is meant by the voltage regulation of a rectifier circuit?
14. Describe the function of a bleeder resistor in a rectifier circuit.
15. Explain the operation of a zener diode as a voltage-regulation device.
16. What is the purpose of a voltage-divider network?
17. Draw a schematic diagram of a diode rectifier circuit used for three-phase rectification.
18. Explain the operation of the silicon controlled rectifier.
19. Define detection.
20. Describe a selenium rectifier.

UNIT 20. Transistors

The word "transistor" is derived from the phrase "transfer of resistance." How a transfer of resistance takes place in a transistor circuit will be discussed in the following unit, which is devoted to the transistor as an amplifier device. In this unit, the general features of the typical junction transistor will be discussed, together with the conduction of current in the basic transistor circuit. This will provide the information that is necessary before beginning the study of the transistor in various kinds of amplifier circuits.

When compared with the conventional electron tube, the transistor has several advantages. It is, first of all, more efficient in respect to the power output it can deliver with reference to a given power input. This is largely because the transistor requires no filament or heater power and can be operated at considerably lower voltages than those applied to the plate and the screen grid of an electron tube. Being a solid-state device, the transistor is more rugged than the electron tube, which is subject to damage from jarring and vibration as a result of its "hollow" construction.

The transistor is also smaller than the electron tube and, unlike the tube, does not require

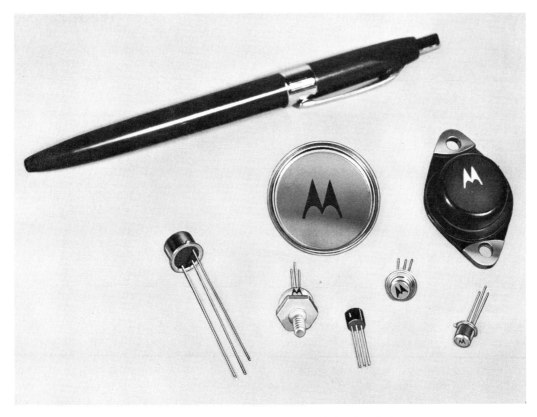

Fig. 20-1. Typical n-p-n and p-n-p transistors. (*Motorola Semiconductor Products, Inc.*)

a "warm-up" period before beginning normal operation (Fig. 20-1). The physical and operational characteristics of the transistor cause it to be especially well adapted to the miniaturization or more compact construction of electronic products. This is true not only because of the smaller size of the transistor itself, but also because the components (resistors, capacitors, transformers, etc.) in a transistor circuit can usually be made smaller than the corresponding components found in electron tube circuits.

Before beginning the study of the following paragraphs, it is recommended that Unit 4, "Semiconductors," be thoroughly reviewed. This review is most important, since the theory and operation of transistors cannot be mastered until the information pertaining to the structure, the conductivity, and the p-n junction combination of semiconductor materials is completely understood.

THE JUNCTION TRANSISTOR

The triode junction transistor, which has almost entirely replaced the earlier point-contact transistor, consists basically of three sections of a doped semiconductor material. These sections make up what are, in effect, two p-n junctions.

Fabrication (construction). One method of transistor fabrication produces what is known as the *diffused-junction* transistor. The fabrication of this transistor begins with a base wafer (crystal) of a doped semiconductor material such as germanium or silicon (either p-type or n-type). If the wafer is p-type, it is placed into a heated space containing a gaseous form of a donor impurity. The impurity then spreads (diffuses) into the base wafer to a predetermined depth, thus forming n-type sections on both sides of the wafer. This results in the fabrication of an

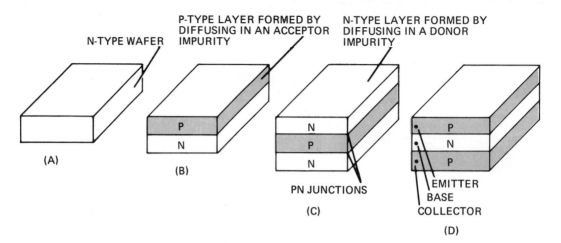

Fig. 20-2. Basic steps in the fabrication of diffused-junction transistors.

n-p-n transistor, which can be described pictorially as shown in Fig. 20-2A, B, and C.

If the base wafer consists of n-type material, an acceptor impurity is diffused into it, producing what is known as a p-n-p transistor (Fig. 20-2D).

Different methods of transistor fabrication result in the alloy-junction and the related diffused-junction drift-field transistors (Fig. 20-3). Each of these transistors has operational characteristics that make its use desirable under certain circuit conditions.

Although both germanium and silicon transistors are popular, the silicon transistor does have the advantage of being able to withstand somewhat higher operating temperatures without being damaged. However, since both of these devices operate in very much the same way, a presentation of the germanium transistor will also serve as an aid to the understanding of the silicon transistor.

In addition to germanium and silicon, a relatively new semiconductor material, gallium arsenide, has been developed for use in fabricating transistors and diodes. This material appears to be well adapted for a broad range of semiconductor applications, since it possesses the desirable characteristics of both germanium and silicon.

Electrodes. The triode junction transistor consists of what are, in effect, two diodes placed "back-to-back" that share a common (center) electrode known as the *base*. The two remaining (end) electrodes of the transistor are the *emitter* and the *collector* (Fig. 20-4A). In low-power transistors, the electrodes are usually connected to thin wire leads that are "brought out" at the base of the plastic or the hermetically sealed enclosure that houses the electrode assembly (Fig. 20-4B).

Many transistors packaged in metal cases are equipped with a fourth (shield) lead that is connected to the case (Fig. 20-4C). This lead is connected to the ground or common point of a circuit and allows the case to shield the transistor from extraneous magnetic and electric fields.

In other transistors, particularly those used as power amplifiers, the emitter and base electrodes are extended from the base by means of

Fig. 20-3. Structural features of alloy-junction and diffused-junction drift-field transistors. (RCA)

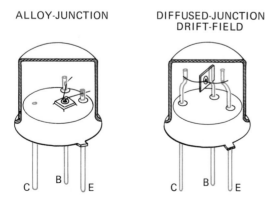

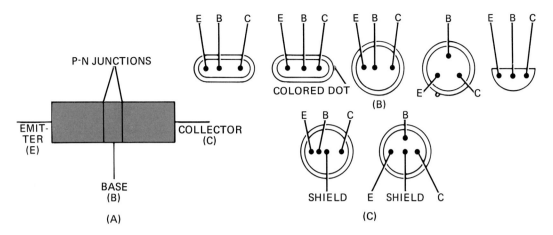

Fig. 20-4. Electrodes and base-lead configurations of common-type, low-power n-p-n and p-n-p transistors.

Fig. 20-5. Power transistors: (A) transistor in TO-3 case, (B) transistor in TO-36 case, (C) transistor mounted upon heat sink.

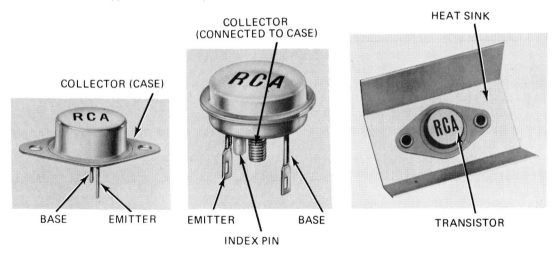

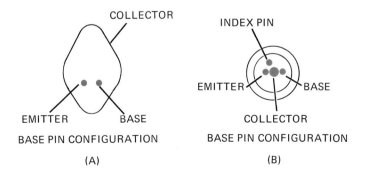

197

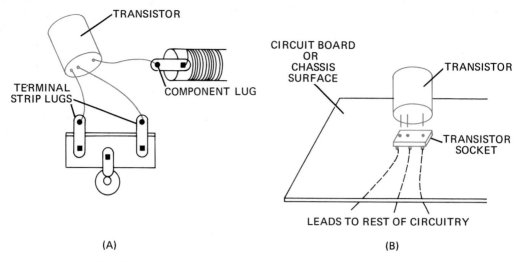

Fig. 20-6. Methods of connecting transistors to associated circuits: (A) "soldered-in" transistor, (B) transistor socket.

pins that are insulated from the metallic case that serves as the collector terminal (Fig. 20-5A and B). This type of construction allows the transistor to be conveniently mounted upon a metallic surface that serves as a heat sink (Fig. 20-5C).

Methods of Connection. Wire-lead transistors are connected into a circuit either by leads connected directly to the terminals of the circuit wiring or by prongs that fit into a socket (Fig. 20-6). While the direct-connection method results in less costly circuit construction, it does make servicing the transistor circuit more inconvenient. At the same time, this method of connection introduces the danger of damaging a transistor because of the heat that is conducted to it during the soldering process.

Type Designation. A large number of transistors are designated as to type by a number-letter combination such as 2N406 and 2N591, or by a series of numbers such as 36547 and 40217. This designation, unlike the electron-tube-type designation, does not provide any indication of operational voltages or general function. These factors, as well as the complete operational data pertaining to any specific type of transistor, can be obtained by referring to any one of a number of transistor manuals that are available (Fig. 20-7).

Comparison With the Electron Tube. Although the construction of the transistor is altogether different from that of an electron tube, these devices are similar in that both control current. In the transistor, the electrode that controls the current is the base. In the tube it is the control grid. Likewise, the emitter of the transistor corresponds to the cathode of a tube, while the

Fig. 20-7. Transistor manual data relating to the type 2N647 alloy-junction n-p-n transistor. (RCA)

TRANSISTOR 2N647

Ge n-p-n alloy-junction type used in large-signal af-amplifier applications in battery-operated portable radio receivers and phonographs. N-P-N construction permits complementary push-pull operation with a matching p-n-p type, such as the 2N217. JEDEC TO-1, Outline No.1. **Terminals:** 1 - emitter, 2 - base, 3 - collector (red dot).

MAXIMUM RATINGS

Collector-to-Base Voltage	V_{CBO}	25	V
Collector-to-Emitter Voltage	V_{CEO}	25	V
Emitter-to-Base Voltage	V_{EBO}	12	V
Collector Current	I_C	100	mA
Emitter Current	I_E	−100	mA
Transistor Dissipation:			
$T_A = 25°C$	P_T	100	mW
$T_A = 55°C$	P_T	50	mW
$T_A = 71°C$	P_T	20	mW
Temperature Range:			
Operating (Ambient)	$T_A(opr)$	−65 to 71	°C
Storage	T_{STG}	−65 to 85	°C

CHARACTERISTICS

Collector-Cutoff Current ($V_{CB} = 25$ V, $I_E = 0$)	I_{CBO}	14 max	µA
Emitter-Cutoff Current ($V_{EB} = 12$ V, $I_C = 0$)	I_{EBO}	14 max	µA
Static Forward-Current Transfer Ratio ($V_{CE} = 1$ V, $I_C = 50$ mA)	h_{FE}	70	

TYPICAL OPERATION IN CLASS B COMPLEMENTARY-SYMMETRY CIRCUIT

DC Collector-Supply Voltage	V_{CC}	6	V
DC Collector-to-Emitter Voltage for driver stage	V_{CE}	2.3	V
Zero-Signal DC Base-to-Emitter Voltage for output stage	V_{BE}	0.14	V
Peak Collector Current for each transistor in output stage	i_C(peak)	70	mA
Zero-Signal DC Collector Current for each transistor (driver and output stage)	I_C	1.5	mA
Signal Frequency		1	kc/s
Input Resistance	R_S	1100	Ω
Load Resistance	R_L	45	Ω
Power Gain		54	dB
Total Harmonic Distortion		10	%
Power Output (input = 20 mV)	P_{OE}	100	mW

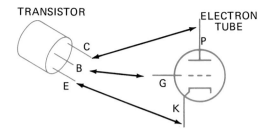

Fig. 20-8. Comparison of the electrodes of an n-p-n or a p-n-p transistor with the corresponding electrodes of a triode electron tube.

collector performs a function similar to that of the tube plate (Fig. 20-8).

THE N-P-N TRANSISTOR

As in the case of the electron tube, the amplifying function of the transistor is possible because this device is able to act as a valve to control the current in the output (collector-base) circuit by means of the voltage that is applied to the input (emitter-base) circuit. To understand this action, it is helpful to first study the conduction of current through a basic circuit utilizing the n-p-n transistor (Fig. 20-9).

In this circuit, the emitter-base (p-n) junction is in a forward-biased condition as a result of the voltage that is applied to it by energy source B_1. At the same time, the collector-base junction is in a reverse-biased condition as a result of the voltage that is applied to it by energy source B_2.

With the bias voltages maintained at a given magnitude, the emitter-base bias voltage overcomes the potential barrier existing across the emitter-base junction. As a result, electrons pass from the negative terminal of B_1 to the emitter, through the emitter-base junction, and into the p-type germanium of the base section. During this process, electrons are said to be injected from the emitter into the base of the transistor. Within the base section, some of the injected electrons combine with holes of the p-type germanium. Each time that such a combination occurs, an electron leaves the base and is attracted to the positive terminal of source B_1. Thus, a small amount of current, the base current I_B, is present in the external base to B_1 circuit.

However, since the base section material is extremely thin, approximately 98 percent of the electrons injected into the base pass through it and into the collector section of the transistor. Here, they come under the influence of the positive terminal of source B_2. As a result, these electrons form the collector current I_C by passing through the collector, from the collector to the positive terminal of B_2, through B_2 and B_1, and

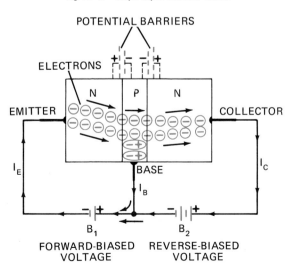

Fig. 20-9. Simple n-p-n transistor circuit.

on to the emitter. Therefore, the emitter current I_E is equal to the sum of the base current and the collector current, or

$$I_E = I_B + I_C$$

Majority Carriers. In the n-p-n transistor circuit, the principal current through the transistor material itself consists of electrons. Hence, in this transistor, electrons are said to be the *majority carriers*, while the holes that diffuse through the base are referred to as *minority carriers*.

Symbol. The symbol for the n-p-n transistor is shown in Fig. 20-10. A comparison of this symbol with the circuit of Fig. 20-9 shows that the arrow attached to the emitter of the symbol points in a direction of electron movement from the emitter to the base in the external circuit. As an aid in remembering the direction of the arrow in this symbol, the three letters n-p-n can be interpreted as n (not) p (pointing) n (in).

THE P-N-P TRANSISTOR

A basic p-n-p transistor circuit and the conduction of current through this circuit are shown in Fig. 20-11. Here, as in the case of the n-p-n transistor, the p-n-p transistor is operated with the emitter-base p-n junction forward-biased, while the collector-base junction is reverse-biased, by sources B_1 and B_2 respectively.

In the forward-biased condition, the voltage of B_1 overcomes the potential barrier existing across the emitter-base junction. This allows the holes within the p-type germanium of the emitter section that are repelled by the positive terminal of B_1 to diffuse through the junction and into the base. During this process, holes are said to be injected from the emitter into the base of the transistor.

Some of the holes injected into the base from the emitter combine within the base section with electrons of the n-type germanium.

Fig. 20-10. Symbol for the n-p-n transistor.

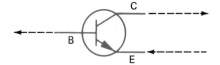

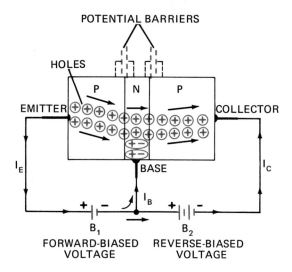

Fig. 20-11. Simple p-n-p transistor circuit.

Each time that such a combination occurs, an electron leaves the negative terminal of B_1 and enters the base. This movement of electrons in the external circuit from B_1 to the base is the base current I_B.

Because, as in the n-p-n transistor, the base section is extremely thin, most of the holes that are injected into the base pass through it and into the collector section. Here, the holes come under the influence of the negative terminal of B_2. As a result, the holes continue moving through the collector section toward the point where the collector is joined to B_2. As holes reach this point, they combine with electrons that leave the negative terminal of B_2 and enter the collector through the external circuit. These electrons form the collector current I_C.

With each combination of a hole with an electron in the collector section, an electron from within the emitter section breaks its electron-pair bond and enters the positive terminal of B_1. Therefore, an electronic current, the emitter current I_E, is maintained in the external circuit from the emitter to the positive terminal of B_1. This current, as in the n-p-n transistor circuit, is equal to $I_B + I_C$. The base portion of the emitter current then passes through B_1 to the base, while the collector portion of the emitter current passes through B_2 and to the collector.

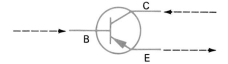

Fig. 20-12. Symbol for the p-n-p transistor.

Majority Carriers. In the p-n-p transistor circuit, the principal current through the transistor material itself consists of holes. Therefore, in this transistor, holes are said to be the majority carriers and the electrons that diffuse through the base are the minority carriers. However, it is important to note that, while holes are the majority carriers inside the p-n-p transistor, the current in the external circuit of this and all other transistors is an electronic (electron) current.

Symbol. The symbol for the p-n-p transistor is shown in Fig. 20-12. Notice that the arrow attached to the emitter of the symbol points in a direction opposite to the direction of the electron movement from the emitter to the base in the external circuit. As an aid in remembering the direction of the arrow in this symbol, the first two letters of p-n-p can be interpreted as p (pointing) n (in).

THE FIELD-EFFECT TRANSISTOR (FET)

The transistors that have been described previously in this unit are bipolar devices, since their operation depends upon the conduction of two charge carriers of opposite polarity (electrons and holes). In a more recently developed transistor, referred to as the *field-effect* or *unipolar transistor,* the charge carriers are basically either electrons or holes (Fig. 20-13). If the charge carriers are electrons, the transistor is known as an n-channel transistor. Likewise, if the charge carriers are holes, the transistor is known as a p-channel transistor.

Fabrication. In its simplest form, the n-channel field-effect transistor consists of a wafer (the substrate) of n-type silicon with a p-type impurity diffused on both sides of the wafer. In a more practical form, the n-channel transistor consists of a p-type substrate with an n-type impurity diffused on one side only. These diffusion processes result in the formation of three electrodes that are identified as the source S, the gate G, and the drain D (Fig. 20-14A). A comparison of the functions of these electrodes with those of the triode electron tube and the triode bipolar transistor is shown in Fig. 20-14B. The p-channel transistor is fabricated in a manner similar to that illustrated in Fig. 20-14A. However, in this transistor, the substrate consists of n-type material into which a region of p-type material has been diffused.

Operation. If a direct voltage is applied across the drain and the source of the field-type transistor with no bias voltage applied to the gate, the transistor behaves somewhat like an ordinary resistor. Under this condition, charge carriers (electrons) pass through the n-type material (the channel) from the source to the drain (Fig. 20-15A). When a reverse-bias voltage (negative for the n-channel transistor) is applied to

Fig. 20-13. Plastic-encapsulated field-effect transistors. (*Texas Instruments, Inc.*)

201

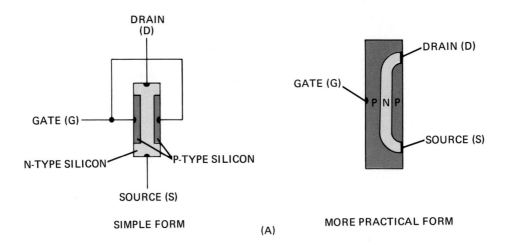

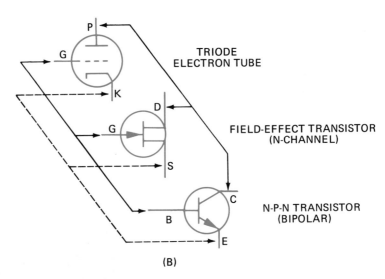

Fig. 20-14. The field-effect transistor: (A) basic structural features, (B) comparison of the electrodes of the n-channel transistor with the corresponding electrodes of an n-p-n transistor and a triode electron tube.

the gate, the bias voltage establishes electrostatic fields, which repel the source-to-drain electrons within areas of the channel that are referred to as *depletion areas* (Fig. 20-15B). Since the depletion areas, in effect, increase the resistance of the channel, the reverse bias produces a decrease of the source-to-drain current. Hence, the bias voltage applied to the gate controls the magnitude of the source-to-drain current through the transistor. In this respect, the field-effect transistor is operationally similar to the electron tube in which the control grid (bias) voltage controls the cathode-to-plate current.

In practice, and in one type of field-effect transistor circuit, the signal voltage to be amplified is applied to the gate. As a result, the effect

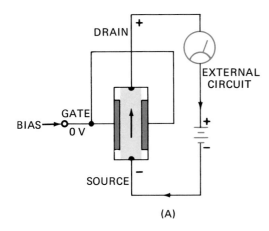

(A)

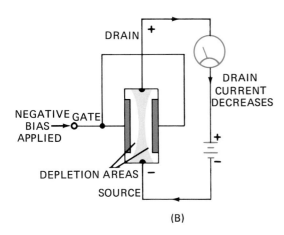

(B)

Fig. 20-15. Basic operation of the field-effect transistor: (A) with transistor unbiased, (B) with reverse-bias voltage applied to gate.

aluminum wire is joined. The ends of the silicon bar form the base one B_1 and the base two B_2 electrodes, and the aluminum is the emitter E electrode (Fig. 20-17A). The aluminum, in contact with the silicon, forms the single p-n junction diode structure of the device.

Operation. With the emitter open, the ohmic resistance between the B_1 and B_2 electrodes of typical unijunction transistors ranges from approximately 5 to 12 kilohms. In normal practice, with one common type of unijunction transistor, a positive voltage V_{BB} is applied to B_2, and B_1 is connected to the common (ground) point of the circuit. Under this condition, the silicon bar acts as a conventional voltage divider (Fig. 20-17B). Thus, a portion of the total voltage applied between B_1 and B_2 appears between the emitter and B_1. This portion of the total voltage is often referred to as the *intrinsic standoff ratio* and is represented by η, the lower case of the Greek letter eta.

The signal input voltage V_E to the transistor is applied between the emitter and ground. When the signal input voltage applied to the emitter causes the emitter to be less positive with respect to B_1 than to ηV_{BB}, the p-n junction diode is reverse-biased. Under this condition, there is no current in the input circuit, and the current through the silicon bar (between B_1 and B_2) consists almost entirely of electrons. When the input voltage causes the emitter to be more positive with respect to B_1 than to ηV_{BB}, the p-n junction diode is forward-biased and holes are injected (emitted) into the silicon bar. These holes then diffuse toward B_1. During this process some of the holes combine with free electrons within the silicon bar. As a result of these com-

of the gate reverse-bias voltage upon the conduction of the transistor is changed in accordance with the variations in the magnitude of the signal voltage. This action, in turn, establishes a varying voltage drop (the output signal) across the transistor (Fig. 20-16).

THE UNIJUNCTION TRANSISTOR

The *unijunction transistor* (UJT) is sometimes referred to as a *double-base diode*. One common type of unijunction transistor consists of a bar or crystal of n-type silicon to which an

Fig. 20-16. Basic field-effect transistor amplifier stage.

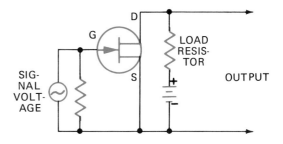

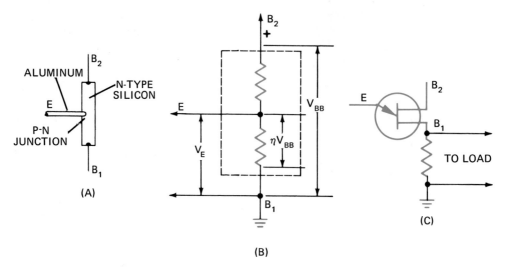

Fig. 20-17. The unijunction transistor: (A) structural features, (B) equivalent voltage-divider circuit, (C) load resistor connected in series with electrode B_1.

binations, electrons are injected into the silicon from B_1 in order to maintain the charge neutrality of the n-type material.

This action, in effect, sharply reduces the resistance between the emitter and B_1, causing a proportional increase of electronic current in the external circuit between B_1 and the emitter. In many unijunction transistor circuits, a load resistor is connected in series with B_1 (Fig. 20-17C). The varying voltage drop developed across this resistor because of the variations of B_1 current is the output voltage of the device. This output voltage is applied to the associated load.

Negative-resistance Characteristic. As previously stated, the injection of holes into the base bar of the unijunction transistor effectively decreases the resistance between the emitter and B_1, causing an increase of current between these electrodes. This action in turn, has the effect of lowering the voltage $E = IR$ between the emitter and B_1. Hence, the unijunction transistor exhibits what is known as a *negative-resistance characteristic*. Under conditions of negative resistance, the current between two points of a device increases as the voltage between these points decreases, or vice versa. Such a current response is, of course, the reverse of that which occurs in accordance with the conventional form of Ohm's law.

The negative-resistance characteristic of the unijunction transistor makes this device capable of being "triggered" into sharply increased conductivity by means of a signal input pulse of the correct polarity. Because of this operational feature, unijunction transistors are very commonly used in control-circuit applications that include timing circuits, time-delay circuits, and switching circuits.

For Review and Discussion

1. State four physical and/or operational advantages of transistors as compared with electron tubes.
2. Define the phrase "solid state."
3. Describe the fabrication of the diffused-junction transistor.
4. Name three semiconductors that are commonly used in the fabrication of transistors.
5. Give the names of the electrodes found on an n-p-n or a p-n-p transistor; also give the names of the corresponding electrodes of a triode electron tube.

6. Describe the two methods used to connect transistors into their associated circuits.

7. Describe the contents of a transistor manual.

8. Draw diagrams illustrating three different methods of positioning the electrodes (pins) upon bases of low-power n-p-n or p-n-p transistors, and identify the electrodes on each of these bases.

9. Draw a diagram representing the "electrode" side of a typical power transistor, and identify the electrodes in this diagram.

10. Draw a schematic diagram showing the bias conditions of an n-p-n transistor, and explain the basic operation of this transistor.

11. What are the majority carriers in an n-p-n transistor?

12. Draw a schematic diagram showing the bias conditions of a p-n-p transistor, and explain the basic operation of this transistor.

13. What are the majority and the minority carriers in a p-n-p transistor?

14. Draw the symbol for an n-channel field-effect transistor, and identify the electrodes on this symbol.

15. Explain the basic operation of a field-effect transistor.

16. Draw the symbol for a typical unijunction transistor, and identify the electrodes on this symbol.

17. Explain the basic operation of a unijunction transistor.

18. For what purpose are unijunction transistors primarily used?

UNIT 21. Transistor Amplifier Stages

The term *amplify* means to enlarge or increase. In the field of electronics, the quantity that is increased by the process of amplification is usually a signal voltage that is present in receiving, transmitting, and sound-reproducing circuits. In these circuits, one or more steps or stages of amplification are used to increase the magnitude of a signal voltage, which may be extremely small, to levels that are adequate for the operation of various circuit devices.

One purpose of this unit is to present the operational characteristics of the triode (n-p-n and p-n-p) junction transistor amplifier stages. A second purpose is to introduce the triode field-effect transistor amplifier stage. The tunnel diode, another semiconductor device that performs the amplification function, is also discussed.

CURRENT CONTROL IN THE TRANSISTOR

In the typical n-p-n or p-n-p transistor amplifier stage, the magnitude of the collector (output) current depends primarily upon the magnitude of the emitter-base current. The emitter-base current is, in turn, dependent upon the bias voltage applied to the emitter-base junction. Therefore, any condition that causes the emitter-base bias voltage to vary will also cause the collector current to vary in direct proportion.

The emitter-base bias voltage of the transistor amplifier stage is varied by introducing the input-signal voltage into the emitter-base circuit. The variations in the signal voltage produce changes in the emitter-base current that controls the collector current. Thus, the transistor can be considered as a current-controlled device, since the current in the input circuit controls the current in the output circuit.

Both n-p-n and p-n-p transistors can be used in amplifier stages. The only difference in the

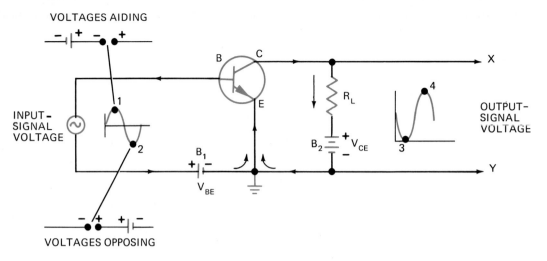

Fig. 21-1. Common-emitter n-p-n transistor amplifier stage.

structure of the associated circuits is that the polarities of the bias voltages used with the n-p-n transistor are opposite to those used with the p-n-p transistor.

When used in an amplifier stage, a transistor can be connected into a circuit in either of two different ways (or circuit configurations). These configurations, defined in terms of that electrode which is common to both the input and the output circuits, are the common-emitter circuit and the common-collector circuit. Each of these circuits has individual operational characteristics that affect its performance as an amplifier.

THE COMMON-EMITTER CIRCUIT

The common-emitter (CE) configuration is used in most transistor amplifier circuit applications because of its relatively high current, voltage, and power gains. In this circuit, the input signal is introduced in series with the base-emitter circuit, while the output signal is taken from the collector-emitter circuit.

N-P-N Transistor. A basic common-emitter circuit using an n-p-n transistor is shown in Fig. 21-1. Notice that the input or base-emitter p-n junction of the transistor is forward-biased, while the base-collector p-n junction is reverse-biased.

When the sinusoidal input-signal voltage is at the positive peak of its cycle (point 1 of the input waveform), it aids V_{BE}, thus increasing the forward-bias effect at the base-emitter junction. This, in turn, causes a greater number of electrons to be injected into the base of the transistor from the emitter. As a result, the collector current and the IR voltage drop developed across the load resistor R_L are at a maximum. At the same time, the voltage drop across the transistor (from collector to emitter) is at a minimum, since this section of the transistor and R_L are connected in series and across battery B_2, the voltage of which remains constant. Point X of the output circuit is now least positive (most negative) with respect to point Y. This condition, which is said to represent a negative-going output signal, is illustrated by point 3 of the output-signal voltage waveform.

When the input signal is at the negative peak of its cycle (point 2 of the input waveform), it opposes bias voltage V_{BE}, thereby decreasing the forward-bias effect at the base-emitter junction. This causes the collector current and the voltage drop across R_L to be at a minimum. At the same time, the voltage drop across the transistor is at a maximum. Point X of the output circuit is now most positive (least negative) with respect to point Y. This condition, which is said to represent a positive-going output signal, is illustrated by point 4 of the output-voltage waveform.

It is very important to understand that the output voltage of the circuit developed across the transistor is a direct voltage, since current in the collector circuit does not change direction. When the input signal is sinusoidal in nature, the output voltage varies in proportion to the changes in the magnitude and polarity of the input voltage. This varying direct voltage can then be represented by a sine-like waveform that shows the magnitude of the output voltage as the alternating input voltage passes through one cycle.

P-N-P Transistor. An example of the basic common-emitter p-n-p transistor amplifier stage is shown in Fig. 21-2. The operation of this circuit is similar to that of the n-p-n transistor circuit. However, since the transistor is of the p-n-p type, the input-signal voltage indirectly controls the number of holes injected into the base from the emitter. This action controls the base-emitter current, which, in turn, controls the collector current. Notice that the polarities of the bias voltages V_{BE} and V_{CE} are opposite to those in the n-p-n circuit shown in Fig. 21-1. This, of course, produces a change in the direction of the collector current as well as the base and emitter currents.

Phase Relationship. In the common-emitter circuit, a phase inversion occurs between the input and output voltages. This is illustrated by the waveforms of Figs. 21-1 and 21-2, which show these voltages to be 180 degrees out of phase with each other.

Input and Output Impedances. As a result of the forward biasing of the base-emitter p-n junction, the input impedance of the common-emitter circuit is relatively low, usually less than 5,000 ohms. The output impedance of such a circuit is moderate; typical values range from 50 to 50,000 ohms.

Current Gain. The current gain, or current-transfer ratio, of the common-emitter circuit is the ratio of the current in the output circuit to the current in the input circuit. Since the input current is present in the base circuit, the current gain is also the ratio of the collector current to the base current, or I_C/I_B. This ratio is called the *beta (β) characteristic* of the transistor in the circuit.

As an example of beta, assume that in the operation of a given common-emitter circuit, 98 percent of the majority carriers within the transistor diffuse through the base and arrive at the collector. Then, for all practical purposes, 2 percent of these carriers enter into combination with minority carriers within the base to produce a base current. As a result,

$$\beta = \frac{0.98}{0.02} = 49$$

In addition to providing an indication of the current gain in a common-emitter circuit, beta provides an indication of the effect that a change in the base current will have upon the collector current under a given collector-to-emitter voltage condition. Because of this, the current gain of the common-emitter circuit can also be determined by using the following formula:

$$\beta = \frac{\Delta I_C}{\Delta I_B}$$

where ΔI_C = change of collector current
ΔI_B = change of base current

and where the currents are expressed in the same unit (milliamperes, microamperes, etc.).

Fig. 21-2. Common-emitter p-n-p transistor amplifier stage.

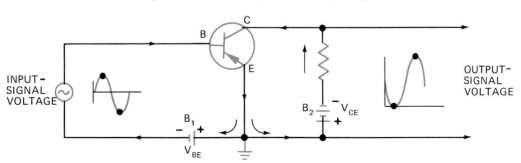

Cutoff Frequency. The cutoff frequency of a transistor is that input-signal frequency at which the value of beta decreases to 0.707 of the beta when the input signal has a frequency of 1 kilohertz. This information provides an indication of the effective frequency range of a transistor when it is used in an amplifier stage.

VOLTAGE GAIN AND AMPLIFICATION

The voltage gain of the common-emitter circuit is the ratio of the input-signal voltage to the output-signal voltage. As indicated previously, a small voltage change in the input circuit will indirectly cause a large current change in the collector (output) circuit. Since the output impedance of the circuit is high, the increase of collector current will, in turn, cause a large voltage to be developed across the output circuit as compared with that of the input voltage. Hence, the input signal has, in effect, been greatly amplified.

Voltage Gain Formula. The voltage gain of a given common-emitter stage can be determined by using the following formula:

$$\text{Voltage gain} = \beta \frac{R_{out}}{R_{in}}$$

where R_{out} = resistance or impedance of output circuit, in ohms
R_{in} = resistance or impedance of input circuit, in ohms

Extrinsic Transconductance. The extrinsic transconductance of a transistor in an amplifier stage may be defined as the quotient obtained when a given small change in collector current is divided by the change in the base-emitter bias voltage that produced it while other bias voltages remained constant. Transconductance is expressed in mhos and is usually given in micromhos.

POWER GAIN

The outstanding advantage of the common-emitter circuit is its high power gain. This is brought about by the combination of the high current gain and the high ratio of the output impedance to the input impedance. Expressed as a formula,

$$\text{Power gain} = \beta^2 \times \frac{R_{out}}{R_{in}}$$

where R_{out} = resistance or impedance of output circuit, in ohms
R_{in} = resistance or impedance of input circuit, in ohms

COLLECTOR CHARACTERISTICS

The circuit response of a transistor is often graphically illustrated by means of a set or family of curves known as *characteristics* curves. In transistor circuit analysis, the most commonly used characteristics curves, known as *collector characteristics,* usually show the relationship between the collector current, the base current, and the collector-to-emitter voltage (Fig. 21-3). The most common transistor letter symbols used with collector characteristics data are defined in Appendix G, page 467.

Current Polarity. On the collector characteristics curves shown in Fig. 21-3, the collector current and the base current are indicated as being negative (−) quantities. In this case, the negative sign is an indication of current polarity. A *negative electronic current* is defined as one that enters or moves into a given electrode. Likewise, a *positive electronic current* is defined as one that leaves or moves out of a given electrode.

During the operation of a p-n-p transistor, the external circuit collector and base currents consist of electrons that enter these electrodes

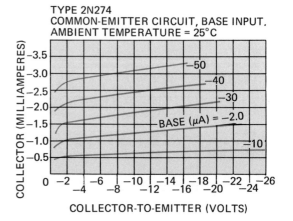

Fig. 21-3. Collector characteristics for the type 2N274 p-n-p transistor. (RCA)

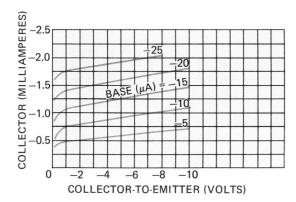

Fig. 21-4. Typical collector characteristics. (RCA)

(see Fig. 21-2). Therefore, in accordance with the definitions stated, these currents are identified with the minus sign to show that they are negative in polarity. The collector-to-emitter voltages on the collector characteristics curves are identified by means of a negative sign because the collector in a common-emitter p-n-p transistor circuit is operated at a voltage that is negative with respect to the emitter.

Problem. The collector characteristics of a p-n-p transistor in a typical common-emitter circuit are shown in Fig. 21-4. From the data given by these curves, compute the current gain of the circuit when the base current changes from 5 to 10 microamperes while the collector-to-emitter voltage remains constant at 8 volts.

Solution. With the collector-to-emitter voltage at 8 volts, a change of from 5 to 10 microamperes in the base current produces a change of from 0.68 to 1 milliamperes in the collector current. Thus,

$$\beta = \frac{\Delta I_C}{\Delta I_B}$$
$$= \frac{1{,}000 - 680}{10 - 5}$$
$$= \frac{320}{5} \text{ or } 64$$

where β = current gain.

THE COMMON-BASE CIRCUIT

An example of the basic common-base (CB) n-p-n transistor circuit is shown in Fig. 21-5. In this circuit, the sinusoidal input-signal voltage is applied in series with the external emitter-base circuit.

Fig. 21-5. Common-base n-p-n transistor amplifier stage.

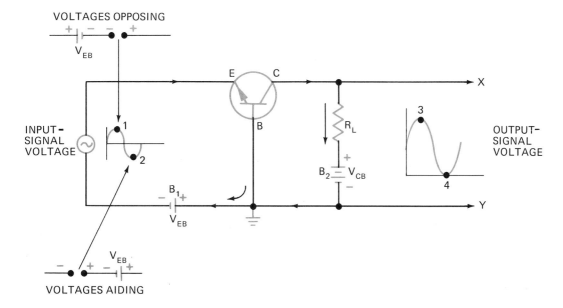

When the input voltage is at the positive peak of its cycle (point 1 of the input waveform), it opposes the bias voltage V_{EB} applied across the emitter-base junction by B_1. As a result, the forward-bias effect at the base-emitter junction is decreased, causing a reduction in the number of electrons injected into the base from the emitter. Because of this, the collector current that passes through the load resistor R_L decreases to a minimum and the IR voltage across the resistor is also at a minimum. At the same time, the voltage drop across the transistor (from collector to emitter) increases to a maximum. Point X of the output circuit is now most positive with respect to point Y. This condition, which is said to represent a positive-going output signal, is illustrated by point 3 of the output-signal voltage waveform.

When the input-signal voltage is at the negative peak of its cycle (point 2 of the input-voltage waveform), it aids the base-emitter bias voltage. Hence, the forward-bias effect at the base-emitter junction is increased, causing a corresponding increase of collector current. Hence, the voltage drop across R_L is at a maximum and the voltage drop across the transistor is at a minimum. Point X of the output circuit is now least positive with respect to point Y. This condition, which is said to represent a negative-going output signal, is illustrated by point 4 of the output-signal voltage waveform.

P-N-P Transistor. A common-base p-n-p transistor amplifier stage is shown in Fig. 21-6. The operation of this circuit is similar to that of the n-p-n circuit, except that the majority carriers within the transistor are holes instead of electrons.

Phase Relationship. In the common-base circuit, the input and the output voltages reach their respective peaks at the same instant as shown by the waveforms of Figs. 21-5 and 21-6. Hence, the input and output voltages are in phase with each other.

Input and Output Impedances. As in the common-emitter circuit, the emitter-base p-n junction is forward-biased, while the collector-base junction is reverse-biased. This causes the input circuit, consisting of the emitter-base junction, to have a comparatively low impedance of approximately 200 ohms in the typical circuit. Since the collector-base junction is reverse-biased, the circuit has a very high output impedance, which may be 800,000 ohms or more.

Operational Characteristics. In the common-base circuit, the forward current transfer ratio is

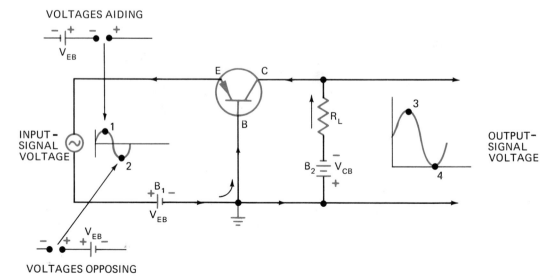

Fig. 21-6. Common-base p-n-p transistor amplifier stage.

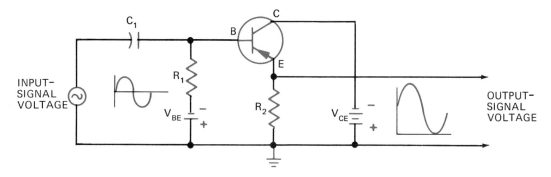

Fig. 21-7. Common-collector p-n-p transistor stage.

the ratio of the collector current to the emitter current or I_C/I_E. This ratio is commonly called the *alpha* (α) *characteristic*. The value of alpha is always less than 1 and averages approximately 0.98. Therefore, there is no current gain in the common-base circuit, because not all the electrons injected into the base of an n-p-n transistor and all of the holes injected into the base of a p-n-p transistor reach the collector.

Since the output circuit of the transistor in a common-base circuit has a high impedance, the load resistor (or inductor) in the output circuit can also have a high resistance or impedance. As a result, the magnitude of the voltage drop across the load is much higher than that of the input voltage. Thus, the circuit is capable of producing a high voltage gain.

In accordance with the power formula $P = I^2R$, a given magnitude of current produces a greater amount of power in a high-resistance (or impedance) circuit than in a low-resistance circuit. The common-base amplifier stage is then capable of producing a significant power gain, since approximately the same amount of current is present in the high-impedance output circuit as in the low-impedance input circuit.

THE COMMON-COLLECTOR CIRCUIT

In the common-collector (CC) circuit, the input signal is applied to the base-collector circuit and the output signal is taken from the emitter-collector circuit (Fig. 21-7). Thus, the collector is common to both the input and output circuits.

Since the input circuit includes the reverse-biased collector-base junction, the input impedance of this circuit is much higher than that of either the common-base or the common-emitter circuit, usually ranging from approximately 100,000 to 300,000 ohms. The output circuit includes the forward-biased base-emitter junction. Because of this, the output impedance of the circuit is relatively low, usually less than 500 ohms.

Gains. The current gain of the common-collector circuit is slightly higher than β and is most often equal to $\beta + 1$. The voltage gain of the circuit is always less than 1 because of the low-impedance load that is used in conjunction with such a circuit. However, because of the high current gain there is a power gain, although it is less than that obtained with either the common-base or the common-emitter circuit. As in the common-base circuit, the input and output signals of the common-collector circuit are in phase.

Application. The common-collector circuit is rarely used as an amplifier stage. Its principal function is that of impedance matching where it is necessary to match the high-impedance output of one stage to the low-impedance input of the following stage. It is also used for isolating one stage from another.

TEMPERATURE EFFECTS

Although the transistor is a rugged, dependable device, it is sensitive to variations in temperature. This is true primarily because thermal energy disturbs the lattice structure and the electron-pair bond arrangement upon which transistor action depends.

As the temperature to which a transistor is exposed (the ambient temperature) increases, there is an increase in the number of electron-pair bonds that are broken and a corresponding increase in the number of free electrons and holes in the transistor material. As a result, there is an increase in the input and the output currents of the transistor circuit with any given magnitude of bias voltage. Thus, the degree of control that the emitter or the base current exerts upon the collector current is reduced. This, in turn, produces a considerable variation in the current transfer ratio and the input and output impedance of the circuit.

If the temperature becomes excessive, the conductivity of the transistor material is increased to a point at which there are a large number of collisions between charge carriers (electrons and holes) and atoms of the transistor material. This action, known as *thermal runaway*, enhances the disturbance of the lattice structure and causes the current through the transistor to increase in magnitude until the transistor may be damaged.

BIASING

The polarity of the bias voltages used in conjunction with transistor amplifier circuit stages provides a forward-bias across the emitter-base p-n junction. The proper polarity of the voltages necessary to establish these bias conditions is conveniently indicated by the first two letters of n-p-n or p-n-p. These letters, which can be interpreted to mean n for negative and p for positive, indicate the normal operating polarities of the emitter and the collector with respect to each other and to the base (Fig. 21-8).

Operating Point. The magnitude of the input circuit (base-emitter) bias voltage to be used in a given transistor amplifier stage is determined by reference to collector characteristics curves. For example, assume that a common-emitter p-n-p transistor circuit is to be operated with a collector-to-emitter bias of 6 volts (Fig. 21-9A). Assume also that the transistor is to operate with a base current I_B of 30 microamperes when there is no signal voltage applied to the input of the circuit. This establishes what is known as the *quiescent operating point* (Fig. 21-9B). The point establishes the practical limits within which an input signal voltage can "swing," or vary.

Notice that with V_{CE} at 6 volts, the operating point is located approximately at the center of the straight (linear) portion of the 30-microampere base-current curve. This condition is desirable, since it will result in circuit operation that will provide for the least amount of distortion or change in the waveform of the output signal as compared with the waveform of the input signal.

Under the conditions specified by the operating point ($V_{CE} = 6$ volts and $I_B = 30$ microamperes), it is, of course, necessary to limit the no-signal base-emitter current to 30 micro-

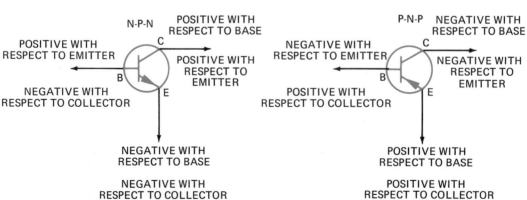

Fig. 21-8. Relative bias conditions used in conjunction with transistor amplifier stages.

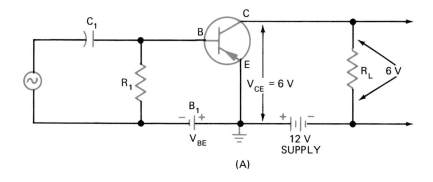

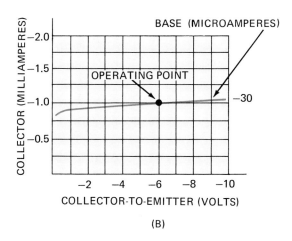

Fig. 21-9. Explanation of the operating point: (A) schematic diagram, (B) graphical representation of the operating point.

amperes. This is done by inserting resistor R_1 in series with the base-emitter B_1 circuit.

Single-source Bias. In the circuits previously mentioned, the transistor was biased by using two separate voltage sources (a cell and a battery). While this method of biasing is effective, a more convenient method involves using only one voltage source that is most commonly a 6-, 9-, or 12-volt battery. This battery provides the necessary collector-to-emitter or collector-to-base voltage and, when used in conjunction with a fixed-bias resistor, it also provides the emitter-base bias current.

A basic single-source bias configuration for a common-emitter p-n-p transistor amplifier stage is shown in Fig. 21-10. Here the 6-volt battery voltage V_{CC} supplies the collector-to-emitter current through load resistor R_L. Bias resistor R_1, which limits the current in the base-emitter circuit, is connected between the negative terminal of the battery and the base of the transistor.

With a given operating-point base current, the value of the bias resistor can be accurately determined by Ohm's law:

$$R = \frac{V_{CC}}{I_B}$$

If the operating-point base current of the circuit is assumed to be 40 microamperes,

$$R_1 = \frac{6}{0.00004} = 150,000 \text{ ohms}$$

An examination of the circuit diagram shows that all the polarity conditions for biasing a p-n-p transistor are satisfied. First, the collector

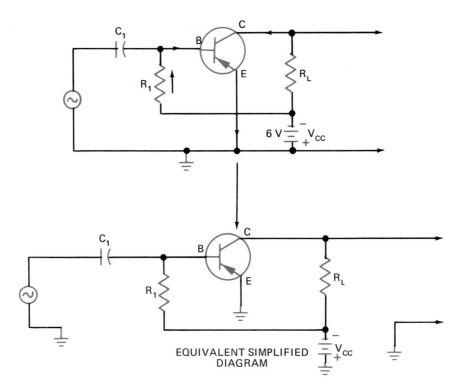

Fig. 21-10. Single-source biasing of common-emitter p-n-p transistor amplifier stage.

is negative with respect to the emitter because of the polarity of V_{cc}. Then, there is usually more voltage drop across R_1 than there is across R_L; therefore, the collector is also negative with respect to the base. Finally, the emitter is positive with respect to the base, because it is connected directly to the positive terminal of the battery.

BIAS STABILIZATION

Although fixed-biasing a transistor does have the advantage of simplicity, it is not the most satisfactory method of providing the bias in a common-emitter circuit. This is true primarily because it is difficult to maintain the base current on which the base-emitter bias depends at a constant value because of operational variations between different transistors and the sensitivity of the transistor to changes in temperature. As a result, the currents vary erratically, thus shifting the operating point, which must remain fixed for efficient operation of an amplifier stage.

Feedback Bias. The bias voltage can be stabilized to a greater extent by means of the feedback bias circuit configuration shown in Fig. 21-11. In this circuit, the bias resistor R_1 is connected between the collector and the base to provide a feedback voltage from the collector to the base.

Fig. 21-11. Feedback-biased common-emitter p-n-p transistor amplifier stage.

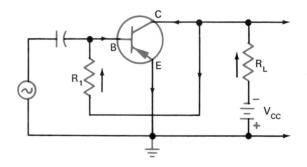

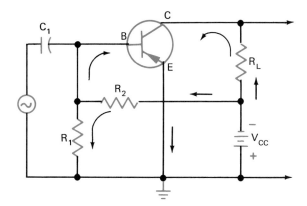

Fig. 21-12. Voltage-divider transistor biasing.

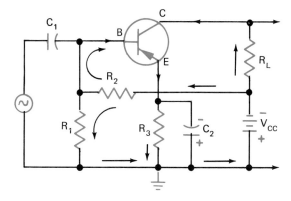

Fig. 21-13. Transistor amplifier stage bias network using voltage divider and emitter stabilizing resistor.

When the base current increases, the collector current also increases, thus increasing the voltage drop across R_L and causing the collector to become less negative. Since the collector is the base bias source, the decrease in collector voltage will cause the base current to decrease. Thus, the increasing base current is opposed and the circuit currents tend to remain constant. This negative feedback from the collector stabilizes the operating point.

Voltage-divider Network. Additional bias stability is provided by using the circuit shown in Fig. 21-12. In this circuit, the voltage-divider network, consisting of resistors R_1 and R_2, establishes the necessary forward-bias condition across the base-emitter junction. The bias voltage is determined primarily by the resistors. Any change in the transistor as the result of a temperature shift will have less effect because of the improved regulation offered by the divider network.

The Emitter Stabilizing Resistor. The bias stability of the common-emitter circuit is improved by using a resistor that is connected in series with the emitter as shown in Fig. 21-13. The voltage drop across resistor R_3 is of a polarity that reduces the forward-bias effect of the base-emitter junction. Therefore, as the emitter current increases, the resulting increase of the voltage drop tends to stabilize the base current and the collector current. Likewise, a decrease of emitter current increases the forward-bias effect, thereby also tending to stabilize the base and collector currents under this condition.

The function of the bypass capacitor C_2 that is connected in parallel with the emitter resistor is to minimize signal reduction or degeneration in the amplifier. In practice, this is usually achieved by using an electrolytic capacitor with a capacitance value of from 30 to 50 microfarads.

Thermistor Bias Stabilization. The bias stability of the voltage-divider network is improved by adding a thermistor to the circuit (Fig. 21-14). This method of bias stabilization is particularly effective in a power amplifier stage where the comparatively large collector current increases the temperature of the transistor to a level that makes the operating point subject to significant variation.

As the temperature of the transistor in the circuit increases, the resistance of the thermistor RT_1 decreases because of its negative tempera-

Fig. 21-14. Thermistor bias stabilization.

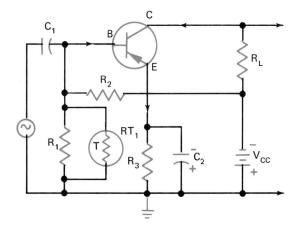

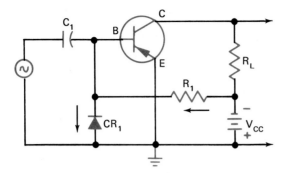

Fig. 21-15. Diode bias stabilization.

ture coefficient. As a result, there is an increase of current through the voltage-divider network. The increased current, in turn, produces a larger voltage drop across resistor R_2. Since this action decreases the forward-bias effect at the base-emitter junction, it also tends to counteract the increase of temperature.

Diode Bias Stabilization. An example of the use of a diode to provide bias stabilization is shown in Fig. 21-15. In this circuit, current passes through the voltage-divider network consisting of resistor R_1 and diode CR_1 in the direction indicated. With an increase of temperature, the resistance of CR_1 decreases, thus allowing more current to pass through the diode and the resistor. Hence, the voltage drop across R_1 increases and there is a corresponding reduction in the voltage drop across CR_1. This action has the effect of reducing the forward bias at the base-emitter junction of the transistor, thereby reducing the collector current, which also tends to increase with an increase of temperature.

THE LOAD LINE

A *load line* is a line used to indicate the relationships between the collector-to-emitter voltage, the collector current, and the base current in a transistor amplifier stage with a given value of load resistance. Because of this feature, load lines are useful in designing amplifier circuits.

The load line is commonly drawn upon the collector characteristics "curves." A basic common-emitter p-n-p transistor amplifier stage and its associated load line are shown in Fig. 21-16.

To draw the load line, at least two points must first be located. One of these points is A of Fig. 21-16B. With the collector current I_C at zero, the voltage drop across the load resistor R_L is also zero. Hence, the total collector supply voltage V_{CC} is applied across the collector and the emitter and $V_{CC} = V_{CE}$.

The second point is located by computing the collector current when it is assumed that the transistor is acting as a perfect conductor from collector to emitter. In this case, the current in the collector circuit is limited only by the resistance of R_L and can be determined by using Ohm's law as follows:

Fig. 21-16. A basic common-emitter p-n-p transistor amplifier stage and the associated load line.

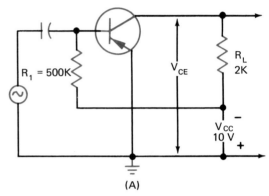

(A)

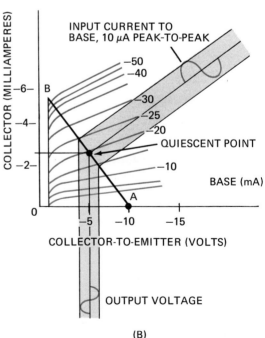

(B)

216

$$I_c = \frac{V_{cc}}{R_L}$$
$$= \frac{10}{2{,}000} = 5 \text{ milliamperes}$$

The second point is then 5 milliamperes on the vertical (collector current) axis. This is identified as point B in Fig. 21-16B. To complete the load line, a straight line is drawn through points A and B.

By using other values of R_L, other load lines for this amplifier stage can be "constructed" in a similar manner. The slant or slope of any given load line is governed by the load resistance. As the load resistance increases, the slope of the associated load line also increases.

The quiescent operating point is found by computing the base bias current. Since R_1 limits this bias current, it is equal to 10/500,000 or 20 microamperes. The operating point is then located where the 20-microampere base-current curve intersects the load line. It can be seen that at this point the collector-emitter (output) voltage is 5 volts and that the quiescent (no input signal) collector current is approximately 2.5 milliamperes. When an input signal is applied to the associated circuit, the base current will, of course, vary above and below the operating point. Thus, the load line makes it possible to determine the output voltage and collector current produced as the result of any variation of base current.

AMPLIFIER STAGE CLASSIFICATION

The operating characteristics of an amplifier stage are often classified in accordance with the bias (operating point) conditions under which transistors are operated. This method of classification is convenient since it indicates what portion of the input signal appears in the output, and the expected efficiency of the stage. The three principal classes of operation are A, B, and AB.

Class A Operation. A class A transistor amplifier stage is biased at approximately the midpoint of the collector characteristics curve. This operating point condition provides for continuous collector current during the complete cycle of the input-signal voltage and also when there is no input signal.

Because of the low distortion factor, class A operation is commonly used in audio voltage amplifier and driver stages. It has the disadvantage, however, of being a comparatively inefficient and performs adequately only when used in relatively low-power applications.

Class B Operation. A class B amplifier stage is one that is biased at or near cutoff. Thus, collector current is present only during one half of each input-voltage cycle. Although class B operation is much more efficient than class A operation, it does result in considerable distortion because of the serious "clipping" of the output signal. For this reason, such operation is not used in conjunction with single-ended (one transistor) amplifier stages, except when distortion can be tolerated.

The principal application of class B operation is found in push-pull output stages, which are discussed in Unit 23. Here, the distortion is reduced, since each transistor of the stage amplifies only half of each input cycle.

Class AB Operation. A class AB amplifier stage is one that is biased for operation in the region between class A and class B. In such a stage, collector current in the output circuit is present for considerably more than half the input-signal voltage cycle, but for less than the complete cycle.

THE EMITTER-FOLLOWER CIRCUIT

The output signal from a given transistor amplifier stage can be "taken" from across an emitter resistor as shown in Fig. 21-17. This re-

Fig. 21-17. The emitter-follower circuit.

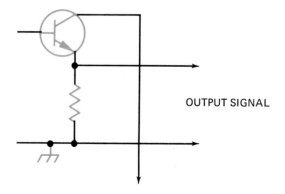

OUTPUT SIGNAL

sults in an output signal with a voltage that is less than the voltage of the input signal, a condition commonly referred to as a *voltage gain of less than one*. The output signal is in phase with the input signal. Because the follower circuit has a comparatively high input impedance and a very low output impedance, it is often used for impedance matching and for isolation purposes.

HIGH-FREQUENCY TRANSISTORS

The frequency response of the typical transistor is limited by two factors: the transit time and the interelectrode capacitance. The transit time is related to the period of time that is required for the majority carriers to diffuse through the base region and pass on to the collector. Since the majority carriers diffuse throughout the base region, the path from the emitter to the collector within the transistor is somewhat longer than a straight line between these electrodes. This condition increases the transit time and thus decreases the efficiency with which the transistor will operate in a high-frequency circuit.

Interelectrode capacitance exists between the emitter-base and the collector-base junctions of the transistor. As the operating frequency is increased, the capacitive reactance decreases sharply, resulting in feedback by means of which a portion of the output signal is internally coupled to the input. This action reduces the stability of the transistor stage by reducing the control that the input signal has over the output.

Efficient high-frequency transistor operation is achieved by special transistor structures that have the effect of preventing the diffusion of majority carriers throughout the base region and reducing the interelectrode capacitance. Among the transistors designed for high-frequency applications are the planar, the drift, the epitaxial, and the mesa types.

In the mesa transistor, for example, the base and the emitter regions are "undercut" as shown in Fig. 21-18. This shape decreases the effective collector area, thereby decreasing the interelectrode capacitance at the collector-base junction. Since this structure also decreases the cross-sectional area of the base region, the majority carriers tend to follow a relatively straight path as they move from the emitter to the col-

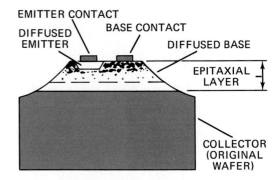

Fig. 21-18. Mesa-type transistor. (*RCA*)

lector. As a result, the transit time is decreased. The net effect of these structural and operational features is a transistor with an efficient frequency-response characteristic that can extend to several hundred megahertz.

THE FIELD-EFFECT TRANSISTOR STAGE

An outstanding feature of the field-effect transistor is its high input impedance. In this respect, the transistor can be considered as the solid-state counterpart of the electron tube, which typically possesses similar impedance characteristics.

The triode field-effect transistor amplifier stage is most commonly operated as either a common-source or a common-gate configuration. For general-purpose amplifier applications, the common-source configuration has proved to be the most useful. In the common-source circuit, the input-signal voltage is applied to the gate and the output-signal voltage is "taken" from the drain-source circuit, the source being common to both the input and output circuits. A basic example of the common-source, n-channel amplifier stage is shown in Fig. 21-19.

In this circuit, battery B_1 applies a voltage across the drain and the source, which makes the drain positive with respect to the source. As a result, electrons move through the transistor from the source to the drain in the direction indicated. Under this condition, the upper end of bias resistor R_2 is positive with respect to the lower end. Hence, the gate-to-source bias voltage V_{GS} is of a polarity that causes the gate to be negative with respect to the source. It is

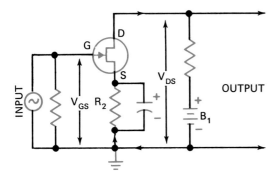

Fig. 21-19. Common-source, n-channel field-effect transistor amplifier stage.

interesting to note that these operating-voltage polarities are identical to those existing across corresponding electrodes in the operation of the typical electron-tube amplifier stage.

A preamplifier circuit utilizing a common-source, p-channel field-effect transistor stage Q_1 in conjunction with an n-p-n transistor stage is shown in Fig. 21-20. In this circuit, the drain of Q_1 is negative with respect to the source, and electrons move through the external output circuit of Q_1 from the source to the drain. Under this condition, the upper end of bias resistor R_2 is negative with respect to the lower end, thus causing the source to be negative with respect to the gate. Conversely, the gate is positive with respect to the source. It is important to notice that these operating-voltage polarities are the reverse of those existing across identical electrodes in the n-channel field-effect transistor stage.

THE TUNNEL DIODE

In its basic form, a tunnel diode is a p-n junction semiconductor device made of silicon, germanium, or gallium arsenide that is doped with a relatively high concentration of impurities (Fig. 21-21). Hence, the p-type material contains a high concentration of holes, while the n-type material contains a high concentration of electrons. This condition results in the formation of a very narrow junction depletion region, thus making it possible for these carriers to move across the junction by means of what is referred to as a *quantum-mechanical action*, also commonly known as "tunneling."

Characteristic Curve. The characteristic current-response curve of a typical tunnel diode under a forward-bias condition is shown in Fig. 21-22.

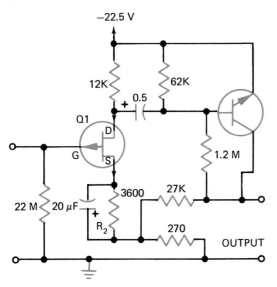

Fig. 21-20. Preamplifier circuit utilizing a p-channel field-effect transistor and an n-p-n transistor. *(Siliconix, Inc.)*

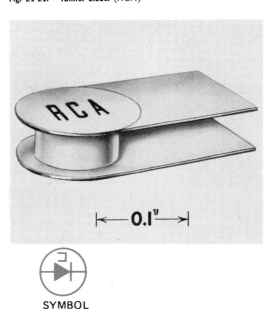

Fig. 21-21. Tunnel diode. *(RCA)*

SYMBOL

The curve shows that the diode begins to conduct current at an extremely low applied forward-bias voltage of only several millivolts. This conduction response is quite different from that observed in the operation of conventional diodes, which do not begin to conduct until their applied forward-bias voltage is approximately 0.3 volt for the germanium diode and 0.8 volt for the silicon diode.

The curve also shows another very interesting and important current-response characteristic. When the bias voltage is first applied, the current increases rather sharply from zero to point A. As the bias voltage is further increased from V_1 to V_2 on the horizontal reference (voltage) axis, the current sharply decreases to point B on the curve. Therefore, the diode exhibits a negative-resistance characteristic between points V_1 and V_2, since the current through it decreases as the applied voltage increases.

Basic Amplifying Action. When used as an amplifier stage, the tunnel diode is usually biased so that its no-signal-input operating point is at a level that is approximately in the center of the linear portion of the negative-resistance slope of the characteristic curve (point C of Fig. 21-22). As a result, the variations in the magnitude of the signal voltage will affect the bias voltage in a manner that causes the current through the device and its associated load to vary significantly above and below the operating point. This action, in turn, produces an amplification of the

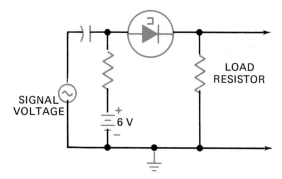

Fig. 21-23. Basic tunnel diode amplifier stage.

signal voltage because the peaks of the varying voltage developed across the load resistor are greater than those of the signal voltage (Fig. 21-23).

The tunnel diode has several outstanding features. It is, first of all, capable of responding to extremely high frequencies. Because of this, tunnel diodes are commonly used in microwave amplifier circuits that operate at frequencies of 10,000 megahertz or more. The device also has the property of very low power dissipation, which makes its application desirable in circuits that must be contained within a small enclosure. An additional advantage of the tunnel diode is its relatively high resistance to nuclear radiation. This feature allows it to be used in equipment units that are exposed to levels of radiation that could cause serious malfunctions in transistorized circuitry.

The principal disadvantage of the tunnel diode is the difficulty encountered in stabilizing the operating-point current. This is primarily because the current along the negative-resistance slope of the characteristic curve tends to vary somewhat during the course of normal operation.

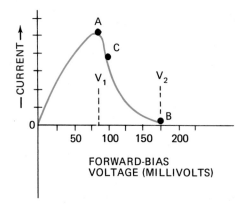

Fig. 21-22. Characteristic current-response curve of a typical tunnel diode.

For Review and Discussion

1. Explain why a transistor operates as a current-controlled device.
2. Draw a schematic diagram illustrating a basic n-p-n or p-n-p common-emitter transistor amplifier stage.

3. Describe the phase relationship between the input and output voltages of a common-emitter circuit. Explain why this occurs.
4. State the input and output impedance characteristics of a common-emitter circuit.
5. Define beta and state the operational characteristic of a transistor that it identifies.
6. What relationships are given by the collector-characteristics data of a transistor?
7. In transistor terminology, what is meant by a negative current with respect to a given electrode?
8. Describe the phase relationship between the input and output voltages of a common-base circuit.
9. Define alpha and explain why it is always less than 1.
10. For what purposes is a common-collector transistor stage used?
11. What is the effect of an increase in temperature upon the operation of a transistor? Explain why this occurs.
12. Define what is known as the operating point of a transistor stage.
13. Define bias stabilization.
14. Draw a schematic diagram of a common-emitter p-n-p transistor amplifier stage in which the single-source method of biasing is used.
15. Illustrate the feedback method of biasing.
16. Describe diode bias stabilization.
17. What is a load line?
18. Define class A and class B amplifier stage operation.
19. Describe an emitter-follower circuit.
20. State one outstanding operational feature of the field-effect transistor.
21. Draw a schematic diagram illustrating a basic n-channel field-effect transistor amplifier stage.
22. Describe the constructional features of a typical tunnel diode.
23. Explain what is meant by the negative-resistance characteristic of a tunnel diode.

UNIT 22. Electron Tubes

Although electron tubes have generally been replaced by solid-state devices, they are found in various kinds of old products and are still being used in the circuits of some new products. This unit presents the structure and the theory of operation relating to the most common types of tubes.

THE DIODE TUBE

The typical diode electron tube consists basically of two electrodes: the cathode and the anode or plate. These electrodes are sealed within an enclosure (the envelope) from which, as in most types of tubes, most of the air has been removed (Fig. 22-1). The cathode serves as the electrode from which electrons are emitted by the process of *thermionic emission*. In order to perform this function, the cathode may be heated either directly or indirectly (Fig. 22-2).

To function as a rectifier, the plate and the cathode of the tube are connected to an alternating voltage in a manner similar to the semiconductor diode (see Fig. 19-6). The tube will conduct current only during the time that the plate is positive with respect to the cathode. In this condition, the electrons emitted from the cathode are attracted to the plate. The tube does not conduct during the time that the plate is negative with respect to the cathode.

In the full-wave or duo-diode tube, there are two separate plates and two cathodes. Hence, one pair of elements will conduct during each half-cycle of the alternating input voltage.

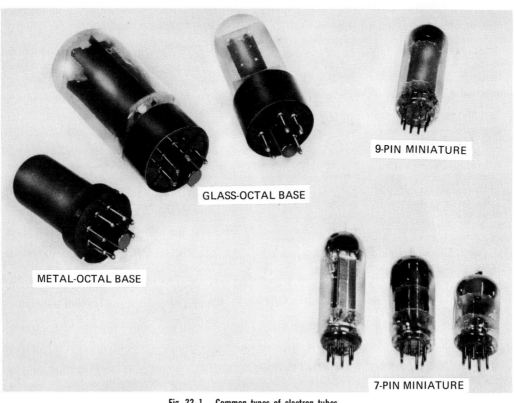

Fig. 22-1. Common types of electron tubes.

Fig. 22-2. Electron-tube cathodes: (A) directly heated, (B) indirectly heated.

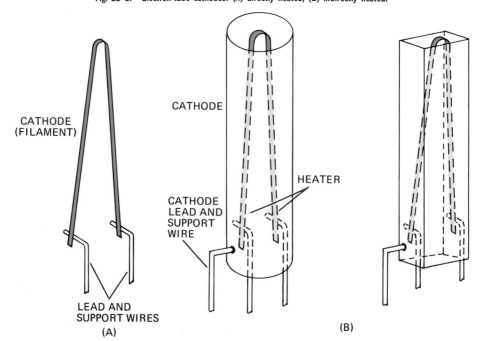

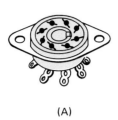

(A)

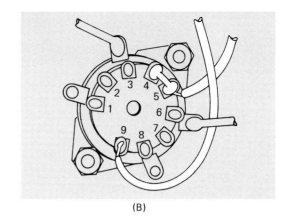

(B)

Fig. 22-3. Tube socket: (A) unwired, (B) wired.

TUBE BASES AND SOCKETS

The electrodes within an electron tube are connected into a circuit by means of a tube base pin and tube socket assembly, with the actual circuit wire connections being made to terminals of the tube socket (Fig. 22-3). The pattern of pin placement on three common types of tube bases is shown in Fig. 22-4. The pins are numbered in a clockwise direction starting with the "key" and with the tube viewed from its base end.

Fig. 22-4. Electron-tube base-pin configurations.

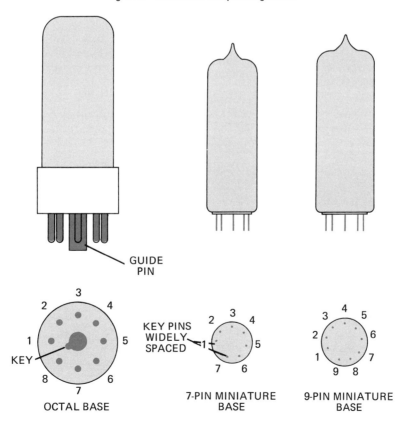

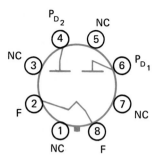

Fig. 22-5. Basing diagram of a type 5U4GB full-wave vacuum rectifier tube.

THE BASING DIAGRAM

A basing diagram shows (in schematic form) the electrodes within an electron tube and the base pins to which each electrode is connected (Fig. 22-5). The letters NC appearing on many tube basing diagrams indicate that the pins do not have a connection to any electrode within the tube. Note that pins are numbered clockwise from the key.

TUBE TYPE DESIGNATION

Vacuum rectifier electron tubes, as well as other so-called "receiving" tubes used in radio receivers, amplifiers, and other relatively small-size products, are identified by means of a system of numbers and letters. Common diode (or duo-diode) tubes are the 5U4G, the 35W4, and the 35Z5GT.

The first number (or numbers) of the tube type designation indicate the voltage at which the heater is operated. Thus, the 5U4G tube is operated with a heater voltage of 5 volts, while the 35Z5GT tube is operated with a heater voltage of 35 volts.

The letters between the numbers of a tube type designation have no standardized meaning. However, the letters near the end of the alphabet usually indicate that a tube is a rectifier tube, while other letters indicate an amplifier or some other tube function.

The last number of a tube type usually indicates the number of pins on the tube base that are actually connected to electrodes within the envelope. Thus, while a 5U4G tube, which is an octal base type, may have eight pins on its base, only four of these are connected to electrodes. These are pins 2, 4, 6, and 8. Any remaining pins are often identified on a schematic basing diagram as NC (not connected). When counting tube base pins, it is important to note that a pin position is counted whether or not a pin is actually located at this position on the base.

The letter G or the letters GT appearing at the end of an octal tube type designation indicate a glass tube, GT meaning glass-tubular shaped. Therefore, the 5U4G is a glass tube. If this same tube is identified as a 5U4GA, the letter A indicates that it is a revised model of the earlier 5U4G type. Likewise, the 5U4GB tube is a revised model of the 5U4GA. All 7- and 9-pin miniature tubes are glass tubes, although the letter G does not appear in the tube type designation.

THE TUBE MANUAL

The tube manual is a publication that lists the descriptions, operating characteristics, and ratings, and gives the basing diagrams of a wide variety of electron tubes. In addition, most of these manuals also present useful information pertaining to electron-tube circuits and circuit design.

THE TRIODE TUBE

The typical triode tube contains a plate, a directly or an indirectly heated cathode, and a

Fig. 22-6. The triode tube.

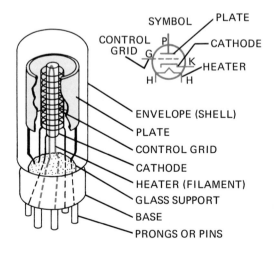

224

third electrode known as the *control grid* (Fig. 22-6). This grid, which is often made of molybdenum, Nichrome, or tungsten, consists of several turns of wire wound upon supporting structures and a lead that connects it to a base pin.

Function of the Control Grid. The function of the control grid is to provide a means for the electrostatic control of the unidirectional current from the cathode to the plate. This is accomplished by connecting the grid to the signal voltage output of the preceding stage or portion of the circuit (Fig. 22-7A). When the signal voltage is of a polarity that makes the grid negative with respect to the cathode, the electrostatic repulsion existing between the negative grid and the electrons restricts the flow of electrons through the tube (Fig. 22-7B).

When the signal voltage applied to the grid is of a polarity that makes the grid positive with respect to the cathode, electrons are attracted toward the grid. However, since the grid allows most of these electrons to pass through it and on to the more highly charged positive plate, there is an increase of plate current in the output circuit of the tube (Fig. 22-7C). Thus, the control grid acts as an electric gate or valve that can control plate current.

Bias Voltage. The control grid will be more effective in controlling the current through a tube if it is operated at a voltage that makes it constantly negative with respect to the cathode. This is achieved by introducing a bias voltage. The bias voltage is applied to the grid by connecting the grid to the negative terminal of a voltage-divider network or by utilizing the voltage that is developed across a resistor in the cathode circuit of the tube.

If the negative bias voltage is increased to a certain magnitude while the positive voltage applied to the plate remains constant, the current from the cathode to the plate (the plate current) will be reduced to zero. This condition is referred to as the *cutoff point* of the tube, and the bias voltage at this point is known as the *cutoff bias*. If, on the other hand, the bias voltage is such that, at a given plate voltage, all the electrons emitted from the cathode reach the plate, the tube is said to be operating at its *saturation point*.

THE AMPLIFICATION PROCESS

The process of amplification in a triode tube occurs as the result of the control that the signal voltage applied to the control grid has upon the plate current. Because of this control, small variations in the magnitude of the signal voltage can cause large variations in the plate current. In order to utilize these variations in current in the form of an output voltage, the plate of the tube is connected in series with the primary winding of an interstage transformer, a plate load resistor, or an inductor.

Fig. 22-7. Function of the control grid.

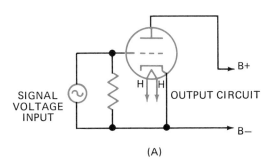

(A)

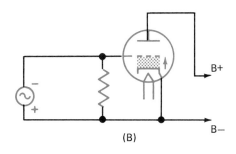

(B)

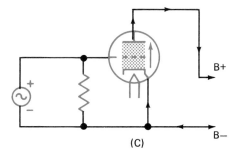

(C)

THE TETRODE TUBE

The tetrode or four-electrode tube is designed to minimize the effect of the interelectrode capacitance between the plate and the control grid and between the plate and the cathode. In this tube, an additional grid, the *screen grid,* is located between the control grid and the plate. On a tube basing diagram, the control grid of a tetrode tube is identified as G_1 and the screen grid is identified as G_2.

In practice, the screen grid is most often operated at a positive voltage that is somewhat lower than the voltage applied to the plate. This enables the grid to act as an electrostatic shield between the control grid and the plate of the tube.

THE PENTODE TUBE

A major disadvantage of the tetrode tube is that its screen grid is capable of accelerating electrons to a degree that produces a significant amount of secondary emission from the plate. Because of this, many of the electrons that are emitted secondarily from the plate are attracted back to the screen grid, causing a serious reduction of plate current (unless the plate is operated at an abnormally high voltage).

To overcome the undesirable effect of secondary emission, a pentode, or five-electrode, tube is often used. In the pentode, a third grid, known as the *suppressor grid,* is located between the screen grid and the plate (Fig. 22-8). On a basing diagram this grid is identified as G_3. The suppressor grid is usually operated at the cathode voltage by connecting it directly to the cathode within the tube or by connecting it to the cathode externally.

Since the suppressor grid is operated at the cathode voltage, it is, of course, highly negative with respect to the plate. However, since this grid is a widely spaced structure, electrons emitted from the cathode readily pass through it to the plate because of the influence of the screen grid and the plate voltages. At the same time, the suppressor grid has a significant effect upon the secondary emission electrons, causing them to be repelled back to the plate.

MULTIUNIT OR MULTISECTION TUBES

A multiunit tube is one in which two or more individual electrode structures are combined within a single envelope. In addition to the duo-diode rectifier tube previously discussed, there are several other multiunit tubes in common use. These include various electrode combinations such as the twin diode-triode, the duo or twin triode, and the triode-pentode (Fig. 22-9).

TUBE SHIELDING

In many high-gain amplifier circuits, it is often necessary to shield the electron tubes of the various stages to prevent excessive capacitance and magnetic field interaction between tubes. When metal tubes are used, the shell itself serves

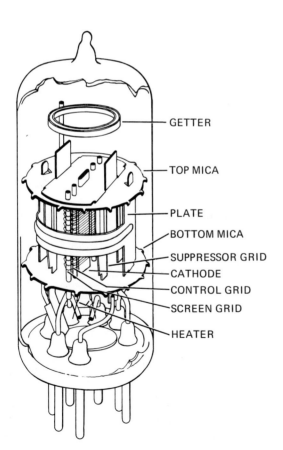

Fig. 22-8. Constructional features of a 7-pin miniature pentode tube. (*General Electric Co.*)

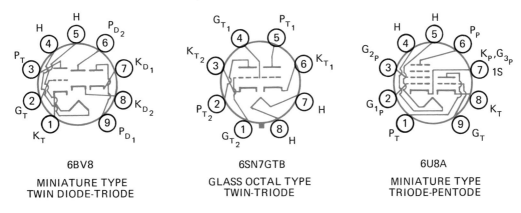

Fig. 22-9. Basing diagrams of common multiunit electron tubes. (RCA)

as the shield. The shell is usually connected to a base pin of the tube, which is identified as S upon the tube basing diagram. Shielding is then achieved by connecting the corresponding terminal of the tube socket to the chassis or to the common (or ground) point of the circuit. Some glass tubes are equipped with an internal shield that is connected internally to a base pin identified as I_s upon the basing diagram.

When shielding is necessary for a glass tube that is not equipped with an internal shield, a metallic tube shield is slipped over the tube. The shield is then connected to the chassis with a wire lead or by using a tube socket that is equipped with a shield base (Fig. 22-10).

THE THYRATRON TUBE

The thyratron is a gas-filled triode tube containing a grid located between the indirectly heated cathode and the plate. This grid controls the magnitude of the ionization voltage that is required before the tube will conduct current.

If the voltage applied to the grid is negative, the resulting electrostatic field about the grid will counteract the positive charge of the plate, thus reducing the tendency of the gas to ionize. Under this condition, ionization of the gas within the tube will occur only if the plate voltage is increased. Therefore, the magnitude of the grid voltage determines the plate voltage that is required to "fire" the tube and start the conduction of current through it. However, when conduction does take place, the grid loses its controlling effect, since it is then surrounded by positive ions of gas. As a result, the tube will not stop conducting until the plate voltage is reduced to a certain magnitude, which is referred to as the *extinction point* or the *extinction voltage*

Fig. 22-10. External tube shield and tube socket equipped with a shield base. (Heath Co.)

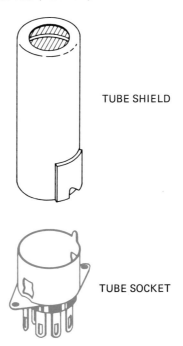

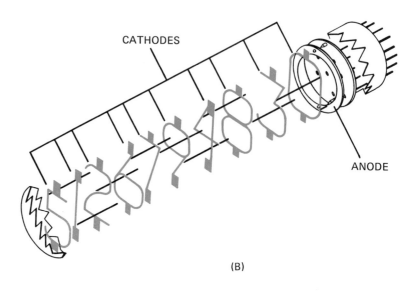

Fig. 22-11. Indicator tubes: (A) typical tube, (B) exploded view of a tube structure that permits a reading of the digits from the "top" or "end" of the envelope. (Burroughs Corp.)

Although thyratron tubes are not commonly used as rectifiers, they are found in a variety of circuits where the electronic control of a relatively large current is desirable. Examples of such circuits are motor controls, welder controls, and sawtooth waveform oscillators.

INDICATOR TUBES

Indicator tubes are used to provide a "visual readout" of a quantity indicated by digital-type measuring instruments, counters, and computers. A common type of cold-cathode, glow-lamp indicator tube is shown in Fig. 22-11A.

In such a tube, a "stacked" assembly of ten cathodes, each shaped in the form of a digit (9 through 0), and a common anode are contained within a glass envelope that is filled to a low pressure with a gas such as neon (Fig. 22-11B). During typical operation, the anode is operated at a positive voltage with respect to the cathodes. When a control voltage is applied to a given cathode, this cathode is made more negative with respect to the anode. As in other types of glow tubes, the voltage between the anode and the cathode is sufficient to produce ionization of the gas within the tube. As the resulting positive gas ions strike the negative cathode, their energy is released in the form of a glow discharge surrounding the cathode, thereby making it visible.

For Review and Discussion

1. Describe the construction of a typical diode rectifier tube.
2. For what purpose is the heater in an electron tube used?
3. What is a tube basing diagram?

4. What do the letters NC on a basing diagram mean?
5. Explain the meaning of a tube-type designation such as 5U4GB.
6. State the function of the control grid in a triode tube, and explain how the grid is able to perform this function.
7. What is meant by an electron-tube bias voltage?
8. Why is a screen grid used in a tetrode tube?
9. State the purpose of the suppressor grid in a pentode tube.
10. Describe a multiunit tube.
11. For what purpose is a tube shield used?
12. Explain the basic operation of a thyratron tube.
13. For what purpose are indicator tubes used?
14. Describe the construction of a typical cold-cathode, glow-lamp indicator tube.

UNIT 23. Amplifier Systems

Amplifier systems usually consist of two or more amplifier stages, associated control circuits, and the coupling networks used to transfer the signal from one stage to another. The amplifying system may be contained within an individual unit called an *amplifier* or it may be an integral part of a complete circuit system such as that used in conjunction with a receiver or a transmitter.

The process of amplification must, of course, be used in both audio-frequency (AF) and radio-frequency (RF) circuit systems. In this unit, those amplifiers which operate at audio frequencies will be emphasized. Radio-frequency amplifiers and their application in radio receivers, television receivers, and transmitters are discussed in later units that are devoted to receiving and transmitting equipment.

THE AUDIO AMPLIFIER

The most commonly used individual amplifier system is the audio amplifier, which is designed to operate at the audio range of frequencies that extends from approximately 15 to 20,000 hertz (Fig. 23-1). These amplifiers are most often used in conjunction with signal sources such as microphones, record players, tape recorders, and radio tuners.

Fig. 23-1. Solid-state stereo amplifier. (*Lear Siegler, Inc., Bogen Communications Division*)

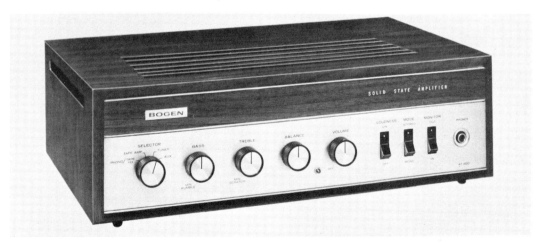

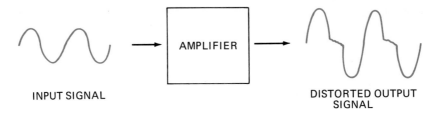

Fig. 23-2. Distorted amplifier output signal.

Distortion. Distortion in an amplifier results when the variations of the amplifier output signal do not closely resemble the variations in the waveform of the input signal (Fig. 23-2). This condition produces a loudspeaker sound response that is said to be distorted or of poor quality, since the sound waves are not true reproductions of the input signal.

Fidelity. Fidelity is the degree to which an amplifier is capable of amplifying a given input signal with the minimum amount of distortion. Hence, a high-fidelity amplifier is one that is capable of faithfully reproducing input signals over the entire range of audio frequencies.

In the low-fidelity amplifier, the signal response is often limited (distorted) at the upper end of the audio-frequency range. This condition, known as *frequency distortion,* usually results because the voltage gain of the various amplifier stages is less at higher frequencies than at lower frequencies.

Although a high-fidelity amplifier response is desirable for music appreciation, it is not always important. In a public-address system, for example, it is most often necessary to amplify only voice frequencies that can be adequately interpreted, even though the sound reproduction is not of high-fidelity quality. In certain industrial applications, amplifiers are used in conjunction with signals that are intended only for circuit control purposes. Here, the primary requirement of the amplifier is to increase the magnitude of the signal input to a degree that will allow the control of a device such as a relay or a solenoid-plunger mechanism.

Compatibility. In a high-fidelity system, the various units, which consist of the signal source, the amplifier, and the associated loudspeaker or loudspeakers, must be compatible. This means that each of the units must be capable of handling the entire range of frequencies that are to be reproduced. Thus, high fidelity cannot be achieved with a microphone that has an upper frequency-response limit of 8,000 hertz, even though the amplifier and the loudspeaker are high-fidelity units. Likewise, high fidelity cannot be achieved with a good quality signal source and amplifier if the loudspeaker has a high-frequency response limit of only 8,000 hertz.

CAUSES OF DISTORTION

The extent to which an audio amplifier is capable of reproducing the signal-input voltage is dependent upon the design of its circuit and the quality of the components used. The quality factor is of particular importance with regard to transistors, electron tubes, bypass capacitors, coupling capacitors, and transformers. If these devices are not capable of wideband (entire audio-frequency range) response, the fidelity of the amplifier will be limited, even though its circuit is well designed.

Distortion in the output of an amplifier can be caused by a number of conditions within the amplifier. The most common of these conditions are discussed in the following paragraphs.

Defective Transistors and Tubes. In transistor circuits, distortion often results when a transistor has become damaged from overheating. In electron-tube circuits, distortion is often produced by tubes that are gassy and therefore not able to respond to input signal variations satisfactorily.

Microphonics. A microphonic condition occurs in an electron tube when its electrodes become loose because of faulty construction or service "wear." As a result, the spacing between the electrodes is changed when the tube is subjected to vibration. Since this causes the tube to conduct current erratically, the tube acts as a "microphone" to produce a distortion noise in the loudspeaker output.

Defective Capacitors. Capacitors that are shorted or have excessive leakage often cause a change in the bias of a transistor or an electron tube. This condition, in turn, causes the transistor or the tube to introduce a distorted signal into its output circuit.

Defective Resistors. The overheating of a carbon-composition-type resistor can cause it to either "burn out" or undergo a significant change of resistance. As with defective capacitors, each of these conditions can cause a transistor or electron tube bias variation that produces a distorted output signal.

Overloading. Overloading occurs when an input signal having an excessive magnitude is applied to a transistor or electron-tube amplifier stage. In common practice, the input signal should cause a single audio amplifier transistor or tube stage to operate along the linear portion of its characteristic curve. If the magnitude of the input signal is excessive, the amplifying device will be "overdriven," causing the output signal to exceed the linear portion of the curve. As a result, the output signal is distorted, since it is not a true reproduction of the input signal. This condition is commonly referred to as *amplitude distortion* (Fig. 23-3).

COUPLING

One of the most important considerations in the design of an amplifier circuit is the coupling that exists between the input signal and the first stage and the interstage coupling between the various stages. To be effective, the coupling network must provide the necessary impedance match between associated circuits; it must also perform the coupling function without undue loss of signal strength or power.

In typical amplifiers, interstage coupling is most commonly achieved by three methods: resistance-capacitance (*RC*) coupling, transformer coupling, and direct coupling. The resistance-capacitance and transformer methods of signal coupling are illustrated by the schematic diagram of the basic record-player amplifier shown in Fig. 23-4. In this circuit, capacitor C_1 and resistor R_1 form an impedance-matching network that serves to couple a high-impedance crystal or ceramic phono cartridge to the relatively low-impedance base-emitter input circuit of transistor Q_1.

Resistance-Capacitance Coupling. The signal output of the first-stage transistor Q_1 is coupled to the second-stage transistor Q_2 by means of the resistor-capacitor network consisting of resistor R_4, capacitor C_2, and resistor R_5. Resistor R_4 acts as the collector load resistor for transistor Q_1. Capacitor C_2 is the dc blocking capacitor that prevents the direct voltage applied to the collector of Q_1 from appearing at the input of Q_2. Resistor R_5 is a current-limiting resistor through which the necessary base-emitter bias is applied to Q_2.

As the collector current of transistor Q_1 varies in proportion to the variations of the input signal, a varying voltage drop is produced across transistor Q_1. As a result, capacitor C_2 charges and discharges through the circuit of R_5. The effect of this action produces a varying voltage drop across R_5, which is in turn applied to the input circuit of Q_2.

Transformer Coupling. Interstage coupling between transistor Q_2 and transistors Q_3 and Q_4 in the circuit of Fig. 23-4 is accomplished by transformer T_1. The primary winding of this

Fig. 23-3. Amplitude distortion.

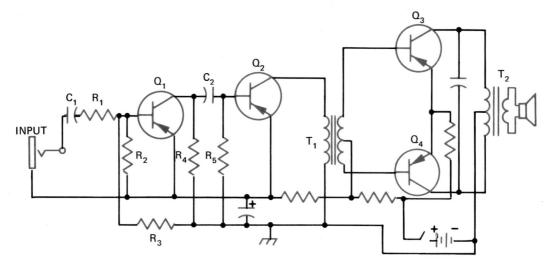

Fig. 23-4. Schematic diagram of a low-power audio amplifier utilizing resistance-capacitance and transformer interstage coupling.

transformer provides the necessary load impedance for the collector (output) circuit of Q_2. The center-tapped secondary winding of the transformer inductively couples the output signal of Q_2 to transistors Q_3 and Q_4; it also acts as the circuit through which the base-emitter bias is applied to these transistors.

Transformer coupling is also used between the output stage (transistors Q_3 and Q_4) and the loudspeaker voice coil. Here the output transformer T_2 acts to match the relatively high impedance of the transistor output circuit to the low impedance of the voice coil. Since this is a step-down transformer, its coupling action produces a low voltage across the secondary winding. As a result, the magnitude of the current in the secondary-winding circuit is sufficient to energize the voice coil to the level necessary for loudspeaker operation.

Direct Coupling. In the direct-coupled amplifier circuit, the output of one stage is connected directly to the input of the following stage (Fig. 23-5). This method of coupling is used primarily with stages that are designed to amplify signals down to a frequency of zero hertz, or direct current.

Because of its design, the direct-coupled amplifier system offers the advantage of wide frequency response. However, this method of coupling has two outstanding disadvantages. First, the number of amplifier stages that can be

Fig. 23-5. Direct-coupled audio amplifier stages.

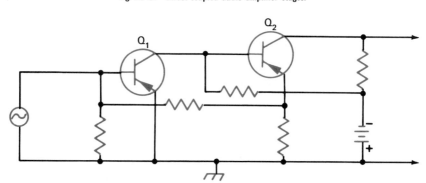

efficiently operated by means of direct coupling is limited. The second disadvantage results from the fact that in a direct-coupled transistor amplifier circuit, any variation of a bias current brought about by an increase of temperature is amplified by all the following stages. Thus, the circuit tends to be rather difficult to stabilize.

PUSH-PULL OPERATION

The output or final (power-amplification) stage of the amplifier circuit shown in Fig. 23-4 represents what is known as a *push-pull stage*. In push-pull operation, two transistors or two tubes are used in one stage. When an amplifier circuit contains only one amplifying device in the final stage, that stage is, by comparison, referred to as a *single* or *single-ended stage*.

The input signals to a push-pull stage are developed across the ends of the center-tapped secondary winding of an interstage transformer that couples the stage to the preceding or "driver" stage. As shown in Fig. 23-6, the signal voltages applied to the input circuits of the push-pull stage from the secondary winding of the interstage transformer T_1 are 180 degrees out of phase with each other. The transistor output currents of the stage are then also 180 degrees out of phase.

In this condition, the current from end 1 to the center tap of the primary winding of the output transformer T_2 is increasing, while the current from end 2 of the winding to the center tap is decreasing. Thus, the associated magnetic field generated about one-half of the primary winding is expanding while the magnetic field about the other half is collapsing. The effects of these magnetic fields are additive and the voltage induced across the secondary winding of the output transformer is the result of the combined action of both magnetic fields.

Advantages. One important advantage of the push-pull amplifier stage is that for a given signal input, it will deliver approximately twice the power output that is obtained with the use of a single-ended stage (which utilizes an equivalent transistor or electron tube). Another advantage of push-pull operation is that it tends to cancel the even-order (second, fourth, etc.) harmonics of the input signal that may be developed in the amplifier circuit. If these harmonics are not cancelled, they often produce a harmonic distortion in the output signal. Finally, push-pull operation results in a reduction of the

Fig. 23-6. Push-pull amplifier stage.

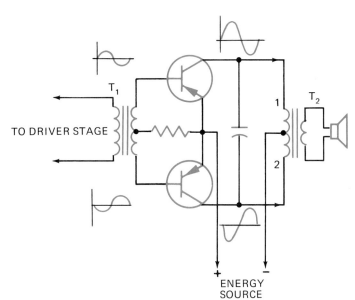

humming "noise" that may be produced in the output of an amplifier because of supply voltage fluctuations. While this feature of push-pull operation is not particularly significant in battery-powered circuits, it is highly desirable in circuits that are operated in conjunction with a rectifier circuit.

Defective Component. If one of the transistors or tubes of a push-pull stage becomes defective for some reason other than an interelectrode short circuit, the stage will continue to operate as a typical single-ended stage. However, as might be expected, the stage will then deliver only approximately one half the power that is produced by the push-pull stage under normal conditions. Also, distortion may increase to a high level.

RC AND TRANSFORMER COUPLING CHARACTERISTICS

The principal advantage of the resistance-capacitance interstage coupling system is that it provides a high gain over a relatively wide band of frequencies (Fig. 23-7). However, a study of this frequency-response curve will show that the response of the conventional RC network is significantly reduced at frequencies below approximately 100 hertz and at frequencies above approximately 10 kilohertz. Additional advantages of the RC coupling network are simplicity of construction, low cost, and the small size of the components that are required.

The principal advantage of transformer coupling is that it provides a highly efficient means of matching the output impedance of a given amplifier stage to the input impedance of the following stage. This condition makes it possible to achieve maximum power gain between stages.

The principal disadvantage of transformer coupling is the limited voltage-gain frequency response that is obtained, as compared with the wider band frequency response of the RC coupling network (Fig. 23-8). At the lower operating frequencies, the reactance of the primary winding of the coupling transformer is significantly reduced. As a result, there is also a reduction in the gain of the signal between the stages that are coupled by the transformer. At the higher operating frequencies, the frequency response of the coupling transformer is limited by the distributed capacitance of its windings. Since the reactance of this capacitance at high frequencies is sharply reduced, the effect of the capacitance causes a portion of the signal energy to be bypassed from the windings. Thus, the signal gain at high frequencies decreases rapidly.

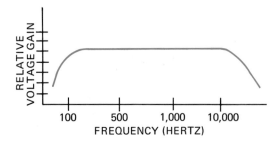

Fig. 23-7. Frequency-response curve of typical resistance-capacitance-coupled amplifier stages.

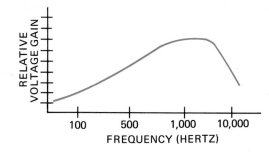

Fig. 23-8. Frequency-response curve of typical transformer-coupled amplifier stages.

THE PHASE-INVERTER CIRCUIT

The phase-inverter circuit is used in an amplifier when, for reasons of economy and improved frequency response, it is desirable to operate a push-pull output stage without the use of a center-tapped secondary-winding coupling transformer. In the phase-inverter circuit shown in Fig. 23-9, resistance-capacitance coupling is used between the driver and push-pull stages. Here, the output current of driver stage transistor Q_1 passes through the collector load resistor R_3 and the emitter load resistor R_2. Resistor R_1 is the base-emitter bias resistor.

Since transistor Q_1 conducts a varying direct current because of variations of the input signal,

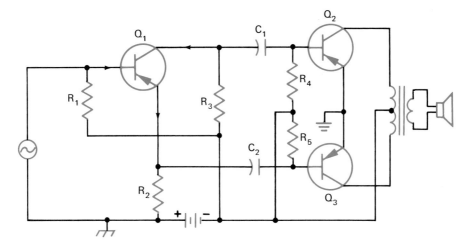

Fig. 23-9. Phase inverter circuit used to couple a driver stage to a push-pull amplifier stage.

the collector and emitter voltages both change with respect to ground (chassis) because of the load resistors R_2 and R_3. As the transistor conducts more heavily, the collector becomes less negative with respect to ground because of the decreased voltage drop across the transistor and the increased voltage drop across the load resistors. At the same time, the emitter becomes more negative with respect to ground. The result is two out-of-phase output voltages that are utilized to drive the push-pull power amplifier stage consisting of Q_2 and Q_3. The signals are coupled from the collector and emitter of Q_1 to the base of Q_2 and Q_3 by coupling capacitors C_1 and C_2. Resistors R_4 and R_5 set the bias current for the push-pull amplifier stage.

THE COMPLEMENTARY-SYMMETRY STAGE

A basic form of an audio-amplifier output stage using what is referred to as a *complementary-symmetry circuit* is shown in Fig. 23-10. In this circuit, an input signal has opposite effects upon the conductivity of transistors Q_1 (an n-p-n transistor) and Q_2 (a p-n-p transistor). As a result, the voltage developed across coupling capacitor C_2 is essentially an alternating voltage which, in turn, causes an alternating current to pass through the voice coil of the loudspeaker.

The complementary-symmetry stage provides the desirable operational characteristics of a conventional push-pull amplifier stage; in addition, it makes possible direct coupling to the loudspeaker. This, of course, eliminates the need for an output transformer.

Fig. 23-10. Basic complementary-symmetry output stage.

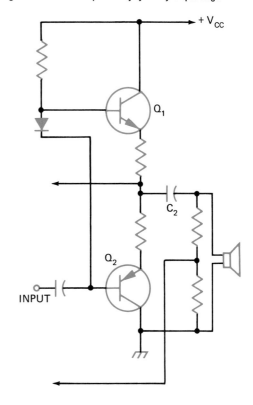

VOLUME CONTROL

The volume control of an amplifier circuit usually takes the form of a potentiometer-type variable resistor that is inserted into the input circuit of an amplifier stage (Fig. 23-11). The volume control then acts as a voltage divider by means of which a smaller or larger portion of the voltage developed across the fixed resistance portion of the potentiometer is applied to the stage.

As the shaft of the potentiometer shown in the accompanying circuit is adjusted so that the sliding arm moves upward, a greater portion of the voltage developed between points 1 and 2 of the control is applied to the input circuit of transistor Q_3. As a result, the volume or output sound level of the amplifier circuit is increased. Likewise, as the sliding arm is moved downward between points 1 and 2, the volume decreases.

Taper. Volume-control potentiometers used in audio-frequency amplifiers are most often logarithmic or nonlinear taper types. This feature makes it possible to vary the volume evenly throughout the total range of control. Thus, the control of the volume at the "low-volume end" is much less critical than would be the case if a linear taper control was used.

Pads. In addition to the volume control that is an integral part of an amplifier circuit, it is often desirable to provide a separate means of con-

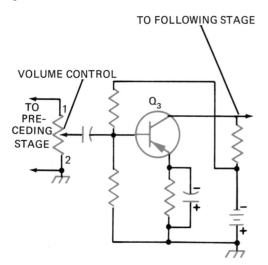

Fig. 23-11. Volume control circuit.

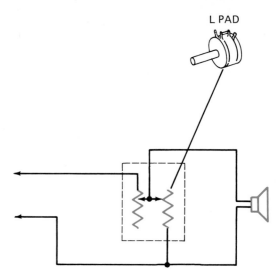

Fig. 23-12. L pad and associated circuit.

trolling the output of the associated loudspeaker or loudspeakers. For this purpose, variable resistor networks known as *T pads* and *L pads* are used (Fig. 23-12). These controls offer a very satisfactory method of controlling loudspeaker volume, since they do not alter the impedance of the total load that is connected to the amplifier.

THE TREBLE CONTROL

A treble or tone control is, in effect, a variable filter that permits adjustment of the frequency response of an audio amplifier. Hence, the control provides a means whereby the low (bass) frequencies or the high (treble) frequencies are emphasized or deemphasized. In this regard, the treble control circuit can be considered as a network that allows a degree of frequency distortion to be deliberately inserted into the output of an amplifier.

A simple type of treble-control circuit consists of a capacitor and a variable resistor (potentiometer) that are connected in series across the output of a given amplifier stage (Fig. 23-13). In this circuit, the capacitor C_2 has a capacitance value that causes it to present a low reactance to the higher frequency signals appearing in the output circuit. The capacitor tends to bypass these high-frequency signals to ground, thus preventing them from reaching the

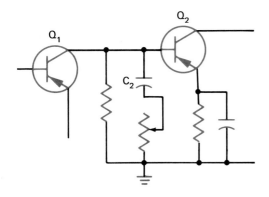

Fig. 23-13. Basic treble or tone control circuit.

loudspeaker. This action, of course, deemphasizes the high frequencies.

The extent to which the bypassing effect of the capacitor takes place is limited by the setting of the potentiometer. As the potentiometer sliding arm moves upward, a greater and greater resistance is inserted in series with the capacitor. As a result, the bypassing effect of the capacitor is decreased and more and more high frequencies appear in the loudspeaker output to increase the treble response.

THE PREAMPLIFIER

A preamplifier is a single-stage or multistage amplifier that is connected between a low-output-level transducer, such as a magnetic phono cartridge or a dynamic microphone, and the main or basic audio amplifier. The purpose of the preamplifier is to increase the signal source (transducer) output voltage to a level that will allow the basic amplifier to be operated (driven) to its full power output with the minimum amount of distortion. The circuit of the preamplifier may be contained within a separate unit or it may be incorporated within the design of the basic amplifier circuit.

The preamplifier may or may not be equipped with volume and treble controls. In high-fidelity preamplifiers, the amplifier stages are usually operated in conjunction with filter networks that provide for frequency compensation or equalization. These equalizing networks are designed to deemphasize (attenuate) high-frequency signals and to emphasize (boost) low-frequency signals. This is necessary since, as the result of the recording process, the voltage levels of the high-frequency signals obtained from a record are greater than those of low-frequency signals.

THE STEREO SYSTEM

The monoaural sound-reproducing system, which consists basically of a single-channel (one input and one output) amplifier and its associated loudspeaker, produces sound that does not

Fig. 23-14. Typical placement of microphones for stereo recording.

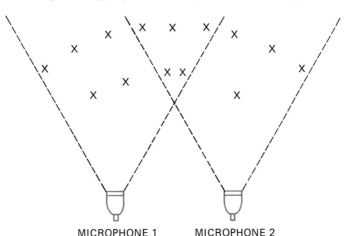

have a directional characteristic. As a result, the sound does not have the true quality that exists when the sound produced by a musical group, for example, is heard "live."

To overcome this loss of the directional characteristic, the stereo (stereophonic) system of sound recording and reproduction has been developed. In its simplest form, the stereo recording process involves the use of at least two microphones placed some distance apart and in front of the origin of the sound that is to be recorded (Fig. 23-14). The resulting "two-part" sound signal is then recorded upon either a stereo record or a stereo magnetic tape.

Stereo Records and Cartridges. The grooves on the typical stereo record consist of "walls" that are cut at right angles to each other and at an angle of 45 degrees with respect to the surface of the record (Fig. 23-15). Each of the walls is then varied in accordance with the sound "pattern" that has been recorded upon the record. Sounds from one direction (one channel) are recorded by variations upon one wall of the groove. Sounds from the other direction (the second channel) are recorded by variations on the remaining groove wall.

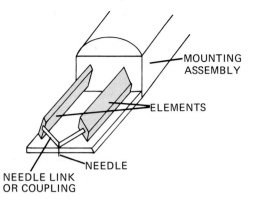

Fig. 23-16. Principal constructional features of a ceramic-type stereo phono cartridge.

The stereo phono cartridge, which is often of the ceramic or the magnetic type, consists of one needle and two separate elements that are coupled to the needle. The principal constructional features of such a cartridge are shown in Fig. 23-16.

Stereo Magnetic Tape. Stereo recording is accomplished by passing a multitrack magnetic tape over a recording head consisting of two separate coils, one for each sound channel to be recorded (Fig. 23-17). When recording upon four-track tape, tracks 1 and 3 of the tape are first magnetized. The tape is then reversed (reels reversed) and the recording is done upon tracks 2 and 4. On playback, a similar procedure is employed except, of course, that the playback head is now energized. Many tape recorder/reproducer units use an eight-track tape. This allows the recording and playback of four separate stereo "programs." The basic principles of magnetic tape recording were previously discussed in Unit 14.

The Stereo Amplifier. To reproduce the stereo sound, an audio amplifier consisting of two separate amplifier circuits or channels is used (Fig. 23-18). To retain the stereo effect, a loudspeaker is connected to the output of each channel. These loudspeakers should have similar operational characteristics if the best quality of sound reproduction is to be obtained.

The balance control of the stereo amplifier adjusts the output level of the channels so that they are approximately equal. As a result, the

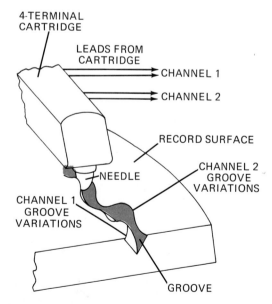

Fig. 23-15. Example of the groove pattern on a stereo record produced by lateral recording.

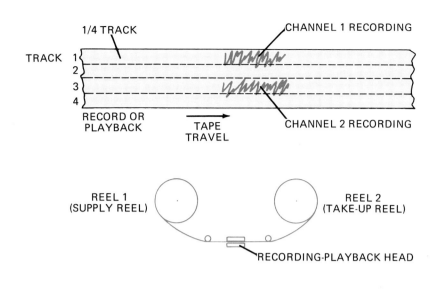

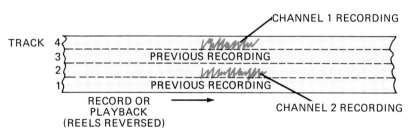

Fig. 23-17. Stereo recording upon four-track magnetic tape.

loudspeakers form a truer reproduction of the sound that was originally recorded. If the balance control is not properly adjusted, the output of one of the channels will dominate and the stereo effect will be significantly reduced.

LOUDSPEAKER CONNECTIONS

Two or more loudspeakers may be connected to the output terminals of an amplifier, either in series or in parallel. The parallel connection is usually preferred in a permanent installation, since it will prevent an open-circuit condition in the voice coil of one loudspeaker from affecting the operation of the remaining loudspeakers.

Fig. 23-18. Simple diagram of a stereo sound system.

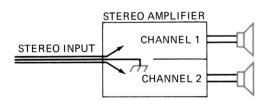

239

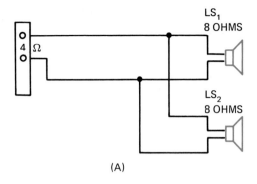

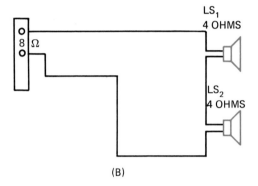

Fig. 23-19. Loudspeaker connections: (A) parallel, (B) series.

Parallel Connection. To minimize power loss and distortion, the connection of loudspeakers in parallel must provide a total load impedance that equals or nearly equals the output impedance of the amplifier. For example, two 8-ohm loudspeakers operated in parallel would be connected to the 4-ohm output terminals of an amplifier (Fig. 23-19A).

When loudspeakers having equivalent impedances are used, an equal amount of power will be delivered to each loudspeaker from the amplifier. However, when loudspeakers having unequal impedances are connected in parallel, the loudspeaker with the lowest impedance will have the greatest amount of current passing through its voice coil. Hence, the amplifier will deliver more power to this loudspeaker than it delivers to each of the remaining loudspeakers. If it is assumed that all the loudspeakers have identical power ratings, then the loudspeaker having the lowest impedance will produce the greatest volume of sound.

Series Connection. The proper impedance match between the output terminals of an amplifier and the loudspeaker load must also exist when loudspeakers are operated in series. Thus, two loudspeakers, each with an impedance of 4 ohms, would be connected to the 8-ohm terminals of an amplifier (Fig. 23-19B). When loudspeakers having unequal impedances are connected in series, the highest voltage will be developed across that loudspeaker which has the highest impedance. As a result, the greatest amount of power will be delivered to this loudspeaker by the amplifier.

Semiparallel Connection. When an amplifier is equipped with multiple output-impedance terminals, loudspeakers having unequal impedances may be connected to it by means of a semiparallel connection (Fig. 23-20). This method of connection often simplifies the installation of loudspeakers, since it eliminates the necessity of computing the total impedance of the loudspeaker load.

Phasing of Loudspeakers. The phasing of loudspeakers means that they are connected so that at any instant of time a voltage of a given polarity applied across their voice coils will produce cone motions in the same direction. This "in phase" condition is important, since sound cancellation will often occur if adjacent loudspeakers or closely spaced loudspeakers are not in phase with each other.

The phasing of loudspeakers can be accomplished in the following manner: Temporarily disconnect the loudspeakers from the amplifier output; connect a dry cell to one of them and notice the direction in which its cone moves.

Fig. 23-20. Connection of loudspeakers having different impedances.

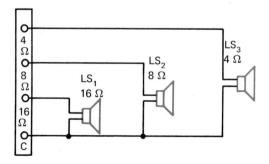

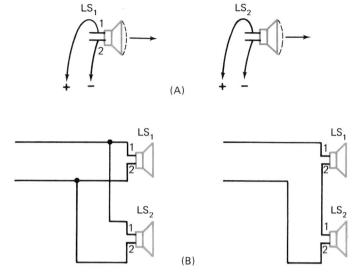

Fig. 23-21. Phasing of loudspeakers.

Identify the terminal of the loudspeaker that is connected to the positive terminal of the cell as terminal 1 and identify the other terminal as terminal 2. Next, connect the cell to each of the remaining loudspeakers, one at a time, in a polarity that will cause their cones to move in the same direction. Finally, identify the terminals of each of these loudspeakers in the manner previously described for the first loudspeaker (Fig. 23-21A).

If the loudspeakers are to be operated in parallel, the proper phasing condition will result when all the #1 terminals are connected together and all the #2 terminals are connected together. When the loudspeakers are operated in series, they are properly phased if "opposite" terminals are connected together (Fig. 23-21B).

CONSTANT-VOLTAGE SPEAKER SYSTEMS

In addition to the various output impedance terminals (4 ohms, 8 ohms, etc.), most public address (PA) audio amplifiers are also equipped with what are generally referred to as constant-voltage outputs. The two most commonly used constant-voltage outputs apply a voltage of either 25 volts or 70.7 volts across the transmission line to which the loudspeakers are connected.

The use of a constant-voltage speaker system makes it possible to provide conveniently for the correct impedance matching of loudspeakers to the amplifier, regardless of the number of loudspeakers used. However, it is important to note that the number of loudspeakers used in any system is limited by the output power rating of the associated amplifier.

When loudspeakers are connected to either the 25- or the 70.7-volt terminals of an amplifier, each loudspeaker is operated in conjunction with a line-to-voice coil transformer (Fig. 23-22).

Fig. 23-22. Constant-voltage loudspeaker system.

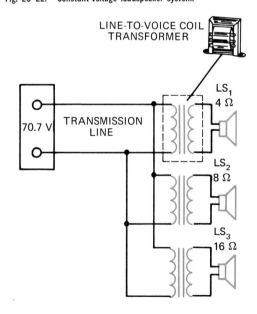

241

Since most of these transformers are of the universal type (have various impedance taps), loudspeakers having different impedances can be conveniently connected to the same audio transmission line.

Some line-to-voice coil transformers are also equipped with different output-level terminals. This arrangement makes it possible to vary the volume of any individual loudspeaker by simply connecting its voice coil to those terminals of the transformer that provide the desired volume.

THE CROSSOVER NETWORK

The crossover or frequency-divider network is a filter circuit that is often used in a high-fidelity system to channel the output of an amplifier into loudspeakers with different frequency-response characteristics. In a two-loudspeaker system, that loudspeaker which responds most favorably to low-frequency signals is called the *woofer* and the high-frequency loudspeaker is called the *tweeter*. In most cases, the use of separate loudspeakers provides a more desirable sound response than can be obtained by the use of a single, full-range loudspeaker that is designed to reproduce signals of all frequencies within the audio range.

The schematic diagram of a basic crossover network is shown in Fig. 23-23. Here, the woofer loudspeaker is coupled to the secondary winding of the amplifier-output transformer through inductor L, while the tweeter loudspeaker is coupled to the winding by capacitor C.

During the operation of this network, low-frequency signals of approximately 20 to 400 hertz are channeled into the woofer, since at these frequencies the reactance of L is comparatively low, while the reactance of C is high. As the frequency of the signal increases above approximately 400 hertz, the reverse condition occurs. Now, the reactance of L increases significantly, while the reactance of C decreases. As a result, signals having a frequency above 400 hertz tend to be channeled into the tweeter.

The frequency at which the major portion of the signal passes from one loudspeaker of the network to the other is referred to as the crossover frequency. At the crossover frequency which, in the typical network, is usually between 400 and 1,200 hertz, $X_L = X_C$. Thus, at the crossover frequency, both loudspeakers receive the same amount of signal energy.

Problem. Compute the crossover frequency of the crossover network shown in Fig. 23-23.

Solution. At the crossover frequency, $X_L = X_C$. Therefore, the crossover frequency F will be equal to the resonant frequency of a circuit consisting of L and C:

$$F = \frac{1}{2\pi \sqrt{LC}}$$

$$= \frac{1}{6.28 \sqrt{0.004 \times 0.00003}}$$

$$= \frac{1}{6.28 \sqrt{0.00000012}}$$

$$= \frac{1}{6.28 \times 0.000346}$$

$$= \frac{1}{0.00217} = 462 \text{ hertz}$$

THE COAXIAL LOUDSPEAKER

The base and treble (woofer and tweeter) loudspeakers used in a crossover or other high-fidelity sound system are often combined within a single unit that is referred to as a *coaxial loudspeaker* (Fig. 23-24). Such a loudspeaker employs a single magnet but is equipped with two separate cones, each of which is designed to operate primarily at a given band of audio frequencies.

Fig. 23-23. Loudspeaker crossover network.

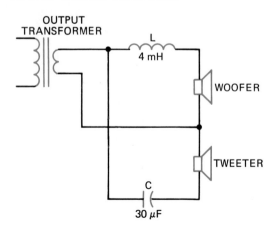

Fig. 23-24. Coaxial loudspeaker with a frequency response of from 30 hertz to beyond audibility. (*Minneapolis Speaker Co.*)

For Review and Discussion

1. Define distortion and fidelity as related to the operation of an amplifier.

2. Explain the compatibility requirement of a high-fidelity amplifier system.

3. Define microphonics as related to the operation of an electron tube.

4. State four causes of amplifier distortion.

5. Draw a schematic diagram illustrating a basic example of resistance-capacitance (*RC*) signal coupling.

6. At what signal frequencies is direct coupling most effective?

7. Give two desirable and two undesirable characteristics of direct coupling.

8. Describe a push-pull amplifier stage.

9. State three desirable operational characteristics of the push-pull amplifier stage.

10. What is the principal advantage of resistance-capacitance coupling as compared with other coupling systems?

11. State the principal advantage that is gained by the use of transformer coupling. Also, state the principal disadvantage of this coupling system.

12. For what purpose is a phase-inverter circuit used in an amplifier system?

13. What component, often found in the output stage of an audio amplifier, is eliminated by the use of a complementary-symmetry output stage?

14. For what purpose are pads used in an amplifier-loudspeaker system?

15. Define bass and treble frequencies.

16. Explain the operation of a basic treble control.

17. What is a preamplifier and for what purpose is it used?

18. Give a basic explanation of the stereo sound system.

19. What is meant by the phasing of loudspeakers and why is this done?

20. State the desirable feature of the constant-voltage loudspeaker system.

21. Describe a crossover network, and give the reason for using such a network.

22. Describe a coaxial loudspeaker.

UNIT 24. Oscillators

The term "oscillate" means to "swing" forward and backward or to vibrate in a manner similar to that which occurs during the operation of a pendulum. In electronics terminology, an oscillator is a circuit that is capable of producing oscillations in the form of an alternating voltage at a frequency determined by the specific components used. The oscillator can then also be thought of as a type of generator that transforms direct voltage into alternating voltage, since the components of an oscillator circuit are operated by the application of a direct voltage.

The outstanding advantage of the electronic oscillator is its ability to produce alternating voltages having frequencies far above those that can be obtained by using a rotating (electromechanical) type of generator. In addition, the oscillator circuit is more compact than the rotating generator circuit. Because of these advantages, oscillator circuits are used in many applications where a controlled alternating voltage at a fixed or variable frequency is desired. The most common of these applications are the generation of the carrier waves produced by transmitter circuits and necessary for the radio (electromagnetic) transmission of energy, the generation of the mixing or heterodyning signal that is used to produce the intermediate frequency in various kinds of receiving circuits, and the generation of the signals that are produced by certain units of test equipment.

The basic theory relating to the oscillatory response of a tuned (tank) circuit has been previously discussed in Unit 18. It is recommended that this material be thoroughly reviewed before proceeding with the study of the sinusoidal oscillator circuits that are presented in the following paragraphs.

LC OSCILLATOR CIRCUIT REQUIREMENTS

Before either a typical transistor or an electron-tube circuit will function as an *LC* oscillator, it must satisfy the three fundamental requirements illustrated in Fig. 24-1. These are:

Amplification. The transistor or the tube in the circuit must be capable of amplifying a signal that is applied to its input electrode.

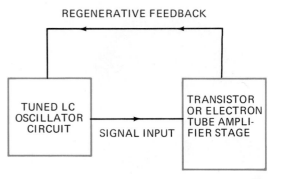

Fig. 24-1. Fundamental requirements for an *LC* oscillator circuit.

Tuned Circuit. The circuit must be equipped with a tuned *LC* (tank) circuit that is designed to oscillate at or to be resonant to the desired frequency. The tank circuit also forms a part of the input circuit to the transistor or the electron tube.

Feedback. The circuit must contain a coupling system that allows a portion of the amplified output signal of the transistor or the tube to be fed back to the input circuit. This signal must be regenerative or in phase with the input signal and of sufficient magnitude to replace any energy losses that occur in an *LC* circuit as a result of its resistance. If the feedback signal is out of phase with the input signal, it will prevent oscillation within the tuned *LC* circuit by a process that is known as *degeneration*.

THE HARTLEY OSCILLATOR

A basic Hartley oscillator circuit is shown in Fig. 24-2. In this circuit, the first surge of current through coil L_2 in the collector circuit of the transistor generates an expanding magnetic field about the coil. This magnetic field induces a voltage across coil L_1, thus shock-exciting the tuned circuit consisting of L_1 and capacitor C_1 into oscillation. A portion of the induced voltage (the feedback voltage) is coupled to the base of the transistor through C_2. The polarity of this voltage is such that it aids the emitter-base transistor bias, thereby causing the transistor to be driven rapidly into the maximum-conduction or

OSCILLATOR COILS

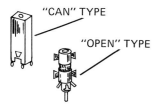

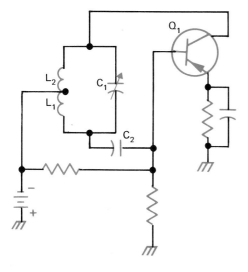

Fig. 24-2. Hartley oscillator circuit.

saturation state. In this condition, the collector current is constant and the induced voltage across L_1 drops to zero.

With the additional (aiding) emitter-base bias removed, the transistor begins to conduct less and the magnetic field about L_2 begins to collapse. This causes a voltage to be induced across L_1 that is opposite in polarity to the voltage that was originally induced across the coil. As this voltage is coupled to the base of the transistor, the transistor continues to conduct less until it reaches the nonconducting or cutoff state. Hence, the magnetic field about L_2 ceases to exist, and the circuit returns to its original state to complete a cycle of operation. As the cycle is repeated, the transistor alternately operates between saturation and cutoff at a frequency determined primarily by L_1, L_2, and C_1. As a result, a sinusoidal voltage (and current) is present in the tuned (tank) circuit.

THE COLPITTS OSCILLATOR

A basic transistor Colpitts oscillator circuit is shown in Fig. 24-3. In this circuit, regenerative feedback is obtained from the voltage that is developed across capacitor C_3 in the tank circuit. The voltage is then applied to the emitter that serves as the input to the common-base p-n-p transistor configuration.

SIGNAL COUPLING

The oscillating signal within the tank circuit of a Hartley or Colpitts oscillator can be coupled to an associated circuit by either capacitive or transformer coupling (Fig. 24-4). However, in many practical circuits, these methods of coupling produce a "loading" condition that causes the tank circuit to deliver an excessive amount of energy to the circuit to which it is coupled. This condition, in turn, often tends to decrease the frequency stability of an oscillator circuit.

THE BUFFER AMPLIFIER

To minimize the undesirable effects of "loading," an oscillator circuit is often coupled to a buffer amplifier stage that precedes the first power amplifier stage in a transmitter circuit. In addition to amplifying the relatively weak signals from the oscillator, the buffer amplifier effectively isolates the oscillator circuit from the remaining portion of the transmitter circuit.

Fig. 24-3. Colpitts oscillator circuit.

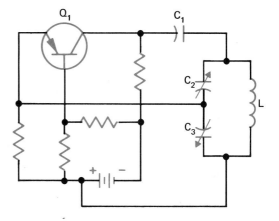

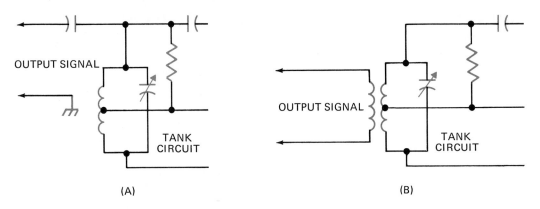

Fig. 24-4. Oscillator signal coupling: (A) capacitive coupling, (B) transformer coupling.

CRYSTAL-CONTROLLED OSCILLATOR

Certain crystalline materials (crystals) such as quartz, tourmaline, and Rochelle salts generate a voltage when they are subjected to mechanical vibration or stress (Fig. 24-5). This action is known as the *piezoelectric effect*. These crystals also possess the opposite property. That is, they will be excited into physical vibration when a voltage is applied to their surfaces (Fig. 24-6). The frequency of these vibrations is equal to the frequency of the applied voltage. However, the vibrations will be greatest at one particular frequency of the applied voltage. This frequency is referred to as the *resonant frequency* of the crystal.

In a crystal-controlled oscillator, the feedback voltage is utilized to excite a crystal into vibration. As a result, the crystal generates a voltage at that frequency to which it is resonant. This voltage, in turn, controls the frequency at which the oscillator circuit operates. Since the frequency of the voltage that is developed by a crystal under this condition is practically constant, a high degree of frequency stability is obtained. For this reason, crystal-controlled

Fig. 24-5. Quartz crystals: (A) natural crystal form, (B) crystal "blanks" cut from the natural form. (*International Crystal Mfg. Co., Inc.*)

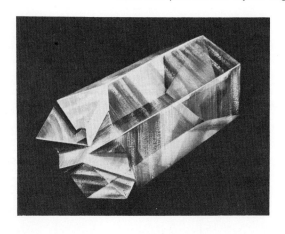

(A)

(B)

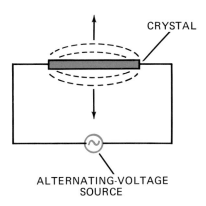

Fig. 24-6. Physical vibration of a piezoelectric crystal resulting from the application of an alternating voltage to the crystal.

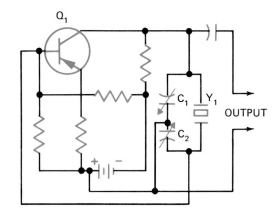

Fig. 24-7. Pierce oscillator circuit.

oscillators are very commonly used where precise control of the output frequency is required, such as in radio and television broadcasting, radar, amateur radio, and telemetry, and in citizen's band communications units.

A popular form of transistor crystal-controlled oscillator is the Pierce type, a modification of the basic Colpitts oscillator circuit (Fig. 24-7). The resistors in this circuit provide the necessary bias and operating-point stabilizing conditions for the common-emitter configuration.

After the first surge of collector current, the crystal is activated and begins to generate a voltage. Capacitors C_1 and C_2 act as a voltage-divider network, and the crystal voltage that is developed across C_2 is applied (fed back) to the base of the transistor. This action maintains oscillations within the crystal at a frequency that is determined by the resonant frequency of the crystal and the total series capacitance of C_1 and C_2.

THE RC OSCILLATOR

A schematic diagram of the RC or phase-shift oscillator is shown in Fig. 24-8. As the circuit begins to operate, the voltage across the transistor (collector-to-emitter) begins to decrease

Fig. 24-8. RC oscillator circuit.

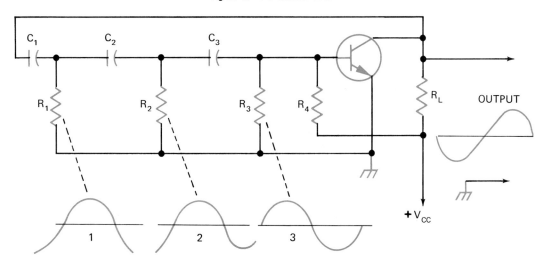

as a negative-going voltage. This voltage is applied to or fed back across C_1 and R_1. Because of the phase-shifting effect of the C_1-R_1 combination, the voltage developed across R_1 leads the voltage across the transistor by 60 degrees, as indicated by waveform 1 associated with the diagram. This voltage is then applied across the C_2-R_2 combination, where the phase-shifting effect produces a second 60-degree phase shift, causing the voltage developed across R_2 to be 120 degrees out of phase with the voltage across the transistor (refer to waveform 2 associated with the diagram). A third phase shift of the voltage produced by the combination of C_3 and R_3 results in a voltage across R_3 that is 180 degrees out of phase with the voltage across the transistor (refer to waveform 3 associated with the diagram). This voltage is of a polarity that aids the base-emitter bias of the transistor, and the transistor is driven into saturation.

With the transistor at saturation, the voltage applied to C_3 begins to decrease, causing a corresponding decrease of current through R_3 and the voltage developed across this resistor. In this condition, the base-emitter bias of the transistor also decreases, causing the collector current to decrease and the voltage across the transistor to increase or become positive-going.

With a positive-going voltage across the transistor, the RC network causes a 180-degree phase shift of this voltage so that a negative-going voltage is fed back to the base of the transistor until the transistor is cut off. Now, the voltage across the transistor is maximum in a positive-going direction. In the transistor cutoff condition, the collector voltage stabilizes, the discharge current of C_3 decreases, and the voltage drop across R_3 also decreases. As a result, the forward bias applied to the base-emitter junction increases and the transistor begins to conduct once again. During this time, the voltage across the transistor has gone from maximum negative-going to maximum positive-going to complete one cycle of oscillation. As the circuit continues to operate, a continuous sinusoidal voltage is present across the transistor.

NONSINUSOIDAL OSCILLATORS

The oscillators previously discussed produce a sinusoidal waveform oscillating action. However, not all oscillations produced by electronic circuits are sinusoidal in nature. Instead, the output-voltage waveforms may be sawtooth, square, or pulsating (peak) shaped.

Nonsinusoidal oscillators are very often referred to as *relaxation oscillators*. As a general rule, these circuits operate by means of a regenerative feedback circuit that is used in conjunction with an RC (resistance-capacitance) or an RL (resistance-inductance) network.

The most common relaxation type of oscillator is the *multivibrator*. Multivibrators are used in a variety of applications for the purpose of timing and current control in circuits that include television receivers, transmitters, radar systems, and computers.

NEON LAMP OSCILLATOR

The neon lamp oscillator, although impractical for many purposes other than experimentation, does illustrate the basic operation of a relaxation oscillator and its application for purposes of time control. In this circuit, the direct supply voltage is applied to a series RC circuit and the neon lamp (load) is connected in parallel with the capacitor (Fig. 24-9A).

Because of the comparatively large amount of resistance in the circuit, the capacitor charges rather slowly. When the voltage across the capacitor (and the lamp) becomes equal to the starting or firing voltage of the lamp (approximately 60 volts), the capacitor discharges rapidly through the lamp, causing it to flash. With the capacitor discharged, the entire charge-discharge cycle is repeated.

The period of time required for the capacitor to become charged to 63.2 percent of the supply voltage is determined by the time constant t of the RC circuit. If the values of R and C are as indicated on the diagram (Fig. 24-9A),

$$t = RC$$
$$= 1{,}000{,}000 \times 0.000001 = 1 \text{ second}$$

Thus, after one second of time, the voltage developed across the capacitor is equal to

$$E_c = 100 \times 0.632 = 63.2 \text{ volts}$$

Since this magnitude of voltage is sufficient to cause the neon lamp to "fire," the lamp will flash at intervals of one second. The time between flashes can be adjusted by changing the

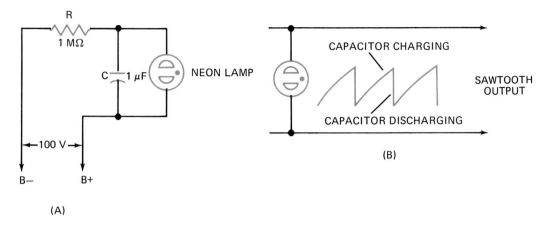

Fig. 24-9. Neon lamp relaxation oscillator: (A) schematic diagram, (B) sawtooth output waveform.

values of R and C so that a different capacitor-charging time-constant period is obtained.

The response of the RC circuit during the time that the capacitor is charging results in a gradual "build-up" of the voltage that is applied to the neon lamp. However, the voltage across the capacitor decreases very rapidly as the capacitor discharges through the lamp at the end of each time-constant period. These conditions, therefore, produce an oscillator output voltage that has a sawtooth waveform (Fig. 24-9B).

THE MULTIVIBRATOR

The multivibrator is essentially a relaxation type of oscillator circuit that is most commonly used to produce square or rectangular voltage waveforms. These output voltages are very often used for timing and signal control in several different applications, including television, radar, and computer circuits.

Multivibrator oscillator circuits are either free-running or driven (triggered). In the free-running oscillator, the oscillating action begins when the operating voltage is applied to the circuit and continues for as long as the circuit is in operation, at a frequency determined by the values of the circuit components. In the driven multivibrator, the performance of each cycle is triggered by a signal input from an external source. As a result, this oscillator operates at a frequency that is determined by the frequency of the external control signal.

The Free-running (astable) Multivibrator. The free-running multivibrator is basically a nonsinusoidal, two-stage oscillator in which one stage conducts while the other stage is at the cutoff (nonconducting) point. This operation continues until a circuit stage is reached that causes the action to be reversed. The stage that was at cutoff then begins to conduct, while the previously conducting stage is at cutoff.

The schematic diagram of a free-running multivibrator circuit is shown in Fig. 24-10. Here, the output of each common-emitter amplifier stage is resistance-capacitance coupled to the input of the other stage, thus providing the necessary in-phase regenerative feedback required for oscillation.

In this circuit, slight variations in the values of the components will cause one transistor to conduct before the other or to conduct a greater amount of current than the other when the circuit is placed into operation.

If it is assumed that the application of the supply voltage V_{cc} causes transistor Q_1 to conduct more current than is conducted by transistor Q_2, the operation of the circuit can be described as follows:

As the collector current of Q_1 increases because of its increasing conductivity, the voltage across R_2 increases, causing a proportional decrease of the Q_1 collector voltage. As a result, the collector of Q_1 becomes less negative with respect to ground (chassis). Since the collector of Q_1 is coupled to the base of Q_2 through

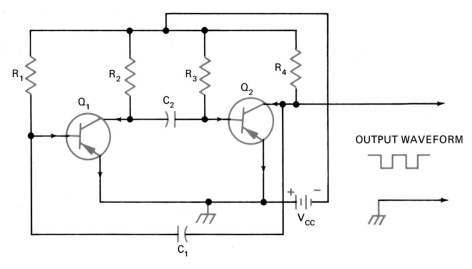

Fig. 24-10. Free-running multivibrator oscillator.

capacitor C_2, a reverse-bias voltage is applied to the base-emitter circuit of Q_2, causing Q_2 to be driven to the cutoff point. Capacitor C_1 then charges through the path provided by resistor R_4 and the base-emitter circuit of Q_1. Since the charging current is in a direction that aids the forward bias of the Q_1 input circuit, the transistor is driven to saturation (maximum conduction).

As C_2 becomes fully discharged, the voltage across it decreases, and therefore the base of Q_2 becomes less positive. Now, a forward-bias current begins to pass from the battery through R_3, and through the base-emitter junction of Q_2, and the transistor begins to conduct. In this condition, the voltage developed across Q_2 is applied to the base-emitter circuit of Q_1 in a polarity that applies a reverse-bias to the circuit, causing Q_1 to be driven to the cutoff point. Now, capacitor C_2 recharges through R_2 and the base-emitter circuit of Q_2, thereby increasing the forward bias at the Q_2 input. This causes transistor Q_2 to be driven to saturation, thus completing one cycle of oscillation. The oscillations are then repeated at a frequency that is primarily determined by the values of C_1, C_2, R_1, and R_3.

The Driven (monostable) Multivibrator. The schematic diagram of a basic monostable multivibrator is shown in Fig. 24-11. This is also commonly referred to as a *one-shot multivibrator*. In the absence of a control-triggering pulse at the input circuit, the base-emitter junction of transistor Q_1 is forward-biased and the transistor conducts. At the same time, the base-emitter circuit of transistor Q_2 is not forward-biased, causing this transistor to be at cutoff.

When a positive trigger pulse is applied to the base-emitter circuit of transistor Q_1, the forward bias established by resistors R_1 and R_2 is overcome and Q_1 is cut off. As a result, the collector voltage of transistor Q_1 rapidly increases in a negative direction. This negative-going voltage is coupled by capacitor C_1 to the base of Q_2 and turns Q_2 on by forward-biasing its emitter-base junction. In the absence of a trigger pulse, capacitor C_1 will discharge through R_6, and transistor Q_2 will return to its nonconducting state.

The output of the monostable multivibrator is taken from the collector circuit of transistor Q_2 (Fig. 24-11B). The waveform at this portion of the circuit then represents the response of Q_2 during any given complete cycle. The output voltage waveform is essentially square, since each input trigger pulse causes transistor Q_2 to temporarily conduct during the time that capacitor C_1 is discharging.

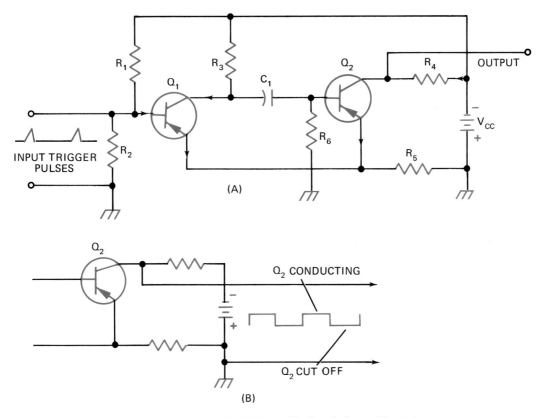

Fig. 24-11. Driven (monostable) multivibrator: (A) schematic diagram, (B) output circuit and waveform.

THE BISTABLE (FLIP-FLOP) MULTIVIBRATOR

The schematic diagram of a bistable multivibrator is shown in Fig. 24-12. This multivibrator is similar to the free-running multivibrator; however, this circuit has two stable states. A trigger pulse causes the circuit to switch from one state to the other. A second trigger pulse causes the circuit to switch back to its original state. Hence, the name *flip-flop*.

Assume that in the circuit of Fig. 24-12 transistor Q_2 conducts more heavily than transistor Q_1. This will cause the voltage across Q_2 (collector-to-emitter) to be less than the voltage across Q_1. As a result, the voltage across the series circuit consisting of R_3 and the base-

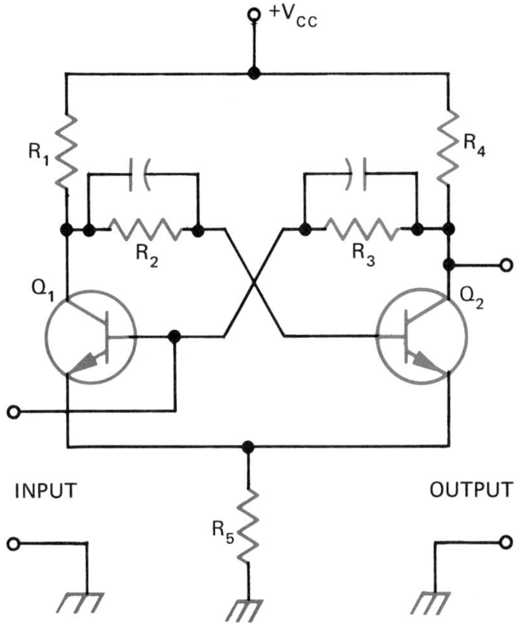

Fig. 24-12. The bistable (flip-flop) multivibrator.

emitter junction of Q_1 also decreases, causing the forward bias of Q_1 to decrease. The voltage across Q_1 now increases, and the voltage across the series circuit consisting of R_3 and the base-emitter junction of Q_2 also increases. This increases the forward-bias of Q_2, which is rapidly driven to saturation. In this condition, the voltage developed across R_5 is greater than the voltage across the base-emitter junction of Q_1. Since this results in the application of a negative base-emitter bias to Q_1, the transistor is cut off and the circuit stabilizes, with Q_2 conducting and Q_1 cut off.

A positive trigger pulse applied to the base of Q_1 will overcome the reverse bias of this transistor and cause it to begin to conduct. The voltage across Q_1 now decreases, thus decreasing the voltage applied across R_2 and the base-emitter junction of Q_2. This decreases the forward bias of Q_2, causing it to conduct less and causing the voltage across the transistor to increase. As a result, the voltage applied across R_3 and the base-emitter junction of Q_1 also increases, thus increasing the forward-bias of Q_1 and causing this transistor to be driven to saturation. With the voltage across Q_1 at a minimum, the voltage applied across R_2 and the base-emitter junction of Q_2 decreases, causing the voltage across Q_2 to become less than the voltage across R_5. This reverse-biases Q_2, and drives it to cutoff. The circuit remains in this state, Q_1 conducting and Q_2 cut off, until a negative trigger pulse is applied to the input.

Since the input pulses in this circuit are applied to the base of Q_1, a positive input pulse when this transistor is cut off will cause the circuit to switch. And when Q_1 is at saturation, a negative input pulse will also cause the circuit to switch. Therefore, alternate positive and negative input pulses will cause the voltage across the output to vary from minimum to maximum in what is essentially a square waveform.

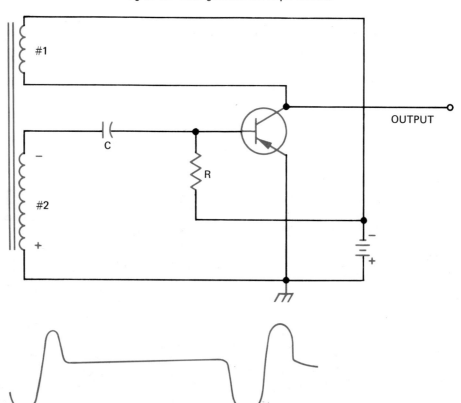

Fig. 24-13. Blocking oscillator and output waveform.

The Blocking Oscillator. The blocking oscillator, which may be the free-running or triggered type, is a form of nonsinusoidal oscillator that operates for a short period and then is cut off or "blocked" for a much longer period. The time period between oscillation pulses is determined by the time constant of capacitor C and resistor R in the basic circuit of the oscillator shown in Fig. 24-13.

When the circuit is energized, the collector current increases rapidly because of the forward base-emitter forward-bias voltage applied to the transistor. As a result, a voltage of the polarity indicated on the circuit diagram is induced across winding 2 of the transformer by the expanding magnetic field about winding 1. This voltage charges capacitor C through the low forward resistance of the base-emitter junction, thereby aiding the forward-bias condition and driving the transistor to saturation.

At saturation, the collector current is, of course, zero, and therefore there is no longer a voltage induced across winding 2 of the transformer. Now, the capacitor begins to discharge through winding 2 and resistor R, making the upper end of the resistor positive with respect to the lower end. This reverse-biases the base-emitter junction, causing the base current and the collector current to decrease toward zero. The collapsing magnetic field about winding 1 of the transformer now induces a voltage across winding 2 in a polarity that aids the reverse-bias condition, and the transistor is driven to cutoff.

The transistor is held at cutoff until the capacitor discharge current through R decreases to a magnitude at which the voltage drop across the resistor is no longer able to reverse-bias the base-emitter junction to cutoff. At this time, the transistor begins to conduct once again, thus beginning another oscillation cycle.

THE SCHMITT TRIGGER

The Schmitt trigger circuit is similar to the basic monostable multivibrator in appearance but is bistable in operation. It is very commonly used as a switching circuit to provide the function of waveshaping or changing sinusoidal and other waveforms into square waveforms for the operation of certain digital computer circuitry. A basic Schmitt trigger circuit is shown in Fig. 24-14.

Fig. 24-14. The Schmitt trigger.

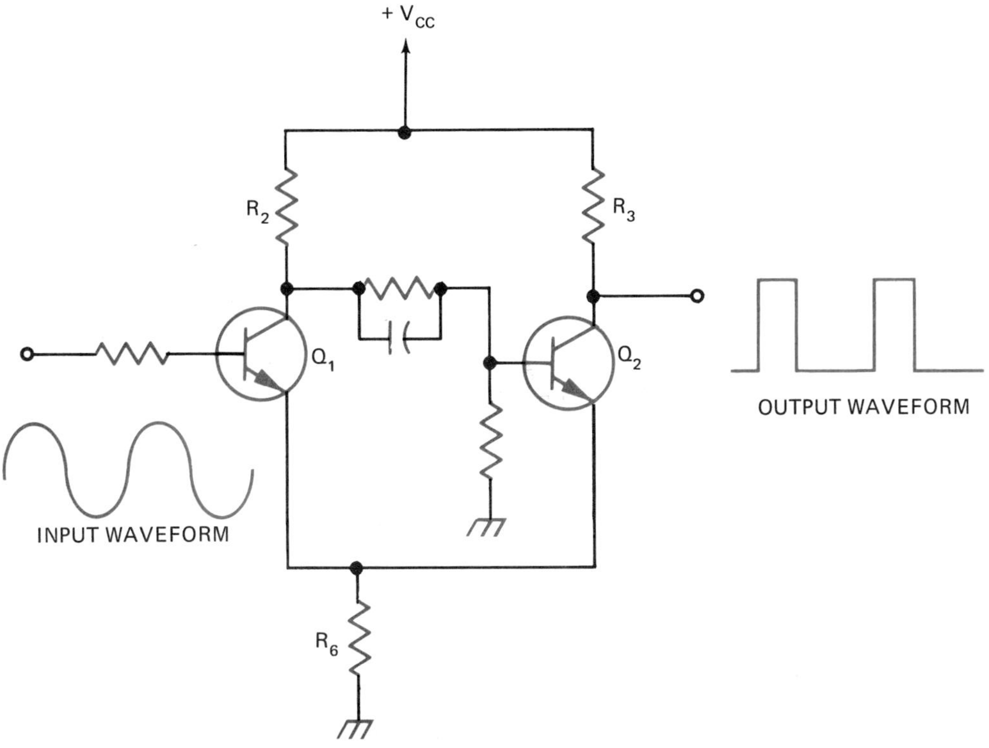

Assume that in the circuit of Fig. 24-14 there is no signal input. In this condition, transistor Q_2 is on and Q_1 is off because there is no effective base-bias current path for transistor Q_1. Transistor Q_1 remains cut off because of the polarity of the voltage developed across R_6.

To switch the circuit, a sinusoidal voltage greater in magnitude than the voltage across R_6 must be applied to the input. As a result, Q_1 begins to conduct. At the same time, Q_2 begins to turn off very rapidly, primarily because the conduction of current through Q_1 causes the voltage across R_6 to increase, thus making the emitter of Q_2 more positive. As Q_2 turns off, Q_1 continues to turn on very rapidly because the voltage across R_6 now decreases, producing a corresponding increase of Q_1 base-emitter forward-bias.

When the signal input voltage decreases, the circuit returns to the original state, with Q_2 again conducting and Q_1 off. By varying the values of R_2 and R_3, the output can be controlled to provide the necessary output-signal pulse width.

TRANSISTORIZED POWER CONVERTERS

The transistorized (solid-state) power converter is used to change a direct voltage to an alternating voltage, or to change a given direct voltage to a higher direct voltage. These converters are commonly referred to as *dc-to-ac* and *dc-to-dc units*. The dc-to-ac converter is also known as an *inverter* (Fig. 24-15). Because of its high degree of reliability as a result of the absence of mechanically moving parts, the transistorized power converter has replaced the rotary- and vibrator-type converters in many power-supply applications.

Operation. The power converter circuit is basically a multivibrator-type oscillator circuit in

Fig. 24-15. Solid-state dc-to-ac inverter. *(Electro Products Laboratories, Inc.)*

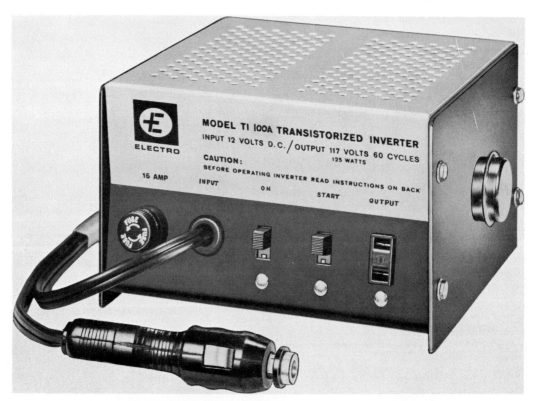

As the collector current of Q_1 decreases, the decreasing current through winding 2 induces opposite polarity voltages in the other windings and decreases the flux density within the core so that the core is no longer saturated. Now, the voltage induced across feedback winding 4 is of a polarity that causes transistor Q_2 to begin conduction.

When the current through Q_2 reaches the saturation point, almost the entire battery voltage is applied across winding 3. The resulting current through this winding produces a flux density that saturates the core of the transformer in the opposite magnetic polarity. Because of the core saturation, the induced voltages across the other windings decrease, driving Q_2 to cutoff and causing Q_1 to conduct, thus beginning another cycle of oscillation.

Output Voltage. The waveform of the voltage developed across the secondary winding of the power transformer in the circuit just described (Fig. 24-16) is essentially a "spiked" square wave (Fig. 24-17). When the circuit is to be used as a dc-to-ac inverter, a conventional choke coil-capacitor filter is connected across the secondary winding. This filter circuit tends to remove the

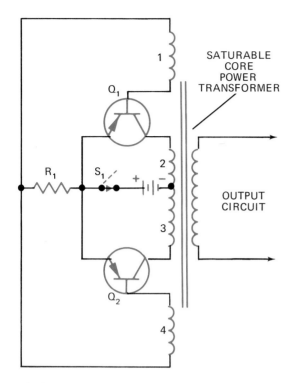

Fig. 24-16. Power converter (inverter) circuit.

which the switching action of transistors is utilized to produce a change in the direction of current through the primary windings of the associated power transformer. The schematic diagram of such a circuit is shown in Fig. 24-16. Assume that a slight imbalance of the circuit components causes transistor Q_1 to conduct more current than transistor Q_2 when switch S_1 is turned on. As the increasing Q_1 collector current passes through winding 2, it induces a voltage across "feedback" winding 1. This voltage increases the negative voltage applied to the base of the transistor, thereby causing the transistor to be driven to saturation. Simultaneously, the voltage induced across feedback winding 4 causes transistor Q_2 to be driven to cutoff.

Under this operating condition, almost the entire battery voltage appears across winding 2, producing a current that is sufficient to drive the core of the transformer to magnetic saturation. Therefore, the voltages induced across the feedback windings (windings 1 and 4) are sharply decreased, and as a result, transistor Q_1 is driven toward cutoff.

Fig. 24-17. Spiked waveform.

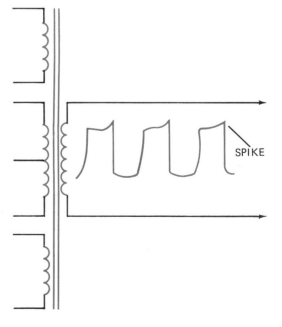

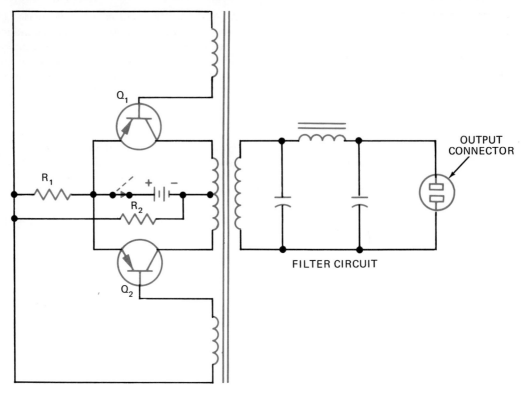

Fig. 24-18. Schematic diagram of inverter circuit equipped with "starting" resistor.

spikes from its input waveforms, thereby producing a circuit output voltage which, for all practical purposes, approximates a sine waveform.

When the circuit is to be used as a dc-to-dc converter, a conventional rectifier-filter circuit is connected to the secondary winding of the power transformer. The output voltage is then a direct voltage, the magnitude of which depends mainly upon the primary-to-secondary winding turns ratio of the transformer.

Starting. The basic transistorized converter (or inverter) circuit will often fail to start (begin to oscillate) readily under full-load conditions. To assure starting, the circuit is modified as shown in Fig. 24-18. Here, the addition of resistor R_2 causes the transistors to be biased to a point that allows oscillations to begin immediately after the circuit is energized.

For Review and Discussion

1. Define an oscillator circuit.
2. State three fundamental requirements of an *LC* oscillator circuit.
3. Explain "feedback" as it relates to an oscillator circuit.
4. Draw a schematic diagram of a basic Hartley oscillator and explain the operation of this circuit.
5. For what purpose is a buffer amplifier often used in conjunction with an oscillator stage?
6. Describe the property of a piezoelectric crystal that makes it suitable for the frequency control of an oscillator.
7. State the outstanding operational characteristic of the crystal-controlled oscillator.

8. Explain the basic operation of an *RC* oscillator.

9. Define a nonsinusoidal oscillator.

10. Draw a schematic diagram of the neon lamp oscillator and explain the operation of this circuit.

11. For what purposes are multivibrators used?

12. Name the three principal types of multivibrators and describe the operational characteristics of each.

13. For what purpose is a Schmitt trigger used?

14. Describe the function of a "power" converter.

SECTION 5
TESTS AND MEASUREMENTS

Tests and measurements are important in designing, maintaining, troubleshooting, and servicing all types of electrical and electronic products and circuit systems. While it is possible to detect some circuit actions and defects by visual inspection, in electricity and electronics we are concerned with operational characteristics that are not always visible.

To detect these characteristics (voltage, current, resistance, etc.), it is necessary to transform an electrical quantity or condition into a visible or aural indication. This is done with the help of meters, cathode-ray tubes, and loudspeakers (or headsets) that are used in conjunction with many different items of test equipment. The ability to use these instruments, coupled with a sound theoretical knowledge of electronic behavior, is essential to the technician in analyzing and troubleshooting various types of circuits logically and efficiently.

UNIT 25. Meters

The meter is a basic measuring device that usually indicates the magnitude of quantities such as current, voltage, and resistance by the position of a pointer that moves over an appropriate scale (Fig. 25-1). While the meter serves as the final indicator, it may or may not be used alone. In many of its applications, a meter is used in conjunction with a circuit configuration, making it possible to vary the range of measurements and to use the same meter to measure different electrical quantities. Thus, the meter is a highly versatile instrument that can be used to perform a wide variety of tests and measurements.

There are several different kinds of meter mechanisms or movements in use. However, in this unit only the most popular type, the D'Arsonval meter, will be discussed.

THE D'ARSONVAL METER

Most common direct-current meters used in electrical and electronics work are of the D'Arsonval type. The D'Arsonval meter movement consists of a pointer that is usually attached to a rectangular aluminum frame around which a lightweight coil is wound. The coil assembly is pivoted so that it is free to rotate about a stationary cylindrical permanent-magnet core (Fig. 25-2). A coiled hairspring located at one end of the coil provides a highly flexible electrical connection between the coil and the meter terminals. The spring also holds the pointer at zero when the meter is inactive, and dampens or slows down the pointer so that it does not oscillate (move back and forth) when coming to a stop.

Additional pointer damping action is produced because of the aluminum frame. The frame, acting as a one-turn coil, moves into the permanent-magnet field when the meter is in use. The current that is induced within the frame then generates another magnetic field that tends to oppose the movement of the coil. This action also tends to prevent the pointer from oscillating as it comes to a stop.

The adjustment screw on the meter mechanism is used to change the tension of the hairspring. The screw can be turned in either direc-

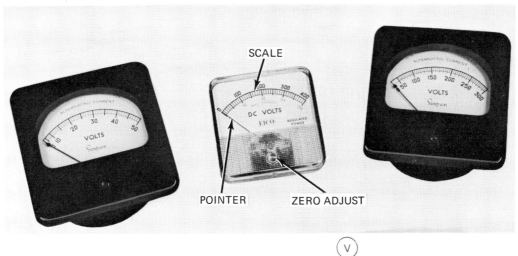

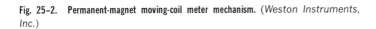

Fig. 25-1. Panel-type voltmeters. (*Gail E. Henderson*)

Fig. 25-2. Permanent-magnet moving-coil meter mechanism. (*Weston Instruments, Inc.*)

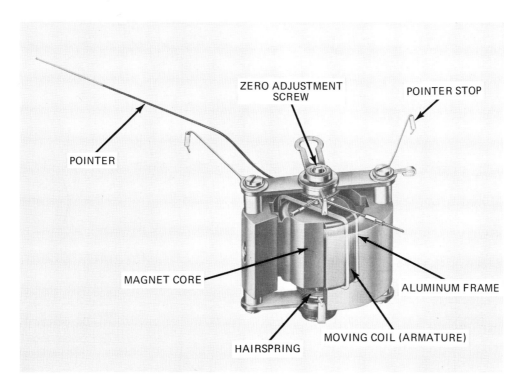

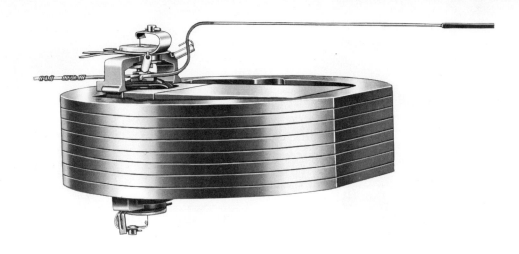

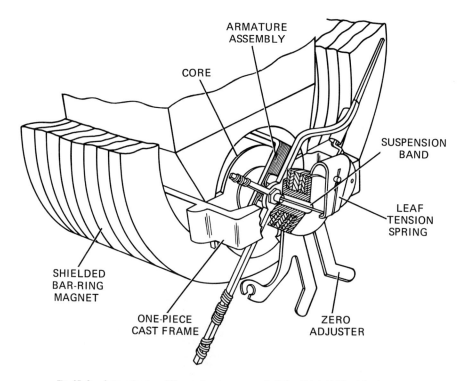

Fig. 25-3. Suspension-type D'Arsonval meter movement. (*The Triplett Electrical Instrument Co.*)

tion to move the pointer slightly to the right or left of zero on the meter scale. Hence, the pointer can be reset to zero if for some reason it does not return to this position after the meter has been used.

A suspension-type D'Arsonval meter movement is shown in Fig. 25-3. This mechanism has proved to be highly rugged and particularly well adapted to the measurement of extremely small currents.

Meter Action. The D'Arsonval meter is an electromagnetic device that is actuated by the passage of current through the movable coil. The current, in turn, generates a magnetic field that interacts with the field of the permanent magnet to produce a "motor" action. As a result, the pointer moves away from the zero position on the meter scale to a distance that is proportional to the magnitude of the current.

Although the movement of the pointer is a direct result of the magnetic effect produced by current, its position upon an appropriate scale can be used to indicate the magnitude of almost any electrical quantity. A particular scale is calibrated (marked) in terms of the quantity to be measured, and the meter then translates a current into an indication of the quantity being measured.

Since the direction in which the pointer moves is determined by the direction of the current through the movable coil, the D'Arsonval meter is a direct-current instrument. If alternating current is used, the pointer will attempt to follow each reversal of the current. The result is a rapid vibration of the pointer, which does not provide a readable indication upon the meter scale.

When the meter is used to measure alternating current or voltage, it must be operated in conjunction with a rectifier circuit (Fig. 25-4). This circuit permits current to pass through the meter only during the half-cycle that the polarity of the current (or voltage) under test is positive and CR_1 is conducting. When the polarity of the voltage reverses during the following half-cycle, rectifier CR_1 does not conduct and the current is shunted or bypassed from the meter through rectifier CR_2. Under the operating condition indicated in Fig. 25-4, the movable coil is, of course, energized with direct current. The pointer then moves to a position on the meter scale that is usually calibrated to indicate the rms value of the alternating current or voltage of the circuit to which the rectifiers are connected.

Sensitivity. The sensitivity of a basic meter movement is most often expressed in terms of the magnitude of current that is necessary to produce a full-scale deflection (from zero to maximum) of the meter pointer. This characteristic of the meter movement is determined primarily by the resistance of the movable coil and the field strength of the permanent magnet. Examples of the sensitivity of common meters used in electrical/electronics testing and measuring are 50, 200, and 500 microamperes, and 1 milliampere.

Accuracy. The general-purpose D'Arsonval meter is accurate to approximately 2 percent of its full-scale value when used to measure direct current, and to approximately 3 percent of its full-scale value when used to measure alternating current. When a meter is used in conjunction with a test-instrument circuit, its accuracy is affected by the tolerance of the components (mostly resistors) in the circuit. For this reason, standard test instruments are equipped with precision resistors, which are usually wire wound.

Care. Meters are delicate instruments and must be handled and used with care to avoid serious or permanent damage. The most sensitive part of the meter is the movable coil. Because of its low resistance, the coil can quite easily be burned out if it is severely overloaded for even a short period of time. Although the coil may safely withstand a very brief pulse of current that is 100 percent greater than the full-scale deflection current, an overload often produces a violent motion of the pointer that may cause it to be bent as it strikes the pointer stop.

Another very sensitive part of the meter is the movable coil assembly and the pivot points that hold it in place. If the meter is dropped or otherwise severely jarred, this assembly may be knocked from its bearings, causing serious and perhaps permanent damage.

Fig. 25-4. Rectifier assembly used in conjunction with a D'Arsonval meter.

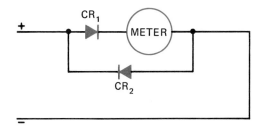

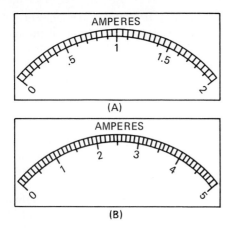

Fig. 25-5. Linear meter scales.

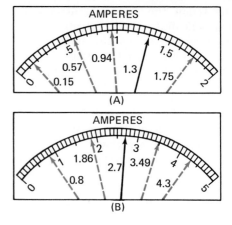

Fig. 25-6. Reading the linear meter scale.

READING THE METER

To use a meter correctly, two things are necessary. First, the meter must be connected to the circuit or the device under test in the proper manner. Second, the meter must be "read" by interpreting the position of the pointer upon the scale.

The two most common types of meter scales are the linear and the nonlinear. On the linear scale, the divisions of the total range of measurement are all the same distance apart. On the nonlinear scale, the divisions are not the same distance apart.

The Linear Scale. To read any scale, it is first necessary to determine the value of each minor subdivision on the scale. This is done by dividing the value of each major division by the number of its subdivisions. The subdivisions on a linear scale will, of course, all have the same value.

On the ammeter scale shown in Fig. 25-5A, for example, the scale is marked off into four major divisions, each having a value of 0.5 ampere. The value of each subdivision on the scale is then equal to 0.5/10, or 0.05 ampere.

By following a similar procedure it will be seen that the value of each minor subdivision on the meter scale shown in Fig. 25-5B is equal to 1/10, or 0.1 ampere.

When the pointer is located between two minor subdivisions on any scale, it is necessary to approximate the reading of the meter. With practice, this can usually be done with a reasonable degree of accuracy. A study of the readings shown in Fig. 25-6 will be helpful in providing some practice in this procedure.

The Nonlinear Scale. To read a nonlinear scale, care must be taken to determine the correct value of each minor subdivision, since all of them do not have the same value (Fig. 25-7). Since some of the major divisions on this type of scale may not be subdivided, special care must be taken to obtain a close approximation of the reading. A careless reading of such a scale can easily result in a significant error in measurement.

The Multirange Scale. Some instruments are designed to measure different ranges of one or more electrical quantities. The multirange scale is subdivided into a number of individual scales,

Fig. 25-7. Nonlinear meter scale. (*The Triplett Electrical Instrument Co.*)

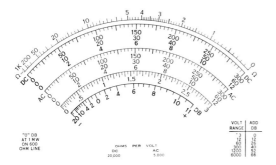

Fig. 25-8. **The multirange meter scale.** (*The Triplett Electrical Instrument Co.*)

each pertaining to a given range and quantity (Fig. 25-8). When reading a multirange scale, it is important to use the proper individual scale and to determine the correct value of its subdivisions.

The Viewing Angle. When reading any meter scale, it should be viewed from an angle perpendicular to the surface of the scale at the location of the pointer. Failure to do this will result in parallax error, or a significant difference between the apparent reading and the actual reading that would be obtained if the scale were viewed from the correct angle.

To decrease the possibility of parallax error, some meter scales are equipped with a small mirror or other reflecting surface, usually located just below or above the scale markings (Fig. 25-9). When this type of scale is being read, the pointer is "lined up" with its reflection on the mirror until both pointer and reflection are in the direct line of sight. As a result, parallax error is reduced to a minimum.

For Review and Discussion

1. Describe the D'Arsonval meter movement and explain how this meter operates to produce a deflection of the pointer.

2. For what purpose are rectifiers (diodes) sometimes used in conjunction with the basic circuit of a D'Arsonval meter?

3. What value of a voltage (or current) is indicated by a D'Arsonval meter and by other common types of meters?

4. Define full-scale deflection current.

5. What is the difference between a linear and a nonlinear meter scale?

6. Define parallax error.

Fig. 25-9. **Milliammeter equipped with scale mirror.** (*Weston Instruments, Inc.*)

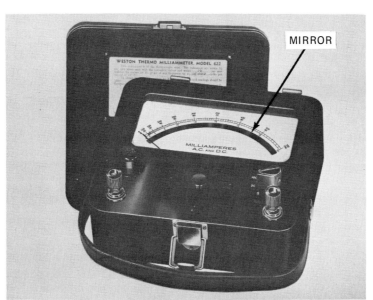

UNIT 26. Measuring Current, Voltage, and Resistance

A basic meter movement, such as the D'Arsonval type, can be used in several different kinds of test instruments, including the ammeter, the voltmeter, and the ohmmeter. Each instrument may have a separate movement, or a single movement can be used within a multipurpose instrument called a multimeter. The multimeter is an instrument that contains the circuitry necessary to measure either current, voltage, and resistance or just voltage and resistance. A knowledge of the various kinds of instruments and the procedures for using them is essential to understanding and performing tests and measurements.

THE AMMETER

Whether an ammeter is used to measure current in amperes, milliamperes, or microamperes depends upon the sensitivity of the particular meter movement and the circuit arrangement with which it is operated (Fig. 26-1). Some typical panel-type direct-current ammeters using the D'Arsonval movement can measure small currents of only 0 to 10 microamperes, while others measure larger currents of 0 to 20 amperes or more. Direct-current ammeters designed to measure extremely small quantities of current are very often referred to as *galvanometers* (Fig. 26-2).

Using the Meter. An ammeter is always connected in series with one conductor of the circuit under test (Fig. 26-3A). Since the ammeter is a low-resistance device, it will be seriously damaged if it is connected across a line or between any two points in a circuit with a significant voltage existing across it.

A direct-current ammeter is a polarized device and must therefore be connected into a circuit in the proper polarity (Fig. 26-3B). If this is not done, the pointer will "kick back," or move in the wrong direction, thus making a current measurement impossible and, at the same time, increasing the danger of the pointer being bent as it hits against the pointer stop.

Meter Shunts. The range of a D'Arsonval-type ammeter can be extended by using a resistor that

Fig. 26-1. Panel-type microammeters and milliammeter. (*The Triplett Electrical Instrument Co.*)

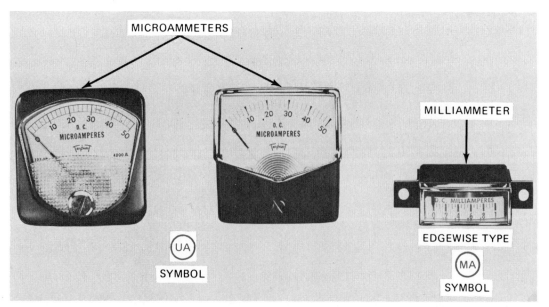

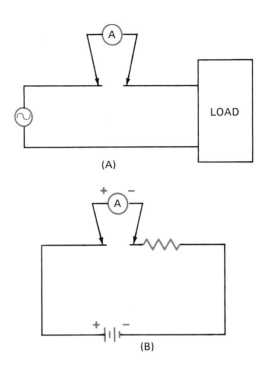

Fig. 26-3. Connecting an ammeter into a circuit.

SYMBOL

Fig. 26-2. Laboratory-type galvanometer. (*Weston Instruments, Inc.*)

Fig. 26-4. Typical meter shunts: (A) plug-in type, (B) terminal-screw type. (*The Triplett Electrical Instrument Co.*)

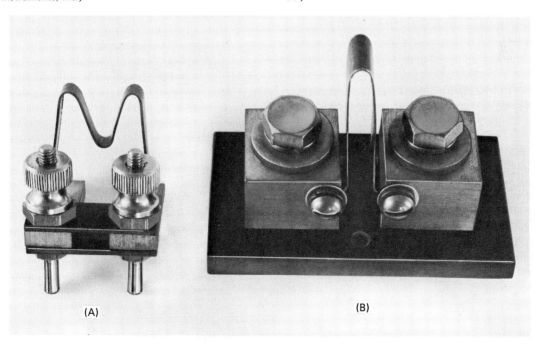

is connected in parallel with or across the terminals of the meter. This resistor is called a *shunt* (Fig. 26-4). When used with the proper shunts, one meter can be made into a multirange unit that will measure currents much larger in magnitude than the full-scale deflection current of the meter itself.

To compute the resistance value of the shunt resistor, it is first necessary to know the full-scale deflection current and the internal resistance of the meter movement. If the full-scale deflection current of a particular ammeter is unknown, it can be determined by using the circuit shown in Fig. 26-5. Before the dry cell is connected to this circuit, rheostat R_1 should be adjusted so that its full resistance exists across terminals A and B. It should then be adjusted until the meter under test, M_1, indicates a full-scale reading. The full-scale deflection current of this meter is then equal to the current indicated by meter M_2.

The direct-current resistance of a D'Arsonval-type ammeter can be determined by an indirect procedure. In this procedure, rheostat R_1 is first adjusted to obtain a full-scale deflection of the meter pointer. R_2 is then added to the circuit and adjusted to produce a half-scale deflection of the pointer. Finally, R_2 is removed from the circuit and its resistance is measured with an ohmmeter. The internal resistance of the meter is equal to the resistance obtained.

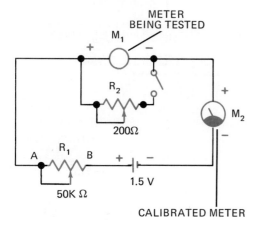

Fig. 26-5. Circuit used to determine the full-scale deflection current and the direct-current resistance of an ammeter.

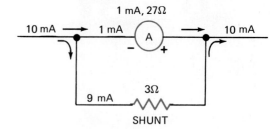

Fig. 26-6. Ammeter shunt resistance circuit.

The resistance of a D'Arsonval meter should never be measured directly with an ohmmeter. If this is done, the voltage of the cell or battery within the ohmmeter circuit will be applied across the meter, thus causing an excessive amount of current to pass through the movable coil.

Computing Shunt Resistance. Since the full-scale deflection current of a given meter movement does not change, the shunt resistor used to extend the range of a meter must "carry" the excess current passing through the meter circuit when it is in use. For example, assume that a meter having a full-scale deflection current of 1 milliampere and an internal resistance of 27 ohms is to be converted into a meter having a range of 0 to 10 milliamperes.

To produce a full-scale deflection of the meter pointer at this range, 9 milliamperes or 0.009 ampere of current must pass through the shunt resistor (Fig. 26-6). The shunt current is then nine times the current that passes through the meter itself. Therefore, the resistance of the shunt must be one-ninth of the resistance of the meter. Hence,

$$R_{shunt} = \frac{27}{9} = 3 \text{ ohms}$$

The resistance of the shunt used with any meter can be determined by using the following formula:

$$R_{shunt} = \frac{\text{Internal resistance of meter}}{(\text{Max. value of desired range}/\text{Full-scale deflection current of meter})-1}$$

Problem 1. A given D'Arsonval-type ammeter has a full-scale deflection current of 2 milliamperes and an internal resistance of

45 ohms. Compute the shunt resistance necessary to extend the existing 0- to 2-milliampere range to 0 to 20 milliamperes.

Solution.

$$R_{shunt} = \frac{45}{(0.02/0.002) - 1}$$

$$= \frac{45}{10 - 1}$$

$$= \frac{45}{9} = 5 \text{ ohms}$$

Reading the Shunted Meter Scale. When the range of an ammeter is extended by the use of a shunt, the original readings of the scale are no longer appropriate. To read the scale correctly, a multiplier must be applied to any reading on the scale. If, for example, a 0- to 1-milliampere meter is converted into one having a range of 0 to 10 milliamperes, any reading on the scale must be multiplied by 10. If the meter is converted into one having a range of 0 to 20 milliamperes, the multiplier would be 20, and so forth.

THE VOLTMETER

The voltmeter is used to measure the voltage in volts or in units of the volt. Typical panel-type direct-current voltmeters using the D'Arsonval meter movement have ranges extending from 0 to 1.5 volts to 0 to 1,000 volts or more.

Using the Meter. A voltmeter is always connected in parallel with or across the two points under test (Fig. 26-7). The voltmeter will not be damaged if it is connected in series with a circuit conductor, however. A direct-current voltmeter is a polarized device, and must therefore be connected to the test points in the proper polarity (+ to + and − to −).

The Multiplier. Since the D'Arsonval meter has a low resistance, it would be damaged if connected directly to two points across which a significant voltage exists. To prevent damage, a resistor referred to as a *multiplier resistor*, or *multiplier*, is connected in series with the D'Arsonval meter movement when it is used as a voltmeter. The amount of resistance necessary will depend upon the voltage measurement range for which the meter circuit is designed. The multiplier may be located inside the meter case or mounted externally.

Assume, for example, that a D'Arsonval meter having a full-scale deflection current of 1 milliampere and an internal resistance of 20 ohms is to be used as a direct-current voltmeter having a range of 0 to 5 volts. To produce a full-scale deflection of the meter pointer (an indication of 5 volts), the resistance of the meter circuit must be

$$R = \frac{E}{I}$$

$$= \frac{5}{0.001} = 5,000 \text{ ohms}$$

Since the internal resistance of the meter itself is 20 ohms, a multiplier of 5,000 − 20 or 4,980 ohms must be connected in series with it (Fig. 26-8).

When a single meter is to be used to measure voltages in several different ranges, it is often

Fig. 26-7. Connecting a voltmeter to a circuit.

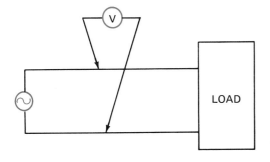

Fig. 26-8. Voltmeter multiplier resistor.

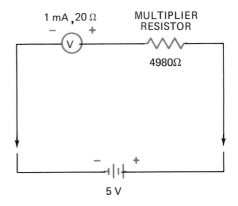

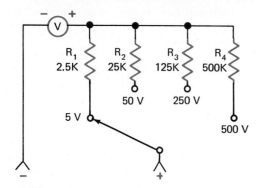

Fig. 26-9. Example of a multiplier resistor network used with a multirange voltmeter.

operated in conjunction with a multiplier network and a range selector switch. This makes it possible to switch different multipliers or combinations of resistors into the circuit to provide the various desired ranges of measurement (Fig. 26-9). The scale of the meter is then calibrated so that a voltage reading within any of the ranges can be readily interpreted.

Problem 2. A D'Arsonval-type meter has a full-scale deflection current of 2 milliamperes and an internal resistance of 27 ohms. Compute the resistance of the multiplier resistor that must be connected in series with this meter if it is to be used as a voltmeter having a range of 0 to 100 volts.

Solution. At the maximum voltage of the desired range (100 volts), the total resistance of the meter circuit must be

$$R = \frac{E}{I}$$

$$= \frac{100}{0.002} = 5,000 \text{ ohms}$$

Therefore, since the resistance of the meter movement is 27 ohms, the resistance of the multiplier must be 5,000 − 27, or 4,973 ohms.

Sensitivity. The sensitivity of a voltmeter is expressed in terms of ohms per volt (Ω/V). This factor is determined by dividing the total resistance of the meter circuit by the maximum voltage that the meter will measure at a given range.

If, for example, the total resistance of a 0- to 5-volt voltmeter is 5,000 ohms, its sensitivity is 5,000/5, or 1,000 ohms per volt.

Typical panel-type voltmeters have a sensitivity of from 1,000 to 2,000 ohms per volt, while the meters used in high-quality test instruments may have a sensitivity of up to 1 million ohms per volt. This higher sensitivity makes possible more accurate voltage measurements. The sensitivity of a multirange voltmeter will be the same for every range, since the sensitivity is determined by the meter movement used.

Loading Effect. Since a voltmeter is connected in parallel with the points across which a voltage is to be measured, the meter circuit itself acts as a load through which current passes. As in any parallel circuit, this current adds to the total circuit current, and therefore it causes an increase in the voltage drop across components such as resistors. This loading effect, in turn, can cause a significant error in measurement of a voltage.

For example, consider the circuit shown in Fig. 26-10A. Here, two resistors, R_1 and R_2, are connected in series across a voltage of 220 volts. This condition results in a circuit current of 0.002 ampere and voltage drops across the resistors as indicated on the diagram.

Now, assume that a 0- to 300-volt, 1,000-ohms-per-volt voltmeter is used to measure the voltage across R_2 (Fig. 26-10B). This, in effect, places a 300 × 1,000 or 300,000-ohm resistor in parallel with R_2. As a result, the total resistance of the circuit is reduced to

$$10,000 + \frac{100,000 \times 300,000}{100,000 + 300,000} = 85,000 \text{ ohms}$$

and the total circuit current that passes through R_1 is increased to 220/85,000, or 0.00258 ampere.

By using Ohm's law, the voltage drop across R_1 is now found to be 25.8 volts instead of the 20 volts appearing across this resistor before the meter was placed into the circuit. Because the supply voltage between A and B has not changed, the voltage across R_2 is 220 − 25.8, or 194.2 volts. This is the voltage the meter will indicate.

Compared with the original 200-volt drop across R_2, this condition represents a considerable error in measurement, which is, of course,

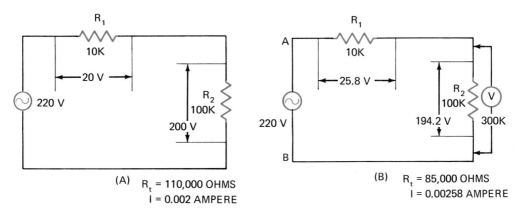

Fig. 26-10. Circuits illustrating the loading effect of a voltmeter.

undesirable. To reduce this error, it would be necessary to use a voltmeter having a higher ohms-per-volt sensitivity. Because of its higher resistance, this meter would load the circuit to a lesser degree, thus making it possible to obtain a more accurate indication of the voltage that should exist across R_2. To prevent significant loading in any given situation, the resistance of a voltmeter circuit should be at least 20 times greater than the resistance of the circuit to which the meter is connected.

The Digital Voltmeter. The digital voltmeter indicates the magnitude of the measured voltage by means of digits (numbers) appearing upon the "face" or dial of the instrument (Fig. 26-11). The obvious advantage of such a voltmeter is that any given voltage may be determined more rapidly and more accurately than is possible with the conventional pointer-type meter. However, the comparatively high cost of the instrument limits its application to highly specialized military and industrial processes.

THE OHMMETER

An ohmmeter is used to measure electrical resistance in ohms (Fig. 26-12). In almost all cases, the meter used in an ohmmeter circuit is of the D'Arsonval type having a full-scale deflection of from 50 microamperes to 1 milliampere. Typical ohmmeters have ranges extending from 0 to 100 megohms or more.

Unlike the ammeter or voltmeter, the ohmmeter circuit does not receive the energy necessary for its operation from the circuit under test. In the ohmmeter, this energy is supplied by a self-contained source of voltage, such as a cell or battery. When the meter is connected to the device or circuit under test, current passes

Fig. 26-11. Digital voltmeter. (Hewlett-Packard Co.)

SYMBOL

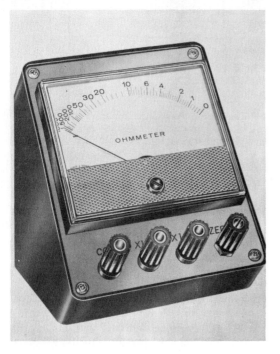

Fig. 26-12. Laboratory-type ohmmeter. (*The Triplett Electrical Instrument Co.*)

the value of the resistance. As a result, the meter scale can be calibrated so that the value of the resistance under test can be read directly from it.

Series Ohmmeter. A basic series-ohmmeter circuit is shown in Fig. 26-13A. Here, the meter is used in conjunction with series resistors R_1 and R_2. These resistors, together with the internal resistance of the meter, limit the current through the circuit to the full-scale deflection value of 1 milliampere when the test prods are shorted (in contact) with each other. A portion of the circuit resistance R_2 is in the form of a variable resistor that is used to compensate for a normal decrease in cell voltage from usage. This resistor corresponds to the zero or ohms adjustment control of a practical ohmmeter that "zeros," or sets the pointer at the zero position of the scale before the meter is used.

When the test prods of the meter are shorted together, the maximum (full-scale deflection) current passes through the meter circuit and the pointer moves to the right-hand end of the scale. Since the resistance between the test prods is now assumed to be zero, the right end of the meter scale is marked zero. When the test prods are connected to a resistance under test, this resistance is added to the resistance of the meter circuit. Therefore, the current through the meter decreases and the meter pointer "backs off" from zero or moves toward the right a shorter distance (Fig. 26-13B).

through the circuit and the meter itself. Since the magnitude of the current is affected by the resistance under test, the meter pointer moves to a position on the scale that is determined by

With the test prods in an open-circuit (not touching) condition, the meter pointer does not

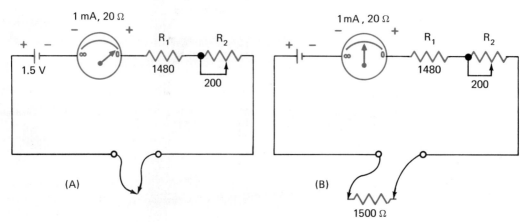

Fig. 26-13. Basic series ohmmeter circuits.

move. This indicates an infinitely high resistance across the test prods. Hence, the left-hand end of the scale on many ohmmeters is marked with the "lazy eight" infinity symbol. At all points between zero and infinity, the meter scale is calibrated to show the resistance values within a given range.

Meter Range. The useful range of the ohmmeter circuit shown in Fig. 26-13 extends from approximately 0 to 30,000 ohms. When resistances above this range are measured, the pointer deflects for only a short distance from the end of the scale, thereby making it difficult to obtain reasonably accurate readings. To increase the range of the circuit, it is necessary to use either a higher voltage (a meter battery instead of a cell) or a more sensitive meter movement that will provide full-scale deflection with less than 1 milliampere of current.

Shunt Ohmmeter. A shunt-type ohmmeter circuit that provides for more accurate low-resistance measurements than are possible with the series ohmmeter is shown in Fig. 26-14. In this circuit, the test prods are connected in shunt or in parallel with the meter. Notice that, unlike the series ohmmeter, the scale of the meter used with the shunt ohmmeter is calibrated to read from left to right.

When the test prods of the meter are "shorted" together, the total circuit current passes through the short circuit and there is no current through the meter. At this time the meter reading is zero. With the test prods "open," or not in contact with each other, the current through the meter is equal to the full-scale deflection current and the pointer moves to the extreme right of the meter scale.

When the test prods are connected to a device or circuit being tested, the total circuit current is divided between the meter "path" and the circuit being tested. As a result, there is less than full-scale deflection current through the meter. The amount of pointer deflection is then proportional to the resistance of the circuit being tested. For example, if the resistance being measured is 20 ohms, or equal to the resistance of the meter movement, the pointer will move to the midpoint of the meter scale.

Multirange Ohmmeters. A three-range ohmmeter circuit is shown in Fig. 26-15. In this circuit, an additional battery is connected in series when the ohmmeter is adjusted to the high ohms range. The resulting higher voltage provides for greater accuracy of high-resistance measurements.

Using the Ohmmeter. An ohmmeter is connected to the terminals of the device or circuit being tested (Fig. 26-16). If the ohmmeter is to be used to make tests of any part of a "powered" circuit, the circuit should be unplugged. In the case of a battery-operated circuit, the battery should be disconnected. Any electrolytic capacitors in the circuit should also be completely discharged before the ohmmeter is connected to any part of the circuit. These procedures are necessary, since an ohmmeter may be seriously damaged if it is connected across a circuit or device in which a voltage already exists.

Before using the ohmmeter to make a resistance test or before making a test after the ohms range has been changed, the meter should be "zeroed." This is done by touching the test prods together and adjusting the zero or ohms adjustment control until the pointer is at the zero position on the scale. A significant error of measurement can easily result if the meter is not zeroed.

If an ohmmeter appears to be in good working condition and cannot be zeroed, it will be necessary to replace the dry cell or battery.

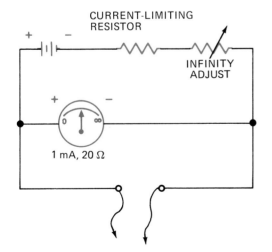

Fig. 26-14. Shunt ohmmeter circuit.

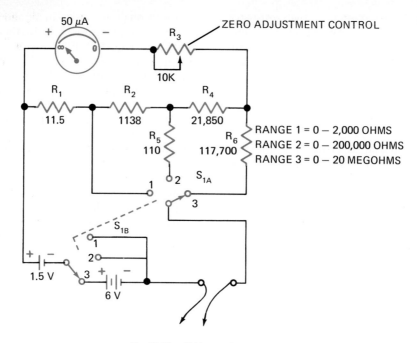

Fig. 26-15. Multirange ohmmeter circuit.

In ohmmeters containing both a cell and a battery, the cell should be replaced if zeroing is impossible on the low ohms range or ranges. Likewise, the battery should be replaced if the meter cannot be zeroed on the high ohms range or ranges.

To avoid erroneous readings when, for example, a resistor in a circuit is being tested, care should be taken to see that components such as other resistors and coils are not connected in parallel with it. A common method of eliminating this condition is to isolate the component being tested from the circuit by disconnecting one of its leads (Fig. 26-17).

To further assure accurate resistance measurements, hands should contact only the insulated portions (handles) of the test prods. If hands are in contact with the metallic portions of the prods, the resistance of the body is placed

Fig. 26-16. Connection of an ohmmeter to a device under test.

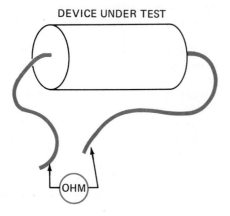

Fig. 26-17. Isolating a component from a circuit to avoid an erroneous resistance measurement.

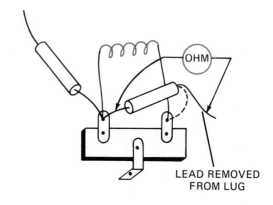

in parallel with the resistance under test, thus causing the resistance measurement to be inaccurate on high ohms ranges.

Whenever possible, resistance measurements should be made on an ohms range that will allow the pointer to move to a position between zero and the midpoint of the scale. This will make it possible to obtain a more accurate reading than that which can be obtained if the pointer stops along the more crowded midpoint-to-infinity portion of the scale.

The ohmmeter should be protected by preventing the metallic portions of the test prods from contacting each other when the instrument is not in use. If the test prods remain in contact and the meter is not turned off, the cell and/or battery will deteriorate and the meter itself may be damaged by chemical cell leakage.

Polarity of Jacks. When using an ohmmeter in making certain tests, it is important to check the polarity of its cell and/or battery because the voltage of these devices appears across the jacks (and, of course, across the test prods when the test leads are plugged into the jacks). Since the ohmmeter is usually supplied with one red and one black test lead, it is easy to assume that, when the red test lead is plugged into its associated jack (marked "Pos.," "V-A," "OHMS," etc.), it is connected to the positive terminal of the ohmmeter cell or battery. However, this is not always the case. In some ohmmeters the red test lead is connected to the negative terminal of the cell or battery when it is plugged into its associated jack. This is not a defective condition; it is simply due to the manner in which the instrument is designed and constructed.

The most convenient way of determining the polarity of ohmmeter jacks is to connect a direct-current voltmeter across them, with the ohmmeter operated at its highest ohms range. If the pointer of the voltmeter moves in the proper direction during this test, the positive terminal of the voltmeter is connected to the positive jack of the ohmmeter.

THE CONTINUITY TEST

Since the ohmmeter contains a cell and/or a battery and a means of connecting either of these devices in series with a meter, it is a very useful instrument for testing continuity, as well

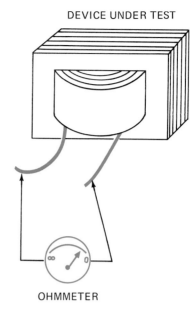

Fig. 26-18. Example of a continuity test made with an ohmmeter.

as for measuring resistance. This means that the ohmmeter can be used to determine whether or not there is a conducting path between any two points of a circuit or across the terminals of a component. Hence, the continuity test often makes it possible to locate open-circuit conditions quickly in many different situations.

To make a continuity test, the ohmmeter is usually operated at the low ohms range. After the meter has been zeroed at this range, its test prods are connected across the circuit or device to be tested. If there is continuity between the test points, the meter pointer will move to a position on the scale that is determined by the resistance of the conducting path (Fig. 26-18). Failure of the pointer to move is an indication that an open-circuit condition exists between the test points.

MULTIMETERS

The most common types of multimeters are the volt-ohm-milliammeter (VOM), the vacuum-tube voltmeter (VTVM), and the transistorized voltmeter (TRVM) (Fig. 26-19). The volt-ohm-milliammeter measures a wide range of voltage, current, and resistance. The vacuum-tube volt-

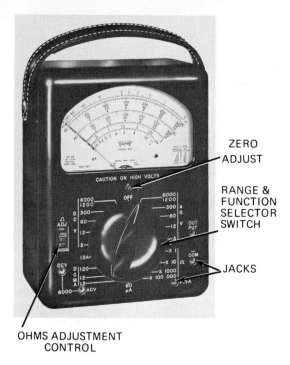

Fig. 26-19. Volt-ohm-milliammeter. (*The Triplett Electrical Instrument Co.*)

meter and the transistorized voltmeter measure voltage and current only.

Because of the high input impedance of the vacuum-tube voltmeter and the transistorized voltmeter as compared with the generally lower input impedance of the volt-ohm-milliammeter, these instruments do not significantly load the circuit under test and are capable of indicating more accurate voltage measurements.

Since multimeters use a direct-current D'Arsonval type of meter movement, they must be equipped with a rectifier (diode) assembly when alternating-current voltages are to be measured. The output of the rectifier, a pulsating direct voltage, is then applied to the meter movement and read on a scale that is calibrated to indicate the magnitude of the input (alternating) voltage.

The instructions relating to multimeters are usually supplied by the manufacturer. They should be carefully studied before attempting to use an instrument for any purpose.

METER PROTECTION

Since all types of meters can be seriously damaged by an excessive current, it is desirable to protect them from the effects of overload. Silicon diodes provide an efficient low-cost method of protecting direct-current meters when extreme meter accuracy is not required. In the circuit shown in Fig. 26-20A, diodes are connected across the D'Arsonval movement of a typical multimeter. The 1N1692 (and similar) low-cost diodes used will not begin to conduct a significant current until the voltage across them is from 0.5 to 0.7 volt, which is also the range of voltage that will cause a beyond-

Fig. 26-20. Silicon-diode-meter-protection circuits.

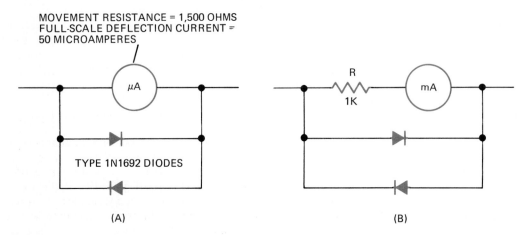

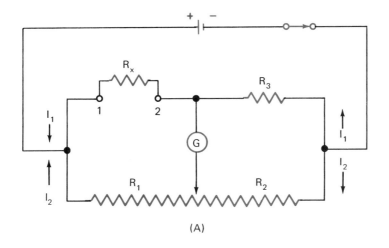

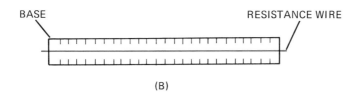

Fig. 26-21. The Wheatstone bridge: (A) basic circuit, (B) resistance wire base assembly.

full-scale reading of the meter. Therefore, when the voltage across the meter (and the diodes) approaches 0.5 volt, the diode that is forward-biased will shunt most of the current from the meter, thus protecting it from the effect of excessive current.

The diode meter protection circuit described is most efficient if it is used with meters having a relatively high movement resistance of 500 ohms or more. If a given meter has a much lower resistance, the use of a series resistor in conjunction with the diodes is recommended (Fig. 26-20B). This resistor will increase the total resistance of the meter circuit so that the voltage developed across the meter will be sufficient to cause the diodes to conduct before the movement is damaged.

THE WHEATSTONE BRIDGE

Although the ohmmeter is a useful instrument, it does not provide an exact measurement of resistance. This is primarily due to the difficulty encountered in reading its scale. When an accurate resistance reading is required, a circuit arrangement known as a *Wheatstone bridge* is employed. Because of its accuracy, the Wheatstone bridge is used in laboratory work and in detecting the location of shorted or grounded cable conductors in communications transmission lines.

One type of Wheatstone bridge consists of a cell (or battery), three resistors, and a galvanometer connected as shown in Fig. 26-21A. In this circuit, resistor R_3 is a wire-wound precision resistor. Resistors R_1 and R_2 are formed by the lengths of a uniform resistance wire on each side of a sliding-type contact and are mounted upon a base that is divided into a number of equal units (Fig. 26-21B). The resistance wire is made of an alloy such as constantan that has a very low temperature coefficient. Thus, the resistance of the wire does not change significantly with moderate changes in temperature.

To operate the bridge, the resistance to be measured, R_x, is connected to terminals 1 and

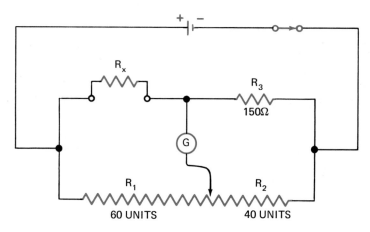

Fig. 26-22. Circuit illustrating Problem 3.

2. With the switch closed, the bridge is balanced by moving the sliding contact until the galvanometer reads zero. In this condition,

$$I_1 R_x = I_2 R_1$$

and

$$I_1 R_3 = I_2 R_2$$

Therefore

$$\frac{I_1 R_x}{I_1 R_3} = \frac{I_2 R_1}{I_2 R_2}$$

and

$$\frac{R_x}{R_3} = \frac{R_1}{R_2}$$

Thus

$$R_x = \frac{R_1 R_3}{R_2}$$

Problem 3. When the unknown resistance R_x is measured with a Wheatstone bridge, the bridge is balanced when R_1 is 60 units in length, R_2 is 40 units in length, and R_3 has a resistance value of 150 ohms (Fig. 26-22). What is the resistance of R_x?

Solution. Since the resistances of R_1 and R_2 are directly proportional to the lengths of wire from which they are formed, R_1 can be expressed as 60 and R_2 can be expressed as 40. Thus

$$R_x = \frac{R_1 R_3}{R_2}$$

$$= \frac{60 \times 150}{40}$$

$$= \frac{9{,}000}{40} = 225 \text{ ohms}$$

The Resistance Box. When the Wheatstone bridge is used to perform a wide range of resistance measurements, a variety of precision reference resistors corresponding to R_3 in Fig. 26-21 must be available. These resistors are often contained within one or more resistance (decade) boxes (Fig. 26-23). The desired resistor is then selected by adjusting the appropriate resistance-selector switches.

For Review and Discussion

1. Draw a diagram showing how a typical ammeter is connected into a circuit.

2. What condition causes the pointer of a meter to "kick back" or move in the wrong direction?

Fig. 26-23. **Resistance box.** (*Ohmite Manufacturing Co.*)

3. State the purpose for which an ammeter shunt is used, and explain how the shunt is able to perform this function.

4. Draw a diagram showing how a typical voltmeter is connected into a circuit.

5. For what purpose is a multiplier resistor used in conjunction with a voltmeter?

6. Define the ohms-per-volt sensitivity of a voltmeter.

7. Describe the loading effect that occurs when a voltmeter having a low ohms-per-volt sensitivity is connected to a circuit.

8. What is a digital voltmeter?

9. Draw a schematic diagram illustrating a basic series ohmmeter circuit, and explain how the circuit operates to measure resistance.

10. Why is it important not to connect an ohmmeter to any two points across which a voltage exists?

11. How is an ohmmeter "zeroed"?

12. What ohmmeter condition often makes it impossible to zero the meter?

13. Describe the procedure to be followed in ascertaining the polarity of ohmmeter jacks.

14. For what purpose is a continuity test made?

15. What is the outstanding advantage of the vacuum-tube voltmeter and the transistorized voltmeter as compared with the volt-ohm-milliammeter?

16. What is the purpose of the rectifier assembly in a multimeter?

17. Explain how the diodes used as meter protection devices operate to protect a meter from an excessive current.

18. Explain the operation of a basic Wheatstone bridge.

UNIT 27. Other Test Instruments

In addition to the measurement of current, voltage, and resistance, a large number of other tests and measurements are made in connection with electrical and electronics activities. In this unit, several test instruments that perform a variety of functions are discussed. These instruments are used to test the operation of circuits and the condition of circuit components, and to aid in adjusting circuits for proper operation.

THE OSCILLOSCOPE

The cathode-ray oscilloscope is an extremely useful and versatile instrument that is closely related to the television receiver in both function and circuit design (Fig. 27-1). It is used to provide a visual representation of a waveform that is produced by an electron beam striking

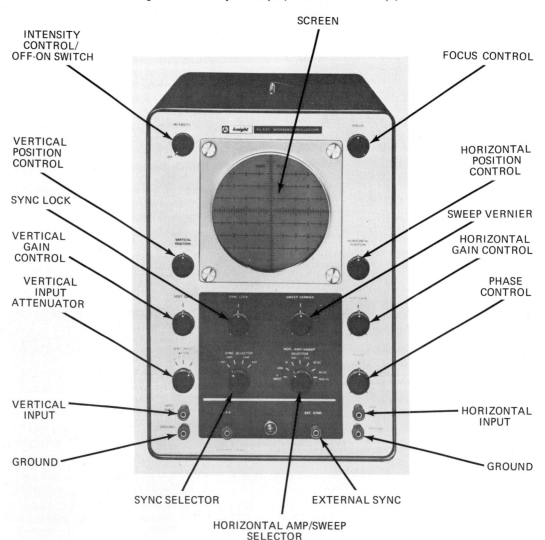

Fig. 27-1. Cathode-ray oscilloscope. (*Allied Electronics Corp.*)

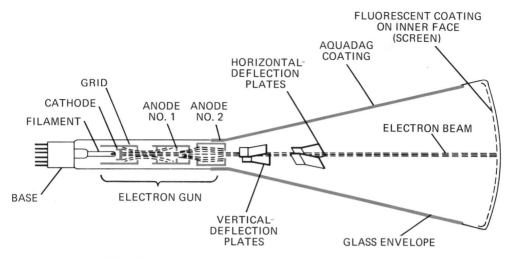

Fig. 27-2. Principal constructional features of the cathode-ray tube used in oscilloscopes.

the fluorescent screen of a cathode-ray tube (Fig. 27-2). The electron beam is automatically controlled to produce only a horizontal sweep that is controlled by the sawtooth output voltage of a sweep generator. This action produces a line-like horizontal trace of the beam when a voltage is not applied to the vertical input of the oscilloscope (Fig. 27-3A). When an external "test" signal is applied to the vertical input of the instrument and the horizontal gain control is turned "down," the beam is deflected vertically (Fig. 27-3B).

In addition to the observation of waveforms, the oscilloscope can also be used to perform a number of other important functions. Among these are voltage comparison, frequency measurement, and phase measurement.

Electrostatic Deflection. The electron beam in the cathode-ray tube of an oscilloscope is usually controlled electrostatically. This is accomplished by applying voltages from the vertical deflection and horizontal sweep circuits to ver-

Fig. 27-3. Horizontal and vertical traces of the electron beam appearing upon the screen of an oscilloscope. (Hewlett-Packard Co.)

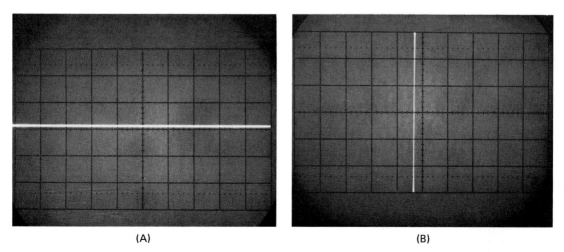

(A)   (B)

279

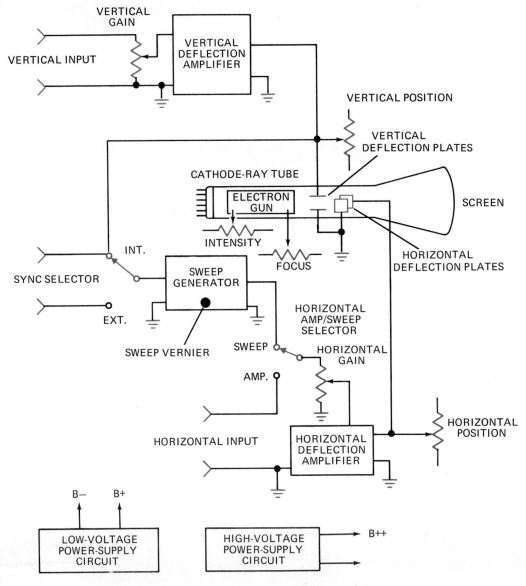

Fig. 27-4. Simplified schematic diagram of the oscilloscope circuit system with control and function data.

tical and horizontal deflection plates within the tube (Fig. 27-4).

Since the electron beam is, in effect, a negative charge, the direction in which it is deflected is determined by the polarity of the voltages that are applied to the deflection plates. For example, if one of the horizontal deflection plates is made more positive than the other, the beam will move in a horizontal direction (Fig. 27-5A). The extent or distance of this movement is directly proportional to the magnitude of the applied voltage. Likewise, if one of the vertical deflection plates is made more positive than the other, the beam will move toward the more positive plate in a vertical direction (Fig. 27-5B).

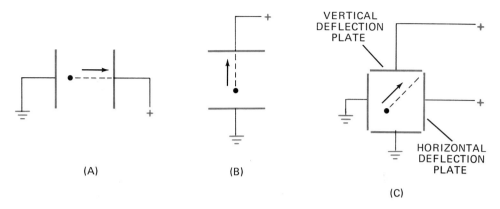

Fig. 27-5. Horizontal, vertical, and angular electrostatic deflection of the electron beam.

Because the vertical deflection and horizontal sweep circuits operate independently of each other, the electron beam can be brought simultaneously under the influence of two separate electrostatic forces that are at right angles to each other. This is necessary if at any instant the beam is to be deflected in a direction that is neither horizontal nor vertical, as required for the representation of the majority of waveforms.

To illustrate this action, assume that identical polarity voltages of equal magnitude are applied to the deflection plates at the same instant. The beam will then move in a straight line at a 45-degree angle from the horizontal (Fig. 27-5C). It is important to note that, in practice, the electron beam can be made to move in any direction by simultaneously varying the control voltages applied to the deflection plates.

Basic Waveform Observation. To observe a waveform, for example, a 60-hertz sine wave, the sync (synchronization) selector control is first adjusted to the line (60-hertz) position. The sweep selector control is then also adjusted to the line position. In this condition, a 60-hertz voltage from within the oscilloscope triggers the horizontal sweep generator circuit so that its sawtooth output voltage has a frequency of 60 hertz. Since the synchronization voltage applied to the vertical plates and the sweep voltage applied to the horizontal plates are of the same frequency, a stationary sine waveform will appear on the screen (Fig. 27-6A).

Variations of the sine waveform can be produced by changing the frequency of the sweep generator output voltage with an adjustment of the sweep selector and the sweep vernier controls. If the sweep frequency is set at 120 hertz, one alternation (half-cycle) of the sine wave is obtained (Fig. 27-6B). If the sweep frequency is 30 hertz, a two-cycle sine waveform will appear on the screen (Fig. 27-6C). In any case, the exact shape of the resultant waveform will be determined by the ratio of the frequency of the horizontal sweep voltage to the frequency of the voltage applied to the vertical plates.

In practice, the oscilloscope is most often used by applying the signal that is to be observed to the vertical input and "ground" jacks. The frequency of the horizontal sweep voltage is then adjusted to the frequency of the input voltage by using the sweep selector (coarse tuning) and the sweep vernier (fine tuning) controls. The input voltage is applied to the vertical plates, and the resulting waveform on the screen of the cathode-ray tube represents the input-voltage waveform.

An example of the waveforms obtained by following such a procedure appearing at various points of a television receiver is shown in Fig. 27-7. These waveforms are very often included with a schematic diagram as an aid in servicing the equipment. In checking a circuit, the technician uses an oscilloscope at the indicated check points to see if the waveform at these points of the circuit is as indicated on the diagram. If the circuit is in proper operating

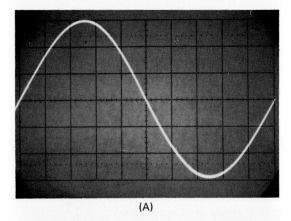

(A)

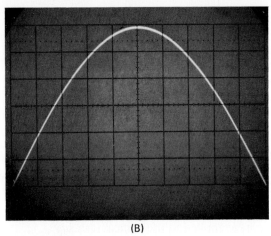

(B)

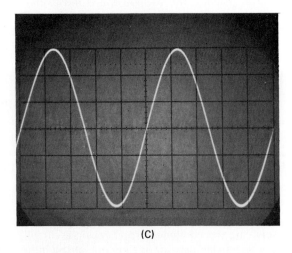

(C)

Fig. 27-6. Sinusoidal waveform traces of the electron beam appearing upon the screen of an oscilloscope. (*Hewlett-Packard Co.*)

condition, the waveform observed at any check point will be similar to the waveform shown on the diagram at the corresponding point.

Sensitivity. The sensitivity of an oscilloscope is a measure of the deflection of the electron beam that results when a given voltage is applied to the input. This is generally expressed in terms of millivolts (mV) per inch. Thus, if the horizontal sensitivity of a given oscilloscope is indicated as being 600 mV rms/inch, it means that an alternating voltage of 0.6 volt (600 millivolts) rms applied to the horizontal plates will cause the beam to move horizontally for a distance of one inch. The vertical sensitivity, which in the typical oscilloscope is much greater than the horizontal sensitivity, is often expressed in the same manner.

Comparing Voltages. The oscilloscope can be used to compare the magnitude of voltages existing at two different points in a circuit. This procedure is often used to determine the voltage gain of amplifier stages. For example, assume that when an oscilloscope is connected across the input of a given amplifier stage, the resulting waveform has a vertical peak-to-peak deflection of 2 on the cathode-ray tube chart (Fig. 27-8A). Also, assume that when the oscilloscope is connected across the output of this same stage with the position of its vertical gain control unchanged, the waveform has a peak-to-peak deflection of 12 (Fig. 27-8B). In each case, the total deflection of the waveform is directly proportional to the magnitude of the peak-to-peak voltage applied to the vertical deflection plates. Thus, the gain of the amplifier stage under this condition is approximately 12/2, or 6.

The outstanding advantage of the oscilloscope in measuring voltage gain and in other applications is that it is capable of responding to higher frequencies than most other indicating instruments. This is because the electrostatically controlled electron beam has negligible inertia and can therefore produce an accurate visual response to any input frequency.

Measuring Frequency. To measure the frequency of a sinusoidal voltage with an oscilloscope, the signal of the unknown (measured) frequency is applied to the vertical input terminals. At the same time, a signal of a known (reference) fre-

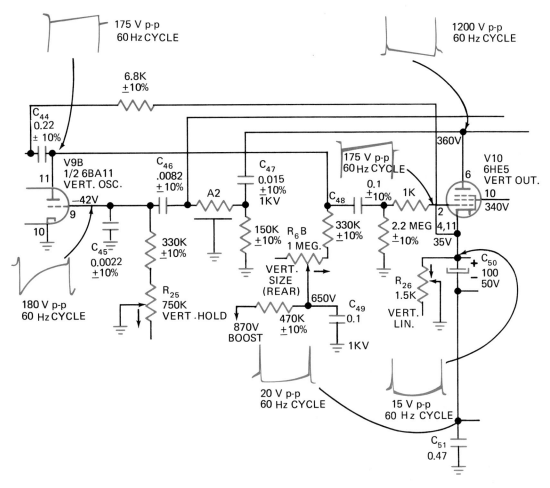

Fig. 27-7. Oscilloscope waveforms appearing at various points of a television receiver circuit.

Fig. 27-8. Using the oscilloscope to compare the magnitudes of two different voltages.

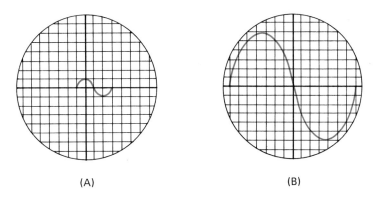

(A) (B)

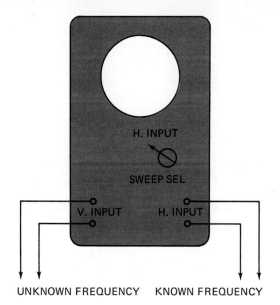

Fig. 27-9. Signal connections to the oscilloscope when measuring frequency.

quency is applied to the horizontal input terminals, with the sweep selector control adjusted to the horizontal input ("AMP") position (Fig. 27-9). This "removes" the sweep generator from the circuit and allows the signal of the known frequency to be applied to the horizontal deflection plates through the horizontal deflection amplifier (see Fig. 27-4).

The known frequency signal used in oscilloscope frequency measurements is usually obtained from the output of a signal generator. Such an instrument is capable of generating voltage outputs over a wide range of frequencies, a specific frequency being selected by adjusting the tuning control to the desired position.

Fig. 27-10. Lissajous figure circle indicating that two frequencies measured with an oscilloscope are equal.

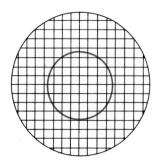

With the operating conditions of the oscilloscope set, the resulting pattern on the screen of the cathode-ray tube provides an indication of the frequency ratio existing between the input signals. For example, when these signals are of equal frequency and 90 degrees out of phase with one another and are applied to the deflection plates with equal magnitude, the pattern, which is referred to as a *Lissajous figure*, is a circle (Fig. 27-10).

When the signal frequencies are not equal, the ratio of the measured frequency to the known frequency can be determined by count-

Fig. 27-11. Lissajous figures showing different frequency ratios.

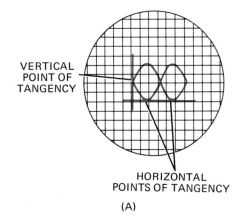

(A)

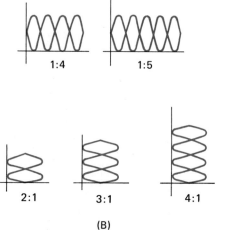

(B)

ing the points of tangency or the points at which the loops of the resulting Lissajous figure touch imaginary horizontal and vertical axes. Expressed as a proportion,

$$\frac{\text{Unknown frequency } F_x}{\text{Known frequency}} = \frac{\text{Vertical points of tangency}}{\text{Horizontal points of tangency}}$$

As an example, assume that the signal (of known frequency) applied to the horizontal input of the oscilloscope has a frequency of 60 hertz and that the resulting Lissajous figure is similar to the one shown in Fig. 27-11A. Then, since this figure has two points of tangency along the horizontal axis and one along the vertical axis, the ratio of the measured (unknown) frequency F_x to the known frequency is 1:2. Therefore,

$$\frac{F_x}{60} = \frac{1}{2}$$

$$2F_x = 60$$

and

$$F_x = 30 \text{ hertz}$$

The frequency of the measured (unknown) signal applied to the vertical input of the oscilloscope is then 30 hertz. Several additional Lissajous figures and the frequency ratios that they represent are shown in Fig. 27-11B.

Measuring Phase. The phase relationship existing between the signals applied to the vertical and the horizontal input terminals of an oscilloscope can also be determined by the use of Lissajous figures. These figures will be in the form of a circle, a straight line, or an ellipse, each representing a definite phase relationship (Fig. 27-12). In each case, the frequencies are equal.

Fig. 27-12. Lissajous figures showing different phase relationships of signals having the same frequency.

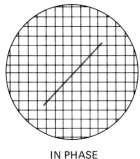

IN PHASE

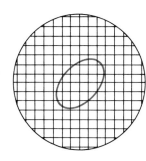

45° OUT OF PHASE

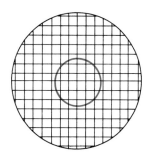

90° OUT OF PHASE

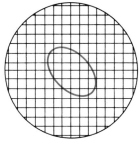

135° OUT OF PHASE

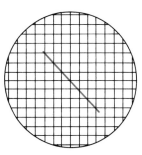

180° OUT OF PHASE

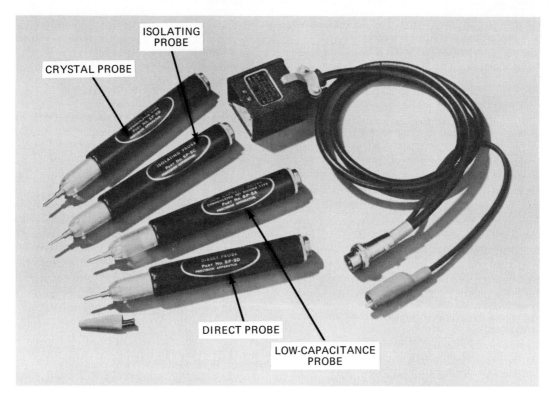

Fig. 27-13. Oscilloscope probes. (*B & K Division of Dynascan Corp.*)

OSCILLOSCOPE PROBES

Several different types of probes are commonly used in conjunction with the oscilloscope to provide a greater versatility in the application of this instrument. A number of typical probes are shown in Fig. 27-13. The crystal probe serves as an amplitude-modulation (AM) detector, thus allowing an observation of the modulation waveforms of both intermediate-frequency (i-f) and radio-frequency (r-f) signals. The isolating probe is essentially a low-pass filter that "sharpens" the response patterns of a variety of different kinds of signals. The low-capacitance probe reduces distortion and loss of gain of a waveform, which may result because of input overloading to the oscilloscope. The direct probe is designed primarily to minimize the effect of "stray" signal inputs resulting from electrical and/or magnetic field radiation.

THE ELECTRON-RAY TUBE

The electron-ray (magic-eye) tube is an indicating device that visually detects small changes in the magnitude of a control voltage. Such a tube consists of a conventional glass envelope containing a cathode heater assembly and several electrodes, including a fluorescent target plate (Fig. 27-14A). When electrons emitted from the cathode strike the target plate, the plate fluoresces (glows). Under different conditions of control voltage, the target plate fluoresces more or less, thus producing a wedge-shaped shadow upon the "face" of the tube (Fig. 27-14B).

SIGNAL GENERATORS

The signal generator is a test instrument that consists of one or more oscillator circuits and the associated controls by means of which the

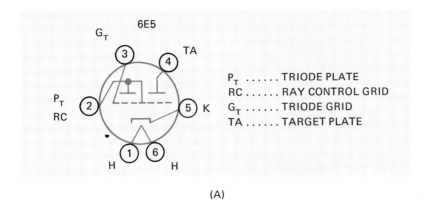

Fig. 27-14. The electron-ray tube: (A) basing diagram, (B) cutaway view of type 6E5 tube and shadow angles.

desired frequency output is selected and adjusted. It is, in effect, a small-size transmitter that generates an alternating-current signal. This signal is applied directly to oscilloscopes, radio and television receivers, audio amplifiers, etc.

When used in conjunction with receivers, the output signal of the generator is applied to various points of the circuit to align or adjust the related tuned circuits for maximum response. When used in conjunction with an amplifier, the signal generator is generally used to check the frequency response and gain characteristics of the various stages.

Signal generators are usually classified as either audio-frequency or radio-frequency units. Each of these is further classified in accordance with the kinds of signals or output waveforms it is capable of producing.

Audio Signal Generator. The audio signal generator is most often designed to produce a continuous signal output, which may have a total frequency range varying from 20 hertz to as high as 1 megahertz. In the typical instrument, the total frequency output range is divided into a number of individual range "steps" that are se-

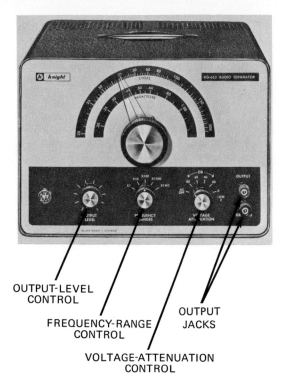

Fig. 27–15. Audio-signal generator. (*Allied Electronics Corp.*)

In some audio signal generators, the oscillator is tuned by means of variable tank capacitors and different feedback coupling networks. In other generators, the output signal is produced by mixing (heterodyning) together the outputs of two separate oscillator circuits. The resulting difference (beat) frequency then forms the output signal.

All general-purpose audio signal generators are capable of producing sinusoidal output signals throughout the total frequency range. More sophisticated instruments are also designed to produce a square-wave output signal at all frequencies or at a certain portion of the total range. In some circuits this is accomplished by operating the final amplifier stage at the saturation point during the positive half-cycles of the sinusoidal input and at cutoff during the negative half-cycles of the input voltage. As a result, the peaks of the output signal are distorted or "clipped" to produce what is essentially a square-wave output signal (Fig. 27–16).

R-F Signal Generator. The radio-frequency signal generator is designed to produce a sinusoidal output signal that may range from 150 kilohertz to 100 megahertz or more. In addition, the generator contains an audio oscillator circuit that is usually designed to produce a signal having a fixed frequency of 400 hertz. The instrument can be operated so that this signal amplitude-modulates the radio-frequency output, thus providing a signal that is extremely useful for

lected by an adjustment of the range control (Fig. 27–15). The exact frequency within each range is then selected by adjusting the tuning pointer to the correct position on the scale or dial.

Fig. 27–16. Producing a square-wave output signal from a sinusoidal input signal.

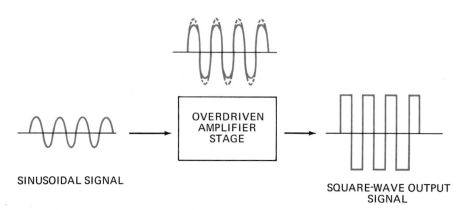

Fig. 27-17. Frequency-response curve of an FM interstage-transformer output to which the signal from a sweep generator has been applied.

aligning all sections of the typical AM receiver circuit. On most generators, the 400-hertz signal is also available at the audio output terminal. It can therefore be used for testing the performance of audio amplifier circuits at this frequency. The radio-frequency signal of the generator can also be modulated by an audio signal from an external source.

The Sweep Generator. The sweep generator is a radio-frequency generator that produces a sinusoidal signal output that continually varies in frequency above and below the center frequency to which the instrument is tuned. The maximum variation of the frequency or width of the "sweep" is usually several megahertz. The rate of the sweep is most often fixed at 60 hertz. This means that the generator produces all the output frequencies during each time period of $\frac{1}{60}$ second. The output is therefore frequency-modulated to an extent that is determined by the setting of the sweep-width or similarly identified control.

The sweep generator is most commonly used to align or tune the radio-frequency and the intermediate-frequency sections of frequency-modulation (FM) radios and the intermediate-frequency section of television receivers. This is generally done by applying the output signal of the generator to a given amplifier stage with an oscilloscope connected to the output of the detector stage. The associated interstage transformer is then aligned to produce the desired frequency response curve on the screen of the oscilloscope (Fig. 27-17).

The Marker Generator. The marker generator is basically a radio-frequency signal generator that is calibrated so that its output frequency can be determined very accurately. It is used in conjunction with a sweep generator and an oscilloscope to identify a specific frequency within the total range of frequencies that is represented by the oscilloscope waveform. This is necessary if any circuit must be aligned in order to have the required operating bandpass characteristics.

In use, the marker generator and sweep generator are connected to the same point of the circuit being aligned. The marker generator is then set to produce a signal having a frequency that is at the lower end of the necessary bandpass of the circuit. As the sweep generator output passes over this frequency, the interaction between the two equal-frequency signals produces a "pip" on the oscilloscope waveform (Fig. 27-18A).

The marker generator is then adjusted to produce a signal having a frequency at the upper end of the bandpass. As the sweep generator output passes over this frequency, a second pip is produced on the waveform (Fig. 27-18B). By marking or otherwise identifying the location of

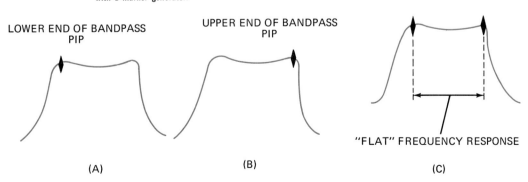

Fig. 27-18. Pips produced on an oscilloscope waveform when a sweep generator is used with a marker generator.

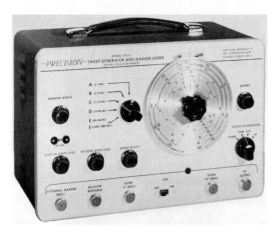

Fig. 27-19. Sweep generator and marker adder. (B & K Division of Dynascan Corp.)

When the grid dip meter has become tuned to a frequency that is equal to the resonant frequency of the circuit to which it is coupled, energy from the oscillator circuit of the instrument is absorbed by the circuit. As a result, the milliammeter pointer dips to indicate a minimum current in the input circuit of the grid dip oscillator. The resonant frequency of the circuit under test is then equal to the frequency to which the grid dip meter is tuned.

Oscillating Detector. As an oscillating detector, the grid dip meter is used to determine the frequency at which an energized radio-frequency-tuned circuit is operating. To use the meter for this purpose, a headset is plugged into the appropriate jacks on the panel and is then

Fig. 27-20. Grid dip meter. (Allied Electronics Corp.)

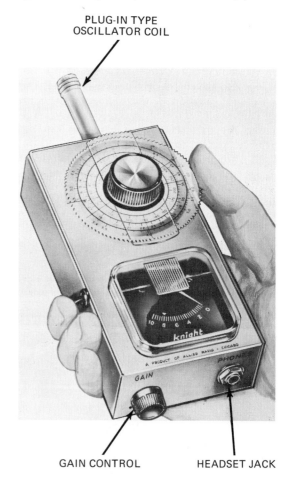

these pips on the oscilloscope screen, the interstage transformers of the circuit can be very accurately aligned to produce the desired "flat" frequency response to the band of frequencies that is represented by the horizontal distance between the pips (Fig. 27-18C).

The sweep generator and the marker generator signals are often mixed together in a marker-adder. A sweep generator with a built-in marker-adder is shown in Fig. 27-19.

THE GRID DIP METER

The grid dip meter consists of a tunable oscillator network, a frequency-calibrated tuning dial, and a sensitive milliammeter that is connected into the input circuit of the oscillator stage (Fig. 27-20). It is a versatile instrument used primarily to determine the resonant frequency of tuned circuits. However, the grid dip meter can also be used as an oscillating detector, an absorption-wave meter, a relative field strength meter, and a signal generator.

Determining Resonant Frequency. To determine the resonant frequency of a circuit, the grid dip meter is first rather closely coupled to the circuit under test (Fig. 27-21). This test is usually performed with the circuit under test in an unpowered or deenergized condition and is begun at the lowest frequency to which the grid dip meter can be tuned.

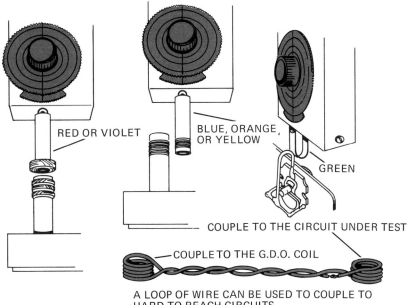

Fig. 27-21. Coupling the grid dip meter to the circuit under test. (*Allied Electronics Corp.*)

coupled to the circuit under test. When the instrument is tuned to a frequency or to a harmonic frequency of the circuit under test, this condition is indicated by a beat note or tone produced in the headset. For safety reasons, a grid dip meter should always be kept some distance away from energized high-voltage circuits.

Absorption-type Wave Meter. When used as an absorption wave meter (frequency meter), the function of the grid dip meter is similar to its use as an oscillating detector. However, for this application, the instrument is operated with a minimal supply voltage or with no supply voltage at all. Hence, the instrument is actuated by the energy absorbed into the grid dip meter tank circuit from the circuit to which it is coupled. Since the maximum absorption of energy occurs when the tank circuit is tuned to the frequency of the circuit under test, the milliammeter will also show a maximum reading at this time.

Relative Field Strength Meter. The absorption of energy into the grid dip meter tank circuit is also made use of when the instrument is employed as a field strength meter to measure the relative intensity of a radiated electromagnetic field. In this application, the coil of the tank circuit acts as an antenna into which a voltage is induced. Thus, the relative intensity of the field strength under test will, at any given time, be indicated by the distance to which the pointer of the milliammeter is deflected.

THE SIGNAL TRACER

The signal tracer is an instrument that is particularly useful in the process of troubleshooting radio receiving equipment (Fig. 27-22). In its basic form, the signal tracer consists of a wide-range amplifier and detector circuit. This device usually provides an aural (audio) indication of the presence of a modulated signal when its probe is applied to a test point in a circuit. A signal generator, which serves as the source of the signal, is first connected to the input of the circuit under test. By connecting the probe of the signal tracer to various points in the circuit,

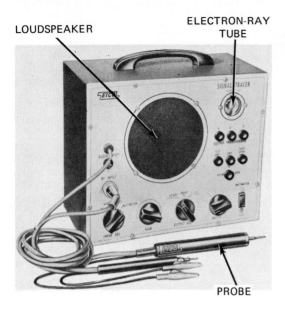

Fig. 27-22. Signal tracer. (EICO, Inc.)

Fig. 27-24B. Here, the voltage of the battery is applied to all electrodes of the transistor. If the transistor is "good," the meter will indicate a current that is from 20 to 300 times greater in magnitude than the base current set by R_b. With this ratio of collector current to base current, the transistor should be capable of performing the function of amplification in a satisfactory manner.

Short and Open Tests. The specific procedures to be followed in making the short and the open tests are usually stated in the instruction manual that is supplied with the instrument. In most cases, however, a short between electrodes is indicated by an excessive deflection of the meter pointer, while an open between electrodes is indicated by failure of the meter to respond at all.

Testing Diodes. When testing semiconductor diodes, the tester battery voltage should be applied to the diode under test in both a forward-

it is possible to follow the progress of the signal through the amplifier, coupling, and detector stages of the circuit being tested.

TRANSISTOR AND DIODE TESTER

The most popular types of transistor testers include a circuit arrangement that makes it possible to test for leakage, gain, open-circuit conditions, and short-circuit conditions (Fig. 27-23). Some transistor testers are equipped to test diodes. These instruments are also available in the form of a combination tube-transistor tester.

Leakage Test. The leakage test is usually made by applying a voltage across the emitter and the collector of the transistor with the base electrode "open" (Fig. 27-24A). If the resulting leakage or reverse current is relatively small, the transistor is in good condition. If, on the other hand, the transistor is defective because of surface or internal structural conditions, the leakage current will be in excess of the maximum limit as specified in the tester instruction manual.

Gain Test. The basic circuit used when making a gain test with a transistor tester is shown in

Fig. 27-23. Transistor and diode tester. (B & K Division of Dynascan Corp.)

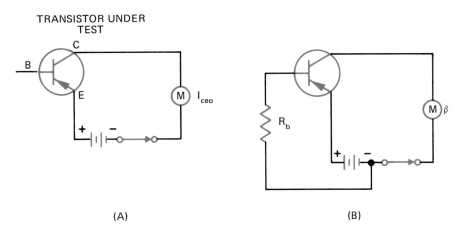

Fig. 27-24. Basic circuits of the transistor tester when making the leakage and gain tests.

and reverse-bias "direction" (Fig. 27-25). To interpret the test, it is then necessary to compare the forward current with the reverse current. If the forward current is from 30 to 50 times greater than the reverse current, the diode is generally considered to be in good condition.

THE TUBE TESTER

The tube tester is an instrument used to check the condition of electron tubes under a given set of conditions. While there are many different types of tube testers, those designed for general-purpose applications usually provide for an indication of emission, undesirable interelectrode contacts (shorts), and the presence of an excessive amount of gas within the envelope. Some tube testers are also equipped with circuits and switching arrangements that allow operational and noise characteristics to be checked.

TUBE SUBSTITUTION TEST

Although the tube tester is a valuable instrument, it is important to note that the tester circuit arrangement subjects the tube under test to operating conditions that may differ from those in the circuit from which the tube was removed. For this reason, the results of a test made with a tube tester may not be valid in all cases. Thus, a given tube that is apparently in good condition on the basis of tester results may not operate properly when inserted into its circuit.

The substitution test should be made in all situations where the actual circuit operating conditions of a tube are in question. This test is performed by replacing the tube with one that is known to operate efficiently. While it is always best to replace a tube with one that is of the exact type, many tubes can be replaced with one or more types whose operational characteristics are similar. A list of the domestic and foreign-made tubes that can be used to replace a particular tube is given in any one of several tube substitution manuals or guides that are available.

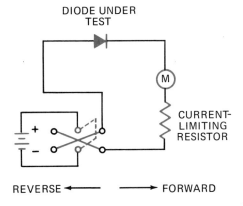

Fig. 27-25. Basic diode test circuit.

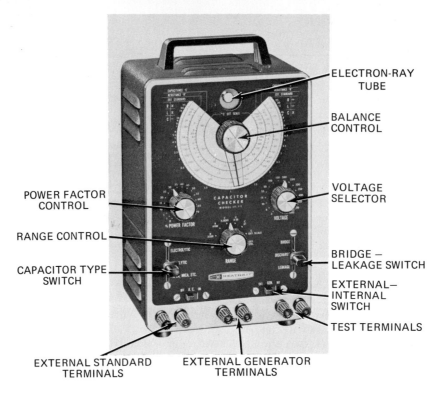

Fig. 27-26. Capacitor checker. (*Heath Co.*)

THE CAPACITOR CHECKER

A typical capacitor checker that is also designed to measure resistance and inductance is shown in Fig. 27-26. The "heart" of this and similar checkers is a bridge circuit, which is energized either by connecting it to an alternating-voltage line or by connecting a separate signal generator to the external generator terminals of the instrument.

A basic form of bridge circuit used for the measurement of capacitance is shown in Fig. 27-27A. Here, R_{1A} and R_{1B} make up the resistance arms of the bridge. The capacitance arms of the bridge include C_s (the capacitance standard inserted into the bridge by means of a range selector switch), and C_x (the capacitance to be measured).

When R_1 is adjusted so that the Wheatstone-type bridge is balanced, the voltage drop across C_x is equal to the voltage drop across R_{1A} and the voltage drop across C_s is equal to the voltage drop across R_{1B}. In this condition, the voltage applied to the triode grid of the electron-ray tube is at a minimum with respect to ground. Hence, the "eye" of the tube is open (has a maximum shadow). The value of C_x is then determined by reading the value of capacitance on the C scale of the checker and multiplying this value by the setting of the range selector switch.

Although an electron-ray tube is used as the indicating device in the bridge circuit described, another type of indicator, such as a meter, could be used by connecting it across points D and E. In any case, the balance control would be adjusted so as to produce a "null" or minimum indication of the indicating device.

When measuring resistance, the capacitors of the bridge circuit are, of course, replaced by resistors (the known standard and the unknown resistance to be measured). As in the measurement of capacitance, the bridge is then balanced and the unknown resistance determined by multiplying the value of the standard resistance by the setting of the range selector switch.

Inductance is measured with most common types of capacitor checkers by connecting an inductance standard to the appropriate (external standard) terminals and the inductance to be measured to the test terminals. The bridge is then balanced by using the checker balance control (Fig. 27-27B). In this condition, the voltage drop across L_x, the inductance to be measured, is equal to the voltage drop across R_{1A}; and the voltage drop across L_s, the inductance standard, is equal to the voltage drop across R_{1B}. Thus, the voltage applied across the indicating device is at a minimum. The value of L_x is then determined by following the instructions accompanying the instrument.

Fig. 27-27. Bridge circuits used for measuring capacitance and inductance.

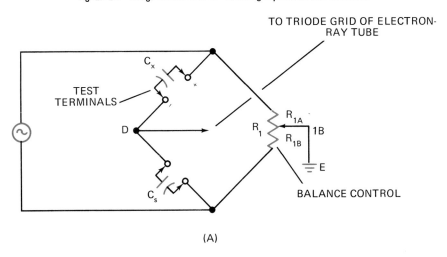

(A)

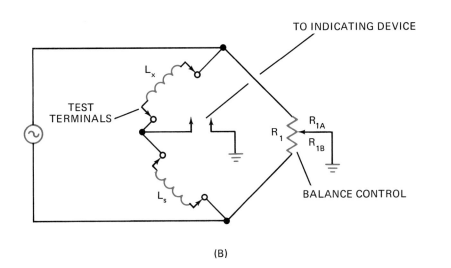

(B)

For Review and Discussion

1. Explain the electrostatic method of deflecting the electron beam in the cathode-ray tube of an oscilloscope.

2. What factor determines the magnitude of the horizontal deflection of the electron beam in the cathode-ray tube of an oscilloscope? What factor determines the magnitude of the vertical deflection of the beam?

3. State the function of the horizontal sweep circuit of an oscilloscope.

4. To which terminals of an oscilloscope is the signal to be observed most commonly applied?

5. What is the function of the sweep selector control of an oscilloscope?

6. How is the sensitivity of an oscilloscope usually indicated?

7. Explain how an oscilloscope is used to compare the magnitudes of voltages.

8. Give the basic procedures to be followed when using the oscilloscope to measure frequency.

9. For what purposes are Lissajous figures used in conjunction with an oscilloscope?

10. State the function of the electron-ray tube and its associated circuit.

11. What is a signal generator?

12. State an important application (use) of the r-f signal generator.

13. Give the frequency of the audio signal with which the output of an r-f signal generator can usually be modulated.

14. State the general characteristic of the output signal produced by a sweep generator.

15. Explain how the sweep/marker generator is used in conjunction with an oscilloscope to align an interstage transformer to the correct frequency-response condition.

16. Describe a grid dip meter and give at least three important uses of this instrument.

17. State the general relationship between the leakage and the gain currents that is indicated by a transistor tester when the transistor under test is in good condition.

18. State the general relationship between the forward and the reverse currents that is indicated by a diode tester when the diode under test is in good condition.

UNIT 28. Troubleshooting Procedures

Troubleshooting involves locating and correcting improper operating conditions in electronic circuits and devices. In this unit, a number of troubleshooting procedures will be discussed, together with several tests and measurements involving the use of the volt-ohm-milliammeter (VOM) or the vacuum-tube voltmeter (VTVM) in determining the condition of individual components and circuit structures.

THE VISUAL INSPECTION

In most cases, a thorough visual inspection of any defective circuit should be made before attempting more detailed troubleshooting procedures (Fig. 28-1). Such an inspection often shows loose connections, short circuits, burned-out resistors, overheated coils, and other conditions that can cause trouble; and will very often save time that might otherwise be wasted by unnecessary testing and measuring.

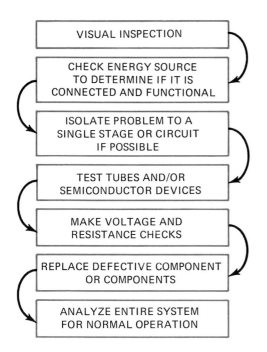

Fig. 28-1. Order of troubleshooting procedures.

ELECTRON TUBES

Following the overall visual inspection, the energy source is checked, and an attempt is made to isolate the malfunctioning circuit. The next step is to test any electron tubes in the circuit. This is strongly recommended because defective electron tubes are a primary reason for incorrect circuit operation.

While the tube tester is the standard instrument for testing tubes, there are three other methods of identifying defective tubes: visual inspection, heater continuity test, and tube substitution.

Visual Inspection. During operation, the heaters of most tubes have a cherry-red glow that is visible through the tube's glass envelope. When heaters are operated in parallel, as they usually are, the lack of a heater glow most generally indicates that the heater of the tube in question is burned out.

When metal tubes are operated in parallel, a burned-out heater can often be detected by feeling the tubes. If, during this "test," a tube is "cold" in comparison with the other tubes in the circuit, there is good reason to believe that its heater is burned out.

A visual inspection of a tube is also very useful in detecting defective tube envelopes into which air has leaked through surface cracks or breaks at the sealed terminal points. An air leak produces a clearly visible milk-white spot on the inside surface of the envelope. Such a spot means that the tube may be inoperative and should be tested.

If a visual inspection shows that the plate of a tube glows cherry-red, the circuit should immediately be turned off. Operation with a red-hot plate for any length of time may cause permanent damage to the tube. This condition most often occurs in rectifier- and power-amplifier tubes as the result of an excessive plate current, and is usually due to a shorted filter capacitor in a rectifier circuit or a shorted grid coupling capacitor in a power-amplifier stage (Fig. 28-2).

Fig. 28-2. Two principal causes of overheated electron tubes: (A) shorted filter capacitor, (B) shorted coupling capacitor.

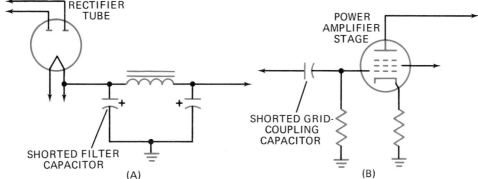

297

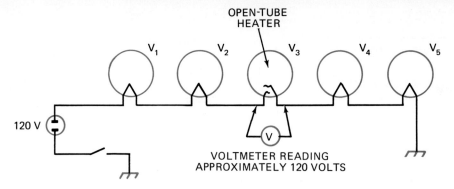

Fig. 28-3. Using a voltage test to locate an "open" electron-tube heater in a series circuit.

Heater Continuity Test. Since burned-out (open) heaters account for approximately 85 percent of the difficulties encountered with electron tubes, the technician is often interested in testing the heaters only. This can be done quickly and conveniently by removing the tube from its socket and connecting a volt-ohm-milliammeter across the heater pins. While performing this test, the ohms-range selector switch should be set at its lowest position. Otherwise, the voltage of the instrument battery may be sufficient to burn out a good tube.

Series-connected Heaters. It is somewhat of a problem to test tube heaters in a circuit where the heaters are operated in series, since any one open heater will affect all other tubes. In this case, each tube can be tested individually. However, an open heater in a series circuit can very often be much more quickly located by measuring the voltages across the heater pins with the circuit in operation. While performing this test, there will be no indication of voltage across those heaters that are "good." On the other hand, when the voltmeter is connected across a heater that is open, it will indicate a voltage that is approximately equal to the total voltage applied to the circuit (Fig. 28-3).

TRANSISTORS

If the visual inspection of a transistorized circuit indicates no obvious defects, it is reasonable to suspect a defective transistor although, as a general rule, these devices cause much less trouble than their electron-tube counterparts. While the transistor tester will provide a more realistic indication of operational efficiency, it is often desirable to "spot-test" a transistor by means of resistance measurements. These tests can be made with the transistor connected into its circuit if there are no low-resistance conducting paths (coils, resistors, etc.) connected across the electrodes. However, it is always good practice to remove a transistor from its circuit before attempting a resistance test, if this can be done without damaging the transistor or its associated circuit structure. A heat sink of some type should always be attached to the lead of any semiconductor device that is being unsoldered from a terminal.

The resistance testing of a triode-junction transistor is based upon the fact that such a transistor, in effect, consists of a combination of two diodes placed "back-to-back" (Fig. 28-4). Therefore, if the transistor is in good condition, each of the "diodes" will have a much lower resistance when biased in the forward direction than it has when biased in the reverse direction.

P-N-P Transistors. When a transistor is being resistance-tested, the ohms-range selector switch

Fig. 28-4. P-N-P transistor represented as a two-diode combination.

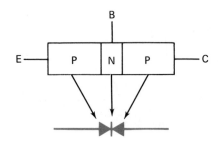

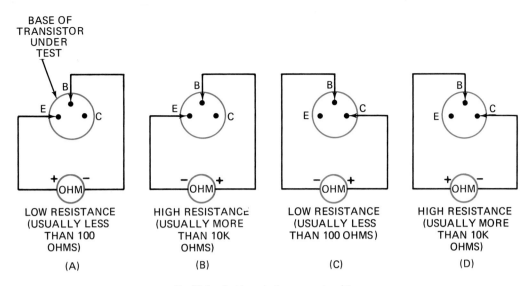

Fig. 28-5. Resistance-testing a p-n-p transistor.

of the volt-ohm-milliammeter should be set at its lowest ohms range. This will prevent the application of an excessive voltage across any pair of electrodes. The negative lead of the instrument is connected to the base and the positive lead to the emitter when testing a p-n-p transistor. This biases the base-emitter junction in the forward direction and the resulting resistance should be relatively low (Fig. 28-5A). When the leads of the ohmmeter are reversed, the junction is reverse-biased; thus the resistance should be considerably higher (Fig. 28-5B).

When it is to be tested, the base-collector junction is first forward-biased by connecting the ohmmeter to the base and to the collector as shown in Fig. 28-5C. The result again should be a low resistance measurement. By reversing the ohmmeter leads to the base and the collector, the base-collector junction is reverse-biased and should show a much higher resistance (Fig. 28-5D).

As a final test, it is desirable to measure the resistance between the emitter and the collector of the transistor. In typical transistors that are in good condition, the resistance between these electrodes should not be less than approximately 1,000 ohms, regardless of how the ohmmeter leads are connected to them.

N-P-N Transistors. N-P-N transistors can be resistance-tested in a similar manner (Fig. 28-6). With these transistors, however, the polarity of

Fig. 28-6. Resistance-testing an n-p-n transistor.

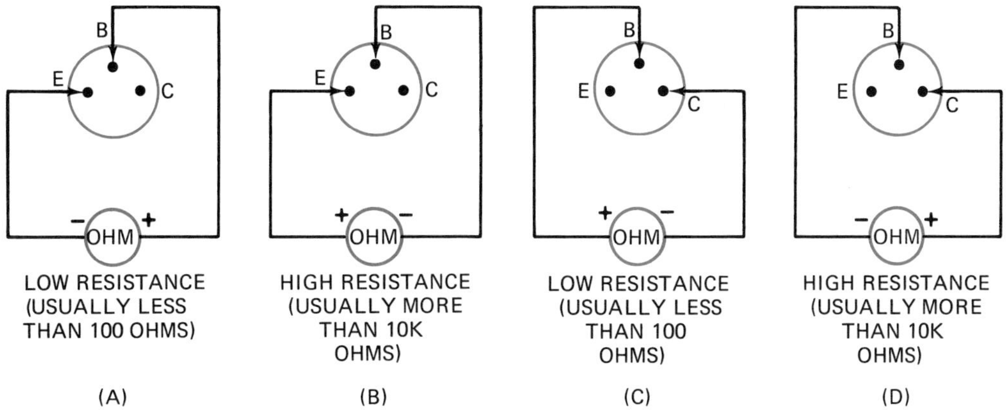

the voltage applied to the electrodes must be opposite to that applied to the electrodes of a p-n-p transistor to obtain the necessary forward- and reverse-bias resistance responses.

Shorts. If, while making a resistance test of a transistor, the resistance between any pair of electrodes is found to be extremely low (less than 1 ohm), there is reason to suspect a short between the electrodes. In this case, a reliable transistor tester should be used to more accurately determine the condition of the transistor.

Change of Operating Characteristics. Although resistance tests of a transistor usually provide a fair indication of its condition, they do not provide an indication of any change in its operating characteristics. These changes sometimes occur as the result of the effects of temperature during normal service life. Hence, a transistor that is designed to operate at radio frequencies may show a marked decrease in efficiency as a radio-frequency circuit device. When there is reason to believe that a change in the operating characteristics of a transistor has occurred, the transistor should, of course, be replaced.

Overheating. Under typical circuit conditions, all transistors except power (output) transistors remain comparatively cool during operation. Hence, if a transistor becomes overheated, the transistor itself may be shorted or an undesirable voltage may be present at an electrode. This condition most often results from the excessive leakage of an associated electrolytic capacitor. If the capacitors in the transistor circuit appear to be in good condition, the transistor should be replaced.

Determining Transistor Type. In performing many troubleshooting procedures, it is important to determine whether a given transistor is of the n-p-n or the p-n-p type. This is particularly true when a schematic diagram of the circuit in question or the pertinent technical data are not available.

The type of a transistor can be conveniently determined by following procedures similar to those given in connection with the resistance testing of transistors. To do this, first connect the negative lead of the ohmmeter to the emitter and the positive lead to the base of the transistor under test. Note the resistance between these electrodes. Next, reverse the leads of the ohmmeter to the emitter and the base and again note the resistance.

If the negative-lead-to-emitter, positive-lead-to-base resistance is significantly lower than the reverse-connection resistance, the transistor is of the n-p-n type; if it is significantly higher, the transistor is of the p-n-p type.

Transistor Substitution. It is always desirable to replace a given transistor with one of the exact type. However, this may not always be possible. In such cases, the transistor should be replaced with one having similar functional and operational characteristics. A satisfactory replacement-type transistor can be determined by referring to a transistor substitution guide, or by comparing the characteristics of transistors as indicated in a transistor manual.

DIODES

All types of common-use diodes can be checked by means of resistance measurements which, as with resistance measurements of transistor diode junctions, involves a comparison of forward resistance with reverse resistance. When testing a crystal diode in this manner, the resistance in the reverse direction should be approximately 100 times greater than the resistance in the forward direction (Fig. 28-7). The ratio of reverse to forward resistance will vary considerably in the case of silicon or germanium diodes, but both of these devices will have a much higher reverse resistance than forward if they are in good condition.

The resistance method of testing a diode with an ohmmeter (VOM or VTVM) can also be used to determine its cathode and its anode terminals. For example, when the ohmmeter is

Fig. 28-7. Resistance-testing a semiconductor diode.

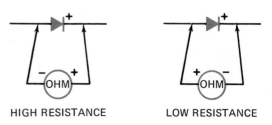

HIGH RESISTANCE   LOW RESISTANCE

connected to a diode in the low-resistance (forward) direction, the negative lead of the ohmmeter is connected to its cathode and the positive lead is connected to its anode.

ANALYSIS OF COMPONENT FAILURE

When circuit components such as electron tubes, transistors, and diodes are found to be defective, the inexperienced technician will often simply replace the defective devices and proceed to test the circuit in an effort to determine whether or not the trouble has been corrected. This procedure is proper only if it can be definitely determined that the failure was due to the termination of the component's normal useful service life. However, an assumption hastily made without a further analysis of the possible reasons for the component's failure may prove to be expensive and time-consuming, since certain circuit conditions will cause a component to be seriously damaged, regardless of its condition. If these conditions are not corrected, the "new" component will also be damaged, thus making another replacement necessary.

An example of "replacement without further investigation" is illustrated in the following paragraphs. Assume that the silicon diode of the rectifier circuit shown in Fig. 28-8 is found to be defective. The solution to this problem may be the simple procedure of replacing the diode. Suppose, however, that the replacement not only fails to correct the circuit difficulty but also results in immediate damage to the new rectifier. Obviously, there is now a real need for reasoning and analysis.

What circuit condition or conditions could cause this trouble? In many cases, further investigation will reveal that C_1, the input electrolytic capacitor to the filter circuit, is shorted. As a result, the rectifier becomes effectively connected directly across the power line, causing it to be overloaded beyond its normal current-delivering capacity. Realization of the possibility of this condition before the first replacement of the rectifier would have saved time and money.

VOLTAGE CHECKS

After testing the electron tubes and transistors and correcting any circuit condition that may have caused their failure, the next step in the troubleshooting process is to check voltages appearing at critical points of a circuit. These checks are essential, since active components (electron tubes and transistors) will not operate satisfactorily unless the voltages applied to their electrodes are of the proper magnitude.

Voltage Designation. The voltages that should appear at the various electrodes of transistors and tubes are usually indicated on the schematic diagram of the associated circuit (Fig. 28-9).

Fig. 28-9. Examples of voltage designations on a schematic diagram: (A) transistor circuit, (B) electron-tube circuit.

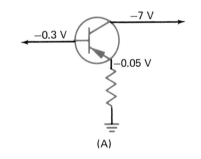

(A)

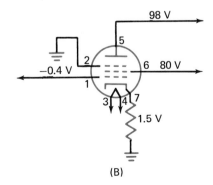

(B)

Fig. 28-8. Circuit used in conjunction with an analysis of component failure.

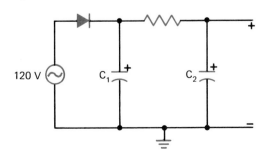

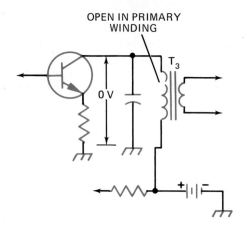

Fig. 28-10. Circuit used in conjunction with an explanation of voltage tracing a circuit.

Unless otherwise indicated, all are direct voltages measured with respect to the common (ground) point of the circuit. Although the voltages can, of course, be measured with a volt-ohm-milliammeter, it is sometimes desirable to use a transistorized or vacuum-tube voltmeter, since the insignificant loading effect of these instruments provides more accurate results.

Voltage-tracing the Circuit. After an improper voltage condition has been discovered, it is necessary to trace the circuit in an effort to determine the cause. This can be done conveniently by making voltage checks along the circuit until the trouble, usually an open or a short, is located.

For example, assume that the collector voltage of an n-p-n transistor stage that should be 8.3 volts actually measures zero (Fig. 28-10). Since the voltage is applied to the collector through the primary winding of output transformer T_3, it is logical to begin the voltage checks along this part of the circuit. If a measurement at the bottom of the winding indicates a voltage of approximately 8.5 volts, there is reason to believe that the primary winding is open. Thus, a voltage tracing check can be used as a continuity test to determine conditions that prevent voltage from being applied to a given point in a circuit.

When making voltage checks in a transistor circuit, it is important to prevent the test prods of the voltmeter from shorting across transistor electrodes. Such a short causes the voltage present at one electrode to be applied to another electrode. This voltage can often cause excessive current to pass through the transistor, and the transistor may be seriously damaged.

RESISTANCE CHECKS

In addition to voltage designations, a schematic diagram or its associated circuit data often indicates the values of resistance that should exist between various points of a circuit (Table 28-1). When this information is given, the point-to-point resistance measurements are very useful

Table 28-1. Example of point-to-point resistance data given with a schematic diagram.

	RESISTANCE MEASUREMENTS*								
	Tube Socket Terminals								
Tube	1	2	3	4	5	6	7	8	9
V1	30K	0	0	0	100K	150K	4M	X	X
V2	3M	0	0	0	120K	150K	200	X	X
V3	220K	500K	950	0	0	210K	0	750	0
V4	500K	0	0	0	100K	100K	200	X	X

*1. Resistance to be measured from each indicated point to ground with antenna disconnected from receiver. 2. All resistance measurements to be made with a VTVM or a TRVM.

in locating opens, shorts, and significant variations in resistance that may occur in the resistors as the result of excessive heating. Resistance checks are always performed with the circuit deenergized and all electrolytic capacitors discharged.

CAPACITORS

In his normal routine, the technician will find that defective capacitors, most often the paper and electrolytic types, are a frequent cause of circuit difficulty. This is largely due to the constructional features of capacitors and their widespread use in a variety of different applications.

The failure of a capacitor can usually be traced to the deterioration of the dielectric material between its plates. This condition can produce a short between the plates, excessive leakage through the dielectric, and a significant change in capacitance. While all these defects can, of course, be detected by using a capacitor checker, it is often preferable to perform on-the-spot tests by means of resistance and/or voltage measurements. These procedures are particularly effective for detecting shorts, opens, and leakage conditions. However, a capacitor checker is always recommended when it is necessary to test a capacitor for a possible change of its capacitance value.

Testing for Shorts. A resistance (continuity) test is used to test a capacitor for a shorted condition. The capacitor should first be isolated from the circuit by disconnecting one of the leads from its terminal point (Fig. 28-11). As a result, the resistance of the capacitor will not be affected by a parallel conducting path that would otherwise cause erroneous resistance readings. As a precaution, the capacitor should first be discharged to prevent possible damage to the meter or a dangerous shock to the technician.

When an ohmmeter is connected to the leads of any good nonelectrolytic capacitor, it will indicate an almost infinite resistance value. A resistance reading that shows continuity between the leads indicates that the capacitor is shorted or leaking and must be replaced.

Because of the chemical nature of its dielectric, an electrolytic capacitor that is not shorted will indicate some degree of continuity

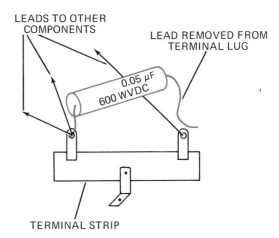

Fig. 28-11. Example of how a capacitor is isolated from a circuit when testing it for a shorted condition.

between the leads. Such capacitors that are in good condition and have a working voltage rating of 100 volts or more will usually have a resistance in excess of 300,000 ohms. Electrolytic capacitors that have a working voltage rating of less than 100 volts will most often have a resistance of from 100,000 to 500,000 ohms.

Since an electrolytic capacitor is a polarized device, its resistance will depend upon the "direction" in which the ohmmeter leads are connected. During a test, the ohmmeter leads are connected to the capacitor first in one way and then they are reversed. While doing this, a technician need not always observe the specific polarity of the ohmmeter connections. Instead, he can simply use the higher resistance reading as an indication of the condition of the capacitor.

The procedure described in the preceding paragraph is entirely satisfactory for resistance-testing an electrolytic capacitor that has a direct-current working voltage (WVDC) rating of 50 volts or more. However, when testing low-voltage capacitors such as those found in transistor circuits, the ohmmeter should be connected to a capacitor in the proper polarity (+ to + and − to −). If this procedure is not followed, the voltage applied to a capacitor from the ohmmeter battery may be sufficiently high to damage the capacitor.

Open Test. A fairly accurate open test of a non-electrolytic capacitor with a capacitance of more

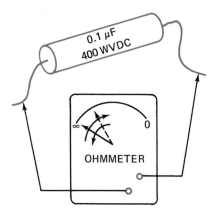

Fig. 28-12. Ohmmeter response indicating the charging and discharging actions of a capacitor that is in good condition.

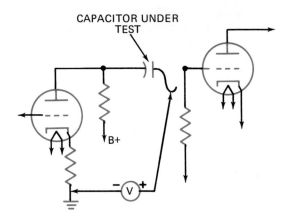

Fig. 28-13. Voltage-testing a capacitor for leakage.

than 0.05 microfarad can be made with an ohmmeter. To do this, the ohms-range selector switch is set at the highest ohms-range position so that the maximum voltage from the instrument battery can be applied across the capacitor.

When the leads of the ohmmeter are connected to a capacitor that is in good condition, the meter pointer will deflect noticeably and then return to the infinite ohms position on the scale (Fig. 28-12). Since this action represents the charging and discharging processes, it is an indication that the capacitor is responding in a normal manner. If the pointer does not move while the test is being performed, there is reason to believe that the capacitor is open.

A similar test of a good electrolytic capacitor will result in a greater deflection of the meter pointer and a final resistance reading of 100,000 ohms or more. If the capacitor is open, the deflection of the pointer will be considerably less than usual and the final resistance reading will approach infinity.

An open condition in an electrolytic capacitor can, of course, occur as the result of an actual break in a lead-to-plate connection. However, this condition most often occurs because the electrolyte dries up from age or from operating under conditions of high temperature. When this happens, the resistance of the electrolyte becomes extremely high, thus causing the capacitance of the device to be partially or completely destroyed.

Testing for Leakage. A capacitor that has an abnormally low resistance is often referred to as "leaking." Excessive leakage can sometimes be detected by means of a resistance test, but usually such a test is not at all valid. This is because in many cases capacitors will not leak when subjected to the relatively low voltage appearing across ohmmeter leads, but will do so under actual operating conditions.

To effectively test for leakage, a voltage test is necessary. To perform such a test, the lead at the "low-voltage" side of the capacitor is first disconnected from the circuit. The circuit is then energized and a direct-current voltmeter is connected between the lead and the ground, or common point, of the circuit (Fig. 28-13). If the capacitor is leaky (or shorted), the meter will indicate a voltage whose magnitude is determined by the extent of the leakage.

A small amount of capacitor leakage can be tolerated in some circuits. In others, however, even a very small amount of leakage seriously disrupts operation. This is particularly true when the capacitor in question is in the input circuit of an amplifier stage, thus significantly altering the bias condition of an electron tube or a transistor.

Checking Variable Capacitors. Variable capacitors sometimes become shorted because of the bending of one or more plates. However, such a capacitor may be shorted at only one position of the rotor plates. For this reason, a continuity test should first be performed with the plates fully meshed (closed). The ohmmeter should then remain connected to the terminals of the capacitor while the rotor plates are slowly unmeshed (opened). By watching the ohmmeter pointer while this is being done, a short between the plates at any position of the rotor plates can be detected.

In some cases, a shorted condition that results because the plates of a variable capacitor are rubbing together can be corrected by simply bending the plates away from each other at the point of contact. In other cases, bending the plates is impractical because of the numerous points of contact that may exist between them.

CHECKING POTENTIOMETERS AND RHEOSTATS

Potentiometers and rheostats often fail to operate correctly because the fixed resistance element is open or because there is improper contact between the sliding arm and the element. To check such a device, it is necessary to perform the following resistance measurement procedures:

1. Measure the resistance of the fixed resistance element by connecting the ohmmeter across the end terminals (Fig. 28-14A). The value of this resistance should be equal to the resistance rating of the device. If the meter pointer fails to move under this condition, the element is open.

2. Rotate the shaft full counterclockwise. Connect the ohmmeter to the center terminal and to that end terminal which produces a reading of direct continuity (Fig. 28-14B).

3. Slowly rotate the shaft in a clockwise direction. The meter should now indicate a gradual increase of resistance up to the maximum value (Fig. 28-14C).

4. Connect the ohmmeter to the center terminal and to the other end terminal. The meter reading should now show direct continuity.

5. Slowly rotate the shaft in a counterclockwise direction. The meter should once again show a gradual increase of resistance up to the maximum value.

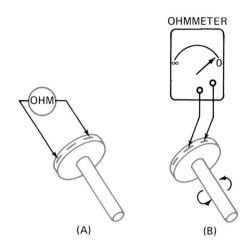

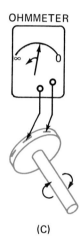

Fig. 28-14. Checking a potentiometer or a rheostat.

CHECKING COILS AND TRANSFORMERS

The most common defects found in coils and transformers are opens and shorts that occur as the result of broken connections to terminals, "burnout" of conductor wires, and insulation damage from deterioration or overheating. When the insulation is damaged, a short circuit can exist between turns and/or between conductors and the core in the case of an iron-core structure.

Coils (inductors). To check a coil for an open, it is necessary to perform a continuity test by connecting an ohmmeter across its terminals. A continuity test can also be used to check for a short between the turns and the core of an

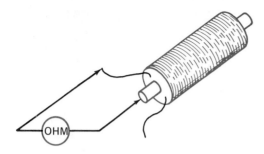

Fig. 28-15. Performing a continuity test to determine whether there is electrical contact between an inductor winding and its core.

iron-core coil. This is done by connecting the ohmmeter to either terminal (or lead) of the coil and the core (Fig. 28-15).

A short between two or more turns of an air-core coil is difficult to detect, since the change in resistance because of the short is usually so small that it cannot be accurately determined by a resistance test. Moreover, the direct-current (ohmic) resistance of most air-core coils is not specified. Thus, the measured resistance value cannot conveniently be compared with the normal resistance. When there is reason to believe that such a coil is shorted, it is best to replace the coil in question with one that is known to be in good condition.

A short between a number of turns in an iron-core coil or a transformer winding can often be detected by the excessive heat that is developed. However, it is important to note that this means of detecting a short is valid only if the current through the circuit is at a normal level.

In some instances when the ohmic resistance of a coil (such as a filter choke) is given, a seriously shorted condition can be detected by a resistance test. If the measured resistance obtained during this test is significantly lower than the specified resistance, the coil should be replaced.

Transformer Continuity Tests. To check a typical transformer, continuity tests are made between the leads of each separate winding and between each winding and the core. There should, of course, be continuity between the leads of each winding and no continuity between any winding and the core or between any two windings. It is important to note that if the device under test is an autotransformer, there will be continuity between all leads. In this case, the only effective check that can be made is a continuity test between the winding and the core.

A continuity test is also valuable for locating the different individual windings of a transformer when these are not indicated by means of a color code. While making this test, those leads across which continuity is observed should be tied or taped together so that they can be quickly identified when the need arises.

Transformer Voltage Tests. To completely "check out" a power transformer, it is often desirable to perform a voltage test across each secondary winding. This procedure is very useful in determining the secondary voltages when these windings are not color-coded.

Before attempting any voltage tests of a transformer, it is essential to properly identify the primary winding. If this is not done, the power line may be connected to a low-impedance (low-voltage) secondary winding, thus causing it to be damaged by the resulting high current. Before the power line is connected to the primary winding, the bare wire ends of each lead should also be examined to see that they are not in contact with each other. When making any voltage test of a power transformer, extreme caution should be observed, since a dangerously high voltage may exist across at least one of the secondary windings (the red leads).

If the transformer is in good condition, the voltage across each secondary winding will be approximately equal to that which is indicated by the color code or the manufacturer's specifications. If the voltage across any secondary winding is much less than normal, there is good reason to believe that a short exists between the turns of the winding in question.

TESTING LOUDSPEAKERS

A loudspeaker can be tested by making a continuity test across its voice coil, with the ohmmeter set at its lowest ohms-range position. Before making a continuity test of a voice coil, however, one of the loudspeaker terminals should be disconnected from its associated output transformer. Otherwise, continuity will be indicated through the secondary winding of the transformer, regardless of whether the voice coil is "good" or "bad."

Distortion in a loudspeaker is often produced when the form around which the voice coil is wound rubs against the pole piece. This most often occurs because of a misalignment of the voice coil or because of dust settling between the form and the pole piece, thus restricting the movement of the cone. To test for this condition, the cone is brought near one ear and gently pressed toward the pole piece with two fingers placed across the voice coil opening. If the voice coil is rubbing against the pole piece, a scratching sound will be heard. Such a condition can be corrected by having the loudspeaker reconed or by replacing it.

TESTING CELLS AND BATTERIES

The operating efficiency of a cell or a battery can be determined quite accurately by means of a voltage test. While such a test is being performed, the device must be connected to the circuit (load) with which it is normally used (Fig. 28-16). This is necessary because the voltage of a cell or a battery under an open-circuit (no-load) condition will be almost equal to the rated voltage, even though it is in a nearly discharged state. However, when such a cell or battery is connected to a circuit, the terminal voltage is greatly reduced because of the relatively large voltage (IR) drop occurring within the device as a result of its high internal resistance.

A cell or a battery can usually be considered to be in good condition if its voltage under load is at least 80 percent of the rated voltage. If the voltage is found to be below this level, the device should be replaced.

Fig. 28-16. Voltage-testing a transistor-radio battery.

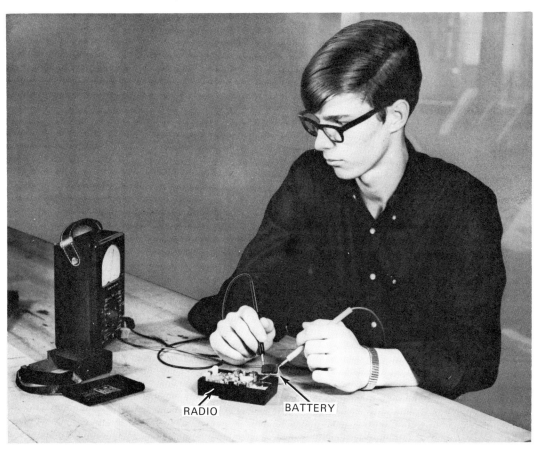

For Review and Discussion

1. What circuit conditions can result in the excessive current that sometimes causes the plate of a rectifier or an amplifier tube to glow cherry-red?

2. Describe the procedure to be followed when checking for an "open" electron-tube heater in a series circuit by means of voltage tests.

3. Give the steps to be followed in resistance-testing a p-n-p transistor.

4. How can a transistor type (p-n-p or n-p-n) be determined by means of resistance testing?

5. Explain the difference between the forward resistance and the reverse resistance of a diode.

6. Describe the procedure to be followed in ascertaining the cathode and the anode of a diode by means of resistance tests.

7. With reference to what point of a circuit are the direct-current voltages appearing at the electrodes of tubes and transistors most commonly measured?

8. Explain why it is necessary to isolate a component from its circuit before testing the component for resistance, for continuity, or for leakage.

9. How is a capacitor tested for an "open" condition by means of a resistance test?

10. Describe the procedure to be followed when testing a capacitor for leakage by means of a voltage test.

11. State the procedure to be followed when testing a variable capacitor for shorts by means of a resistance test.

12. How is a potentiometer or a rheostat checked by means of resistance tests?

13. Between which points of a typical power transformer that is in good condition should continuity be indicated?

14. What physical condition within a loudspeaker commonly causes it to produce a distorted output?

15. Why is it necessary to disconnect at least one terminal of a loudspeaker from the associated output transformer before checking the voice coil for continuity?

16. Explain why it is necessary to voltage-check a cell or a battery while it is connected to its associated load.

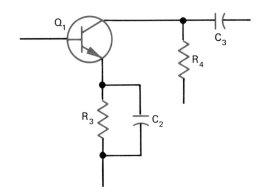

Fig. 29-4. Example of reference designations used on an electronics diagram.

Reference Designations. The symbols used to represent various components on a circuit diagram are most often accompanied by a combination of letters that identify the components but are not themselves a part of the symbols (Table 29-1). When more than one symbol of a specific type of component appears on a diagram, the letter (or letters) are followed by numbers that identify the components. These combinations of letters and numbers are referred to as *reference designations* (Fig. 29-4).

Numerical Values of Components. When details of the type, rating, or value of a particular component are to be given on a diagram, this information is placed adjacent to the symbol or is given by means of notes accompanying the diagram. For example, resistance and capacitance values are indicated as shown in Fig. 29-5. In this scheme, the symbol Ω for ohms and the abbreviations μF and pF for capacitance values are usually omitted and are instead replaced by notes such as the following:

1. All resistors expressed in ohms unless otherwise indicated.
2. All capacitors expressed in microfarads unless otherwise indicated.

A resistance value of 1,000 ohms or more is most often expressed in terms of kilohm (K) or megohm (M) units. Thus, a resistance of 1,500 ohms is written as 1.5K, while a resistance of 220,000 ohms may be written either as 220K or as 0.22M. The comma used when writing a 4-digit number such as 1,500 is not used when such a number is given on a diagram.

Capacitance values of 1 through 9,999 picofarads are usually expressed in picofarad units. Capacitance values greater than 10,000 picofarads should be expressed in microfarad units.

Fig. 29-5. Recommended methods of indicating reference designations and component values.

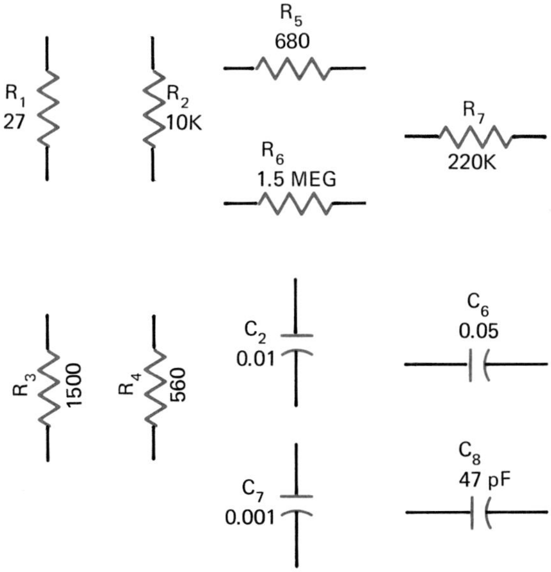

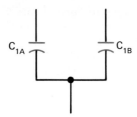

Fig. 29-6. Suffix letters used in conjunction with a capacitor.

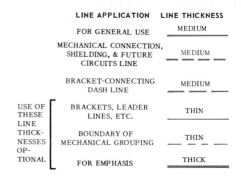

Fig. 29-7. Line conventions used with electrical and electronics diagrams.

DRAFTING PRACTICES

As in other areas of drafting, there is a body of information that is intended to serve as a general guide to some of the procedures that apply to the preparation of electrical and electronics schematic diagrams. The following drafting practices have been adapted from the *USA Standard Drafting Practices, Electrical and Electronics Diagrams* (*USAS Y14.15-1966*), with permission of the publisher, The American Society of Mechanical Engineers.

Suffix Letters. Subdivisions or parts of components often appear as a single, enclosed unit. To identify the separate parts of such a unit upon a diagram, suffix letters are added as shown in Fig. 29-6.

Layout. The layout or form of a diagram should show the main features prominently. The parts of the diagram should be carefully spaced to provide an even balance between blank spaces and lines. Enough blank space should be left in the areas near symbols to avoid crowding any necessary notes or reference information. However, large open spaces should not appear on a diagram, except when it is expected that additions to the diagram will be made in the future.

When planning the layout of a diagram, long connecting lines and line crossings (crossovers) should be kept to a minimum. To do this, the location of the prominent features and symbols (transistors, electron tubes, etc.) of the diagram must often be planned well in advance. This can be done by making a rough sketch of the diagram and then experimenting and modifying until it is certain that all the symbols and parts are in the proper location before the actual "finish" work is started.

Line Thickness. As with other types of diagrams, a schedule of line weights or line conventions is used in drawing electrical and electronics diagrams. The standard line conventions used in the preparation of these diagrams are shown in Fig. 29-7.

Connecting Lines. Lines connecting symbols and other parts on a diagram should, whenever possible, be drawn either horizontally or vertically. As a general rule, no more than three lines should be drawn to any one point on a circuit diagram (Fig. 29-8A). This procedure reduces the possibility of line crowding that could make the interpretation of a diagram more difficult than is necessary.

Fig. 29-8. Connecting lines: (A) recommended and undesirable methods of drawing lines to a point upon a diagram, (B) two "groups" of connecting lines drawn parallel to each other.

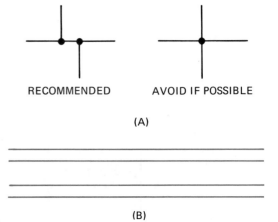

When connecting lines are drawn parallel to each other, the space between them should be at least $\frac{1}{16}$ inch when the diagram is reduced to a final size. Rather long "runs" of parallel lines should be drawn in groups, with double spacing between groups (Fig. 29-8B). It is always good practice to group together those lines that illustrate similar functions on a diagram. Power lines, electron tube heater lines, and antenna lead lines are common examples of similar-function lines.

Interrupted Lines. Connecting lines, whether single or in groups, may be interrupted when a diagram does not provide for a continuation of these lines to their final destination. When a single line is interrupted, the line identification can also indicate the destination (Fig. 29-9A). When groups of lines are interrupted, the destination of the lines is usually given in conjunction with brackets (Fig. 29-9B). In all cases, the identification necessary for use with interrupted lines should be located as close as possible to the point of interruption.

Dashed Lines. Dashed lines (----) are used on schematic and other types of diagrams to show a mechanical linkage between components or parts of components (Fig. 29-10A). Dashed lines are also used to show that a number of components are enclosed within a single container or sealed unit. For example, certain types of interstage transformers and modules may contain a number of components such as coils, capacitors, and resistors. These components are represented on a diagram by using their standard symbols, and the combination is then enclosed with a dashed-line rectangle or square (Fig. 29-10B).

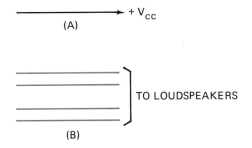

Fig. 29-9. Methods of identifying the destination of single and grouped connecting lines.

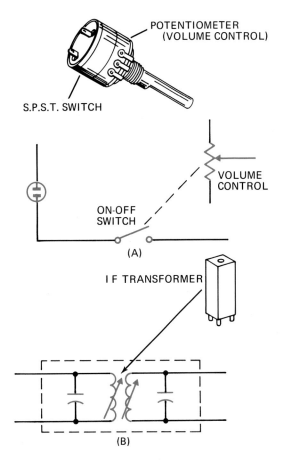

Fig. 29-10. Dashed Lines: (A) used to indicate the mechanical linkage of a potentiometer-switch combination, (B) used to indicate the component content of an enclosed unit.

THE PICTORIAL DIAGRAM

The pictorial diagram is a picture drawn to show the components of a circuit, their location upon a chassis or panel, and how they are connected together (Fig. 29-11). The diagram is usually drawn to a scale that shows the circuit and its related components in a size that is proportional to the finished product. Wire connections are drawn with as little crossover as possible and with as much of the wiring as possible in full view. Circuit components appearing upon the diagram may or may not be identified by name and rating, depending upon the policy of the company that uses the diagram.

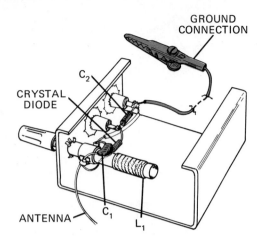

Fig. 29-11. Pictorial diagram.

The outstanding feature of pictorial diagrams is the ease with which they can be read and interpreted. For this reason, such diagrams are widely used for instructional purposes, as guides in assembly line production work, for maintenance activities, and for assembling "do-it-yourself"-type kits of all kinds.

THE SCHEMATIC DIAGRAM

A schematic diagram is one that shows, by means of graphical symbols, the electrical connections and the functions of the different parts of a circuit or a combination of circuits (Fig. 29-12). It does not illustrate the physical size, shape, or chassis location of the component parts and devices. Also, a schematic diagram does not usually show mounting devices or terminals (and lugs) to which wires are connected in constructing the circuit assembly.

The symbols on a schematic diagram are arranged so that the diagram can be "read" from left to right. The symbols are located in a position that will show the order in which each of the components functions in the operation of the circuit. As a result, the schematic diagram provides a convenient way of tracing a signal from input to output or of tracing a circuit operation from start to finish. Because of this feature, schematic diagrams are used in all areas of engineering and technical activities, including service work.

In addition to component values, schematics often show critical point color coding, voltage values, direct-current (ohmic) resistance of inductors, and polarities of electrolytic capacitors and diodes. The notes on the diagram may explain component values, give voltage and resistance measuring information, and indicate the external power supply connections to be made.

Other information often given in conjunction with schematic diagrams includes:

1. Wiring specifications
2. Operating instructions
3. Alignment (tuning) or adjustment procedures
4. Pictorial-type illustrations showing the locations of prominent components and tuning slots or nuts
5. Power and voltage ratings of components

Fig. 29-12. Schematic diagram.

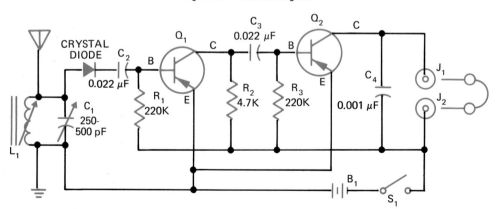

6. Point-to-point resistance values
7. Current or voltage waveforms at certain points of a circuit
8. Assembly and/or disassembly instructions
9. Input and output impedance values
10. Specific component description.

THE SINGLE-LINE OR ONE-LINE DIAGRAM

A single-line diagram shows the component parts of a circuit or circuit system by means of single lines and appropriate graphical symbols (Fig. 29-13). The single lines represent the two or more conductors that are connected between the components in the actual circuit.

The single-line diagram provides the necessary basic information about the sequence of a circuit system, but does not give the detailed information that is found on schematic or other types of "connection" diagrams. As a result, single-line diagrams are particularly suited to illustrating complicated circuits in a simplified manner.

THE FUNCTION DIAGRAM

The ever increasing complexity of circuit systems and the use of modular circuit system structures have brought about the development of a type of single-line diagram commonly referred to as a *function diagram*. In such a diagram, each symbol represents a circuit system module or operational unit (Fig. 29-14). This kind of diagram greatly reduces the time and space required to prepare a working diagram as compared with the time and space that would be required if all components of each module were represented in the conventional schematic manner. Since individual modules of such a circuit are rarely serviced on location, the function diagram fully satisfies the need to illustrate only the functional relationship between the different modules of a given circuit system.

THE BLOCK DIAGRAM

The block diagram is used primarily to show the relationship between the various component groups or stages in the operation of a circuit. It indicates in block form the path of a signal through a circuit or the operational sequence that is performed by a circuit (Fig. 29-15).

The blocks are most often drawn in the shape of squares or rectangles that are joined by single lines. Arrowheads located at the terminal ends of the lines are used to show the direction of the signal path from input to output as the diagram is read from left to right. The information that is necessary to describe components or stages on the diagram is placed within the blocks as a general rule. On some block diagrams, devices such as amplifiers, antennas, loudspeakers, and output meters are represented by standard symbols, instead of by squares or rectangles.

A block diagram shows the functions of the various stages of a circuit in the order of their operation, but does not give any information regarding specific components or wiring connections. As a result, this type of diagram is limited in use but does provide a simplified

Fig. 29-13. Examples of symbols used on single-line diagrams.

AMPLIFIER, GENERAL

BUZZER

HEADSET, SINGLE OR DOUBLE

LOUDSPEAKER, GENERAL

MICROPHONE

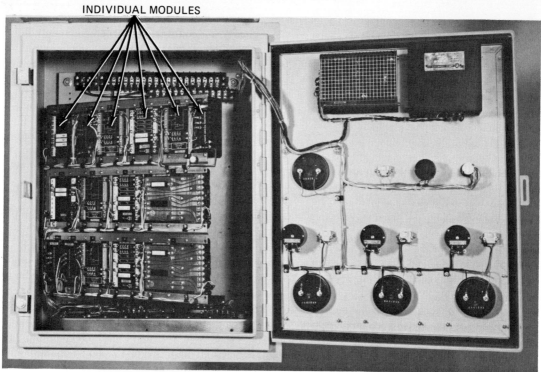
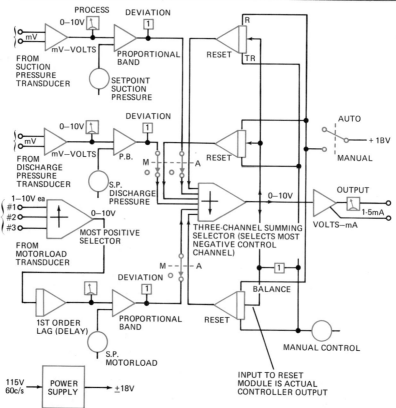

Fig. 29-14. Principal modular units of a liquid pipeline control circuit system, and associated function diagram. (*Consolidated Electrodynamics Corp.*)

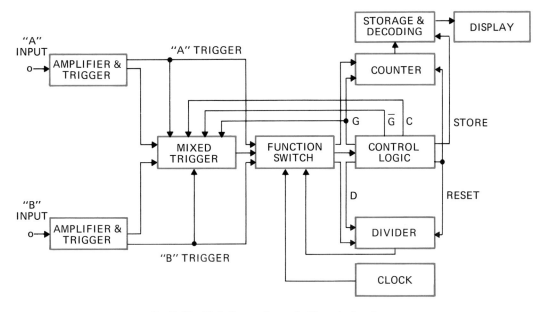

Fig. 29-15. Block diagram of a counter/timer circuit system.

means of illustrating the overall features of a circuit. For this reason, block diagrams are commonly used by engineers and technicians as the first step in designing new equipment. After determining the principal stages of a circuit system and illustrating them on the diagram, which is often in sketch form, the designing process continues on to specific circuit structures. Block diagrams are also widely used in catalogs, advertising copy, and in product description folders.

THE WIRING OR CONNECTION DIAGRAM

The wiring or connection diagram is used to show wiring connections in a simplified, easy-to-follow manner. It may show either internal or external connections or both and is usually drawn as simply as possible to trace out the connections of a circuit system (Fig. 29-16). The components of the circuit are identified by name or are represented by means of pictorial illus-

Fig. 29-16. Automobile interior and trunk lighting wiring diagram.

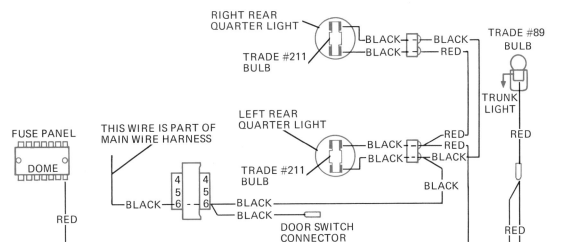

trations that do not follow any well-defined (standard) form. In addition to the component parts, auxiliary devices such as terminal blocks, terminal strips, and fuse mountings are also generally shown on a wiring diagram.

Although a wiring diagram is not drawn to any exact scale, the components are most often arranged to show their relative locations within a given space or area. Since the wiring diagram is primarily a connecting and visual troubleshooting aid, it usually includes the color-coding scheme that is to be followed in preparing such a circuit.

Wiring diagrams are frequently used in manufacturing processes and are widely used in the installation, wiring, maintenance, and modification of equipment. These diagrams are also sometimes used in conjunction with schematic diagrams to provide the additional wiring and connecting information that is necessary to do a particular job.

For Review and Discussion

1. Name four organizations that are responsible for standardizing electronic drafting and symbol usage.
2. For what purpose is a drafting template used?
3. Explain why some of the symbols often found on diagrams do not conform exactly to a typical standard.
4. State five rules for drawing symbols.
5. What is meant by a reference designation?
6. Using appropriate diagrams, show how reference designations and numerical values are correctly given in conjunction with resistors that are drawn horizontally and vertically.
7. Define suffix letters.
8. State the most significant practices followed in laying out a circuit diagram.
9. What is meant by standard line conventions?
10. State three general practices followed when drawing connecting lines on a diagram.
11. Give two examples of similar function lines.
12. For what two purposes are dashed lines used on a diagram?
13. What is the outstanding feature of the pictorial diagram?
14. State several important features of the schematic diagram.
15. Name several supplementary items of information that are often given in conjunction with a schematic diagram.
16. Describe the block diagram and state two of its applications.

UNIT 30. Assembly and Wiring Techniques

The basic steps in constructing a circuit assembly are generally the same, whether the unit to be constructed is complex or simple. These steps include locating and mounting the circuit components, wiring, inspecting, and testing. Because the proper operation of a circuit depends upon the correct location and connection of every conductor, a knowledge of the methods and procedures of producing a neat, accurate, dependable wiring job is essential. Nothing that the technician does creates an impression of his ability more quickly than the appearance of his wiring job. Although a neat job does not guarantee a properly wired circuit, it is most often an indication of good workmanship.

LOCATION OF COMPONENTS

The layout of the components upon or within a chassis is usually best achieved by using a mockup upon which the components are moved

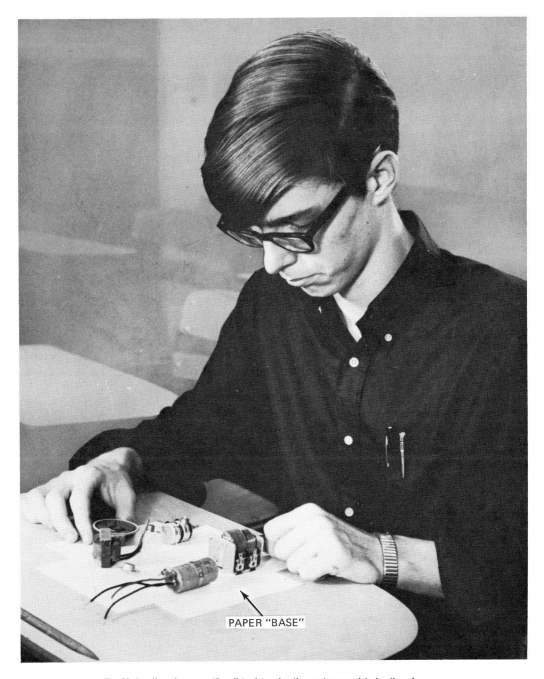

Fig. 30-1. Use of a paper "base" to determine the most appropriate location of components and the size of a chassis.

about until a desirable arrangement is determined. The base of the mockup can be a sheet of paper cut to the size of the mounting surface (Fig. 30-1). Careful planning of the location of the major components and wiring devices such as jacks and binding posts is important since it will be most convenient to drill and/or punch the necessary mounting holes in the flat chassis metal (chassis blank) before it is bent to the desired shape.

When the locations of all components have been determined, they can be transferred to the chassis surface by marking the surface directly or by preparing a chassis layout diagram that will be used as the chassis is punched and formed. Component outlines and other dimensional data should be marked upon a chassis surface with either a soft-lead pencil or a ball-point type of marker. A metallic object such as a scriber will leave scratches that are difficult to remove or to cover completely with paint.

Five basic factors should determine the location of components upon or within a chassis. These are: functional sequence, heat dissipation, radiation of magnetic fields, replacement and test convenience, and safety.

Functional Sequence. A satisfactory component layout usually begins with a consideration of the functions of the major components. Associated components should, insofar as possible, be located near each other in a functional (operationally related) sequence and should be arranged so that the interconnecting leads are as short as possible. This reduces the possibility of undesirable electrical and magnetic interaction between the conductors and also adds to the appearance of the wiring job. Several recommended component-location practices are:

1. Power supply components such as transformers, choke coils, rectifiers, and filter capacitors should be located so as to form a composite unit.

2. Power cords should enter the chassis at a point near the component or components to which they are connected.

3. Antenna leads or terminals should be located near the "front end" circuit components such as tuning capacitors and coils.

4. Amplifier stages consisting of transistors or electron tubes should be arranged so that the signal passes along the most direct route from input to final output, with each stage located as near as possible to the preceding stage.

5. In a radio receiver or an amplifier, the antenna terminal or input jacks should be located near the first transistor or electron-tube amplifier stage.

6. Volume- and tone-control potentiometers should be located near the transistor or tube stage with which they are associated.

7. Output transformers should be located near the final amplifier stage and near the loudspeaker or the output jacks or terminals.

Heat Dissipation. Electron tubes, power transformers, and power resistors develop considerable heat during operation. For this reason, these components should be located as far as is practical from semiconductor devices (diodes and transistors) and from electrolytic capacitors that are subject to deterioration from the effects of heat. Heat-producing components should also be located at a distance from tuning components such as variable capacitors and inductors whose operation is affected by fluctuations in temperature.

Power transistors must be protected from the heat that is developed within them during operation, as well as from the effects of radiated heat. This is done by mounting the component on some form of heat sink. In some circuits the metallic body of the transistor, which serves as the collector terminal, is mounted directly upon the chassis or a portion of the chassis that serves as the heat sink. More often, however, the body of the transistor must be electrically insulated from the chassis by using a mounting arrangement as shown in Fig. 30–2. When this kind of mounting arrangement is used, a silicone grease (compound) should be applied between the transistor and the mica insulator and between the insulator and the chassis. The silicone will allow the maximum conduction of heat from the transistor to the chassis, thereby increasing the ability of the chassis to serve as a heat sink.

Radiation. The currents in the leads extending to power transformers, filter chokes, and power switches are accompanied by relatively strong magnetic fields. If these conductors are placed adjacent to other leads that form a portion of the input circuit to the amplifier stages, the magnetic fields will induce voltages in the input circuits. These voltages are then often passed on to a loudspeaker, thus creating a 60-hertz hum.

To reduce the level of hum noises caused by the radiation of magnetic fields, all power

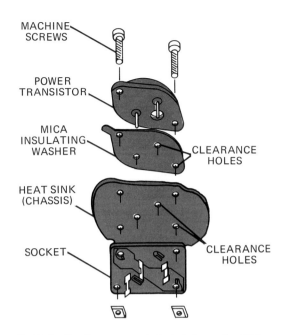

Fig. 30-2. Mounting arrangement of a typical power transistor.

cord, switch, and electron-tube heater leads should be kept as far as possible from other conductors. To further minimize the interference from magnetic fields, these leads should be twisted whenever possible. This effectively cancels the magnetic fields about the leads.

Replacement and Test Convenience. When planning the location of components, some thought should be given to replacing components and servicing circuits in the future. For the sake of convenience, mounting screws used to securely fasten heavy components to a chassis should be located so that they are readily accessible without using special tools and fixtures. Whenever possible, components should be mounted so that they can be removed from the chassis without having to remove adjacent major components or subassemblies.

For convenience in servicing, components should be located so that there is sufficient space to use soldering devices and instrument probes on the terminals and the terminals of adjacent components. Components such as intermediate-frequency transformers, oscillator coils, and variable capacitors that are equipped with adjustment screws or "slug" sockets should be mounted so that they can be readily adjusted without the danger of causing a short circuit to any high-voltage point in a circuit.

Safety. Components such as rectifiers, switches, voltage-dropping resistors, bleeder resistors, and capacitors that operate at line voltages or high, direct voltages should, if at all possible, be mounted under the chassis. This reduces the chances of accidentally contacting their terminals and shorting them with tools and other metallic objects. If it is impossible to place high-voltage components beneath the chassis, they should be enclosed within a vented housing. This will protect them and prevent accidental contact with their terminals.

PREPARING THE CHASSIS

The chassis used in conjunction with typical hand-wired circuits (projects) can be formed from #16 to #20 gauge copper, aluminum, or iron sheet (Fig. 30-3). The "open-end" type of chassis is well suited to the construction of small-size projects. The box chassis should be used whenever heavy components such as power transformers and choke coils are to be mounted upon a relatively large chassis, since this design provides a much stronger supporting surface than that provided by the open-end chassis.

The Chassis Layout. The chassis blank metal should be cut to size only after the appropriate chassis layout diagram has been prepared

Fig. 30-3. Chassis construction: (A) open-end type, (B) box type.

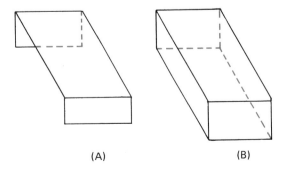

(A)   (B)

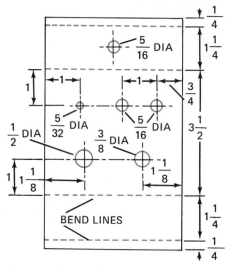

Fig. 30-4. Chassis layout diagram.

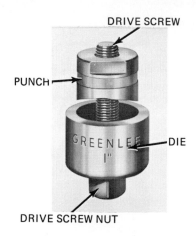

Fig. 30-5. Round chassis punch. (*Greenlee Tool Co.*)

(Fig. 30-4). This practice will reduce the waste of metal and will allow the job to proceed in a more logical and orderly manner. Although the metal can be cut with tin snips, a squaring shear will give a more accurate "straight-edged" job.

Drilling and Punching Holes. After all the holes and other openings have been located on the chassis blank, the next step is to drill or punch them. Small holes of up to $\frac{1}{2}$ inch in diameter can be drilled through the metal. This should be done only after the center of each of the holes has been "center-punched." If the center punch is not used, the twist drill has a tendency to "crawl" over the surface of the metal, missing the correct location of the holes, and marring the metal unnecessarily.

Larger, round holes and other openings that are square or D-shaped are most conveniently punched in sheet metal up to #16 gauge in thickness, using a chassis punch (Fig. 30-5). The steps to be followed when using this device are as follows:

1. Locate and center-punch the center of the hole (or opening).

2. Drill a hole through the chassis that is slightly larger in diameter than the drive screw of the chassis punch to be used.

3. Assemble the chassis punch into the hole as shown in Fig. 30-6.

4. Tighten the drive screw until the punch has cut through the metal at all points along the edge of the opening.

Bending the Chassis. After the holes and/or other openings have been drilled and punched, the next step is to bend the metal to the desired shape. This is most conveniently done by using a box and pan brake (Fig. 30-7). When forming an open-end chassis, the metal can also be bent with a bar folder. When the box and pan brake

Fig. 30-6. Chassis punch-sheet metal assembly.

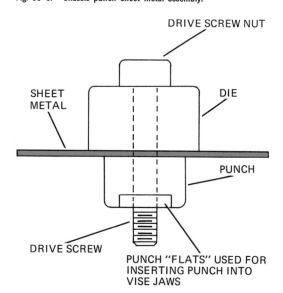

Fig. 30-7. **Hand-operated box and pan brake.** (*The Peck, Stow, and Wilcox Co.*)

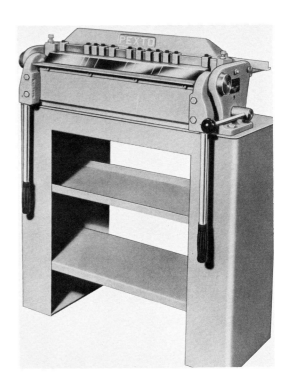

or the bar folder are not available, a small chassis can be formed by using a handy seamer or a wood clamp assembly (Fig. 30-8).

MOUNTING OF COMPONENTS

Heavy components such as transformers and filter chokes must be securely mounted to the chassis, preferably with machine screws (Fig. 30-9A). Plated machine screws should be used to provide both mechanical stability and electrical contact. The plated surface (usually nickel) is much more resistant to corrosion than steel from which ordinary machine screws are made.

The size of a machine screw is given by a combination of numbers indicating the overall diameter of the screw shank, the number of threads per inch on the shank, and the length of the shank. This is commonly indicated as, for example, #6-32 × 1/2". Here, the 6 refers to the

Fig. 30-8. Hand-bending a small chassis.

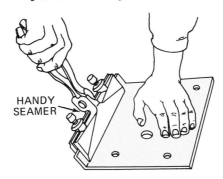

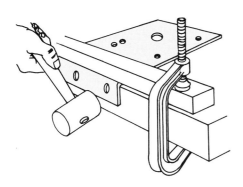

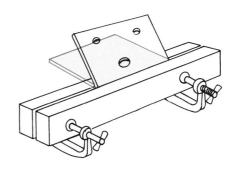

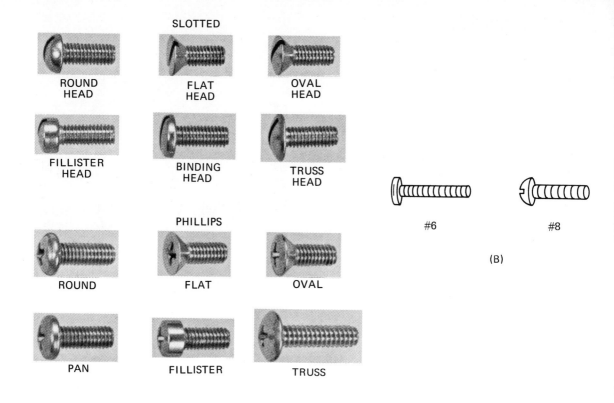

Fig. 30-9. Machine screws: (A) types of machine screws, (B) diameters of #6 and #8 machine screws.

Fig. 30-10. Lock washers: (A) tooth type, (B) spring type, (C) example of how a lock washer is used.

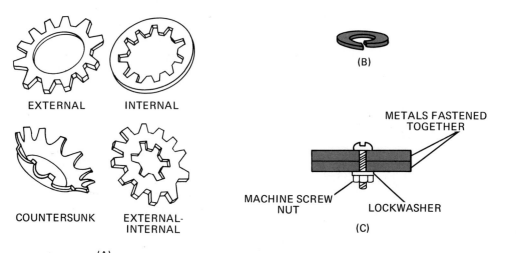

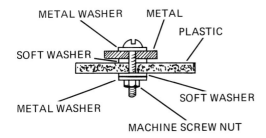

Fig. 30-11. Use of a soft washer when mounting a metallic device upon a plastic surface.

number of the diameter indication increases, the diameter of the screw shank also increases. Thus, a #6 machine screw is smaller in diameter than a #8 screw (Fig. 30-9B).

To prevent a component from becoming loose from vibration or because of the contraction and expansion of metals, the machine screws are often used in conjunction with lock washers (Fig. 30-10). When a metallic device is to be fastened to a plastic surface, soft fiber washers and flat metal washers should be used as shown in Fig. 30-11. This prevents the plastic from being scratched and also reduces the possibility of the plastic being cracked.

In many cases, components are mounted more conveniently with self-tapping (sheet metal) screws (Fig. 30-12). To use such screws,

diameter, the 32 indicates that there are 32 threads per inch upon the shank, and the 1/2 indicates that the shank is 1/2 inch long. As the

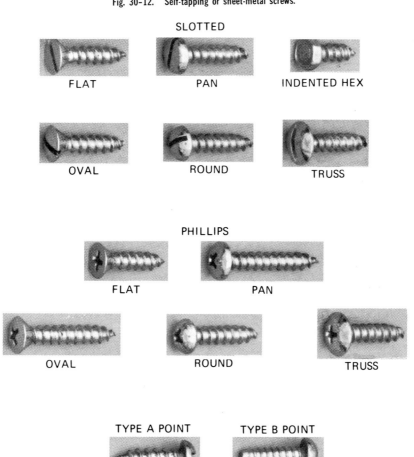

Fig. 30-12. Self-tapping or sheet-metal screws.

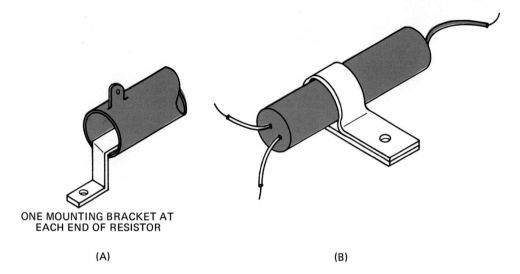

Fig. 30-13. Component mounting brackets: (A) used with power resistor, (B) used with capacitor.

a hole slightly smaller in diameter than the core of the screw is first drilled through the metal. The screw then cuts its own threads into the metal, thereby providing a strong contact between surfaces.

The self-tapping screw is most satisfactorily used with relatively thick metals. When used with thin metals, this type of screw has a tendency to work loose, particularly if it is repeatedly removed and reinserted.

Fig. 30-14. Mounting "can"-type capacitors: (A) mounting wafer, (B) securing capacitor to mounting wafer.

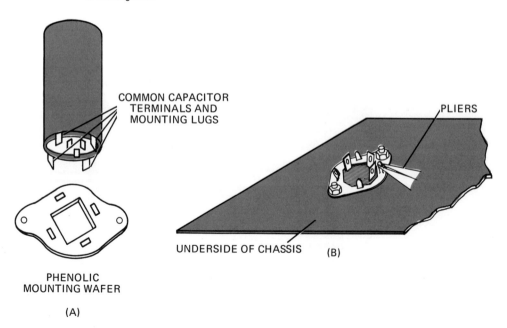

Large Resistors and Capacitors. Mounting brackets should be used to secure large (power) resistors and tubular capacitors firmly to the chassis. They are usually supplied with these components or can be purchased individually (Fig. 30-13). To save space, metal enclosed "can"-type electrolytic capacitors are most often mounted vertically. When the container (negative terminal) of such a capacitor must be insulated from the chassis, a phenolic mounting wafer is used (Fig. 30-14A). The capacitor is then firmly attached to the wafer by twisting each of its mounting lugs approximately one quarter turn (Fig. 30-14B).

Insulators and Terminal Strips. Components that operate at high voltages are often insulated from the chassis by mounting them upon standoff insulators (Fig. 30-15A). More typical components, such as small resistors and capacitors, are secured in place and insulated from the chassis by using terminal strips (Fig. 30-15B). The lugs of terminal strips also provide a very convenient means of terminating wire connections at different points on a chassis. Screw-type barrier terminal strips or terminal boards provide convenient, temporary connections or terminals for connecting leads that extend from a chassis (Fig. 30-16).

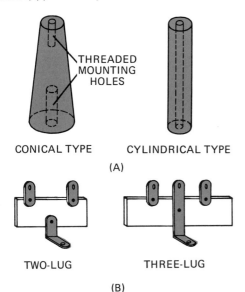

Fig. 30-15. Insulator mounting devices: (A) standoff insulators, (B) terminal strips.

Fig. 30-16. Barrier-type terminal board. (*Kulka Electric Corp.*)

Jacks and Connectors. Phone jacks, banana jacks, tip jacks, and other "in-hole" mounted devices are insulated from a metal chassis by means of shoulder-type, flat fiber washers (Fig. 30-17A). In addition to providing insulation for the flat metallic surfaces of these devices, the shoulder of the shoulder-type washer insulates the threaded portion of a jack body or a mounting screw from the inner surface of the mounting hole. To properly "seat" a shoulder-type washer, the diameter of the mounting hole should be made slightly larger than the outside diameter (O.D.) of the shoulder (flange) of the washer (Fig. 30-17B).

Standard phone jacks are available in several different circuit "make-and-break" types. Among the most common of these are the two-conductor open-circuit (OC) and the closed-circuit (CC) types. The plug associated with the open-circuit jack makes contact with two jack terminals (Fig. 30-18A). Thus, the circuit to which the jack is connected is in an open condition when the plug is removed.

The closed-circuit jack is similar to the open-circuit jack, except that it is equipped with an additional contact point (Fig. 30-18B). When the plug associated with the jack is removed from the jack, the circuit between terminals *A* and *B* is closed.

The most common application of the closed-circuit-type jack is found on radio receivers and amplifiers that can be operated in conjunction with either a loudspeaker or a headset. When the headset plug is inserted into

329

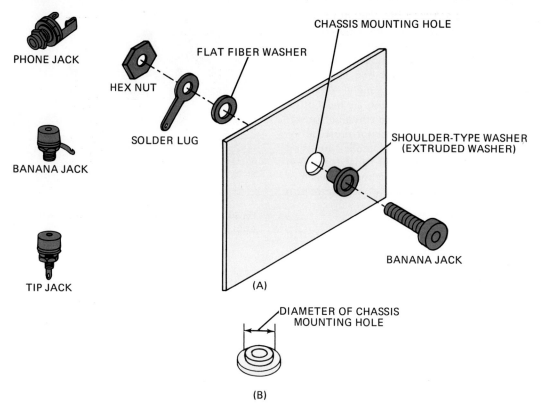

Fig. 30-17. Common types of jacks and how they are mounted on a chassis.

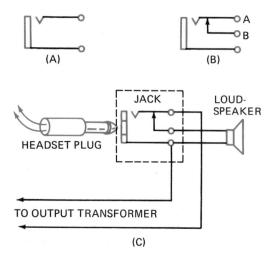

Fig. 30-18. Phone jacks: (A) schematic representation of open-circuit jack, (B) schematic representation of closed-circuit jack, (C) use of closed-circuit jack in loudspeaker circuit.

the jack, the loudspeaker is automatically disconnected from the circuit. When the headset plug is removed from the jack, the loudspeaker is automatically connected into the circuit (Fig. 30-18C).

Potentiometers and Rheostats. Controls such as potentiometers and rheostats are generally not insulated from a chassis. But these devices should be mounted securely so that they will not tend to turn as the shaft is rotated. To prevent them from turning, they are often equipped with mounting tabs or positioning lugs that are inserted into slots cut into the chassis (Fig. 30-19A). To simplify the installation by eliminating the need for chassis slots, the positioning lugs can be turned back or broken off and a lock washer can be used to mount the control (Fig. 30-19B).

When mounting a control, the threaded portion of the control body is usually inserted

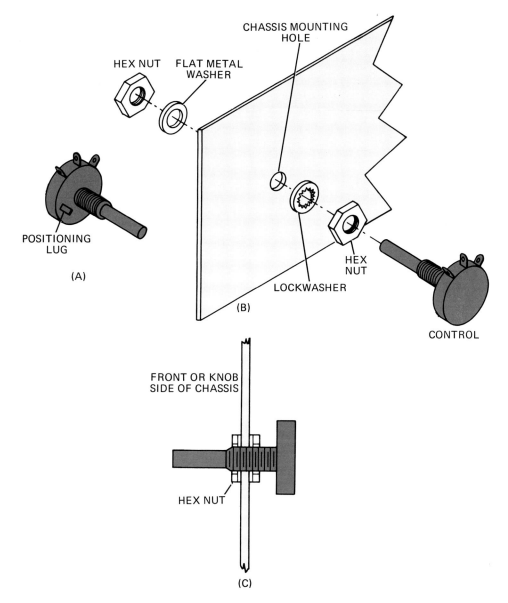

Fig. 30-19. Mounting the potentiometer.

through the mounting hole so that it extends beyond the knob side of the chassis only as far as is necessary to accommodate the front mounting nut (Fig. 30-19C). This procedure improves the appearance of the total job.

WIRES AND WIRING TECHNIQUES

After the major components have been mounted upon the chassis, the circuit is ready for wiring. The general method of routing the wires depends upon the nature of the circuit and the amount of time that is available for the job.

Wires. The specific type of wire to be used in wiring any circuit or part of a circuit is usually selected on the basis of its current-carrying capacity, flexibility, and insulation properties. The most common type of wire (hookup wire) used for interconnecting components is a solid or stranded wire ranging in size from #18 to #24 AWG and insulated with vinyl, thermoplastic, or a fabric braid. The wire itself may be bare copper or pretinned. Pretinned wire is coated with a thin layer of solder. This corrosion prevention coating makes the wire easier to solder to any terminal point.

Solid hookup wire has the advantage of being rather stiff; thus it can be readily bent and routed along a given path from which it will not twist or spring out. However, solid wire can become weakened by nicks and abrasions, and the bending that occurs when wires are unsoldered and resoldered during servicing or circuit changing procedures. Because stranded wire is more flexible and less subject to breakage from bending, it should always be used when connecting components that must undergo serious vibration during operation. Stranded wire should also be used in making semipermanent (frequently changed) connections to terminal points such as jacks and terminal screws.

Shielded Wire (cable). The insulation of ordinary hookup wire will not isolate or shield the wire from extraneous magnetic fields. When a wire such as the input lead to an amplifier stage must be shielded from magnetic fields, shielded wire is used. This wire or cable consists of one or more individually insulated conductors enclosed within a braided shield, usually made of fine copper wire (Fig. 30-20A). Since the shield is generally connected to the ground or common point of a circuit, any voltages induced within it are effectively isolated from the inner conductor(s) (Fig. 30-20B).

Shielded cable is widely used for connecting microphones to amplifiers and for other signal

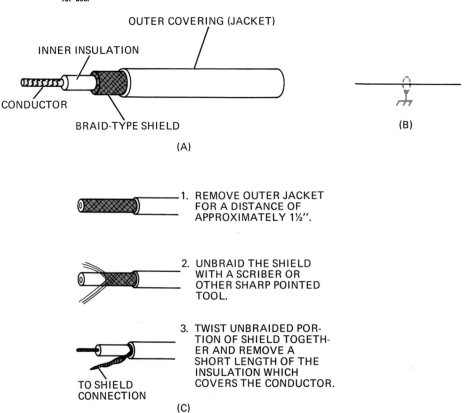

Fig. 30-20. Single-conductor shielded or coaxial cable: (A) structural features, (B) symbol for single-conductor cable with shield connected to chassis, (C) preparing cable for use.

transmission lines. Such cable is also commonly referred to as coaxial cable, since all of its conductors share a common axis.

Shielded cable is prepared for use as shown in Fig. 30-20C. To prevent the internal and the braid conductors from unraveling, they should be tinned or coated with solder after the insulation is removed from them.

Removing Insulation. The removal of insulation from wires before making a connection can be a troublesome chore for the inexperienced person. To make the job easier, several types of insulation stripping tools are available (Fig. 30-21). Although these tools are comparatively simple, they must be operated properly to avoid cutting and nicking the wires.

The insulation from a small wire can be conveniently removed by using the wire cutting jaws of pliers. To do this, the wire is placed between the jaws, which are then brought into gripping contact with the insulation. The insulation is then removed by pulling the wire and the pliers in opposite directions (Fig. 30-22). This procedure often requires a bit of practice before the job can be done without seriously scratching or nicking the wire.

Enameled wire insulation is removed by using sandpaper or emery cloth, or by scraping it with a knife blade or some other metallic edge.

The insulation should be removed from both ends of a wire after the wire is cut to length and before a solder connection is made at either end. If this is not done, the wire may be strained

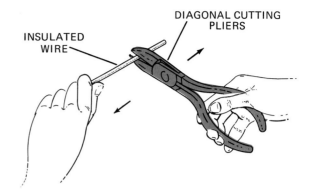

Fig. 30-22. Removing wire insulation with diagonal-cutting pliers.

as the insulation is removed with one end of the wire connected. As a result, a solder connection may be broken or a terminal lug may be bent or pulled out of place.

To prevent the possibility of undesirable contact, no more than the necessary amount of insulation should be removed from the end of a wire. For ordinary wiring purposes, approximately $\frac{3}{8}$ inch of bare wire is sufficient when connecting wires to lugs.

Lugs. Lugs are used to make semipermanent wire connections to screw-type terminals. These devices may be of either the solder or the solderless type (Fig. 30-23). The solderless lug is

Fig. 30-21. Wire-stripping tool. (*Ideal Industries, Inc.*)

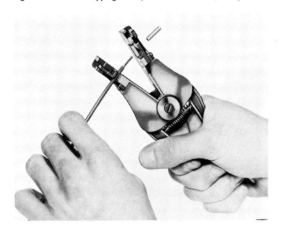

Fig. 30-23. Wiring lugs.

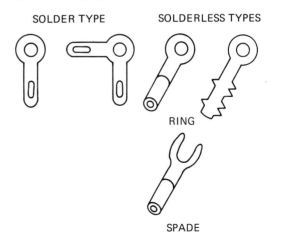

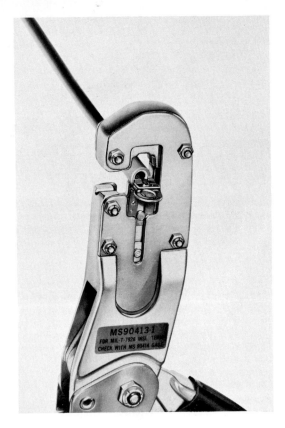

Fig. 30-24. Crimping tool. (*Buchanan Electrical Products Corp.*)

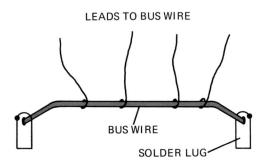

Fig. 30-25. Application of a bus wire.

fastened to the end of a wire by crimping it with what is referred to as a *crimping* or *lugging tool* (Fig. 30-24).

Connections to Chassis. It is very often necessary or desirable to make a wire connection to a metal chassis. While this can usually be done by soldering a wire directly to the chassis, it is much more convenient to use a solder lug for this purpose. To ensure a good connection, both the lug and the surface of the chassis at the point of contact should be thoroughly cleaned before the lug is mounted. The metal at the point of contact should also be smoothly tinned if there is a possibility of its becoming corroded.

Color Coding. Before making any wire connections, some thought should be given to a color code. The standard chassis wiring color code is given in Appendix H, page 467. While this code is useful, it does not provide a color code for all wires.

Although it cannot strictly be called a color code, it is recommended that several colors of wire insulation be made available for a wiring job. Different colors can then be used when several wires are extended through a chassis or in places where wires are grouped together. This practice makes it much more convenient to trace wire connections from one point to another during the wiring process and during testing procedures.

Bus Wire. In some circuit wiring jobs, it is often more convenient to make common-point circuit connections at a specific part of the circuit rather than have these connections made at various points of the circuit assembly. A wire that provides a terminal point for such connections is referred to as a *bus wire,* or simply as a *bus* (Fig. 30-25). This wire is usually a solid, pretinned wire with a relatively large cross-sectional area.

BREADBOARDING

Breadboard wiring involves the use of a baseboard, upon which a given circuit is constructed either by soldering it to terminals or by means of "plug-in"-type component assemblies (Fig. 30-26). This type of circuit construction is widely employed in experimental activities and in circuit design. Since the breadboard technique provides for extreme flexibility of connections, it is also very often used for constructing model or prototype circuits and for correcting operational difficulties in existing circuits. After a cor-

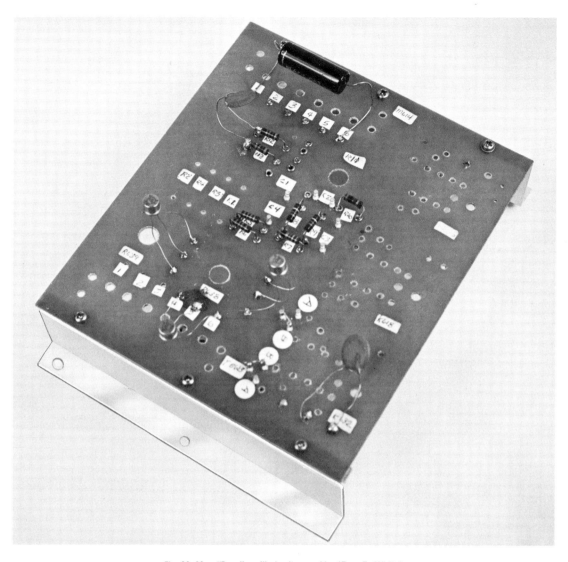

Fig. 30-26. "Breadboard" circuit assembly. (*Ray E. White*)

rect breadboard model is assembled, the circuit is ready for production upon the final chassis form or other assembly structure.

SOLDERING TECHNIQUES

The principal requirements for a good soldering job are: proper solder, a sufficient amount of heat, and clean, tinned surfaces. For general-purpose work, a 50–50 (50 percent lead, 50 percent tin) rosin-core solder with a diameter of $\frac{1}{16}$ or $\frac{3}{32}$ inch is adequate. An acid-core solder should never be used for soldering copper wire or other copper surfaces, since the acid reacts with copper to produce serious corrosion.

The Soldering Iron. The constructional features of a standard-type soldering iron are shown in Fig. 30-27. A 60- to 100-watt soldering iron provides enough heat for ordinary wiring jobs if its tip is kept clean and well tinned. The tinned surface permits the maximum amount of heat to be conducted from the soldering tip to the point at which a soldered connection is being

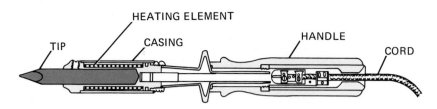

Fig. 30-27. Constructional features of a standard-type soldering iron.

made. It is good practice to tin (melt solder over) only the portion of the tip that will actually be used while soldering. The oxidation coating on the remainder of the tip then acts as a heat insulator to reduce the amount of heat applied to other surfaces that the tip may touch while soldering.

A commonly used, general-purpose soldering iron tip consists of a pure copper base that is iron-clad or nickel-plated to resist corrosion. Such a tip should not be filed, since this destroys the plating and thus shortens tip life. To clean an iron-clad or a plated tip, a tip sponge or a tip cleaner is recommended (Fig. 30-28). After the tip is cleaned it should be re-tinned. The excess solder can then be carefully wiped off with the sponge (or cleaner) or with a dry cloth.

The variable-heat soldering iron is operated in conjunction with a line-voltage isolation transformer, which also serves as a means of varying the voltage applied to the iron (Fig. 30-29). A heat-selector chart on the transformer provides a convenient method of selecting the correct tip temperature necessary to do a particular soldering job.

Soldering Guns. Two types of soldering guns are shown in Fig. 30-30. These devices usually contain an iron-core transformer with the primary winding connected to the power line through a trigger-type on-off switch. The secondary winding of the transformer is a single-turn coil to which the soldering tip is connected. Although the voltage across the secondary winding is harmless, it does cause a significant current to pass through the low-resistance heating element assembly, thereby heating the tip to the proper temperature.

Special Tools and Care. Printed circuit soldering and soldering of semiconductor components requires special tools and care. The procedures relating to these soldering jobs are discussed in later paragraphs of this unit and in the following unit.

Fig. 30-28. Soldering iron tip cleaner. (*American Electrical Heater Co.*)

Fig. 30-29. Variable-heat soldering iron and associated transformer. (*American Electrical Heater Co.*)

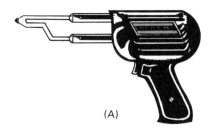

Fig. 30-30. Soldering guns: (A) open-element type, (B) enclosed-element type.

Soldering to Lugs. Before soldering a bare copper wire to a lug-type terminal or to any other device, the wire should first be tinned. Tinning is essential if a good solder connection is to be made with the least amount of effort. Although it is not necessary to tin pretinned wire, it will be more easily soldered if this is done. If the wire is stranded, the strands should be firmly twisted together before tinning.

Most lugs to which wires are to be soldered are sufficiently tinned to allow a good soldering job. However, if the lugs are bare copper, they should always be tinned. When tinning a lug that has one or more wire mounting holes in it, care should be taken to prevent solder from obstructing the unused holes.

The steps to be followed when soldering wires to a lug are illustrated in Fig. 30-31. The wires are crimped to the lug to make a mechanical connection that will prevent the wires from moving while the soldering is being done. Twisting a wire around a lug several times should be avoided in wiring that will not be subjected to severe shock or vibration. This practice not only crowds the lug but also increases the danger of damage to the lug when the wire is unsoldered from it for any reason.

Only enough solder should be applied to the connection to allow the solder to flow freely into all contact areas (Fig. 30-31D). The application of an excessive amount of solder will not improve the connection but will often cause undesirable dripping of solder upon adjacent lugs and wires.

If the correct amount of heat has been applied to the connection, the solder will have a bright silvery appearance after it has cooled. Dull-colored solder most often indicates a cold solder joint which, although it may appear to be good, will frequently result in a loose connection.

Soldering to "Hollow" Plugs and Connectors. Wires must often be soldered into the hollow bodies of various types of plugs and connectors, including phono plugs, auto antenna lead-in

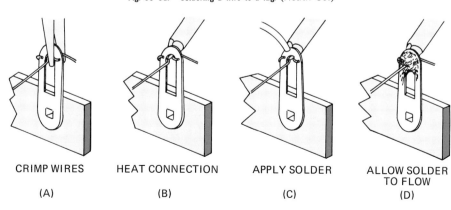

Fig. 30-31. Soldering a wire to a lug. (Heath Co.)

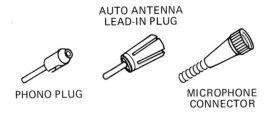

Fig. 30-32. Hollow-body connecting devices.

plugs, and microphone connectors (Fig. 30-32). The steps to be followed in soldering a single-conductor shielded wire to a phono plug are illustrated in Fig. 30-33. After the solder has cooled, it is important to trim excess solder from the tip of the plug. If this is not done, the solder may cause the plug to bind when it is inserted into its associated socket. The connection should then be tested by pulling on the wire to determine whether it has been firmly fastened in place.

Soldering to a Metallic Surface. When a wire is to be soldered to the metal surface of a chassis, it may be necessary to use a larger iron or gun having a wattage rating of 250 watts or more. The larger amount of heat delivered by such a soldering iron or gun is then able to overcome the loss of heat as a result of heat dissipation by the metal. In all cases, both the wire and the metal surface at the point of contact should be cleaned and tinned before the soldering job is attempted. Soldering of this kind may also require the use of a noncorrosive paste flux in addition to the flux that is present within the core of the solder wire.

Metals such as copper, brass, tin plate, and galvanized steel are readily solderable. But solder will adhere to these metals more firmly if the surface to be tinned is first scraped or scratched with a metallic edge.

Soldering to Aluminum. Soldering a wire directly to an aluminum chassis can be a tough job if correct procedures are not followed. Although special-type aluminum fluxes are available for this purpose, the job can be done with ordinary noncorrosive paste flux. The unusual feature of soldering to aluminum is the difficulty of tinning the surface of the metal. After this has been done, the remaining procedures are similar to those followed in soldering to the more readily solderable metals.

To tin aluminum, it is most convenient to use a heavy-duty soldering iron or gun. The surface of the metal to which the wire connection is to be made should first be thoroughly cleaned and rather deeply scratched with the tip of a scriber or some other sharp metallic edge and then covered with a thick layer of noncorrosive paste flux. After a few drops of solder have been applied to this surface, the tip of the soldering iron (or gun) is firmly rubbed over it in a circular motion (Fig. 30-34). This process is continued until a coating of solder is built up on the metal.

SOLDERING SEMICONDUCTOR COMPONENTS

When soldering the leads of semiconductor components such as diodes, transistors, and thermistors, it is desirable to use a light-duty (60

Fig. 30-33. Wiring a phono plug.

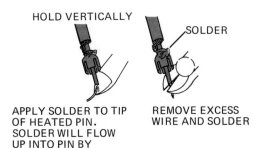

Fig. 30-34. Tinning an aluminum surface.

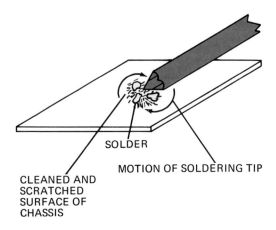

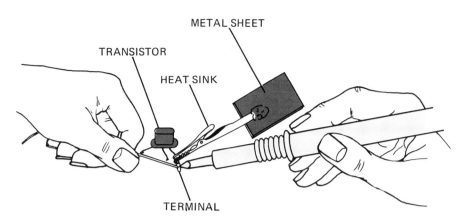

Fig. 30-35. Using a heat sink while soldering a transistor lead.

watts or less) soldering iron. If a larger iron is used, excessive heat may be conducted into the component, causing it to be permanently damaged.

An open-element type of soldering gun is not recommended for soldering semiconductor components. This is because the tip of the gun—actually, the secondary winding of a transformer—is surrounded by a relatively strong magnetic field that can cause damaging electrical interactions within the components.

Heat Sink. When soldering or unsoldering the leads of semiconductor components, a heat sink should always be used to absorb heat that would otherwise be conducted into the components. A very convenient heat sink can easily be made by soldering an alligator clip to a small sheet of copper or brass. In use, the clip is attached to the lead of the component at some point between the component and the terminal to which the lead is being soldered (Fig. 30-35). A convenient, effective heat sink is also provided by long-nose pliers that are used to grasp a bare portion of a wire near the terminal to which the wire is being soldered.

Copper-Wire Tip. A medium- or heavy-duty soldering iron can be converted into a satisfactory soldering tool by wrapping a number of turns of rather large, bare copper wire around its tip as shown in Fig. 30-36. The "open" end of the wire then becomes the "working" tip. In order to cause the maximum amount of heat to be conducted from the tip of the iron to the wire, the tip should be thoroughly cleaned and the wire should be tinned before wrapping.

WIRING FROM A DIAGRAM

It is not at all difficult to route (place) wires and make connections when following a pictorial diagram, since such a diagram usually indicates the layout of the wired circuit. Wiring from a schematic diagram presents an entirely different problem, however, because this type of diagram

Fig. 30-36. Extending the tip of a soldering iron with bare copper wire.

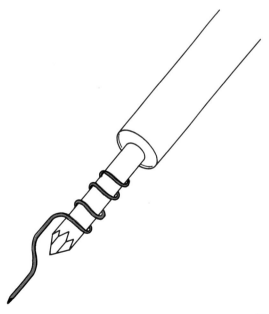

does not show the routing of wires, but only shows how the components are to be connected together.

There are no established rules for the exact procedures to be followed in wiring from a schematic diagram. Before starting it is often helpful to compare the wiring connections as shown by a pictorial diagram with the connections as indicated by the schematic of the same circuit. A careful study and comparison of the two diagrams will result in a better understanding of how a schematic diagram is to be interpreted and transformed into an actual circuit configuration.

A recommended method of starting to wire a circuit from a schematic diagram involves mounting certain key or guide components. As an example, consider the schematic diagram of the two-transistor radio receiver shown in Fig. 30-37A. If the key components selected are

Fig. 30-37. Wiring from a schematic diagram: (A) diagram of the circuit to be wired, (B) mounting of "key" or "guide" components.

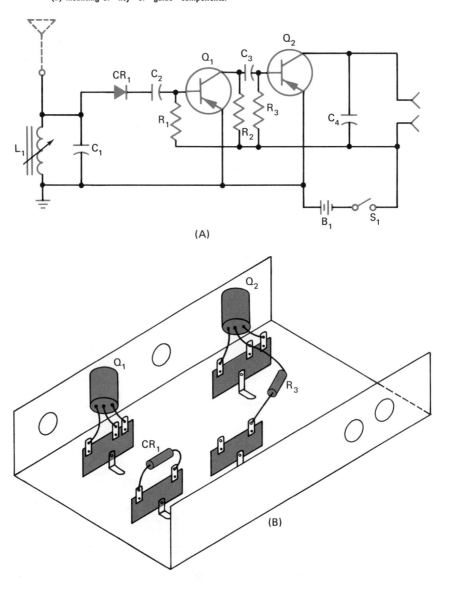

CR_1, Q_1, Q_2, and R_3, these components would first be mounted upon the chassis as shown in Fig. 30-37B. The remaining components would then be connected to the terminal strip lugs to which the key components are connected, or to any other lugs that may be necessary to complete the wiring job.

As the wiring of any particular circuit progresses, more than one wire may be brought to a number of lugs. For the sake of convenience, a single wire should not be soldered to any lug unless it is certain that this wire is the only one that will be connected to the lug in question. After the wiring is completed, all the lugs can be soldered, thus saving time because of the fewer individual solder connections that are necessary.

WIRE ROUTING AND CONNECTING PRACTICES

The proper routing of connecting wires will add to the appearance of the wiring job, improve circuit performance, and use the minimum amount of wire. The following routing and connecting practices can be adapted to almost any wiring job:

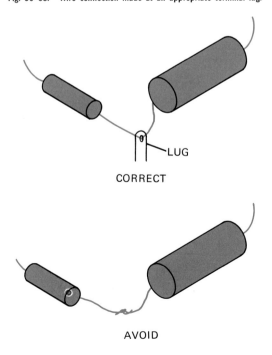

Fig. 30-38. Wire connection made at an appropriate terminal lug.

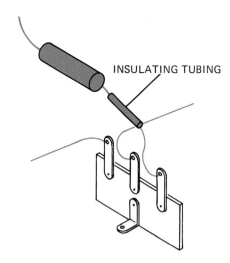

Fig. 30-39. Use of insulating tubing or "spaghetti."

1. Route and connect long, direct wire runs first.

2. When connecting two points of a circuit, use enough wire to allow the wire to be routed along or near the surface of the chassis. Wires that are unnecessarily strained from point to point create a "spiderweb" effect that does not look good. The use of an ample, but not excessive, amount of wire also allows wires to be moved as necessary when components are tested or replaced.

3. Route all wires so that they will not come into contact with any resistor that may develop a considerable amount of heat.

4. Whenever possible, avoid wire runs across components, sockets, terminal lugs, etc. This practice will reduce the unnecessary moving of wires when connections and tests are made at these devices.

5. Always make a wire connection at an appropriate terminal lug (Fig. 30-38). This practice aids in disconnecting wires and adds to the appearance of the wiring job.

6. Cut component leads as necessary to avoid the use of excess wire. When a bare component lead wire or any bare wire is to be placed adjacent to lugs, other bare wires, or metallic surfaces, insulate the wire with plastic or cambric tubing, commonly referred to as "spaghetti" (Fig. 30-39). Always mount a component so that its rating designation, if present, is in full view.

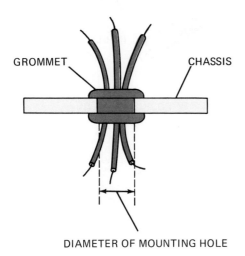

Fig. 30-40. Use of a rubber grommet.

7. Always route wires so that the insulation is protected against abrasion from the metallic edges of the chassis, lugs, mounting screws, etc.

8. Never route wires over the side of any type of chassis.

9. Before wires are passed through holes in a metal chassis, insulate the edges of the holes with rubber grommets (Fig. 30-40). These devices prevent the wire insulation from being damaged by the surface of the metal. Pass a high-voltage wire through a chassis and insulate it from the chassis by using a feed-through insulator (Fig. 30-41).

Fig. 30-41. Feed-through insulator.

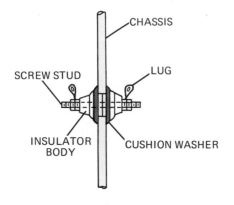

10. Be sure that the power cord enters the chassis at a point near its termination, which is usually a switch or the primary winding of a power transformer. To protect the cord, provide some type of strain relief at the point of its entry into the chassis (Fig. 30-42). A convenient cord strain relief is also provided by the use of an *underwriter's knot* or a simple overhand knot.

11. Use cable clamps when it is necessary to support long wire or cable runs or to hold wires and cables in place (Fig. 30-43). If the cable clamps that are used are not insulated, reinforce the insulation of the wires that they hold with plastic tubing.

THE WIRE HARNESS

In custom wiring where it is necessary to group long runs of wires, the wires are usually either laced together or strapped together with plastic tie straps (Fig. 30-44). This is especially true when wires are grouped together in what is referred to as a *wire harness*. A wire harness is a group of wires that has been prefabricated to suit the requirements of a particular circuit layout (Fig. 30-45A). Although "harness" wiring is not commonly performed in ordinary shop work, it is almost always used in the manufacture of

Fig. 30-42. Installation of one common type of strain relief device. *(Heath Co.)*

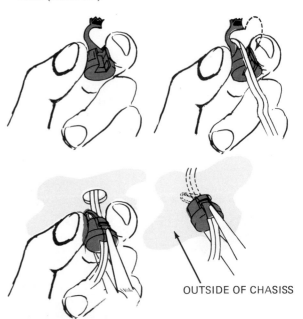

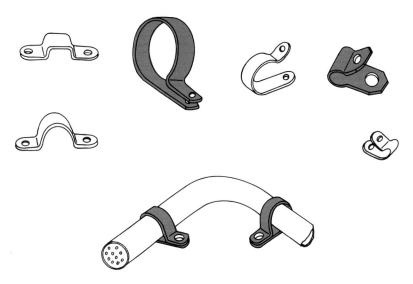

Fig. 30-43. Cable clamps.

Fig. 30-45. The wire harness: (A) basic wire harness assembly, (B) wire harness forming board.

Fig. 30-44. Plastic tie straps.

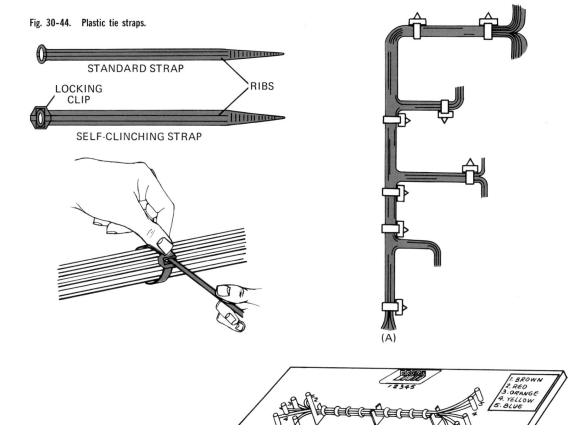

large industrial production units such as transmitters, control consoles, and household appliances.

A wire harness is most often made upon a forming board (Fig. 30-45B). After the wires have been placed in position upon the board, they are strapped together to form a compact unit. The harness can then be installed and wired in as a single assembly. Harness wiring is particularly well adapted to mass-production manufacturing methods, since it eliminates the need to route individual wires.

INSPECTION AND TESTING

After a circuit has been completely wired and before it is placed into operation, the circuit should be carefully checked by means of visual inspection and electrical tests. The purpose of these procedures is to detect any wiring errors that might result in damage to components when the operating voltages are applied to the circuit.

Visual Inspection. During the visual inspection, close attention should be given to those conditions that cause direct-circuit shorts at points where this is not intended. The most common of these conditions are bare wires in contact with each other; solder drippings between bare wires, between adjacent lugs, or between lugs and the chassis; and damage to insulation of wires that rub against metallic surfaces.

The visual inspection should also include checking for bad solder joints, broken terminals, breaks in wires, and correct polarity connections to components such as electrolytic capacitors and transistors. These components will be damaged if voltages of the wrong polarity are applied to them.

Continuity Tests. Although a careful visual inspection will enable obvious circuit defects to be located, it is difficult and often impossible to visually detect all undesirable shorts and missed connections. Because of this, continuity tests should be made at various points in the circuit. Such tests should be made, for example, between the chassis and the metallic portions of jacks. A jack may appear to be properly insulated from the chassis and yet be in electrical contact with it because of improper seating of the insulating washer. A continuity test should also be made to see that all common points in the circuit are connected together. As a final check, a continuity test should be made to see that there is no direct connection between the ungrounded side of a rectifier output and the common point of a circuit.

For Review and Discussion

1. Which components used in typical circuits are subject to deterioration or loss of operational efficiency as the result of being exposed to excessive heat?
2. For what purpose is a silicone compound usually applied between the base of a power transistor and a metallic surface that serves as a heat sink?
3. Describe the procedures to be followed when using a chassis punch.
4. Why is it desirable to use plated screws in conjunction with a mounting job involving an electrical contact?
5. For what purpose are terminal strips used?
6. Describe the electrical contact action of a closed-circuit–type jack.
7. Why are hookup wires often pretinned?
8. Describe a shielded cable and state the purpose of using such a cable.
9. What is meant by a coaxial cable?
10. For what purpose is a bus wire used?
11. Describe the process of "breadboard" wiring and give an important use of this type of wiring.
12. What advantage is gained by using a plated soldering iron tip as compared with an unplated tip?
13. Describe the principal constructional features of a typical soldering gun and explain the operation of this tool.
14. Why is it important to tin all surfaces involved in making a solder connection?
15. State the procedure to be followed when soldering a wire to a "hollow-body" plug or other type of connector.

16. How can common pliers be used as a heat sink in soldering semiconductor components?

17. Describe how a copper-wire tip can be added to the tip of an ordinary soldering iron.

18. What is the recommended procedure when starting to wire a circuit from a schematic diagram?

19. For what purpose is insulating tubing, or "spaghetti," used?

20. For what purpose are rubber grommets used?

21. What is a power cord strain relief device and why is such a device used?

22. Describe a wire harness.

UNIT 31. Printed Circuits

The typical printed circuit assembly consists of components mounted upon a baseboard and soldered to thin strips of copper that are bonded to the base (Fig. 31-1). Although there are several different kinds of base materials used, two very common ones are phenolic and epoxy plastic. In addition to supporting the circuit assembly, the base material must provide adequate insulation and must be able to withstand high soldering and operating temperatures without

Fig. 31-1. Component side of a printed circuit assembly. (Tektronix, Inc.)

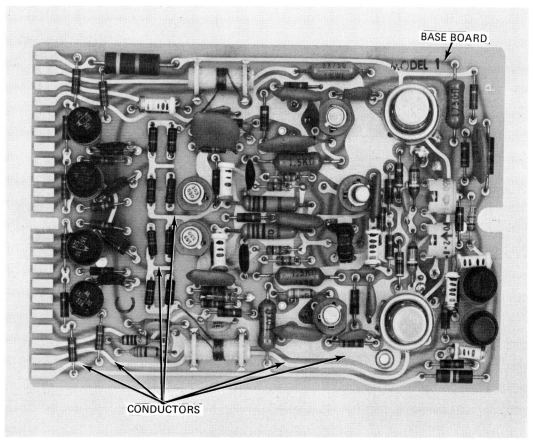

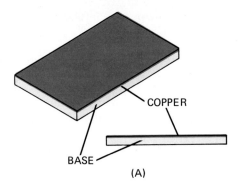

(A)

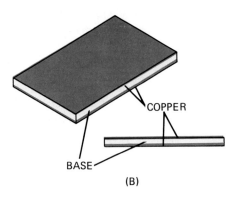

(B)

Fig. 31-2. Structural features of copper clad laminate: (A) copper one side, (B) copper two sides.

significant buckling. In its original form, the entire surface of one or both sides of the base is covered with a thin copper sheet that is bonded to the base at all points. This combination of materials is referred to as *copper clad laminate* (Fig. 31-2).

During the manufacture of the commonly used etched printed circuit board, a *resist* solution is used to transfer the appropriate circuit pattern to the surface of the copper clad. A resist is a material that resists the chemical action of an etching solution. The copper clad is then placed in an etching solution containing a chemical such as ferric chloride or chromic acid. After a short period of time, the copper that is not coated with resist is etched or "eaten" away, leaving only those copper surfaces on the board that were coated with resist. Following this, the resist is washed away with a suitable solvent, exposing the copper strip circuit conductor pattern. After mounting holes are drilled or punched in the board, it is ready for component mounting (Fig. 31-3).

For some circuit applications, it is expedient to plate the copper on a printed circuit board with another metal. This is done by an electroplating process either before or after etching. Plating prevents the corrosion of conductor surfaces, thereby making it easier to make solder connections to the conductors.

Fig. 31-3. Etched and punched printed circuit board as it appears before mounting components. (*Eastman Kodak Company*)

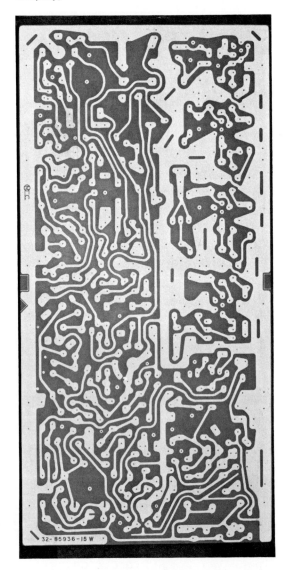

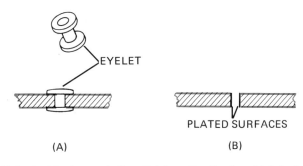

Fig. 31-4. Providing for a conducting path between the sides of a printed circuit board with copper on both sides: (A) using an eyelet, (B) using plated surfaces.

When conductors appear on both sides of a printed circuit board, it is necessary to provide a means of contact between the surfaces at several points of the circuit. This is usually done by means of wire leads or eyelets, or by plating the inner surfaces of holes that are drilled through the board at the points of contact (Fig. 31-4).

Printed circuit wiring has become very popular because it is well adapted to automated mass production manufacturing techniques and to the construction of uniform, small size, lightweight units. Further improvements in the materials used and the ever increasing trend toward more compact (miniaturized) circuit assemblies will no doubt result in an increase in the applications of this process.

THE PRINTED CIRCUIT DIAGRAM

The first step in the manufacture of a printed circuit assembly is to prepare an accurate layout diagram, which is a pattern of the conductors that are to appear upon the baseboard. When it is drawn in ink, such a diagram is usually prepared on a Mylar film or on good quality paper that will not shrink or expand because of humidity or excessive room temperatures. It is very important to use a film or a paper having these properties, since the size of the pattern should remain unchanged over a period of time. The layout pattern is drawn to actual size or to a scale that can be easily reduced, or enlarged photographically to actual size.

Printed circuit diagrams are also often made with tape. In this process, precut strips and pieces of a special tape are assembled into a pattern that conforms with the engineering sketch of a particular circuit (Fig. 31-5).

Drafting Practices. When preparing a printed circuit diagram, the diameter of the holes indicated upon the diagram is usually not less than three-quarters of the thickness of the copper clad to be used. The lines corresponding to the conductors that will appear on the copper clad should not be less than $\frac{1}{32}$ inch wide and should be separated by a distance of at least $\frac{1}{32}$ inch.

Current Carrying Capacity. The maximum current carrying capacity of the printed circuit conductors will of course depend upon the width and thickness of the copper that is bonded to the baseboard. The maximum current carrying capacities of two common thicknesses of copper are given in Table 31-1.

Table 31-1. Typical overload currents for circuits printed upon copper. (*Harris Manufacturing Company*)

Line Width	.00135″ Copper Amps	.0027″ Copper Amps
$\frac{1}{4}''$	23	35
$\frac{1}{8}''$	15	20
$\frac{1}{16}''$	10	15
$\frac{1}{32}''$	5	8

The Component Overlay. Many types of printed circuit boards have the locations of the components that are to be mounted upon them indicated by symbols or other appropriate markings. This is accomplished by means of a component identification overlay prepared for

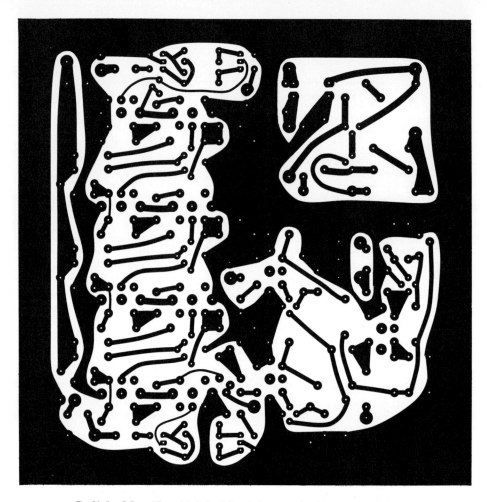

Fig. 31-5. **R-F amplifier printed circuit layout diagram made with tape.** (*Gates Radio Company, Division of Harris-Intertype Corp.*)

use with a given printed circuit layout diagram (Fig. 31-6). After the printed circuit board has been completed, the identification information indicated upon the overlay is stencilled onto it. This identification of components speeds up the assembly of circuit units and helps in servicing the equipment.

TRANSFER OF THE PATTERN

After the printed circuit diagram has been prepared, it must be transferred to the copper surface of the circuit board. Two commonly used production methods of doing this are the *photoengraving* process and the *silkscreen* process.

Photoengraving Process. The photoengraving process produces a very accurate transfer and is used whenever precision and extremely good detail are required. In this process, the diagram pattern is first photographed and a suitable negative is prepared. The negative is then placed in contact with a copper clad board that has been coated with a light-sensitive substance. The circuit board is next exposed through the negative to a high-energy light source such as a carbon-arc lamp or a mercury-vapor lamp

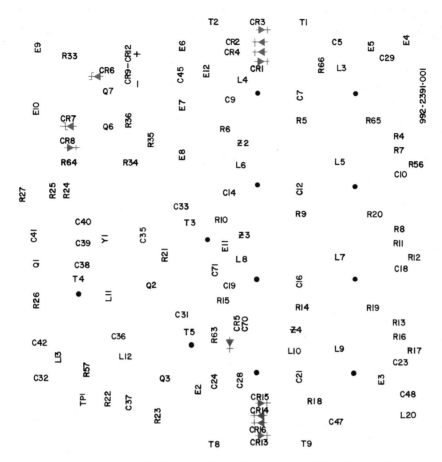

Fig. 31-6. Component identification overlay used in conjunction with the printed circuit layout diagram shown in Fig. 31-5. (*Gates Radio Company, Division of Harris-Intertype Corp.*)

(Fig. 31-7). The light passes through the transparent areas of the negative (the conductor layout) and strikes areas of the copper that are identical in shape to the inked or the taped portions of the layout pattern.

Following exposure, the copper clad is developed. As a result of the development process, an acid-resistant coating, the resist pattern, which is identical to the conductor layout of the printed circuit diagram, is formed on the surface of the copper. Therefore, the copper conductor pattern remaining upon the board after the etching process has been completed is also identical to the conductor layout of the diagram.

Silkscreen Process. The silkscreen process of transferring printed circuit patterns, though less accurate than the photoengraving process, is more adaptable to large quantity, mass-production methods. In the basic silkscreen process, a stencil-type cutout of the circuit conductor pattern is adhered to a silkscreen. The circuit board is then placed under the stencil and a resist compound is applied to the copper through the screen. Because of the stencil, the resist coats only those surface areas of the copper that correspond to the layout pattern. The uncoated copper is then etched from the copper clad.

Fig. 31-7. Photoengraving process of printed circuit board fabrication. (*Eastman Kodak Company*)

Fig. 31-8. Printed circuit components: (A) potentiometer, (B) electrolytic capacitor, (C) i-f transformer.

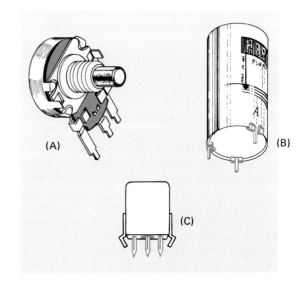

COMPONENT MOUNTING AND SOLDERING

After the printed circuit board is etched and cleaned, unplated conductors are usually coated with a solder-resistant varnish at those points that are not to be soldered. Following this, the components are mounted on the board. The components used with printed circuit wiring are equipped with pin-like terminals to aid in mounting and to contact conductors through holes that are drilled or punched through the board (Fig. 31-8). When the components have been mounted, the assembly is ready for soldering. In mass production of printed circuit assemblies, the soldering is done by the automated process of either *dip soldering* or *wave soldering*.

Dip Soldering. When dip soldering is used, the surface or surfaces of the wiring side of the printed circuit board are first dipped into a liquid flux. The board is then dipped into molten solder to a depth that is sufficient to allow solder to flow freely into all connection points (Fig. 31-9A).

Wave Soldering. In wave soldering (sometimes referred to as *fountain soldering*), the molten solder is pumped up to the level of the printed circuit board in the form of waves (Fig. 31-9B). This method of soldering, as compared with dip soldering, permits more favorable angles of solder insertion, provides better control for the duration of solder contact, and reduces the amount of heat applied to other parts of the assembly.

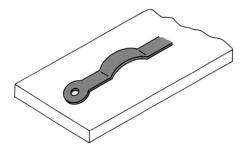

Fig. 31-10. Delaminated printed circuit conductor.

Fig. 31-9. Soldering processes used in the manufacture of printed circuit assemblies: (A) dip soldering, (B) wave soldering.

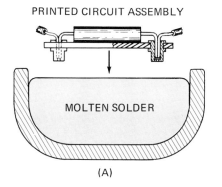

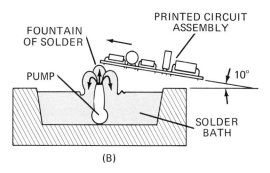

WORKING WITH PRINTED CIRCUITS

The repair and servicing of printed circuit assemblies is not very different from the procedures followed when working with hand-wired circuits. However, because of the nature of the printed circuit and the ease with which it can be seriously damaged by dropping, twisting, overheating, or the excessive application of heat, a more delicate approach is required.

The Soldering Tool. The soldering iron or gun used on a printed circuit board should not have a wattage rating of more than 60 watts. A heavier duty tool, particularly when used by an inexperienced person, produces an excessive amount of heat that may seriously damage the board by burning out conductor strips or by causing the conductors to become separated (delaminated) from the base of the board (Fig. 31-10).

The Solder. Rosin-core solder should be used when working with printed circuits, and the solder should have a comparatively low melting point so that it can be applied with the minimum amount of heat. This condition is satisfied by using a 60-40 (60 percent tin, 40 percent lead) solder with a diameter of approximately $\frac{1}{16}$ inch.

Soldering and Resoldering. When soldering a connecting wire or a component lead to a printed circuit board, it is extremely important not to use any more solder than is absolutely necessary. The use of excessive solder can easily result in shorting out adjacent conductors

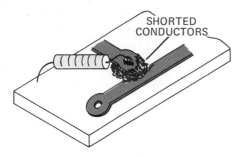

Fig. 31-11. Shorted printed circuit conductors; short caused by use of excess solder.

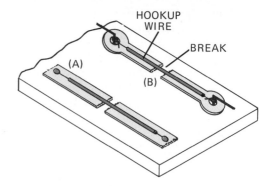

Fig. 31-13. Methods of repairing broken printed circuit conductors.

(Fig. 31-11). Attempts to clear up such a short often result in delamination and/or spreading of the solder so that it contacts a greater length of adjacent conductors.

When soldering a semiconductor component to a printed circuit board, its leads should always be left long enough to allow the use of a heat sink (Fig. 31-12). A heat sink should, of course, also be used when unsoldering the lead of a semiconductor component from the board.

Because of the twisting of a printed circuit board or a defective solder connection, it is sometimes necessary to resolder a connection at some point on the board. If at all possible, this should be done by simply reheating the connection without applying more solder.

Repairing Broken Conductors. A simple, "clean" break of a conductor on a printed circuit board is most conveniently repaired by carefully flowing (applying) solder across the break. When the break is too large to be handled in this manner, the conductor can usually be repaired by soldering a length of bare, tinned wire across the break (Fig. 31-13A). When the terminals at each end of a broken conductor are readily accessible, it is often desirable to connect hookup wire directly to the terminals themselves (Fig. 31-13B).

Replacing Defective Components. In the majority of cases, the most convenient and safest way of replacing a defective component on a printed circuit board is to avoid disturbing the existing solder connections. When replacing lead-type components such as resistors and capacitors, this is done by first cutting the leads on the component side of the board with diagonal cutting pliers (Fig. 31-14A). The leads of the replacement component are then soldered to the projecting "stubs" of the leads that have been cut (Fig. 31-14B).

When replacing multiterminal components such as transformers, potentiometers, and can-type capacitors, it is best to first cut the terminals. The terminal "stubs" can then be removed from the board one by one. In most cases, the safest way to do this is to grip the terminal with a long-nose plier while heating it at the point where it connects to the board (Fig. 31-14C). After the solder is melted, the terminal is pulled through the mounting hole.

If the terminals on a multiterminal component cannot be cut, it will be necessary to unsolder all connections before the component

Fig. 31-12. Pliers used as heat sink while soldering a printed circuit component.

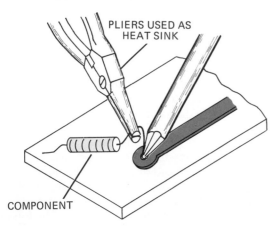

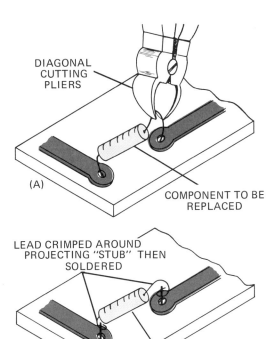

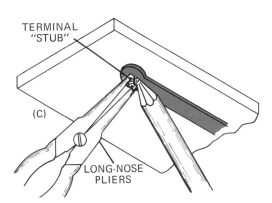

Fig. 31-14. Replacing defective printed circuit components.

shape. Because of the heat dissipation of its comparatively large surface area, the use of a de-soldering tip may require a heavier duty soldering iron than would be advisable if the tip is not attached to the iron.

In situations where it is impossible to unsolder all of the terminals of a multiterminal component simultaneously, the unsoldering must, of course, be done on a one-by-one basis. The principal problem encountered while doing this is to remove all the melted solder from the

Fig. 31-15. Common types of de-soldering tips.

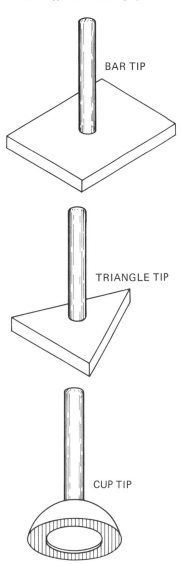

can be removed. When doing this, it is best to de-solder all of the connections at the same time. Since it is usually difficult or impossible to perform this operation with an ordinary soldering tip, special de-soldering tips designed for that purpose are used (Fig. 31-15). If a de-soldering tip for a particular job is not available, one can often be made by cutting and bending a relatively heavy copper sheet into the desired

353

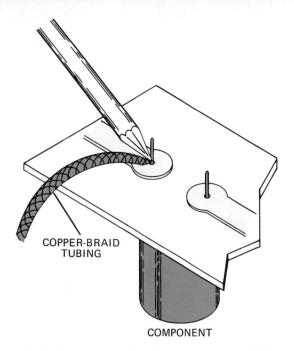

Fig. 31-16. Using copper-braid tubing to remove excess melted solder from a solder connection.

terminal being worked on. If the solder is not removed, the terminal will become resoldered when the soldering iron or gun is applied to another terminal.

A simple, though not always effective, method of removing melted solder from a given terminal connection point is to use the tip of the soldering iron or gun. If the tip is well tinned, it will attract to itself a small quantity of melted solder, which can then be wiped from the tip. By repeating the process, usually all of the solder can be removed from a connection point.

A more effective way to remove melted solder involves using a length of copper-braid shielding "tubing" as shown in Fig. 31-16. The braid, being in contact with the terminal point being unsoldered, becomes heated and attracts solder, which then flows upward into the braid.

The best job of removing melted solder from a printed circuit board terminal is done by using a syringe-type de-soldering bulb or a similar suction device (Fig. 31-17). When the solder at a given terminal has been melted, the finger

Fig. 31-17. Syringe-type de-soldering tool. (Ungar Electric Tools)

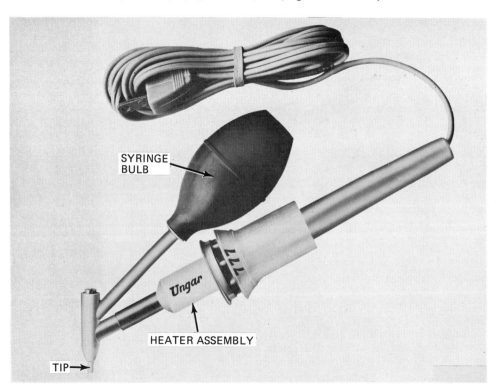

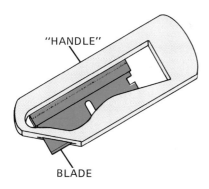

Fig. 31-18. Printed circuit conductor cutting tool.

pressure on the bulb of this tool is released, thereby causing solder to be drawn into its hollow tip.

Isolating Components. In making resistance or continuity tests of components on a printed circuit board, it is often desirable to isolate a given component from the circuit without unsoldering its lead (or leads) from the associated terminals. This is most conveniently done by cutting the printed circuit conductor (or conductors) connected to the component with a cutting tool (Fig. 31-18). After the tests have been completed, the conductor is "repaired" by flowing solder across the break.

FLEXIBLE CIRCUITRY

The use of flexible copper clad laminate allows a printed circuit to be formed into a variety of shapes, thus enabling the design of circuit structures not possible with rigid materials (Fig. 31-19). In addition to its application as circuit structure material, such laminate is also widely used in making flexible flat cable that often provides for a significant reduction in weight and space as compared with more conventional types of cable.

Fig. 31-19. Flexible copper clad polyester laminate used in circuit of telephone handset. (*Schjeldahl, Electrical Products Division*)

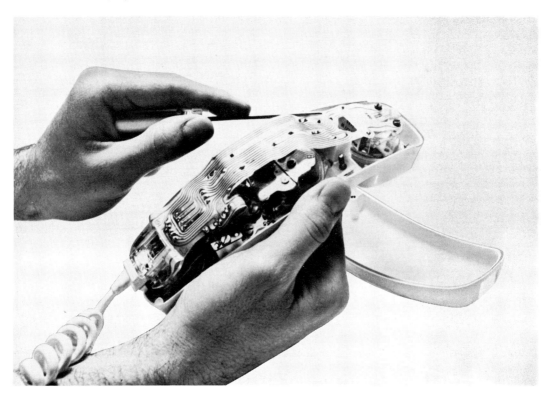

OTHER TYPES OF PRINTED CIRCUITS

Although the etched copper clad laminate printed circuit is the most common type in use today, several additional methods are available for "printing" conductors upon base materials. The scope of this book allows only a brief mention of several of these methods.

The Plated Circuit. The plated circuit board is made by first applying a thin film of a silver compound upon a phenolic laminate plastic base. Those areas of the silver-compound surface that are not to be utilized as conductors are then coated with a resist material. The board is then placed in an electroplating solution, which causes copper (the conductor material) to be plated upon those surfaces not covered with resist.

The Painted Circuit. In the painted circuit process, a liquid containing powdered copper or silver is applied to a base material such as steatite by means of a conductor pattern stencil. This combination of materials is then baked at a temperature which causes the metallic particles to be fused to the base.

Die Stamping. The die-stamped circuit board is produced by first stamping or punching out the desired conductor pattern from a sheet of metal foil. The pattern is then usually attached to a base material by using a suitable adhesive.

Chemical Deposit. In this process of conductor formation, a silver, copper, or other metallic compound solution is deposited upon a base material through an appropriate stencil. The resulting metallic film is then sometimes plated to increase its conductivity.

PRINTED MODULES

An *electronic module* is defined as an assembly of basic components that is used in conjunction with other components to form a complete circuit system. In accordance with this definition, the packaged electronic circuit is one type of module. However, in practice, the module is usually considered to be a more complete assembly that often contains transistors and diodes in addition to the related resistors, capacitors, and inductors. These components are formed by "printing" the essential materials upon individual component or combined component base wafers, which are then interconnected to form the complete unit (Fig. 31-20).

Fig. 31-20. Transistor amplifier module and block schematic diagram of terminal connections. (*Melcor Electronics Corp.*)

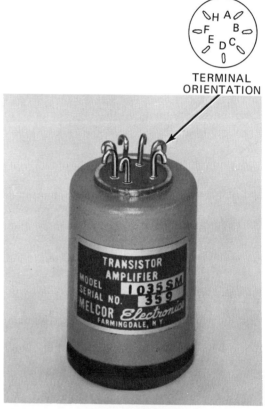

TERMINAL ORIENTATION

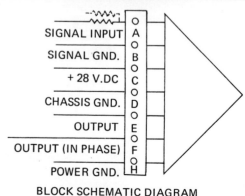

BLOCK SCHEMATIC DIAGRAM

For Review and Discussion

1. What is a printed circuit assembly?
2. Describe the construction of copper clad laminate.
3. State the basic processes used in the manufacture of a printed circuit board.
4. Describe a printed circuit diagram.
5. For what purpose is tape often used in making a printed circuit diagram?
6. Why is a component overlay used in conjunction with a printed circuit assembly?
7. Describe the basic photoengraving process of transferring a printed circuit pattern to a printed circuit board.
8. Give the basic steps followed when using the silkscreen process of transferring a printed circuit pattern to a printed circuit board.
9. Describe the processes of dip and wave soldering.
10. What type of solder is recommended for use with printed circuits?
11. Define delamination.
12. How are broken printed circuit conductors repaired?
13. What is the recommended method of removing lead-type components from a printed circuit board?
14. For what purpose is a de-soldering tip used?
15. State three methods commonly used to remove melted solder from a given terminal point on a printed circuit board.
16. For what purpose is a cutting tool commonly used when working on a printed circuit board?
17. Describe flexible printed circuitry.
18. Define an electronic module.

UNIT 32. Integrated Circuits

The development of transistor, diode, and other semiconductor (solid-state) components, together with the use of printed circuitry, has resulted in the miniaturization of practical circuit assemblies. However, the reduction in size and weight made possible by these component and circuit fabrication methods does not fulfill many of the physical size–weight requirements that have been introduced by the computer and other circuit systems in commercial, military, and space programs.

The electronics industry has moved rapidly into the new era of integrated microelectronics circuitry in order to solve these problems. With this method of circuit fabrication, an astonishing reduction in the size and weight of a circuit assembly is achieved as compared with an identical printed circuit assembly (Fig. 32-1). The reduction in size and the fabrication methods employed in the manufacture of integrated circuits, in turn, provide the advantages of making possible more reliable and speedier circuit operation. All these advantages, coupled with the relatively low cost of production, will no doubt result in the widespread application of integrated circuitry in common consumer products such as radios, television sets, and amplifiers.

The term *integrated* means to form into a whole unit. In the integrated circuit (IC), the whole unit concept is put into practice by several photographic and chemical processes by means of which all the active components (transistors and diodes) and all the passive components (resistors, capacitors, and inductors) are fabricated into or upon the surface of base materials. This results in a combination of components that are inseparably associated within the framework of a circuit structure.

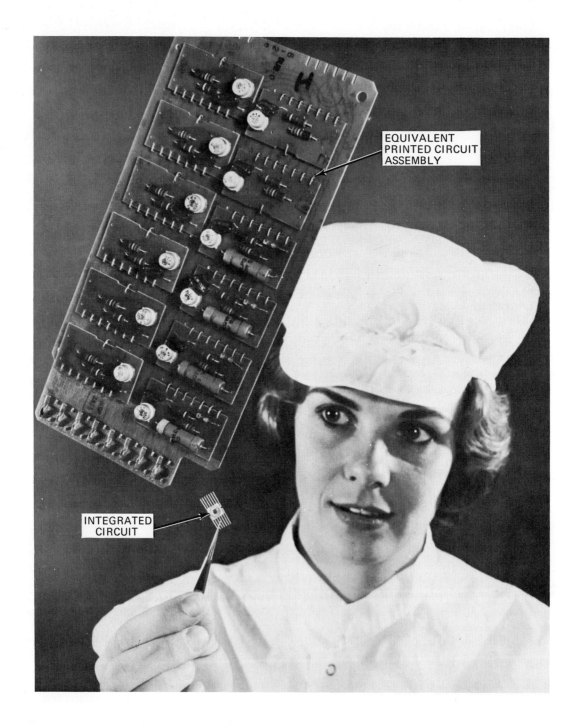

Fig. 32-1. Computer-use "flat-pack" integrated circuit and equivalent printed circuit assembly. (*RCA*)

THE MONOLITHIC CIRCUIT STRUCTURE

The term *monolithic* is derived from the Greek (*mono* for single) and (*lithos* for stone). In the monolithic integrated circuit, the components are fabricated within a single base of a semiconductor material (usually silicon) called a *substrate*. Germanium is seldom used as the substrate since its structure does not react favorably to the fabrication operations required.

In one very commonly used fabrication method known as the *epitaxial diffused system*, the integrated circuit begins with slicing a p-type silicon crystal into thin chips (wafers). Each chip is then ground to the precise thickness desired and cut into a circular, square, or rectangular shape. After being thoroughly cleaned and polished, one surface of each chip is exposed to a heated gas that contains a donor impurity. This operation is referred to as *diffusion*. As a result, the impurity diffuses (spreads) into the substrate, causing a thin layer of n-type silicon, known as the *epitaxial layer*, to be formed (grown) upon the substrate (Fig. 32-2A).

The chips are then placed into a high-temperature oven into which oxygen gas is in-

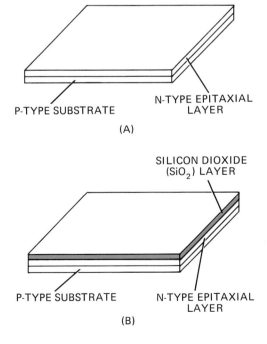

Fig. 32-2. Epitaxial and silicon dioxide layers formed upon an integrated circuit chip.

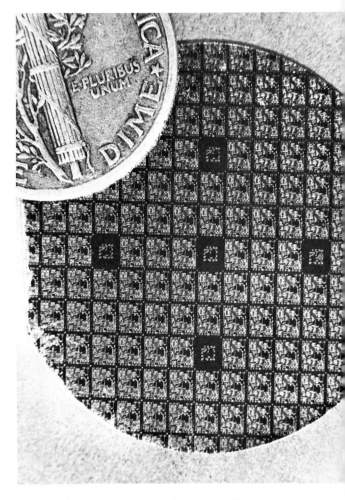

Fig. 32-3. Size of a disk (chip) upon which a large number of integrated circuits have been fabricated compared with the size of a dime. (RCA)

troduced. The chemical process of oxidation that results causes the oxygen to combine with the silicon, thereby forming a protective, insulating layer of silicon dioxide (SiO_2) upon the surface of each chip (Fig. 32-2B).

Normally, large numbers of complete, integrated circuits are simultaneously fabricated into predetermined regions of a given chip (Fig. 32-3). The chip is then scribed (scratched) with a diamond-tipped tool to separate the individual circuits into "pieces" that are referred to as *dice* (plural of *die*).

Each die is then fastened to a ceramic material and mounted upon a suitable overall base

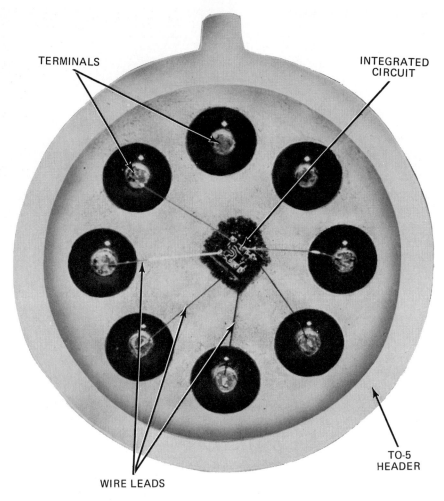

Fig. 32-4. High-frequency video and i-f amplifier monolithic integrated circuit mounted upon a type TO-5 header. (*General Electric Co.*)

called a *header*. The header is equipped with appropriate wire leads and terminals (Fig. 32-4). This assembly may or may not be further encased within a circuit "package" (Fig. 32-5).

Fabrication. For purposes of explaining basic monolithic integrated circuit fabrication, it is best to begin by assuming that components are to be fabricated into a single die. A similar, though greatly expanded, procedure is followed when many individual circuits are to be simultaneously fabricated into an entire chip.

The fabrication of an integrated circuit is begun by coating the silicon dioxide surface of a die with a thin layer of a photosensitive emulsion (photoresist). Those surfaces of the die into which components are to be fabricated are then covered with an isolation mask (a form of stencil) (Fig. 32-6A).

The die is then exposed to ultraviolet light. As the light strikes the unmasked areas of the silicon dioxide in a pattern that corresponds to the clear portions of the mask, the photoresist is transformed into a material that can be removed by etching. As a result of the etching process, the epitaxial layer is also etched away

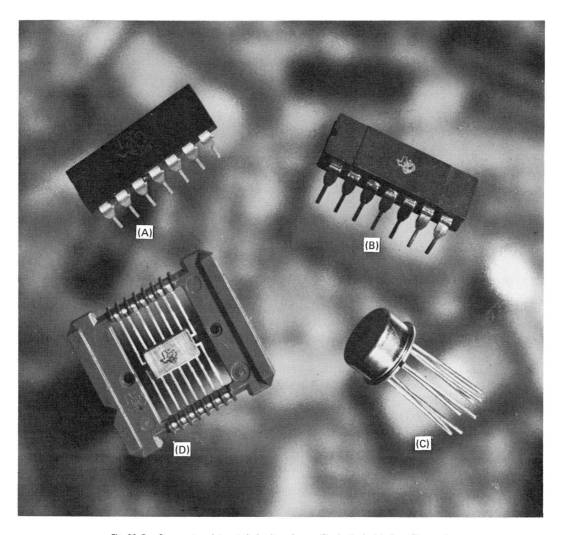

Fig. 32-5. Common-type integrated circuit packages: (A) plastic dual in-line, (B) ceramic dual in-line, (C) TO-5 case, (D) flat pack. (*Texas Instruments, Inc.*)

in accordance with the pattern upon the mask (Fig. 32-6B). Next, a gas that contains an acceptor impurity such as boron is diffused into the etched epitaxial layer channels.

The areas that remain coated with silicon dioxide are now regions of n-type silicon surrounded by p-type silicon (Fig. 32-6C). This effectively "isolates" the n-type regions into which components are to be fabricated, thereby reducing the leakage of current from one component to another during the operation of the completed circuit. After the entire surface of the die is once again coated with silicon dioxide by the oxidation process, the photoresist is reapplied to the die, which is then ready for the fabrication of the necessary circuit components.

The Fabrication Cycle. It is important to note that the photochemical operations involved in the isolation of component areas upon a die consist of four principal processes: oxidation, masking, etching, and diffusion. This oxidation-masking-etching-diffusion cycle is the basis of monolithic integrated circuit structure. In practice, the cycle

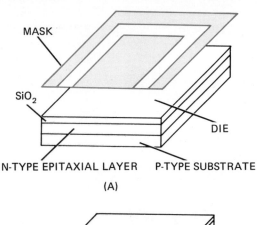
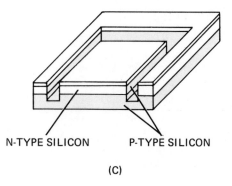
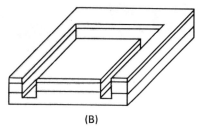

Fig. 32-6. Basic steps in the fabrication of a monolithic integrated circuit.

Fig. 32-7. Fabrication of a monolithic circuit n-p-n transistor.

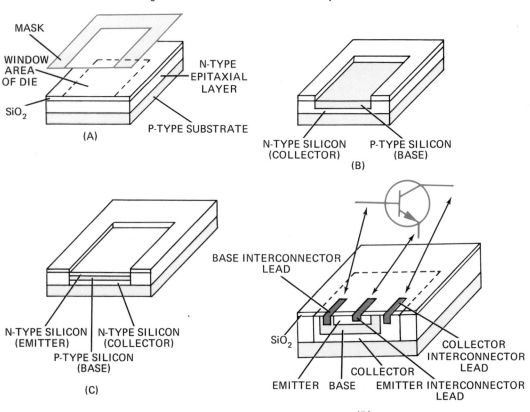

is performed simultaneously by means of automated production methods over all areas of a chip. The cycle of operations is also repeated upon given areas of the chip as necessary to fabricate the desired circuit components within a specified area of each die.

Transistor Fabrication. The fabrication of an n-p-n transistor within a die area of a chip is illustrated in Fig. 32-7. The first step of this process involves masking a "window" area upon the surface of the die (Fig. 32-7A). Following exposure to ultraviolet light and etching, an acceptor impurity is diffused into the window to a predetermined depth. The resulting n-type epitaxial layer immediately below it is the collector of the transistor (Fig. 32-7B). After still another oxidation-masking-etching-diffusion cycle, a donor impurity is diffused into the base p-type material to a given depth. This forms an n-type layer, the emitter of the transistor, above the base region (Fig. 32-7C). A more realistic view of the transistor after the final oxidation cycle has been completed and the electrode interconnector leads have been fabricated is shown in Fig. 32-7D.

Diodes. The diodes of a monolithic integrated circuit usually consist of p-n regions existing within fabricated transistor structures that are not actually used as transistors (Fig. 32-8). Since, in the majority of cases, a circuit that utilizes one or more diodes also contains transistors, this

Fig. 32-8. Monolithic integrated circuit diode formed by using the emitter-base portion of a transistor structure.

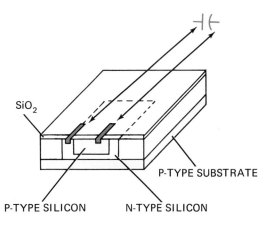

Fig. 32-9. Monolithic integrated circuit capacitor fabrication.

method of providing diodes is more economical than forming individual p-n region units.

Capacitors. One commonly used method of providing capacitors in a monolithic circuit is to utilize the intrinsic (natural) capacitances that exist between p-n junctions because of the ionic charges associated with these junctions. Thus, a p-n junction that is formed as the result of transistor structure fabrication can also be used as a capacitor (Fig. 32-9). In order to increase the capacitance of the resulting junction capacitor, the junction is often operated in a reverse-biased condition.

Resistors. Resistors are most often fabricated into a monolithic integrated circuit structure by diffusing acceptor impurities into the epitaxial layer (Fig. 32-10). Different values of resistance

Fig. 32-10. Monolithic integrated circuit resistor fabrication.

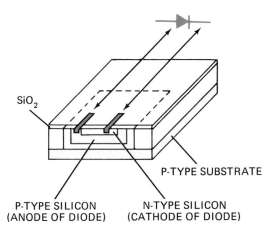

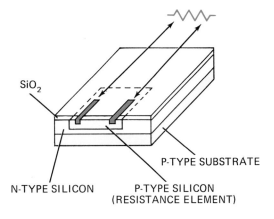

363

are obtained by varying the sizes of the diffusion surfaces and the concentration of the impurities that are diffused into these surfaces.

Inductors. Inductors are by far the most difficult components to fabricate by any method now employed in the manufacture of integrated circuits. Although inductors of very limited Q and inductance ratings are formed by diffusing highly concentrated impurities into a spiral pattern, the manufacture of inductors remains a major problem in integrated circuit technology. Therefore, when these components are required, it is common practice to use small, conventional, high-Q inductor units that are mounted within or near the associated integrated circuit package.

THIN-FILM COMPONENTS

In addition to being fabricated by diffusion, components such as resistors and capacitors are commonly made by utilizing what are known as *thin films*. To form a thin-film component, the desired resistive, metallic, and dielectric materials are first vaporized at a high temperature. The vapors are then allowed to condense upon the specified unmasked areas of a substrate made of glass, ceramic, or oxidized silicon (Fig. 32-11). Since the size of the area on which a film has been deposited in this manner can be very carefully controlled, the thin-film technique makes it possible to fabricate resistors and capacitors that are more precise than those produced by the monolithic diffusion process.

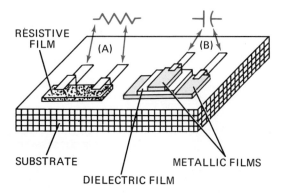

Fig. 32-11. Basic fabrication of thin-film components: (A) resistor, (B) capacitor.

Inductors. The thin-film technique is also used in the fabrication of inductors by forming highly conductive films in a spiral pattern. However, because of their limited inductance, these components are practical only in those circuits that operate at extremely high frequencies.

Transistors and Diodes. The fabrication of practical use transistors and diodes by means of thin films is a highly complex, expensive procedure. Because of this, thin-film transistors and diodes are commonly used only in those circuit applications that are experimental.

THE HYBRID CIRCUIT

The typical hybrid integrated circuit consists of both diffused (monolithic) and thin-film components that are mounted upon a common substrate (base) and interconnected as shown in Fig. 32-12. This method of construction is often used to eliminate certain undesirable operational characteristics that sometimes result from the interaction of components that are actually fabricated into a common substrate. Since the interconnected components are small in size, the entire combination of components can usually be contained in a "package," or module, no larger than would be necessary if the monolithic diffusion process was used to fabricate the entire circuit within a single die.

An additional advantage of the hybrid circuit structure is that it provides for increased flexibility of circuit design. This is because some of the components of the hybrid circuit can be rather easily removed or replaced. For this reason, the hybrid arrangement is often used when a pilot or prototype model of a particular circuit is first constructed. If after evaluation and testing the design of the pilot model is found to be operationally correct, the circuit may then be fabricated in either monolithic or thin-film form.

For Review and Discussion

1. What are the outstanding advantages of the integrated circuit as compared with other types of circuit assembly structures?

2. Define integrated.

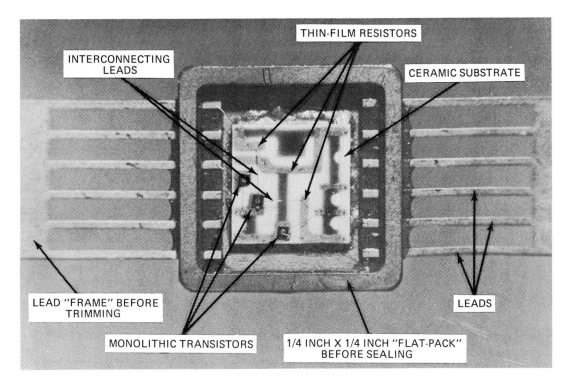

Fig. 32-12. Amplifier hybrid integrated circuit and schematic diagram of the circuit. (*General Electric Co.*)

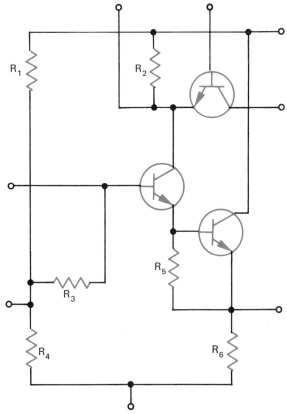

3. Describe the general features of the monolithic integrated circuit.

4. Define a substrate material as related to an integrated circuit.

5. What is an integrated circuit header?

6. For what purpose is the masking step used in the fabrication of an integrated circuit?

7. Describe the basic procedures followed in one method of monolithic integrated circuit transistor fabrication.

8. How are diodes formed from the transistor structure of a monolithic integrated circuit?

9. Describe the process of monolithic circuit resistor fabrication.

10. What is meant by a thin-film component?

11. Describe the fabrication of a thin-film capacitor.

12. What is a hybrid integrated circuit?

SECTION 7
ELECTRONIC COMMUNICATIONS

Electronic communications comprises many forms involving specialized circuit systems that make it possible to communicate by code, voice, pictures, and the printed word. This section presents several very commonly used electronic communications systems, including radio transmitters, receivers, and transceivers; television; radar; and lasers.

The first three units of this section discuss the fundamentals of transmitting, modulating, and receiving electrical energy. The remainder of the section presents information on complete systems and introduces many additional circuits that improve communications.

UNIT 33. Transmitter Fundamentals

The circuit for the radio transmitter shown in Fig. 33-1 contains few electronic components; nevertheless, it represents an electronic system that can transmit intelligence (information) through space by using the energy produced by electricity. The basic electronic circuits in this system are the power supply, the oscillator, the amplifier, the encoder (key), and the radiator (antenna). Input to this system is provided by an operator who moves the key to encode the message.

POWER SUPPLY CIRCUIT

The basic power supply is a 117-volt alternating-current circuit. One half of tube V_1 acts as a half-wave rectifier; the other half serves the dual role of amplifier and oscillator.

The pulsating direct-current output of the rectifier is changed to smooth direct current by filter components C_4, C_5, and CH_1. Capacitor C_6 filters line noise, preventing it from entering the transmitter. It also prevents any radio-frequency output from the transmitter from entering the line.

OSCILLATOR CIRCUIT

The transmitter circuit contains an oscillator circuit that generates the high-frequency radio signal required for transmission. The oscillator circuit consists of the crystal Y_1, R_1, and the pentode section of V_1. Crystal-controlled oscillators are discussed in detail in Unit 24.

AMPLIFIER CIRCUIT

The pentode section of V_1 is also an amplifier that increases the level of the signal produced by the oscillator, thereby greatly increasing the range of the system. Coil L_1 and capacitor C_1 form a tuned circuit at the frequency of the crystal oscillator. Hence, if the crystal is operating at a frequency of 7 megahertz, the circuit formed by L_1 and C_1 will be tuned to 7 megahertz. In some cases, the amplifier serves as a doubler; that is, its tuned circuit is set at the second harmonic of the signal. For example, if the amplifier was serving as a doubler, the tuned circuit would be at 14 megahertz and the crystal in the oscillator section would be a 7-megahertz crystal.

ENCODER CIRCUIT

The encoder circuit in a transmitter is essential to communications. Without it, the electromagnetic waves generated by a transmitter would have no meaning whatsoever to the person receiving the energy.

When two people communicate, they use a language both can understand. Language after all is just a series of sounds that have meaning to the person talking and to the person listening. But unfortunately the voice cannot be projected very far, and so the problem is to invent some kind of system to represent human language (meaning) that the radio transmitter can use.

The *Morse code*, devised by Samuel Morse, is such a language. Each letter of the alphabet and each number is represented by a series of dots and dashes, which in turn can be represented by a series of short or long bursts of radiated energy. And these short and long bursts of energy can be used by the radio transmitter.

If the transmitter's radiated energy can be interpreted according to a predetermined code that is known by both the person sending and the person receiving, then communications between these people can result. This is precisely what happens in radio communications. The trans-

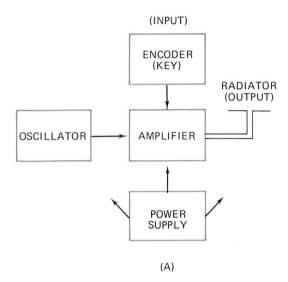

Fig. 33-1. Low-power CW (code) transmitter, 40 meters, 4 watts: (A) block diagram, (B) schematic wiring diagram. (*The Radio Amateur's Journal, CQ*)

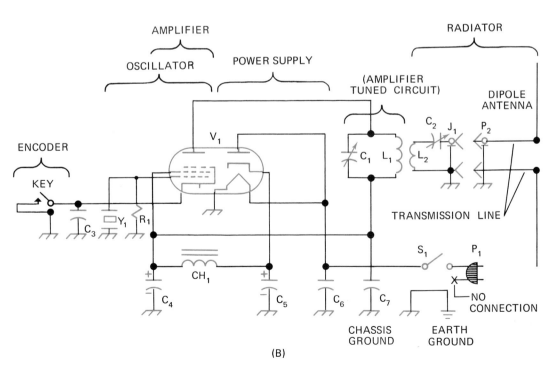

367

mitter is the device that produces the radiated energy and this energy is controlled in a predetermined manner by a device called a *key* or *encoder*, which is really a switch that the person operating the transmitter can control.

In Fig. 33-1B, the encoding circuit is shown with the key (switch) open. In this open position, the circuit is incomplete from the ground through the key to the tube. Therefore, no current flows to the tank circuit and no energy is radiated by the antenna system. Once the key is closed, however, the circuit is completed and the radio transmitter radiates energy into the airways. Capacitor C_3 improves the "make and break" characteristics in the key by providing a capacitive effect as the key is moved. Communication results when the key is used to encode a message in Morse code so that it can be understood by a person at the receiving end. This process is called *radio telegraphy*. The Morse code is also used in the telegraph system. Later units explain how speech and music are encoded (a process called *modulation*) and transmitted.

RADIATION CIRCUIT

Fig. 33-2 shows the basic elements needed to radiate electrical energy into space. The tank circuit contains the energy to be transferred to the radiation circuit. The coil L_2 is called a *link* because it inductively couples the energy from the tank coil to the antenna through the transmission line. The link coil L_2, along with the variable capacitor C_2, forms an antenna coupler that can be adjusted to the resonant frequency of the tank coil. Notice that the antenna coupler is a series-resonant circuit whose current is maximum at the resonant frequency.

Transmission Line. Electrical energy from the series-resonant antenna coupler is fed into a transmission line consisting of wires connected to an antenna coupler at one end and to a radiating antenna at the other. Fig. 33-3 illustrates two types of transmission lines.

Transmission Line Impedance. Because two wires are used in a transmission line, it contains certain electrical characteristics that must be considered. First, the two transmission line wires being parallel to each other form a capacitor. Second, because of the electromagnetic effect of the two lines on each other when the current flows in the transmission line, the wires form an inductor. The metal (usually copper) contains some electrical resistance; hence, the wires also contain some resistance (R). Therefore, the capacitive reactance X_C, the inductive reactance X_L, and resistance R form an impedance to the flow of current in the transmission line.

However, these electrical quantities (X_C, X_L, and R) are not concentrated (lumped) at any one point on the transmission line such as in a regular capacitor or inductor, but are distributed along the line and cannot be separated from one another. In other words, each transmission line contains a specific impedance, known as the *characteristic impedance* of the line (Fig. 33-3C).

The following formula can be used to calculate the impedance of a two-wire transmission line with air as the dielectric:

$$Z_o = \log_{10} \frac{s}{r}$$

where Z_o = characteristic impedance, in ohms
s = center-to-center distance between conductors
r = radius of conductor (expressed in same unit as s)

If another tuned circuit is connected to the output of the transmission line at the characteristic impedance of the line, the source or

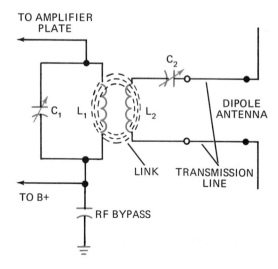

Fig. 33-2. Radiator (antenna) circuit.

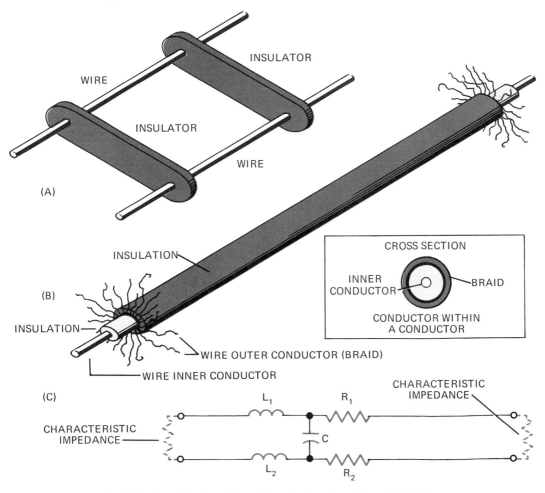

Fig. 33-3. Transmission lines: (A) parallel two-wire (open) line, (B) coaxial cable, (C) electrical equivalent of a transmission line.

input of the transmission line "sees" only that resistance. Therefore, the problem of transferring electric energy from the tank coil to the antenna coupler (antenna link and capacitor), through the transmission line, and finally to the antenna itself is accomplished by matching the impedances of the antenna coupler, transmission lines, and radiator (antenna).

Fig. 33-4A represents the effect of a transmission line terminated in its characteristic impedance. Note that the voltage and current are in phase along the line. When the voltage and current are multiplied to find the power, the resulting waveform is always above the zero axis. This indicates that the line is operating at maximum efficiency and maximum power is being transferred to the antenna.

Standing Waves. The transmission line shown in Fig. 33-4B does not terminate in its characteristic impedance. Thus the voltage and current are out of phase along the line, and when they are multiplied to find the power, the power lobes appear below the zero axis. These negative power lobes represent *reflected energy,* which is in the form of *standing waves* on the line. Reflected energy is energy that is not transferred to the load, but is instead returned to the transmitter. This is inefficient and, in some cases, damaging to the transmitter.

369

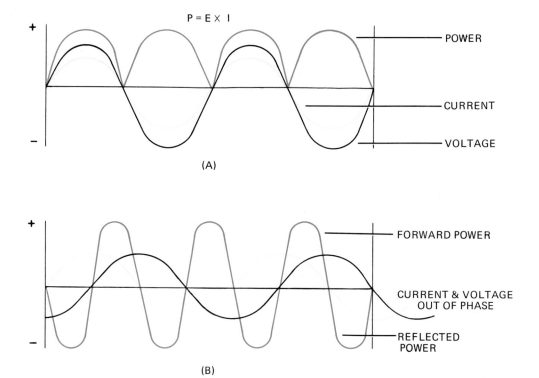

Fig. 33-4. Energy waves on transmission lines: (A) resonant line, (B) nonresonant line.

Radiator (antenna). The radiator (antenna) changes electric currents into electromagnetic waves that radiate out into space, whereas a receiving antenna changes electromagnetic waves into electric currents. Only the former is discussed in this unit.

The radiation of electric energy depends upon several known characteristics of radio-frequency currents. For example, radio-frequency currents produce electromagnetic fields. We have seen that the energy from low-frequency electric currents can be transferred from one transformer circuit to another by the electromagnetic field surrounding the coils of a transformer. These same electromagnetic fields can be used at high frequencies (radio frequencies) to cause electric energy to radiate out into space.

Heretofore, the electrostatic field about a coil energized with low-frequency currents has not been emphasized, but at high frequencies it is an important factor. Scientific investigations indicate that each electromagnetic field has an electrostatic field associated with it. One of the characteristics peculiar to these two fields is that they are perpendicular to each other—just as the electromagnetic field about a current-carrying wire is perpendicular to the direction of the current that produces it (Fig. 33-5).

In studying transmission lines we learned that when the transmission line is matched properly there are no standing waves, and consequently no energy loss. The antenna presents a totally different situation since its purpose is to radiate energy. An antenna wire is a good example of a tuned circuit that is designed with a feed transmission line so that all its energy is "lost," or radiated into space. The lost energy is the "payload" for the transmitter. The more efficient the radiator or antenna, the more efficient the transmitter is. A single wire or antenna arranged in a straight line meets these requirements.

When an alternating-current source of a specified radio frequency is fed into one end of an antenna having a length equal to the wave-

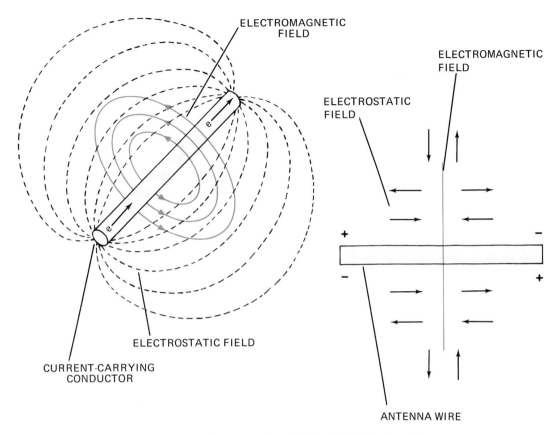

Fig. 33-5. Relationship of electromagnetic and electrostatic fields about an antenna wire.

length of the oscillations, the stream of electrons surges along the antenna, giving rise to a voltage and current waveforms as shown in Fig. 33-6A. Notice that the current and voltage are out of phase. Whenever the current is at peak, the voltage is zero, and whenever the voltage is at peak, the current is zero. Notice, also, that there is a concentration of voltage or charges of electricity at the ends of the antenna.

Although electrostatic and magnetic fields normally exist at right angles to one another, in the case of an antenna, where the current and voltage waves are 90 degrees out of phase, these fields are added together to form a sinusoidally varying, radiated field. This radiated field travels through space at the speed of light (300 million meters per second or 186,000 miles per second).

One wavelength is the distance traveled during the time required to complete one cycle of the radio-frequency current in an antenna. The following formulas can be used to calculate wavelengths:

$$\lambda = \frac{300,000,000}{f}$$

where λ = wavelength, in meters
f = frequency, in hertz

or

$$\lambda = \frac{300}{f}$$

where λ = wavelength, in meters
f = frequency, in megahertz

In many installations, an antenna as long as the wavelength of the oscillating currents would be impractical. For example, some antennas would have to be several miles long. In this case, the length of the antenna can be reduced if it is shortened so that it does not upset the phase relationships between current and voltage. Fig. 33-6B shows a center-fed antenna whose total length is one-half the wavelength. This type of antenna is called a dipole (*di* means *two*), and each of its sides is equal to one-quarter the total wavelength.

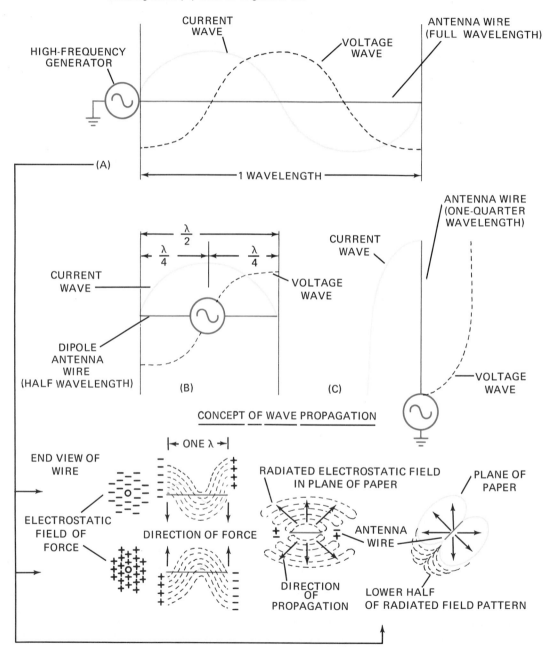

Fig. 33-6. (A) alternating-current (r-f) source fed at one end of antenna, (B) alternating-current (r-f) source fed at center of dipole antenna, (C) vertical antenna, alternating-current (r-f) source fed at grounded end.

Another common type of antenna is the vertical antenna shown in Fig. 33-6C. This antenna needs to be only one-quarter of the wavelength to be effective, because the earth acts as the other half of the dipole.

Antenna Resonant Circuit. Since the main purpose of the antenna is to radiate energy into space, it must have a low "radiation resistance." Antennas, like transmission lines, contain quantities of inductance (L), capacitance (C), and resistance (R) distributed throughout. Thus, the antenna is a tuned circuit whose specific resonant frequency is dependent on various electrical quantities. Antennas, like transmission lines, have their own characteristic impedances, and for maximum power to be delivered by a radiation system, the antenna and transmission line impedances must match. When they do, the radiation circuit is said to be *impedance-matched*.

Directional Beam Antenna. A dipole can be made to be directional; that is, it can direct its major radiation in only one direction like a beam of light. One common directional or beam antenna is the parasitic beam type shown in Fig. 33-7. This antenna has an array consisting of three elements: the reflector, the driven element, and the director. Since the center or driven element of a parasitic beam antenna is a dipole, radiated energy from this element cuts across both the reflector and director elements, causing magnetic fields to be generated. The direction of these fields is opposite from that which generated them. Since these elements (reflector and director) receive their energy from another source, they are called *parasitic elements*.

The physical arrangement of elements in a parasitic beam antenna is called an *array*. To make the antenna directional, the reflector element is made slightly longer than the radiator. This elongation causes increased inductive reactance in the reflector. And, as you recall, an increase of inductance in a circuit causes the current to lag the voltage. This out-of-phase condition causes standing waves to emerge on the reflector of a magnitude sufficient to repel the radiated energy from the reflector element toward the director element.

The director element is made slightly shorter than the driven element, thereby causing a capacitive effect to occur. Since the addition of capacitance in a circuit causes current to lead the voltage, its presence has a leading effect on the radiated energy. Thus, the combined effect of the reflector's inductive reactance "pushing force" and the director's capacitive reactance "pulling force" gives direction to the radiated energy. The distance between the elements also affects the radiation.

Electrical Length of Antenna. It is often impossible to set up an antenna of the exact physical length required for the frequency or band of frequencies to be transmitted. When this situa-

Fig. 33-7. Directional three-element three-band (10-15-20 meters) antenna: (A) director, (B) radiator, (C) reflector.

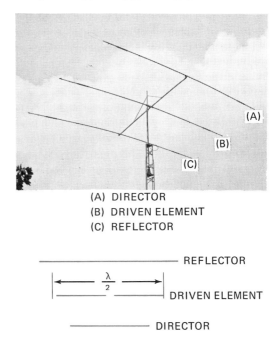

(A) DIRECTOR
(B) DRIVEN ELEMENT
(C) REFLECTOR

REFLECTOR
$\frac{\lambda}{2}$
DRIVEN ELEMENT
DIRECTOR

DIRECTION OF MAJOR RADIATION

tion occurs, the antenna's "electrical length" can be changed by the addition of inductive or capacitive reactance to the circuit. The addition of inductive reactance to a circuit causes the radio-frequency current to lag the voltage and has the effect of adding electrical length to an antenna.

The antenna's length is shortened by the addition of capacitive reactance to the circuit. The addition of capacitive reactance causes the radio-frequency current to lead the voltage, and this has the effect of reducing the electrical length of the antenna.

Adding "lumped" inductance or capacitance to the antenna's circuit is called "loading" the antenna. These quantities are generally added at the point where the transmission line connects to the transmitter or at the feed point to the antenna itself. The process of adjusting the electrical length of an antenna is called *antenna tuning*, and is characterized by a current rise in the antenna circuit, which of course, increases the radiated power.

GROUND AND SKY WAVES

As radio waves leave the antenna, some move outward in contact with the ground and form what is called the *ground wave*, while others radiate upward to form the *sky wave* (Fig. 33-8). The ground wave is used for short-range communications at relatively high frequencies. As it travels along the earth, its energy can be picked up or "sensed" by a receiving antenna at some distant point. However, the radiated energy in the ground wave can be easily absorbed by surrounding hills, ground, and buildings. As it loses its energy and the distance from the antenna increases, the ground wave becomes less effective (energy from high-frequency waves is absorbed by the earth at a faster rate than from low-frequency waves). To minimize this effect, antennas are usually mounted on the highest point in any locale so that the energy they radiate will cover the widest possible area.

Sky waves, on the other hand, are not absorbed so quickly by the earth, since they are radiated upward. Therefore, they can be used to communicate over long distances. These waves leave the antenna and go skyward until they are refracted (bent back) by an ionized layer of gas in the ionosphere (approximately 90-300 miles above the earth). The returning sky waves cover great distances because they "skip around" the earth, making it possible for us to have almost instantaneous around-the-world radio communications.

Fig. 33-8. Radio waves: (A) ground wave, (B) sky wave.

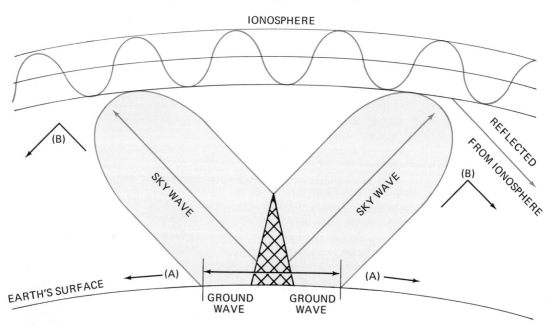

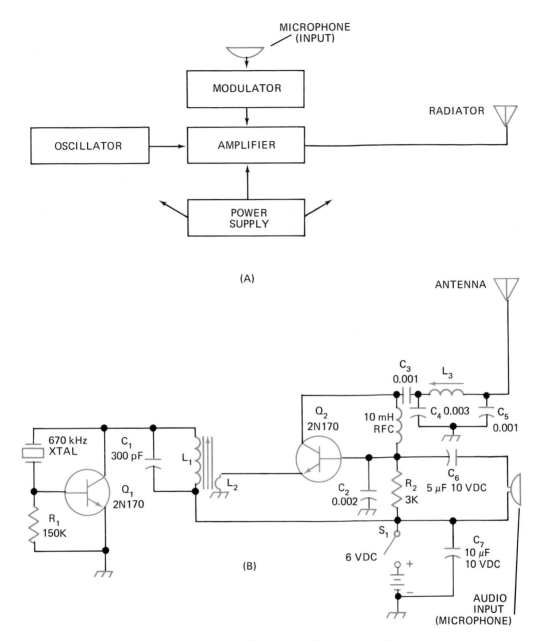

Fig. 33-9. Low-power two-transistor AM transmitter: (A) block diagram, (B) schematic wiring diagram.

LOW-POWER AM TRANSMITTER

A two-stage amplitude-modulation (AM) transmitter is shown in Fig. 33-9. It is designed to transmit a frequency that can be received by an ordinary home-type amplitude-modulation broadcast-band receiver. Although it has low power (400 microwatts), its energy can be picked up by a receiver located 50 to 100 feet away.

This transmitter contains five basic electronic circuits: power supply, oscillator, amplifier, modulator, and radiator. The input for the total system is a voice signal, for example, from a microphone. The output modulated amplitude is a radio-frequency electromagnetic wave. The modulation process is explained in more detail in Unit 34.

Power Supply. The power supply consists of a 6-volt battery.

Oscillator Circuit. The oscillator circuit is composed of resistor R_1, a crystal, Q_1, capacitor C_1, and coil L_1. The movement of the electrons in the 300-picofarad capacitor C_1 and the inductor L_1 forms a parallel resonant "tank" or tuned circuit. It is the action of this tank circuit that produces the continuous sinewave oscillations at a radio-frequency rate, in this case, 670 kilohertz.

Amplifier Circuit. The radio-frequency energy produced by the oscillator is coupled to the r-f amplifier transistor Q_2 through an eight-turn secondary winding L_2. The electron flow in the amplifier-circuit system is from the negative terminal of the battery to the chassis, then to L_2, Q_2, RFC, R_2, and S_1 before returning to the positive terminal of the battery. Practically all the battery power is dropped across the resistor R_2.

Modulation Circuit. The modulation circuit allows intelligence (speech or music) to be superimposed on the radio-frequency signal that can be radiated to distant receivers. This fundamental circuit is discussed in more detail in Unit 34.

Radiating Circuit. The radiating circuit, as shown, consists of capacitor C_3, an antenna matching network (L_3, C_4, C_5), and a three-foot antenna wire. Capacitor C_3 acts as a coupling device for radio-frequency energy into the *pi* network (so named because its schematic topology resembles the Greek letter π). The pi network is a tank circuit that acts like a transformer. This matching method (pi network) allows Q_2 to work efficiently and "loads" up the antenna so that it radiates properly.

Although the radio transmitter shown in Fig. 33-9 uses only two transistors, the principles behind its operation are similar to those of larger and more powerful transmitters. The crystal provides stability. The tank circuits smooth out the energy pulse into a continuous-energy sinewave at a specified frequency. The tank circuit eliminates harmonics and thus reduces the chance of interference. The modulator circuit allows intelligence to be carried to distant receivers. And, the matching pi network loads the antenna properly and increases its efficiency.

For Review and Discussion

1. Identify five basic electronic circuits in a transmitter.
2. Referring to Fig. 33-1, what components make up the oscillator circuit? Explain the electronic action of this circuit.
3. Explain how a transmitter can radiate twice the frequency of the crystal that is used in the oscillator circuit.
4. Define encoder in your own terms.
5. What elements influence the characteristic impedance of a transmission line?
6. Explain what happens when a transmission line is mismatched.
7. Define (a) fundamental frequency and (b) harmonics.
8. Name and describe a three-element beam antenna.
9. Describe the difference between a sky wave and a ground wave.
10. How is the antenna length shortened electrically? Lengthened?
11. Describe how energy is radiated from an antenna.
12. What is the characteristic impedance (in ohms) of a two-wire (parallel) transmission line with air as the dielectric, and whose center-to-center distance between conductors is one inch and wire size #18.
13. Explain the purpose of a pi network as used in a transmitter.
14. Explain how the radio-frequency signal is generated, amplified, and radiated in the transmitter circuit shown in Fig. 33-9.

UNIT 34. Modulation

Radio operators can use a key to turn transmitters off and on in a logical sequence called a *code*. The problem with this system is that it is limited to persons who understand the code. Also, the transmission of information such as music is not possible. This unit deals with how speech and music can be transmitted via radio waves by a process called *modulation*.

AMPLITUDE MODULATION

Fig. 34-1A shows a block diagram of a radio transmitter that is capable of transmitting speech or music. It consists of several basic electronic circuits, including oscillator, buffer amplifier, power amplifier, speech amplifier, modulator, radiator, and power supply circuits. In previous units we have discussed most of these basic circuits in detail. The emphasis in this unit is on speech amplification and modulation.

SPEECH AMPLIFIER AND MODULATOR

In the amplitude-modulated (AM) transmitter, the speech amplifier–modulator increases the power of microphone signals so they can effectively modulate a radio-frequency signal. A diagram of a speech amplifier–modulator is shown in Fig. 34-1C. It consists of three electronic circuits: the first audio amplifier, the second audio amplifier, and the modulator.

First Audio Amplifier. The first audio amplifier, shown in Fig. 34-1C, consists of a mike jack J_1; capacitors C_1, C_2, and C_3; resistors R_1, R_2, and R_4; and half of tube 12AX7 (V_{1A}).

The microphone or input signal enters jack J_1 and passes through capacitor C_1 and resistor R_1. It is then applied to the grid of V_{1A}, where it is amplified approximately 100 times. Capacitor C_1 provides a direct-current block to prevent any external voltages from upsetting the grid bias. Resistor R_1 forms a voltage divider with capacitor C_2. At approximately 160 kilohertz, the reactance of C_2 is 10 kilohms. Hence, stray radio-frequency signals above 160 kilohertz that enter the mike jack dissipate across R_1 and produce very little signal at the grid of V_{1A}. Resistor R_2 serves as a direct-current return for the grid.

Resistor R_4 is the plate load resistor and capacitor C_3 couples the signal from the plate circuit to the grid of the next stage or circuit. Resistors R_5 and R_9 form a voltage-divider network that establishes the plate voltage for both halves of V_1. Capacitor C_7 is a decoupling capacitor with a value large enough to maintain a constant voltage at the junction of R_9 and R_5. It also acts as a bypass to ground for any alternating currents in the power supply.

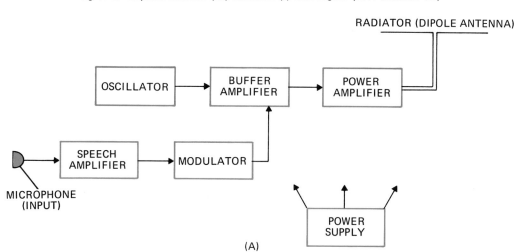

Fig. 34-1. Amplitude-modulated (AM) transmitter: (A) block diagram. (*E. F. Johnson Co.*)

377

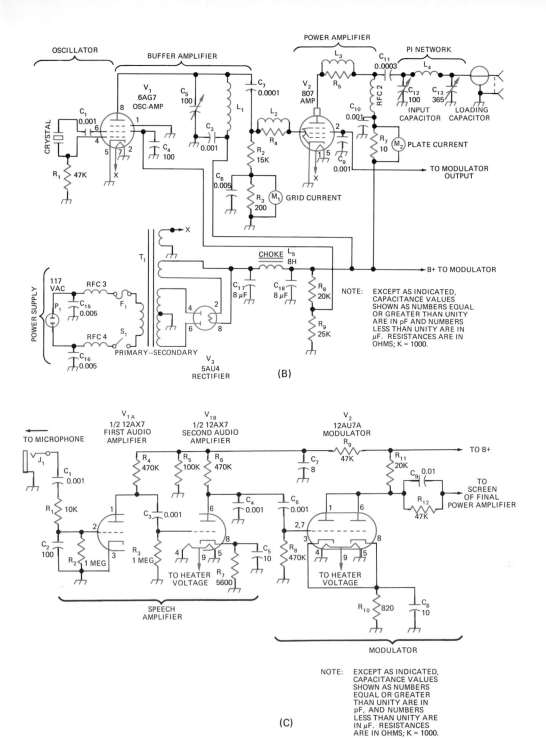

Fig. 34-1. Amplitude-modulated (AM) transmitter: (B) schematic wiring diagram of transmitter circuit, (C) schematic wiring diagram of speech amplifier/modulator. (*E. F. Johnson Co.*)

Second Audio Amplifier. Coupling capacitor C_3 brings the signal from V_{1A} to the grid of V_{1B} for further amplification. Resistor R_3 is the grid return resistor. The main electron flow path for V_{1B} is from the ground into the amplifier chassis, through R_7 into the cathode, out the plate, through R_6 and R_9, and back to the $B+$ terminal of the power supply. In contrast with V_{1A}, V_{1B} uses a cathode bias. Cathode resistor R_7 develops a voltage proportional to the static plate current. The cathode voltage "follows" the top of resistor R_7. As the plate current increases, the voltage at the top of R_7 and the cathode becomes more positive. Having the cathode go positive is identical, of course, to having the grid go negative, because in either event the grid is negative with respect to the cathode.

Capacitor C_5 is large enough to maintain the cathode voltage constant throughout a signal swing. For C_5 to be an effective cathode bypass capacitor, its reactance at the lowest frequency of operation should be about one-tenth the value of the cathode resistor.

Tube V_{1B} amplifies the grid signal by approximately 100 and passes it on to the modulator. Resistor R_6 is the plate load similar to R_4 in V_{1A}. Capacitor C_4 is a high-frequency roll-off component that attenuates frequencies above 3 kilohertz. In communication channels, transmission bandwidths are limited to 3 kilohertz. This represents the optimum combination for intelligibility and the best signal-to-noise ratio. Wider bandwidths offer an improvement in fidelity. Narrower bandwidths improve the signal-to-noise ratio but unfortunately degrade fidelity.

Capacitor C_6 couples the output signal from the plate circuit to the grids of the modulator tube V_2. It is interesting to note that all three coupling capacitors C_1, C_3, and C_6 are 0.001 microfarad. Normally, audio stages use 0.01- or 0.1-microfarad couplers. This amplifier is primarily for voice communications and not for broadcasting music. These coupling capacitors attenuate the low frequencies below 300 hertz, and pass frequencies with high intelligence content.

Modulator. Resistors R_8, R_{10}, R_{11}, and R_{12}; capacitors C_8 and C_9; and tube V_2 constitute the basic electronic modulator circuit. The main electron flow path is from the ground terminal of the power supply into the amplifier chassis through R_{10}, then into both cathodes, out both plates through resistor R_{11}, and back to the positive terminal of the power supply. The dual triodes in the 12AU7 tube can dissipate 2.75 watts each; and when hooked in parallel, as in this case, they can dissipate 5.5 watts of plate power. The modulator circuit or stage develops cathode bias through R_{10} similar to the action of R_7 in stage V_{1B}. The output signal from V_2 develops across R_{11} and feeds through C_9 to the screen grid of the final radio-frequency amplifier. Resistor R_{12} is a screen-dropping resistor to establish the correct direct-current voltage on the power amplifier's screen grid.

MODULATION PROCESS

Fig. 34-2A shows a sketch of several radio-frequency waves. The frequency of these waves is called the *carrier frequency*. Fig. 34-2B illustrates several audio-frequency waves developed by a transmitter's speech amplifier–modulator when a 1-kilohertz tone is applied to the microphone.

The basic idea of the modulation process is that different forces (signal voltages) act upon each other. To illustrate this idea vectors are used. The lines with arrows (Fig. 34-2C) are the vectors. (These represent only a few of the millions of vectors that would be needed to produce an accurate picture of the process.) As you may recall, a vector represents a force that has both magnitude and direction. The arrows indicate direction of the force and the length of the arrows represents the amplitude of the signal voltage. Fig. 34-2D illustrates the modulation process where the a-f and r-f signal voltages act upon each other. Since the carrier and audio waves are sinusoidal, both have positive and negative cycles. Consequently, the modulation process takes place simultaneously in both cycles. It is important to remember that in an amplitude-modulation process, the frequency of the carrier wave does not change. Its amplitude, however, is changed in accordance with the shape of the audio signal superimposed upon it.

Fig. 34-3 shows a complex audio (voice) signal, which varies from approximately 500 hertz to 2.5 kilohertz, modulated on a radio-frequency wave. The shape of the space enclosed by the modulated radio wave is called the *modulation*

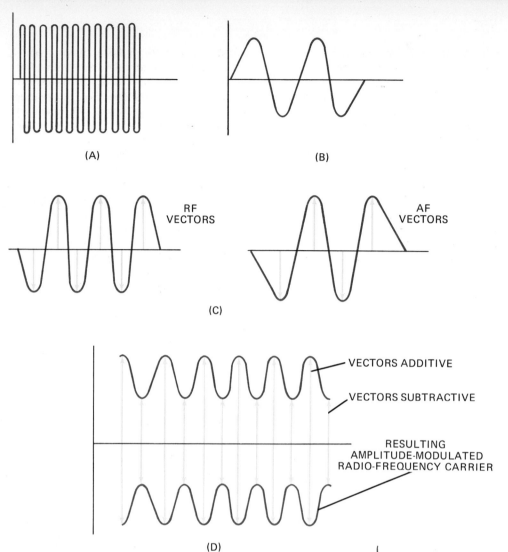

Fig. 34-2. Process of modulation: (A) radio-frequency carrier, (B) audio-frequency intelligence, (C) radio-frequency and audio-frequency vectors, (D) audio tone amplitude-modulating a radio-frequency carrier.

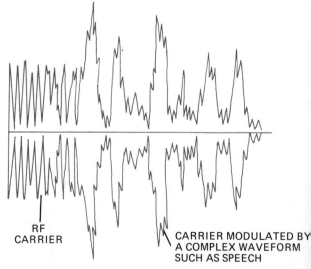

Fig. 34-3. Complex audio (voice) signal.

envelope. Speech and music modulate in the same manner as the single audio tone mentioned previously, but the resultant wave is much more complex.

MODEL AIRPLANE CONTROL TRANSMITTER

The model airplane control transmitter and its companion receiver (see Unit 35) form a com-

380

pact and efficient control system. As can be seen in Fig. 34-4A, the transmitter has six basic electronic circuits: the r-f oscillator, the a-f oscillator, the r-f amplifier, the modulator, the radiator, and the power supply. The inputs to the system are a manually operated switch to control the audio tone signal produced and, of course, the battery. The outputs are a radio wave modulated at a frequency of 26.995 megahertz when the audio tone switch is closed, and a radio wave that is unmodulated when the switch is open.

When the transmitter is turned on by closing switch S_1 shown in Fig. 34-4B, a high-frequency unmodulated radio wave is produced and is radiated from the antenna. Closing tone switch S_2 with the transmitter on starts the tone oscillator, which modulates the radio-frequency signal. The model airplane's receiver, which receives the modulated radio-frequency signal, then actuates an escapement that performs a control function for the airplane, such as moving the rudder to the right or left. The control function continues as long as the tone switch on the transmitter is closed.

The transmitter shown contains two oscillators. One of these is the r-f oscillator that

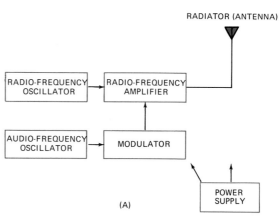

Fig. 34-4. Single-channel radio control model airplane transmitter: (A) block diagram, (B) schematic wiring diagram. (Ace Radio Control, Inc.)

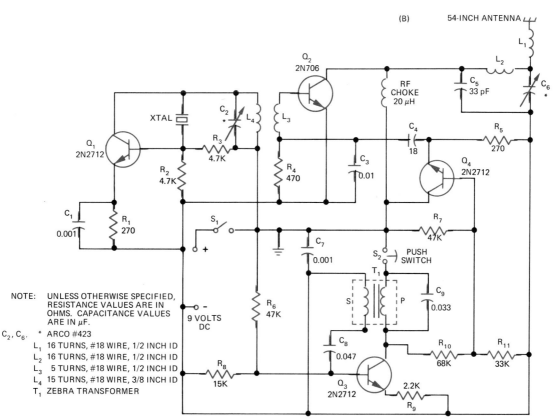

is crystal controlled to provide a precise "station" frequency (26.995 megahertz). The other is an audio oscillator that is on only when a pushbutton switch is depressed (manual input). The process by which modulation is accomplished is discussed below.

R-F Oscillator. The r-f oscillator consists of transistor Q_1, the crystal, capacitors C_1 and C_2, inductor L_4, and resistors R_1, R_2, and R_3, as shown in Fig. 34-4B.

Audio Oscillator. The audio oscillator consists of transistor Q_3, transformer T_1, resistors R_6, R_8, and R_9, capacitors C_7, C_8, and C_9, and the pushbutton switch S_2. Unless the pushbutton switch is closed, Q_3 cannot conduct and therefore cannot oscillate.

Capacitor C_7 provides a radio-frequency and audio bypass around the battery. Capacitor C_8 is a direct-current blocking capacitor used to keep the secondary of T_1 from shorting the bias circuit at the base of Q_3. Capacitor C_9 resonates with the primary of transformer T_1 and determines the tone frequency of oscillation. The tone signal's frequency is between 400 and 450 hertz and can be used with most single-channel model airplane receivers for controlling an airplane.

The primary and secondary windings of T_1 are phased to provide a positive feedback to the base of Q_3. When an electron flow is leaving the collector of Q_3 and moves up through the primary of T_1, electron current also flows up through the secondary. With the bottom of the secondary coupled to the base of Q_3 through capacitor C_8, the induced secondary current "pulls" electrons out of the base. This increase in base current causes the collector current to increase, and the increased collector current through the primary of T_1 further increases the base current to the point where Q_3 saturates. After the saturation point is reached, the primary current levels off. Since a secondary current cannot exist without a changing primary current, the base current in Q_3 decreases. This decrease causes a corresponding decrease in collector current, which induces an electromotive force across the secondary of T_1. Now Q_3 is driven toward cutoff. Transistor Q_3 oscillates continuously between saturation and cutoff.

Modulator Driver. Resistors R_7 and R_{11} are series-connected across the battery and form a voltage divider that forms the base bias of modulator driver Q_4 at about -5.5 volts with respect to ground. Modulator driver Q_4 is connected as an emitter follower and R_5 is the load resistor. Resistor R_{10} couples the audio tone from the oscillator and also attenuates the level of the signal to prevent overloading Q_4. Capacitor C_4 is a direct-current block and coupling capacitor that applies the powered-up audio tone to the base circuit of Q_2.

R-F Amplifier. In the r-f amplifier, two things occur. First, the r-f oscillator power level is boosted about 100 times. Second, the audio tone signal from the modulator driver amplitude modulates the radio-frequency signal. The net result is a moderate-power radio-frequency signal with tone modulation leaving the r-f amplifier. Remember, however, that tone modulation occurs only when the pushbutton switch in the audio oscillator circuit is closed.

The r-f amplifier consists of transistor Q_2, resistor R_4, capacitor C_3, inductor L_3, and the 20-microhenry radio-frequency choke (RFC). The main electron flow path for Q_2 is from the negative terminal of the battery into the emitter of Q_2, out the collector, through the RFC, and back to the positive terminal of the battery. There is no direct-current bias for Q_2, since the base returns to negative through L_3 and R_4. With no forward bias, Q_2 will not conduct until the signal from the oscillator drives its base positive. Transistor Q_2, therefore, serves as a class B amplifier.

Fig. 34-5 shows the r-f amplifier's modulation waveforms. The radio-frequency input to the amplifier is constant and, by varying the voltage across R_4 (Fig. 34-5A), more or less of the input signal appears at the output. Fig. 34-5B shows the input radio-frequency signal. Fig. 34-5C, D, and E show the output waveforms when the voltage across R_4 is zero volts, positive, and negative, respectively. For this transmitter, the voltage developed across R_4 is the tone signal coming from the modulator driver, and the modulated output looks like Fig. 34-5F.

The model airplane control transmitter just described is a basic form of a guidance system. Its control signal, when received in the model airplane, trips a relay or some other device. This causes the airplane to take some specific action

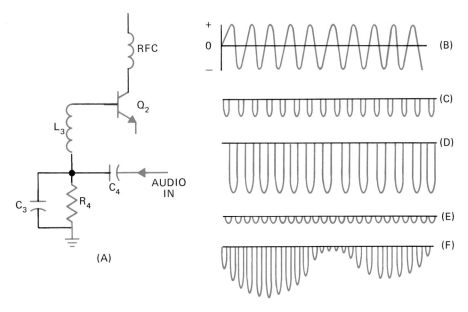

Fig. 34-5. Amplifier and modulation waveforms: (A) radio-frequency amplifier, (B) input signal, (C) output (when voltage across R_4 is zero), (D) output (when voltage across R_4 is positive), (E) output (when voltage across R_4 is negative), (F) modulated output.

such as to turn left or turn right. Some model airplane transmitters have ten or more different tone signals that can be transmitted. The receiver can, of course, differentiate between these tones, thus causing different actions to occur in the plane.

LOW-POWER SPEECH MODULATION

Fig. 34-6 shows a modulation circuit used in a low-power transmitter. The direct-current electron flow for this circuit is from the chassis through L_2, to Q_2, through the radio-frequency choke, R_2 and S_1, back to the positive side of the battery. The complete circuit for the transmitter is shown in Fig. 33-9.

Audio signals are fed from the microphone in and out of the 5-microfarad coupling capacitor. The value of this capacitor is so large that a sizable electron flow is required to change its voltage; thus the audio-signal voltages are passed on to the resistor R_2 and the 0.002-microfarad capacitor at the base of Q_2.

When the audio-signal voltage swings positive at the top of the input (microphone) terminal, a voltage is developed across R_2. This voltage is positive at the top of the resistor and negative at the bottom, and of a polarity that adds to the battery voltage. This action increases the bias voltage on Q_2, resulting in increased electron flow. Since the collector flow consists of radio-

Fig. 34-6. Low-power transistorized speech modulator.

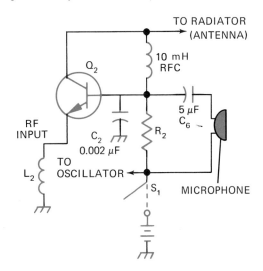

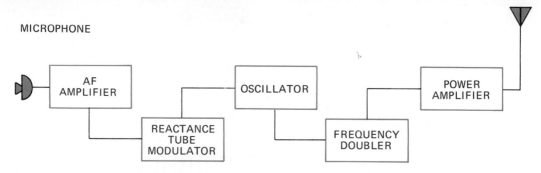

Fig. 34-7. Block diagram of an FM transmitter.

frequency energy, it increases its amplitude at the audio rate and a higher amplitude is radiated.

When the signal voltage goes negative at the audio input, Q_2 "sees" a reduced voltage at its base terminal and a lower amplitude of r-f energy is radiated at the antenna. Thus the audio input modulates the amplitude of the radio-frequency energy being radiated.

FREQUENCY-MODULATED TRANSMITTER

Frequency modulation (FM) is a method of transmitting a message on a radio-frequency wave (carrier) by varying the frequency of the carrier. Since a transmitted frequency-modulation signal does not rely on amplitude changes to send information, the FM system is more immune to noise and interference than is the AM system. This is because static noise and interference, such as the disturbance created by a car's ignition, is basically an amplitude-modulation signal, and FM receivers can be made immune to amplitude-modulation signals.

Fig. 34-7 shows a block diagram of an FM transmitter. Note that the output from the audio amplifier is used to control the frequency of the r-f oscillator, while in the AM transmitter the audio was used to control the amplitude of the radio-frequency carrier in the final power amplifier of the system. Another important difference in the FM transmitter is the use of a frequency doubler (multiplier). This doubles the frequency of the oscillator by tuning the doubler to the second harmonic of the oscillator frequency. The frequency doubler also doubles the amount of frequency deviation produced by the reactance (impedance) tube modulator. For example, if the reactance tube modulator causes the oscillator frequency to swing from 50 to 55 megahertz, the total deviation will be 5 megahertz. When the signal emerges from the doubler, however, the swing is from 100 to 110 megahertz—a total deviation of 10 megahertz. In an FM transmitter, a frequency doubler doubles both the output frequency and the frequency deviation.

REACTANCE TUBE MODULATOR

Fig. 34-8 shows a schematic diagram of a reactance tube modulator. Notice that the reactance tube is connected in parallel with the oscillator tank circuit. Thus, any shift in the reactance of the modulator tube would alter the frequency of the tank circuit.

Output from the audio amplifier is coupled to the cathode of the reactance tube. The audio

Fig. 34-8. Reactance tube modulator.

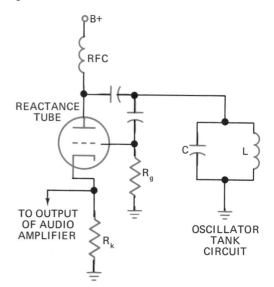

signals present on the cathode cause a change in the instantaneous capacitance of the reactance tube. The frequency of the oscillator is then a function of the audio input to the microphone.

With no audio input, the oscillator frequency is steady at the "resting" frequency. When the positive audio voltage appears at the cathode of the reactance tube, however, the reactance tube current decreases. This occurs because making a cathode positive is equivalent to making the control grid negative. This action decreases the capacitive reactance of the modulator, and increases the oscillator frequency. When the audio signal swings negative, the reactance tube current increases, thus increasing the capacity value and decreasing the oscillator frequency. The frequency of the audio signal determines the rapidity of oscillator frequency change; the amplitude of the audio signal determines the extent. Fig. 34-9 illustrates a frequency-modulated carrier.

DEGREE OF MODULATION AND BANDWIDTH

In an AM modulation system, 100 percent modulation is achieved when the amplitude of the carrier varies between zero and twice its unmodulated value. In an FM system, 100 percent modulation has a different meaning. Since the level of carrier does not change, a constant amount of power reaches the antenna. Thus, in the FM system, 100 percent modulation is achieved when the carrier deviates the maximum permissible amount. In the frequency-modulation broadcast band (from 88 to 108 megahertz), 100 percent modulation is defined as a carrier deviation of 75 kilohertz above and below the resting frequency of the carrier. This amounts to a total frequency swing of 150 kilohertz. A 50 percent modulation, then, would entail a carrier deviation of 37.5 kilohertz above and below the resting frequency.

In other FM systems, a frequency deviation of ±15 kilohertz is considered to be 100 percent modulation. Such a system is sometimes considered a narrowband FM system.

The bandwidth requirements for a standard AM system are twice the highest modulating frequency. If the highest modulating frequency is to be 10 kilohertz, then the bandwidth required would be 20 kilohertz. In an FM system, the bandwidth requirement is greater than twice the frequency deviation of the carrier by an amount approximately equal to twice the highest modulating frequency. Thus, FM systems generally require more bandwidth than AM systems.

OTHER MODULATION METHODS

The modulation process is actually a heterodyning, or mixing, of two different signals, and many methods can be used to achieve it. As mentioned earlier, the amplitude-modulation process changes the amplitude of the carrier wave. Another method of amplitude modulation is called single-sideband suppressed carrier modulation.

Single-sideband Generation. During the heterodyning (mixing) process, there are two important

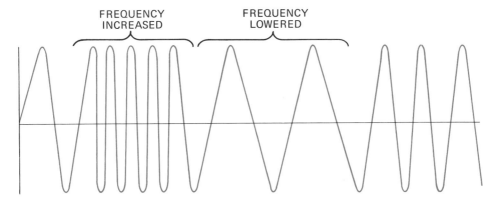

Fig. 34-9. Frequency-modulated carrier.

frequencies generated—one is called the *sum* frequency, the other the *difference* frequency. For example, if a 7,100-kilohertz carrier frequency was transmitted, and amplitude-modulated with voice frequencies ranging up to 3,000 hertz, the result would be a maximum sum frequency of 7,103 kilohertz (7,100,000 *hertz* + 3,000 *hertz*). The minimum difference frequency would be 7,097 kilohertz (7,100,000 hertz − 3,000 hertz). The range of frequencies generated by the audio signals above the carrier frequency is called the *upper sideband*. The range of frequencies generated below the carrier frequency is called the *lower sideband*. Since both sidebands were generated by the same audio frequencies and exist on both sides of the carrier frequency, they naturally contain the same information. Consequently, only one sideband really needs to be transmitted.

Single-sideband generation is essentially a method of transmitting information through the use of only one of the sidebands developed by the modulation process. Hence the term "single sideband." The transceiver described in Unit 36, page 405, makes use of this method of modulation.

There are several advantages to using single-sideband transmission. First, the same message served by the transmitter can be accomplished with a substantially smaller power rating. In a standard AM transmitter having 100 watts of carrier power, each sideband contains 25 watts (assuming 100 percent modulation). The total power required is 150 watts (25 watts upper sideband plus 25 watts lower sideband plus 100 watts carrier). Since only one sideband is needed to convey the information, a 25-watt sideband transmitter alone will provide the same coverage as a 150-watt transmitter, assuming the carrier is not used. Second, since single-sideband transmission uses less space in the frequency spectrum, this method of transmission makes it possible to double the number of voice channels within a given frequency range, as compared with those of a conventional AM transmitter. Third, single-sideband transmission makes it possible to realize savings in size, weight, and cost of transmitting equipment. Fourth, when the carrier is suppressed in single-sideband transmission, it reduces carrier heterodyning, which often produces squeals in standard AM systems.

Angle Modulation. When the modulation process changes in frequency, it is called *frequency modulation* (FM). This method of modulation actually falls under the category of *angle modulation* because the time derivative of the phase angle changes in accordance with the amplitude of the modulating signal. Another method of angle modulation is *phase modulation*. In this method, the instantaneous phase angle of the carrier is changed in accordance with the amplitude of the modulating signal.

Pulse Modulation. Still another method of modulation is called *pulse modulation*. In pulse modulation, one or more aspects of the pulse are varied in accordance with the modulating signal. An example is that in which the pulse-width varies, but the spacing between the pulses remains the same. The Morse code is based on this method.

For Review and Discussion

1. Name the basic electronic circuits found in a speech amplifier and modulator (see Fig. 34-1C).
2. Identify three functions of the speech amplifier and modulator.
3. Define the term "heterodyning."
4. Describe the r-f amplifier and modulation process used in the radio control model airplane transmitter (see Fig. 34-5).
5. Explain the basic difference between the AM and FM radio waves.
6. Identify the basic electronic circuits found in an FM transmitter.
7. Explain how vectors can be added or subtracted and demonstrate this by means of a diagram.
8. Draw two vectors, one representing a current and the other representing a voltage that is 90 degrees ahead of the current. Use any convenient scale.
9. How are sum and difference frequencies generated during the modulation process in an AM transmitter.
10. List three advantages of the single-sideband method of modulation.

UNIT 35. Receiver Fundamentals

Fig. 35-1A shows a block diagram for a three-transistor radio. The input to the system is a modulated radio wave; the output is sound (speech and/or music). The six basic electronic circuits involved are the sensor (antenna), the tuner, the detector, the amplifier (audio frequency), the reproducer, and the power supply. The block diagram shows the signal moving from left to right. The three-transistor radio is essentially a tuned-radio-frequency (trf) receiver followed by a transistor audio amplifier.

SENSOR CIRCUIT

The device used to carry the energy from a modulated radio wave into a receiver is called the *external sensor* or *antenna*. The wire antenna provides a path for the electric currents, which are set in motion by the energy of the modulated radio wave cutting across the antenna. These weak currents flowing through L_1 to ground and back establish voltages called *radio-frequency* (r-f) *signals* at the junction of L_1, C_1, and CR_1 as shown in Fig. 35-1B. At this junction, the radio-frequency signal appears as a modulated alternating voltage similar to that shown in Fig. 35-2. Notice that the sensor circuit shown

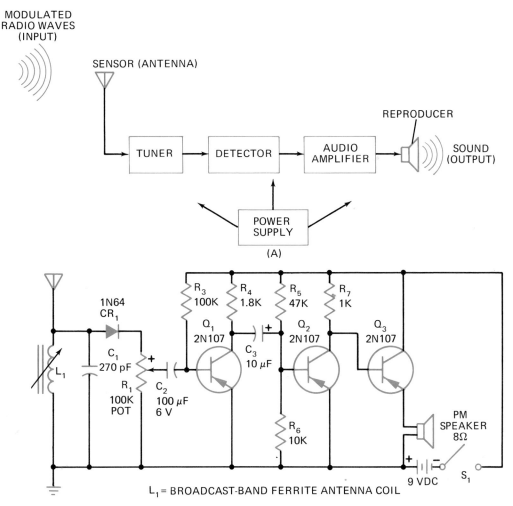

Fig. 35-1. Three-transistor AM radio receiver: (A) block diagram, (B) schematic wiring diagram. (*Popular Electronics*)

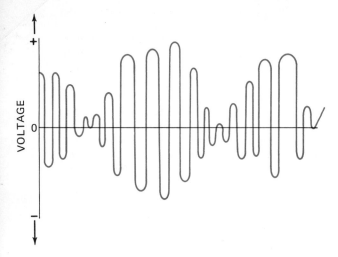

Fig. 35-2. A typical modulated radio-frequency (voltage) signal (artist's conception—greatly enlarged).

in the schematic diagram provides a complete path from antenna to ground.

TUNER

The operation of the tuner depends on three things: the weak electric currents set into motion by the energy from the modulated radio wave, the effect of inductive reactance of the coil on the electric currents, and the effect of capacitive reactance of the capacitor on the electric currents.

When the coil L_1 and the capacitor C_1 are connected as shown in Fig. 35–1B, a *tuner* is formed, whose function is to select the desired station from all others within the range of the radio receiver. A particular station is selected because the combined effect of its electrical characteristics causes the electrical currents in the tuner to oscillate in step with the incoming modulated radio wave. When this occurs, the tuner is said to be *in resonance* with the incoming radio-frequency signals. At its resonant frequency, the tuner offers a low impedance to the currents established by the modulated radio wave. At frequencies other than the resonant frequency, however, the tuner offers a high impedance. Therefore, the electrical values of the tuner determine the radio frequency that will develop a significant voltage across it. As will be noted, the components of the tuner are similar to the components of a transmitter's tank circuit.

The selection of a desired station is accomplished by adjusting the electrical values of the tuner's coil and/or capacitor. For example, a tuner having an adjustable antenna coil with an electrical value ranging from 40 to 300 microhenrys and a 270-picofarad capacitor can select any station frequency within the broadcast band of approximately 550 to 1,600 kilohertz. A specific station is selected by moving the slug (ferrite core) contained within L_1 a slight amount. This changes the electrical value of L_1 so that the circuit formed by it and C_1 will resonate at a specific station frequency; for example, 1,440 kilohertz. In the above example, the electrical value of the capacitor remained unchanged. However, either component can be fixed or adjusted.

DETECTOR

The detector circuit is used to separate speech or music from the modulated radio wave. This process is called *demodulation*. A detector allows the electric currents caused by the modulated radio wave to pass through it easily in one direction, but offers high resistance to the currents in the opposite direction. In effect, the detector rectifies the alternating current comprising the modulated radio wave and changes it into a pulsating direct current (Fig. 35–3).

As the diode CR_1 conducts on the positive half-cycles of the modulated radio-frequency signal, voltages appear at the top of the volume-control potentiometer R_1 (Fig. 35–1B). Although

Fig. 35-3. (A) Pulses of a modulated radio wave after detection, (B) direct-current component (average) pulsating current at an audio-frequency rate.

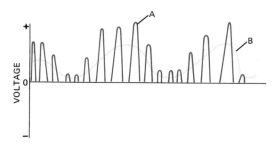

the detected signal consists of a direct-current component (the average level), there are also several radio-frequency components, which come from modulated radio-frequency signals, harmonics, and stray radio-frequency signals. These signals are above the upper cutoff frequency of the 2N107 transistors, and thus are not amplified.

In addition to the incoming energy from the modulated radio wave, this radio uses a 9-volt direct-current battery. With the on/off switch S_1 closed, there are several parallel paths for electron flow. In one path, electrons flow from the battery down through resistor R_3 to Q_1, and then return to the positive terminal of the battery. This action puts a forward bias on the base-emitter junction of transistor Q_1. Electrons also flow down through the collector load resistor R_4, through the collector-emitter circuit of Q_1, and on to the positive terminal of the battery. These two paths establish the direct-current operating point of transistor Q_1.

The direct-current operating point of transistor Q_2 is established in a manner similar to that of Q_1. When electrons flow down through R_5 to Q_1 and return to the battery, a forward bias is established on the base-emitter junction of transistor Q_2. The proper bias causes the transistor to conduct, and collector current flows down through R_7, through the collector-emitter circuit of Q_2, and on to the positive terminal of the battery. The addition of resistor R_6 creates a voltage-divider network consisting of R_5 and R_6 at the base of transistor Q_2. This network stabilizes the direct-current operating point of Q_2 by limiting the voltage that may exist on the base.

Some of the electrons flowing through resistor R_7 forward-bias the base-emitter junction of transistor Q_3 as they return through the reproducer (loudspeaker) to the positive terminal of the battery. When transistor Q_3 conducts, a large number of electrons flow through the collector and emitter circuit of Q_3 to the loudspeaker, and on to the positive terminal of the battery.

AMPLIFIER (AUDIO FREQUENCY)

Referring back to the detected audio-frequency signals or voltages appearing at the top of the potentiometer R_1 it can be seen that the audio-frequency signals are coupled through capacitor C_2 to the base of transistor Q_1, which is connected in a common-emitter circuit. The potentiometer is a voltage-divider circuit that acts as a volume control. Although this circuit is not needed, it does provide a convenient means of controlling the signal strength applied to the amplifier circuits.

When a positive-going signal is applied to C_2, the bias current in the transistor varies about an average or no-signal value. This action causes the current in the direct-current collector circuit to vary through the load of Q_1. In this case, the load for Q_1 is R_4. A positive-going input-signal voltage opposes the base-emitter bias and decreases the base-emitter current. This action decreases the collector current and voltage drop across the load resistor, resulting in an increase in collector-to-ground output voltage. Because the collector is negative with respect to ground, a positive-going input signal will result in a negative-going output signal. Therefore, a phase shift of 180 degrees occurs between the input and output signals.

This negative-going signal voltage is coupled through coupling capacitor C_3 to the base of Q_2, also connected as a common-emitter circuit. This negative-going signal voltage increases the number of electrons flowing into the base of transistor Q_2. The transistor action of Q_2 causes an increase in the electrons flowing into the collector and results in an amplified positive-going signal at the collector of transistor Q_2. The positive-going voltage is directly coupled with the base of transistor Q_3, reducing the number of electrons into the base of Q_3. The transistor action of Q_3 then reduces the number of electrons flowing in the collector-emitter and speaker circuit.

REPRODUCER

Notice that the speaker is connected as the load in an emitter-follower circuit of Q_3. And the output signal is developed between the emitter and the ground. Any change in the current through the speaker coil results in a movement of the speaker cone, which causes an audible sound to be heard.

When the audio signal at the base of Q_1 reverses itself and becomes a negative-going signal voltage, the entire process just described reverses, including the direction of the speaker cone movement. As the cycle is repeated over and over at varying audio rates and amplitudes, intelligible speech or music can be heard.

SINGLE-CHANNEL RC RECEIVER

Because transistors are much smaller and lighter and require less battery power than their tube counterparts, they are used successfully in model airplane radio-controlled (RC) receivers. For these same reasons, transistors are also used in satellites, and in other electronic equipment as well. The single-channel radio-controlled model airplane receiver described in this unit is a superregenerative type. One of the main advantages of this type of receiver is its high sensitivity. The receiver tucked away inside the airplane works in conjunction with a transmitter operated on the ground to provide remote control of the model airplane (see Unit 34, p. 380). Further study of this receiver, relating it to the total radio-control system, shows that the principles used are also found in the more sophisticated satellite communications systems.

Fig. 35–4A shows the block diagram of a radio-controlled model airplane receiver designed to operate on a 26.995-megahertz (radio-frequency) signal that is modulated by a 600-kilohertz tone (radio-frequency) signal. In addition to the relay and the power supply, the basic electronic circuits that make up this radio are the sensor (antenna), the tuner, the detector (superregenerative), the audio amplifier, and the relay driver. Input is provided by a tone-modulated radio-frequency signal and, of course, a battery. The output of the receiver is an electromagnetic force that controls a relay switch. In the model airplane, this switching action causes another circuit to operate, which makes the airplane perform some function; for example, moving the rudder of the airplane to the left or right.

The block diagram shows the signal moving from left to right. First, the sensor, or antenna, picks up the tone-modulated radio-frequency signal and passes it on to the tuner and the superregenerative detector that amplifies and detects the tone signal. The audio amplifier further amplifies the tone signal and passes it on to the relay driver circuit. The relay driver circuit boosts the signal enough to operate a relay. Therefore, when the tone signal is transmitted, the relay is closed and when transmitting ceases, the relay is open.

Sensor (antenna) and Tuner. The sensor, or antenna, and tuner provide the same functions in this receiver as those discussed in connection with the radio receivers. The sensor or antenna picks up the radio-frequency energy and the tuner selects the desired station. In this case, the tuner (C_3 and L_1 in Fig. 35–4B) is adjusted to resonate at 26.995 megahertz. Tuning is accomplished by adjusting the ferrite slug in L_1.

Superregenerative Detector. One of the unique features of this radio receiver is its superregenerative detector circuit. In electronics the term *regeneration* means to take an output signal from an amplifier and feed it back into the input in phase so it adds to the input.

Oscillators (Unit 24) are amplifiers that supply their own input via regenerative feedback. A regenerative amplifier is very similar, except that it is not supposed to oscillate. If this occurs, the amplifier is no longer capable of providing the desired amplification of external signals. A regenerative amplifier should be designed and adjusted so that the regenerative feedback is not quite enough to cause the amplifier to oscillate, but it should significantly increase the gain and sensitivity of the amplifier. Some regenerative receivers have a regeneration control that must be decreased slowly until the amplifier just stops oscillating. At that point, the amplifier usually exhibits its best sensitivity and selectivity.

In the radio-controlled receiver shown in Fig. 35–4B, no regeneration control is shown. In this case, the control of oscillation is automatic, with transistor Q_1 constantly shifting from an oscillating state to a nonoscillating state. Transistor Q_1 serves as an amplifier only when it is not oscillating. This is not a drawback, however, since it serves as an amplifier often enough to perform the desired function.

The superregenerative detector consists of transistor Q_1; resistors R_1, R_4, and R_5; capacitors C_1, C_2, C_3, C_7, C_8, and C_9; inductors L_1 and L_2; and transformer T_1. The main electron flow path is from the negative terminal of the battery through the primary of T_1 to inductor L_1, then into the collector of Q_1, out through the emitter, through L_2 and R_5, and back to the positive

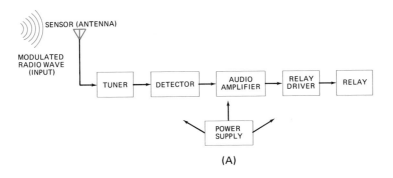
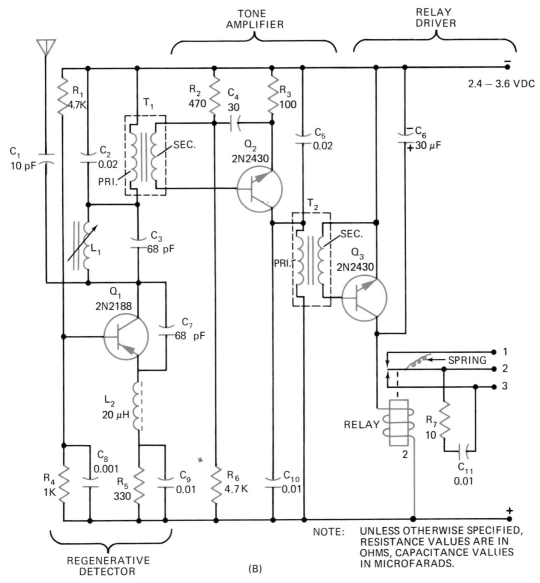

Fig. 35-4. Single-channel radio control model airplane receiver: (A) block diagram, (B) schematic wiring diagram. (*Ace Radio Control, Inc.*)

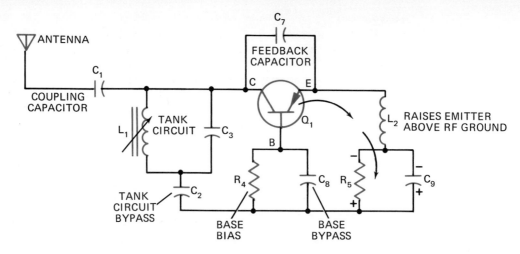

Fig. 35-5. Equivalent regenerative detector circuit.

terminal of the battery. Transistor bias current flows down through R_1 and R_4 (a small amount goes into the base of Q_1) and back to the positive terminal of the battery.

Capacitor C_1 couples the radio-frequency energy received from the sensor (antenna) into the regenerative circuit. The value of C_1 must be large enough to make an efficient transfer of the signal, yet small enough to prevent loading transistor Q_1 unnecessarily. Capacitor C_8 is a radio-frequency bypass for the base circuit of the transistor. Capacitor C_2 is the radio-frequency bypass for the "cold" side of the collector tank circuit. Capacitor C_9 is a bypass. Resistors R_1 and R_4 form a voltage divider for the base, and R_5 is the emitter resistor for direct-current stabilization. Capacitor C_7 is the feedback capacitor that sets up the regeneration in transistor Q_1. Inductor L_2 acts as a radio-frequency control for the emitter. As mentioned before, L_1 resonates with C_3 at the signal frequency.

Fig. 35-5 shows the schematic wiring diagram of the basic regenerative detector circuit. The tank circuit is formed by inductor L_1 and capacitor C_3, and feedback is accomplished by capacitor C_7. When transistor Q_1 begins to oscillate, some of the electron flow from the emitter charges capacitor C_9. The voltage on the top plate of C_9 goes negative, which in turn causes the emitter of Q_1 to go negative with respect to ground. Eventually, the emitter becomes more negative than the base and this turns off the transistor and stops the oscillations. When this occurs, capacitor C_9 no longer has a charging source, so it begins to discharge through resistor R_5. Soon Q_1 is no longer biased and begins to oscillate. The process then repeats itself. In this respect, Q_1 acts as a common-base blocking oscillator.

When a signal appears at the antenna, it couples to Q_1 through C_1 as shown in Fig. 35-4. During the periods when Q_1 is not oscillating, the signal is amplified. Transistor Q_1 acts as a detector as well as an amplifier by rectifying the radio-frequency signal. The detected audio signal develops across capacitor C_2 and the primary of transformer T_1. Transformer T_1 couples the audio tone to the audio amplifier, transistor Q_2.

Audio Amplifier. The audio amplifier consists of transistor Q_2; resistors R_2, R_3, and R_6; capacitors C_4, C_5, and C_{10}; the secondary of transformer T_1; and the primary of transformer T_2. The main electron flow path is from the negative terminal of the battery through resistor R_3, into the emitter of Q_2, out the collector, through the primary of transformer T_2, and back to the positive terminal of the battery.

Capacitor C_4 bypasses resistors R_2 and R_6 for signal purposes; thus the entire audio signal from the secondary of transformer T_1 appears between the emitter and base of Q_2. Transistor Q_2 amplifies the audio tone approximately 100 times and couples it to the relay driver transistor

Q_3. Capacitors C_5 and C_{10} provide filtering to smooth out the signal, and they also provide a radio-frequency bypass for the battery. The output of Q_2 develops across the primary of T_2 and the secondary of T_2 couples the signal directly into the base of Q_3.

Relay Driver. The relay driver circuit consists of transistor Q_3, capacitor C_6, the secondary of T_2, and the relay. The main electron current flow is from the negative terminal of the battery into the emitter of Q_3, out through the collector, through the relay, and back to the positive terminal of the battery. Transistor Q_3 operates at zero bias, since the base returns directly through the secondary of T_2 to the emitter. Theoretically, the current through Q_3, with no input signal, is zero. In practice, not all transistors are perfect and some leakage current exists. If the leakage current gets too high in this case, it could slow the release of the relay. The signal at the secondary of T_2 is the audio-control tone generated by the transmitter. But, with zero bias, Q_3 clips the negative part of the signal and amplifies the positive-going portion. Without capacitor C_6, the output would be a series of pulses as shown in Fig. 35-6. Capacitor C_6 assumes the role of a filter and provides holding current for the relay between pulses. This relatively smooth current operates the relay.

When the transmitter sends out a tone-modulated signal, the electromagnetic pull of the relay coil moves the movable arm of the relay. This causes the contact points to change.

When the tone signal is not transmitted, the relay contacts are returned by spring tension to their original position. Relay connections 1, 2, and 3 are wired to the control mechanism (motor) used in the model airplane or boat. Generally, a connection is made between contacts 2 and 3, which are normally open. Resistor R_7 and capacitor C_{11} provide arc suppression for the contact points.

Thus, via radio signals, the remote control of an object (in this case, a model airplane) is accomplished. The total communication system must have a transmitter that generates a tone-modulated radio-frequency signal at a given frequency, and a receiver tuned to that frequency.

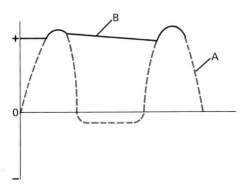

Fig. 35-6. Signal rectification of relay driver: (A) output without filter, (B) output with filter action.

For Review and Discussion

1. Identify six basic electronic circuits commonly found in radio receivers.
2. Describe the functions of each of these circuits.
3. Identify the components of a tuner that are also found in a transmitter tank circuit. Explain the function of the common elements.
4. Explain how the tuner can select one of several broadcasting stations' signals that "strike" the antenna of a receiver.
5. Discuss a method used to raise and lower the volume of a radio receiver.
6. What is meant by the term "superregenerative" as applied to a radio receiver?
7. Describe how the direct-current operating point is established in the audio amplifier of Fig. 35-1B.
8. Explain the process of detection used in the circuit shown in Fig. 35-1B.
9. When a positive-going signal is applied to C_2, what action occurs in the collector circuit of Q_1? (Refer to Fig. 35-1B.)
10. Describe the purpose of the relay driver in Fig. 35-4.

UNIT 36. Transmitter and Receiver Systems

The previous three units discussed several basic electronic circuits necessary for radio communications. This unit discusses additional electronic circuits that are used in modern radio transmitters and receivers to make them highly selective and sensitive.

A MODERN TRANSISTORIZED RADIO RECEIVER

A block diagram for a modern transistorized radio receiver is shown in Fig. 36-1. The input signals to this radio are weak amplitude-

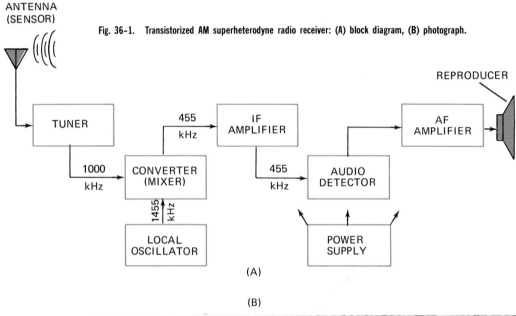

Fig. 36-1. Transistorized AM superheterodyne radio receiver: (A) block diagram, (B) photograph.

modulated radio waves. The basic electronic circuits include: antenna (sensor), tuner, converter (mixer), local oscillator, i-f amplifier, audio detector, a-f amplifier, reproducer, and power supply.

This radio is a superheterodyne radio. That is, it uses the heterodyne principle whereby the input voltages (signals) from the antenna are mixed with voltages generated by a local oscillator in a converter system. The word *heterodyne* is a combination of two Greek words: *hetero* (mixing) and *dyne* (force). In a superheterodyne receiver, two forces (voltage signals) are mixed together.

The output of the converter system is equal to the difference between the input-signal frequency and the local oscillator frequency. This difference is called the *intermediate frequency* (i-f). The principle advantage of the heterodyne technique is that it makes possible the design of amplifier stages that operate at only one frequency, the intermediate frequency, regardless of the input frequency to which a receiver is tuned. As a result, the intermediate-frequency amplifier stages operate more efficiently than amplifier stages designed to operate over a relatively broad band of frequencies.

Consider a station X that is broadcasting at 500 kilohertz. The standard intermediate frequency for broadcast receivers is 455 kilohertz. Therefore, to receive station X, the receiver's local oscillator is tuned to 955 kilohertz. The difference between station X and the local oscillator is 955 kilohertz − 500 kilohertz = 455 kilohertz (the output of the converter). Consider a second station Y at 1,000 kilohertz. If the local oscillator is tuned to 1,455 kilohertz, the output from the converter will again be 455 kilohertz. By adjusting the frequency of the local oscillator, a wide range of different frequencies can be converted into the intermediate frequency.

The audio component of the intermediate-frequency signal is next detected and amplified. Finally, an audio-frequency amplifier is used to drive the reproducer–loudspeaker. Another circuit is also used in this radio receiver. It is called the *automatic volume control* (AVC) and is a convenience to the listener. This auxiliary circuit maintains a constant signal output (volume) from the loudspeaker under conditions of fluctuating signal strength from the transmitter. A 9-volt direct-current battery supplies the necessary electric energy to operate the modern superheterodyne radio shown in Fig. 36-1. The schematic wiring diagram of a typical "superhet" radio is shown in Fig. 36-2.

Input and Tuner. Because of the radio's high sensitivity it does not need an external antenna. The input (sensor) and tuner circuits consist of a ferrite rod antenna (coil L_1), tuning capacitor C_1, and coupling capacitor C_2. Coil L_1 also acts as a matching transformer to prevent the relatively low base impedance of Q_1 from seriously lowering the Q of the antenna circuit. Tuning capacitor C_1 is a variable capacitor with a tuning range of about 20 to 150 picofarads; thus it will resonate with L_1 over the broadcast band of 535 to 1,605 kilohertz.

The antenna circuit has the dual function of selecting the desired radio station frequency and rejecting "image" frequencies. The image frequency problem is a slight shortcoming of superheterodyne receivers. Consider the case where station X operates on 500 kilohertz. To receive a broadcast from this station, a radio's local oscillator would have to be tuned to 955 kilohertz. However, if there is a station Y operating on 1,410 kilohertz (500 kilohertz + 2 × 455 kilohertz = 1,410 kilohertz); the difference between 1,410 and 955 kilohertz is also 455 kilohertz. Therefore, unless the receiver has selectivity before the mixer, it is quite possible to receive interfering image stations.

It is the prime function of L_1 and C_1 to provide this selectivity by tuning to 500 kilohertz at the same time the local oscillator is tuned to 955 kilohertz, and to "track" accordingly through the entire broadcast band. To ensure tracking, tuning capacitors C_1 and C_4 are physically mounted in the same unit and the capacitor rotors are "ganged" together on the same shaft. The dotted line on the schematic (Fig. 36-2) illustrates this arrangement. Mounting the two capacitors this way causes the rotor plates of each capacitor to move simultaneously—thus aligning them exactly with each other.

Capacitor C_2 is a direct-current block to prevent the base bias of Q_1 from being shorted through the lower turns of L_1 that go to ground. The impedance of C_2 at 500 kilohertz is about 2 ohms, so it efficiently couples the weak signals to the base of Q_1.

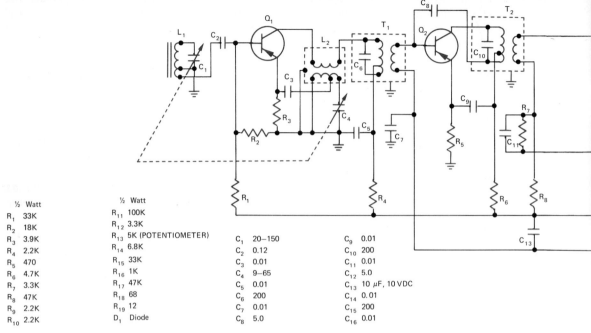

½ Watt		½ Watt					
R_1	33K	R_{11}	100K				
R_2	18K	R_{12}	3.3K				
R_3	3.9K	R_{13}	5K (POTENTIOMETER)	C_1	20–150	C_9	0.01
R_4	2.2K	R_{14}	6.8K	C_2	0.12	C_{10}	200
R_5	470	R_{15}	33K	C_3	0.01	C_{11}	0.01
R_6	4.7K	R_{16}	1K	C_4	9–65	C_{12}	5.0
R_7	3.3K	R_{17}	47K	C_5	0.01	C_{13}	10 μF, 10 VDC
R_8	47K	R_{18}	68	C_6	200	C_{14}	0.01
R_9	2.2K	R_{19}	12	C_7	0.01	C_{15}	200
R_{10}	2.2K	D_1	Diode	C_8	5.0	C_{16}	0.01

Fig. 36-2. Transistorized superheterodyne radio receiver schematic wiring diagram. (Lafayette Radio Electronics Corp.)

Converter (oscillator-mixer). In the converter stage, incoming signals are changed from their original frequencies to another totally different frequency—namely, the intermediate frequency of 455 kilohertz (Fig. 36-3). Coil L_2, capacitors C_3 and C_4, and transistor Q_1 make up the local oscillator. Capacitor C_4, which is ganged to C_1, resonates with the secondary winding of L_2 and establishes the frequency of oscillation. The primary winding of L_2 couples energy from the collector circuit of Q_1 to the tuned secondary of L_2 and back to the emitter by way of C_3. This is the feedback path that sustains the oscillations of Q_1.

While Q_1 is oscillating, the radio signals coupled through C_2 to the base of Q_1 vary the current flow through the transistor, which in turn varies the amplitude of the oscillation signal. Mixing the two signals produces two new frequencies. One is the sum of the radio and local oscillator frequencies, that is, 500 kilohertz + 955 kilohertz = 1,455 kilohertz. Since the intermediate-frequency transformer or "strip" (as it is often called) only amplifies 455 kilohertz, the higher frequency (1,455 kilohertz) is "lost" in the converter stage.

The primary coil of T_1 and capacitor C_6 resonate at 455 kilohertz and have the function of picking out the desired intermediate frequency (455 kilohertz) from all other frequencies within the converter stage. The center tap on the primary of T_1 is at signal ground through the bypass action of capacitor C_5. By putting the signal ground at the center of T_1, less stray noise is picked up in the early intermediate-frequency stage, and the possible detuning effect from the collector capacitance of transistor Q_1 is reduced. This latter point is especially important at the higher frequencies.

Intermediate-frequency (I-F) Amplifiers. The components that make up the i-f amplifiers are shown in Fig. 36-4. Transistor Q_2 is commonly called the first i-f amplifier and Q_3 the second. The primary windings of T_1, T_2, and T_3 resonate with C_6, C_{10}, and C_{15}, respectively. These tuned circuits determine the selectivity of the intermediate-frequency strip. The turns ratios of the i-f transformers are designed to match the low base input impedance to the high impedance of the tuned transformer primary.

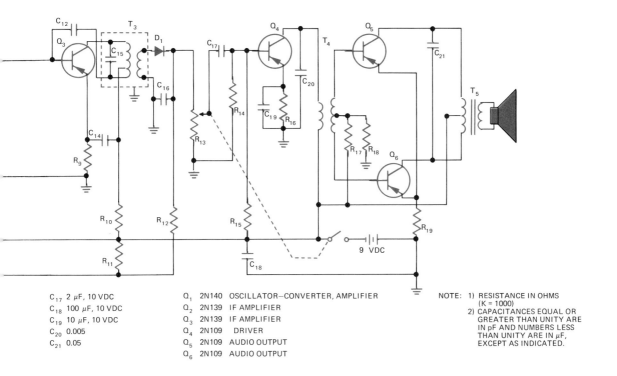

C_{17} 2 µF, 10 VDC
C_{18} 100 µF, 10 VDC
C_{19} 10 µF, 10 VDC
C_{20} 0.005
C_{21} 0.05

Q_1 2N140 OSCILLATOR–CONVERTER, AMPLIFIER
Q_2 2N139 IF AMPLIFIER
Q_3 2N139 IF AMPLIFIER
Q_4 2N109 DRIVER
Q_5 2N109 AUDIO OUTPUT
Q_6 2N109 AUDIO OUTPUT

NOTE: 1) RESISTANCE IN OHMS (K = 1000)
2) CAPACITANCES EQUAL OR GREATER THAN UNITY ARE IN pF AND NUMBERS LESS THAN UNITY ARE IN µF, EXCEPT AS INDICATED.

When the intermediate-frequency signal developed by the converter appears across the tuned primary of T_1, it is fed into the base of Q_2 through the secondary winding of T_1. Transistor Q_2 amplifies the signal by a factor of about 100 and passes it on to the second i-f transformer T_2. It is then amplified by Q_3. By the time the intermediate-frequency signal appears at the secondary of T_3, it is 10,000 times stronger than it was at the primary of T_1.

Fig. 36-3. Converter (oscillator-mixer).

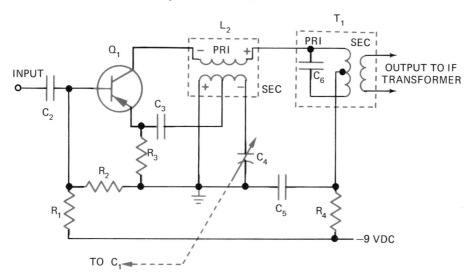

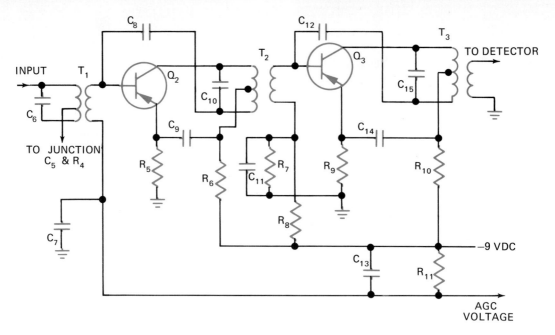

Fig. 36-4. Intermediate-frequency amplifiers (i-f strip).

When a transistor is used in common-emitter configuration (emitter placed at signal ground), the internal capacity between the collector and base creates a feedback path that can cause trouble at radio frequencies. Capacitors C_8 and C_{12} are neutralizing capacitors that are added to prevent the transistors from breaking in oscillation. Capacitor C_{bc} in Fig. 36-5 shows this feedback capacity.

If the transistor works into an inductive load, the phase of the feedback signal becomes 360 degrees as required for oscillation. The addition of C_n, which is connected to the bottom of the tapped primary winding, produces another path that feeds back a 180-degree signal. The two feedback signals then cancel out at the base of the transistor; hence, no feedback voltage and no oscillation.

Capacitors C_7, C_9, C_{11}, and C_{14} shown in Fig. 36-4 have a low impedance at 455 kilohertz and function as bypass capacitors for all the biasing resistors.

Audio Detector and AVC Circuit. Fig. 36-6 shows the components that make up the audio detector and the auxiliary automatic volume control (AVC) circuits, which maintain a constant volume even though input signals may vary.

Fig. 36-6. Audio detector and automatic volume control (AVC) circuit.

Fig. 36-5. Amplifier neutralization capacitor (C_n).

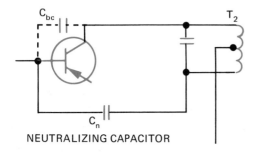

NEUTRALIZING CAPACITOR

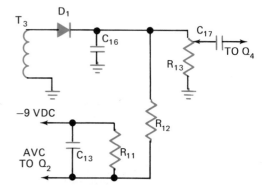

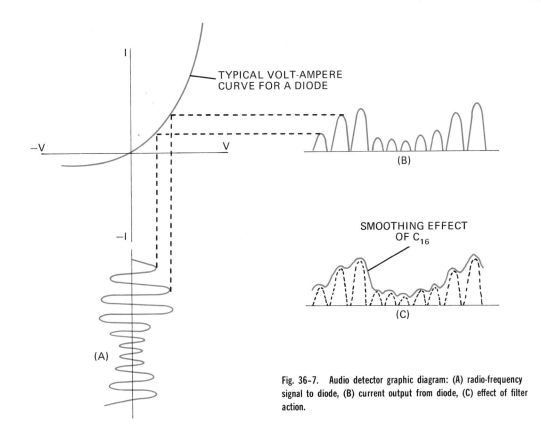

Fig. 36-7. Audio detector graphic diagram: (A) radio-frequency signal to diode, (B) current output from diode, (C) effect of filter action.

The detector D_1, a semiconductor diode, removes the negative-going portions of the intermediate-frequency signal coming from T_3. The output of the diode is a series of half-cycle intermediate-frequency pulses varying in amplitude according to the information modulated originally on the radio wave.

Fig. 36-7A shows the intermediate-frequency signal across T_3. If capacitor C_{16} were removed from the circuit (Fig. 36-6), the voltage across R_{13} would be as shown in the waveform of Fig. 36-7B. Fig. 36-7C shows the smoothing effect C_{16} has on the diode circuit.

Referring back to Fig. 36-6, resistors R_{12}, R_{11}, and capacitor C_{13} form the automatic volume control bias network. The detected signal waveform (Fig. 36-7C) passes through R_{12} and adds charge to C_{13}. Fig. 36-8 shows the sluggish action of C_{13} in changing its charge. Capacitor C_{13} is large, so that the automatic volume control circuit reacts to long-term changes, but not to each audio cycle.

When the receiver is not tuned to a radio station, the diode D_1 does not detect any signal to pass on to C_{13}. In this case, the voltage at the junction of R_{11} and R_{12} is about -0.5 volt and the voltage across C_{13} is about -8.5 volts. This automatic volume control bias sets the no-signal static operating point for Q_2. When a strong radio signal reaches the receiver and develops a $+0.5$ volt signal at the diode, the voltage across C_{13} begins to rise. As the capacitor voltage rises, the automatic volume control bias slips down the load line. Therefore, for weak

Fig. 36-8. Integrating action of capacitor C_{13}, the automatic voltage control circuit.

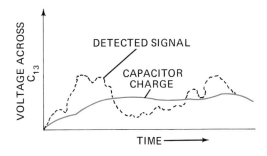

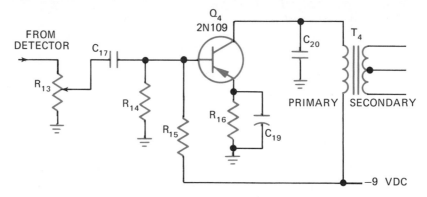

Fig. 36-9. Audio driver circuit.

signals, Q_2 has a high gain and for strong signals, Q_2 has a low gain. Because of the action of C_{13}, the gain does not change instantly.

The static operating point (SOP) moves down the load line as the automatic volume control action lowers the gain. For this reason automatic volume control must be limited to the early amplifier stages where the signal swing is low. When the static operating point drops near the 0 base current line, a large signal swing drives the transistor into cutoff and distorts the intermediate-frequency signal.

Audio Amplifier Circuit. The audio amplifier circuit in this radio consists of an audio driver and a push-pull audio amplifier. Fig. 36-9 provides a schematic representation of the audio driver. Resistor R_{13} is the volume control that taps off the desired amount of the detected signal and feeds it through C_{17} into the base of the driver Q_4. Transistor Q_4 amplifies the audio signal and passes it on to the interstage audio transformer T_4. This transformer couples the driver output to the push-pull amplifier. The secondary of T_4 drives the bases of the push-pull output amplifier transistors Q_5 and Q_6.

Fig. 36-10 shows the push-pull audio output amplifier. When the signal voltage goes negative at the top of the T_4 secondary winding, Q_5 amplifies it and passes it on to T_5 and to the loudspeaker. During this time, the bottom of the T_4 secondary is positive and this cuts off transistor Q_6. However, on the next half-cycle of the signal voltage, the bottom of the T_4 winding becomes negative and Q_6 amplifies while Q_5 is cut off. Since the amplifier delivers energy to the load on each half-cycle, it is considered to "push" or "pull" on the alternate half-cycles of current.

Fig. 36-10. Push-pull audio amplifier and reproducer (loudspeaker).

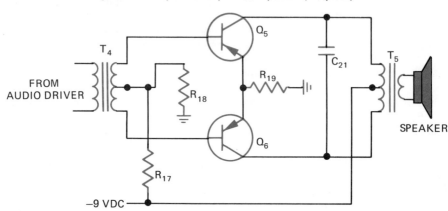

The main advantage of the push-pull type of amplifier is its efficiency. Ideally, it is somewhere around 70 percent efficient, which means that 70 percent of the direct-current power put into the stage comes out as signal power. Another factor that is desirable is that this type of amplifier draws little current when there is no audio-input signal, since Q_5 and Q_6 are biased near cutoff.

The main electron flow path for Q_5 is from the negative terminal of the battery up through T_5, into the collector of Q_5, out the emitter and through R_{19} to ground, and back to the positive terminal of the battery. The electron flow path is the same for Q_6, except the flow is down through the primary of T_5 and into the collector of Q_6.

The bias current path is from the negative terminal of the battery, through R_{17}, into the center tap on the secondary winding of T_4, out each end, and into the bases of Q_5 and Q_6. The electron flow combines again at the emitter junction and flows through resistor R_{19} to ground and back to the positive terminal of the battery. The capacitor C_{21} attenuates the high frequencies to compensate for the response of the small loudspeaker that favors the higher audio frequencies.

Resistors R_{18} and R_{17} develop just enough bias to eliminate crossover distortion, which occurs at the time the amplifying function is being switched from one transistor to another. Fig. 36-11 shows the effects of crossover distortion on a sinewave.

Reproducer System. The output of the audio amplifier is fed into a loudspeaker that changes the audio signals into sound. Transformer T_5 shown in Fig. 36-2 matches the low impedance of the loudspeaker (for example, 3.2 ohms) to the output impedance of the amplifier (typically, 150 ohms).

FIVE-TUBE SUPERHETERODYNE RADIO RECEIVER

The basic principles of heterodyning discussed for the transistorized superheterodyne radio receiver apply also to the five-tube superheterodyne radio receiver—except that now the principles are applied to vacuum tubes.

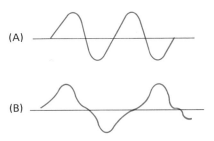

Fig. 36-11. Crossover distortion: (A) normal signal, (B) signal with crossover distortion caused by push-pull amplifier transistors biased at cutoff.

The schematic diagram for a typical vacuum tube superheterodyne receiver is shown in Fig. 36-12.

Power Supply and Filter Circuit. When the on/off switch S_1 is turned on, 120-volt alternating-current voltage is applied across the series string of vacuum tube filaments. It is also applied to the power supply and filter circuit, which is made up of the following components: plug P_1, resistors R_9 and R_{10}, rectifier tube V_5, and capacitors C_{17} and C_{18}.

Resistor R_9 acts as a current-limiting resistor. The filtering action of R_{10} and C_{17} removes practically all the ripple voltage.

Sensor, Tuner, and Oscillator. The radio-frequency signals are picked up by the loop antenna L_1 (sensor), and the desired station is selected by tuning the variable capacitor C_1. The capacitor C_1 and loop antenna together form the tuner. It is a parallel-tuned circuit that can resonate at the desired station frequencies; for example, 1,000 kilohertz. From the tuner the selected station radio frequency is fed into control grid 2 (pin 7) of the pentagrid converter tube V_1.

The tube V_1 is called a *pentagrid converter* because it has five grids. Within the tube, direct-current flow is controlled by the incoming radio-frequency and local oscillator signals. The direct-current plate current flow in the tube is from the chassis ground through oscillator coil L_3, through the cathode and plate of V_1, the primary winding of T_1, filter resistor R_{10}, and back to rectifier tube V_5. The lower screen grid

401

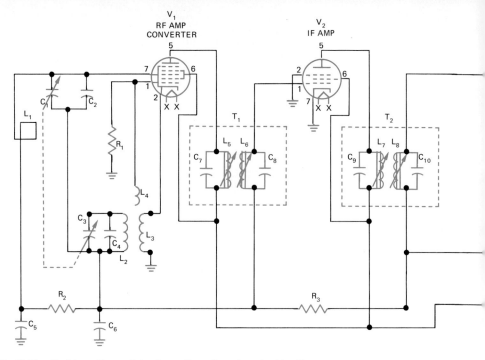

Fig. 36-12. Five-tube ac/dc superheterodyne radio receiver schematic wiring diagram. (Grob and Kiver, Applications of Electronics, 1st ed., New York, McGraw-Hill Book Co., 1960.)

(pin 6) of tube V_1 forms a plate for the oscillator circuit and the upper screen grid (pin 6) forms the screen grid of the tube itself. Note that these grids have a positive charge applied to them and serve two functions.

The local oscillator circuit shown in Fig. 36-12 is made up of variable capacitors C_3 and C_4; coils L_2, L_3, and L_4; resistor R_1; and the oscillator section of V_1. Both the main tuner capacitor C_1 and the oscillator capacitor C_3 are ganged together on the same shaft. The output of the oscillator is fed into control grid 1 (pin 1) of the tube V_1. Capacitors C_2 and C_4 are parallel trimmer capacitors used to make fine adjustments for accurate tracking, dial calibrations, and maximum sensitivity.

The mixing or heterodyning of the local oscillator and station signal frequencies occurs within the tube. This results in the formation of several new frequencies at the plate of V_1. Because of this, the tube is called a *converter* or *mixer*. The frequency of primary interest is the difference frequency between the local oscillator and the desired station radio frequency. The difference frequency, which is 455 kilohertz, is the receiver's intermediate frequency.

I-F Amplifier. The output of the converter tube is fed into the i-f transformer T_1 (Fig. 36-12), which selects the desired intermediate frequency from all the radio-frequency signals in the converter stage. The i-f transformers are often called *doubled-tuned coupling transformers*.

The intermediate-frequency signal, coupled through transformer T_1, is applied to the control grid (pin 1) of the i-f amplifier tube V_2. The amplified signal at the plate of V_2 is coupled through i-f transformer T_2 to the detector circuit.

Detector, AVC, and First A-F Amplifier. The output of the i-f amplifier is fed into the diode (pin 5) section of the duo-diode triode tube V_3 (Fig. 36-12). This tube performs three functions: It detects the audio signal, provides automatic volume control, and amplifies the detected audio signal. Combining the functions within one tube is economical and saves space.

The diode plate and common cathode of V_3 allow current to flow between these elements only during the positive half-cycles of the amplified modulated 455-kilohertz intermediate-frequency signal. It has been seen that this

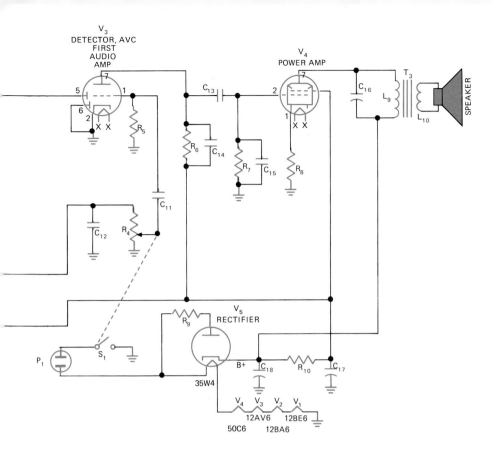

process rectifies the intermediate-frequency signal so that the audio-frequency component can be utilized. This process is called *demodulation* or *detection*.

Capacitor C_{12} filters out the intermediate-frequency components from the detected signal, leaving a negative direct-current voltage varying at an audio rate at the top of the volume control potentiometer R_4.

Resistors R_2 and R_3 and capacitors C_5 and C_6 form the automatic volume control circuit. This circuit controls the receiver gain and keeps the output volume constant even though the incoming modulated radio wave at the sensor (antenna) varies in intensity.

Audio Amplifier. Audio amplification occurs by amplifying the relatively weak audio voltages appearing on the grid of V_3 (see Fig. 36-12). These voltages control the plate current that passes through the plate load resistor R_6. The audio-signal voltage is coupled through C_{13} to the grid (pin 2) of the power-amplifier tube V_4. The capacitor C_{14} is a bypass that reduces the gain of V_3 at the high audio frequencies to improve the tonal qualities of the signal appearing at the speaker.

Power Amplifier. The audio-signal input to the power-amplifier circuit appears on the control grid of V_4 (pin 2) (Fig. 36-12). Grid resistor R_7 is bypassed by capacitor C_{15} to further attenuate the higher audio frequencies. The combination of coupling capacitor C_{13} and grid resistor R_7 also attenuates the lower audio frequencies below 70 hertz. Resistor R_8 provides both bias for V_4 and negative feedback for stability and decreased distortion.

Reproducer. The power-amplifier output is developed across the primary of the output transformer T_3. The transformer delivers the power from the output of V_4 to the speaker's coil, and matches the impedances for maximum power transfer. Capacitor C_{16} attenuates the high audio frequencies in order to obtain a properly balanced combination of base and treble responses at the reproducer (speaker).

Fig. 36-13. A modern single-sideband transceiver. (*Sideband Engineers, Raytheon Co.*)

Fig. 36-14. Inside view of transceiver, showing selected parts. (*Sideband Engineers, Raytheon Co.*)

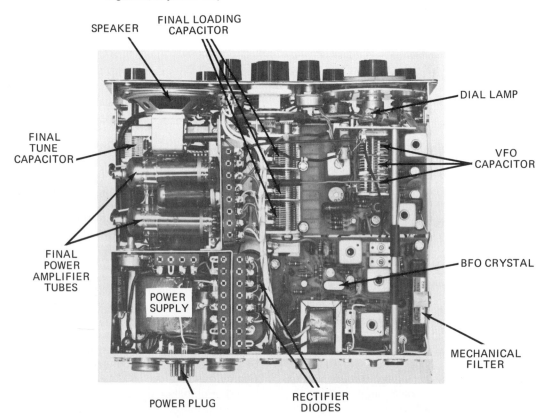

SINGLE-SIDEBAND TRANSCEIVER

A modern transceiver is shown in Figs. 36-13 and 36-14. This transceiver has several features not discussed previously, some of which are single-sideband transmission, mechanical filter, bilateral circuits, and a unique method of band switching.

The single-sideband transceiver operates on four amateur bands: 80, 40, 20, and 15 meters. It requires only a microphone, an antenna, and a source of energy. Transistors are used throughout, with the exception of the radio-frequency driver and final amplifier stages. Exceptional selectivity is provided by a mechanical filter. Bilateral amplifiers and mixers use common circuits to transmit and receive. This is done by controlling the direction of amplification in these circuits. By using these bilateral stages, a saving in size and cost is realized.

A block diagram of a single-sideband transceiver is shown in Fig. 36-15. In the transmit mode, the audio signal is amplified by the microphone amplifier Q_4, and applied to the ring-balanced modulator. The ring-balanced modulator has two functions. First, it mixes the carrier oscillator's frequency with audio frequencies. Second, it suppresses the carrier frequency to a point where it does not appear in the output signal. Therefore, only a double-sideband suppressed-carrier output is impressed across transformer T_2 and amplified by Q_5. Transistor Q_6 is biased off in transmit, and therefore does not interfere with the amplifier operation.

Upon receiving the amplified double-sideband signal from Q_5, the mechanical filter passes the lower-sideband frequency (456.38 kilohertz minus the modulating frequency) and suppresses the upper-sideband frequency (456.38 kilohertz plus the modulating frequency). In the upper-sideband mode, the signal appearing at the output of the filter is subtractively mixed at the high-frequency mixer CR_9 with an injection of 2,738.28 kilohertz from T_5 to produce a resultant frequency of 2,281.9 kilohertz plus the modulating frequency. In the lower-sideband mode, the filter output is additively mixed at CR_9 with an injection of 1,825.52 kilohertz from T_5 to produce a resultant frequency of 2,281.9 kilohertz minus the modulating frequency.

The 2,281.9-kilohertz upper- or lower-sideband output from T_3 is applied to the transmit mixer Q_7. Transistor Q_8 is biased off, since it is not being used in the transmit mode.

Also applied to Q_7 is an injection from the variable-frequency oscillator (VFO), which is variable from 5,456.9 to 5,706.9 kilohertz. This injection translates (changes) the incoming (2,281.9 kilohertz) signal to a range variable from 3,175 to 3,425 kilohertz. This range of frequencies can be transformed by T_6 and passed to the high-frequency mixer Q_9. At this point, an injection from a high-frequency oscillator converts the signal to the final operating frequency. The signal from L_2 is then applied to the grid of the V_1 driver, which in turn drives the final power amplifiers V_2 and V_3. The power-amplifier output is then matched to the transmission line and antenna by means of a pi network.

In the receiver mode (with the microphone button released), a signal from the antenna is loosely coupled from the transmitter pi section network to the top of coil L_3. Diodes CR_{14} and CR_{15} conduct only when extremely strong signals are present, otherwise they do not affect the circuit. The signal at L_3 is coupled to the emitter of Q_{11}, which operates as a common-base amplifier. The amplified signal from Q_{11} is applied to the high-frequency mixer Q_{10} through L_2. The injection from the high-frequency oscillator converts the signal to a frequency within the range of 3,175 to 3,425 kilohertz. The converted signal is then mixed at Q_8 by an injection from the variable-frequency oscillator to 2,281.9 kilohertz. The converted signal is again mixed by CR_9 to the 456.38-kilohertz frequency in a manner exactly opposite to that described for this stage during the transmit function. Output from the mechanical filter is then amplified by Q_6 and coupled through T_2 to the ring-balanced modulator CR_4-CR_7.

With injection from the carrier oscillator, the ring-balanced modulator now functions as a detector and produces an audio output that is applied to the base of audio amplifier Q_2. Output is taken from the collector of Q_2 and coupled to the audio driver Q_1. This amplifier, in turn, drives the audio-power amplifier Q_{20}, which increases the signal strength to operate the speaker.

Ring-balanced Modulator. The ring-balanced modulator produces two output frequencies: the

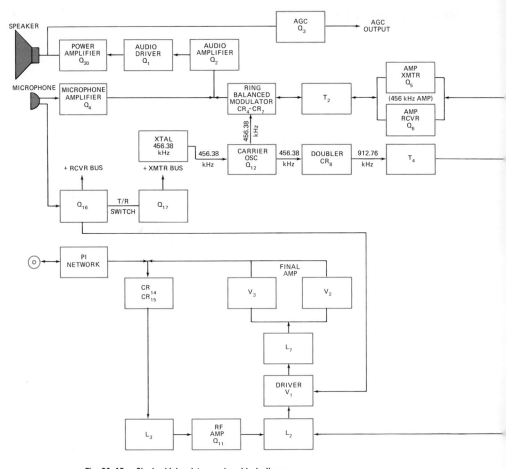

Fig. 36-15. Single-sideband transceiver block diagram.

carrier frequency plus the modulating frequency, and the carrier frequency minus the modulating frequency. The carrier is suppressed by cancellation in the circuit shown in Fig. 36-16.

With no modulation applied to the bridge circuit R_{25}-R_{28}, no signal appears across T_2. The 456.38-kilohertz signal from the carrier oscillator is injected into the bridge through C_{22} to the center arm of R_{25}. Notice that the common point of CR_2 and CR_3 is grounded directly, and the common point of CR_1 and CR_4 is grounded at carrier frequency by C_{17}. On positive carrier half-cycles, CR_2 and CR_4 conduct and current flows through them in a loop that includes the resistive arm across the "ring" and its common ground return. Assuming that CR_2 and CR_4 have forward resistances, a small voltage would appear at both sides of the primary of T_2. These voltages are matched in amplitude by R_{25} and in phase by C_{23}, so that when they arrive at T_2, they are cancelled.

On the negative half-cycle, CR_1 and CR_3 conduct while the other diodes are turned off. Cancellation of the carrier takes place as before.

When an audio signal (controlled by the microphone gain potentiometer and amplified by Q_4) is applied to the ring-balanced modulator, it unbalances the circuit by biasing the diodes in one path, depending on the instantaneous polarity of the audio, and some radio frequency appears at the output. The radio frequency appearing at T_2 is a double-sideband suppressed-carrier signal. In the receiver mode, the process is reversed and the balanced modulator functions as an audio detector.

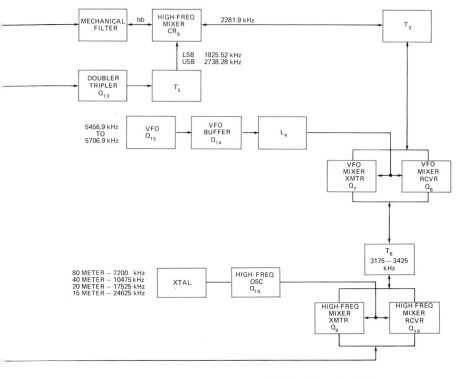

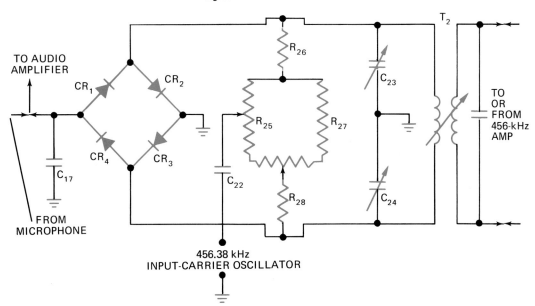

Fig. 36-16. Balanced modulator.

407

Bilateral Amplifier. A typical bilateral amplifier circuit is shown in Fig. 36-17. In the receiver mode, the base-bias resistor of Q_5 is returned to control line A (+ receive), which carries a +12 volt potential. Under this condition, Q_5 is reverse-biased and cannot conduct. The base-bias resistor of Q_6, however, is returned to line B (+ transmit), which is essentially at ground potential, and Q_6 conducts and is able to amplify. Thus, a signal appearing from the mechanical filter will be amplified and delivered to T_2.

In the transmit mode with the microphone button pressed, control lines A and B are reversed (by switching) in polarity. Therefore, Q_6 is biased off and Q_5 is conducting. Under this condition, a signal appearing at T_2 will be amplified and delivered to the mechanical filter.

Mechanical Filter. The purpose of the mechanical filter is to suppress the unwanted upper sideband generated in the ring-balanced modulator. This is accomplished by placing the passband of the filter to one side of the center frequency of the 456.38-kilohertz amplifier (Fig. 36-18). The filter is designed to pass only those frequencies that appear between the points on the curve.

Notice that the reference frequency (the 456.38-kilohertz carrier frequency) and all frequencies above appear outside the passband of the filter. If the modulating frequency, for example, is a 1-kilohertz tone, the output from the balanced modulator will contain the upper- and lower-sideband frequencies (457.38 and 455.38 kilohertz). These frequencies are amplified and passed on to the mechanical filter, which will pass the lower sideband (455.38 kilohertz) and reject the upper sideband (457.38 kilohertz).

Fig. 36-19 summarizes the various frequency conversions or changes for the transmit mode. Notice that the carrier oscillator has a rather low radio frequency (456.38 kilohertz). This low frequency is used because the mechanical filters function better at the lower frequencies. However, to achieve the proper high frequencies necessary for transmitting (either upper or lower sideband) on the proper bands, the carrier frequency is doubled or tripled and separate high-frequency oscillators are used for each transmitting band. The variable-frequency oscillator provides the ability to change a frequency for transmitting within a band.

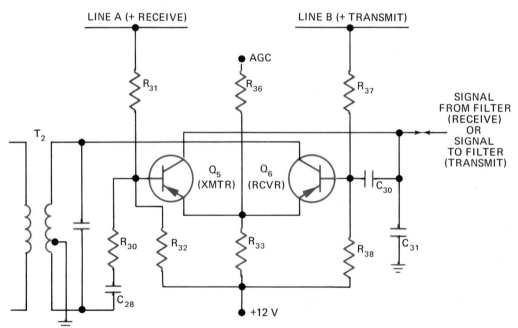

Fig. 36-17. Typical bilateral amplifier circuit (456.38 kilohertz).

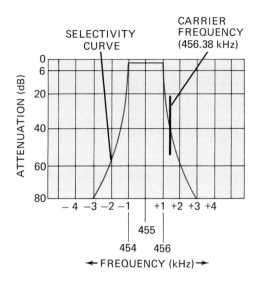

Fig. 36-18. Mechanical filter selectivity curve. (*Sideband Engineers, Raytheon Co.*)

For Review and Discussion

1. List eight basic electronic circuits found in a modern transistorized superheterodyne radio receiver.

2. Name and describe the functions of two auxiliary electronic circuits commonly found in superheterodyne radio receivers.

3. Why is impedance matching important?

4. Identify six characteristics needed by a p-n-p transistor when it operates as an audio amplifier.

5. What is meant by the term "heterodyning"?

6. What is the main advantage of using the intermediate-frequency circuits in a superheterodyne radio receiver?

Fig. 36-19. Varied frequency conversions for transmitting with a single-sideband transceiver.

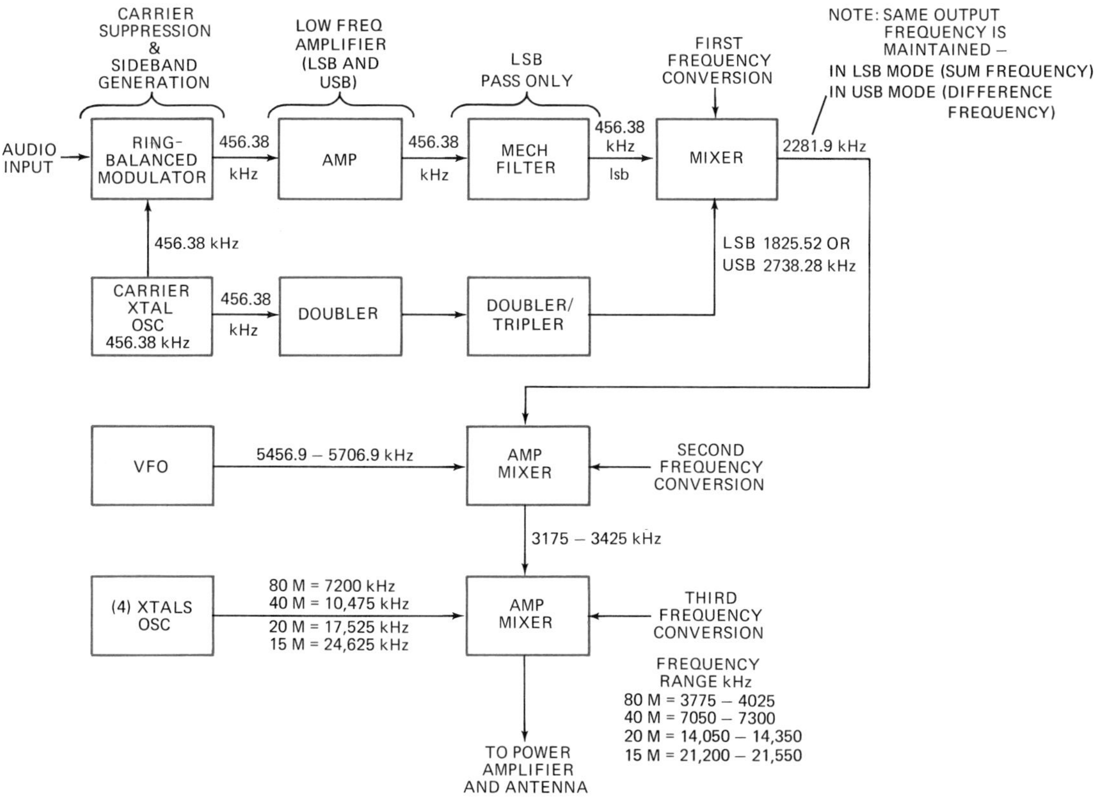

409

7. Draw the block diagram of a superheterodyne radio receiver.

8. Explain the feedback process used in an automatic volume control circuit.

9. Describe the fundamental processes that take place in the converter circuit of a superheterodyne radio receiver.

10. Identify five tubes that are found in a typical superheterodyne radio receiver, and describe their functions.

11. Contrast the main differences between a common transistor and an electron tube-type superheterodyne radio.

12. Give at least five advantages of single-sideband transmission.

13. Explain briefly what is meant by the expression, "suppressed carrier."

14. Describe the function of a ring-balanced modulator.

15. Describe the various frequency conversions used in transmitting with a modern single-sideband transmitter.

UNIT 37. Television

Most of the principles of radio communications that have already been studied also apply to television. However, in television, both picture (video) and sound (audio) signals are transmitted and received simultaneously. Because of the introduction of the picture or video signals, new applications of electronic principles are discussed in this unit. These new applications are reflected in such topics as orthicon tube, cathode-ray tube, scanning, vertical and horizontal sweep, synchronizing pulse, video or "pic" signal, and high-voltage anode. In discussing color television, topics such as chrominance, hue, saturation, and burst appear.

TELEVISION SYSTEMS

There are two types of television (TV) systems in common use: One is the *closed-circuit television* (CCTV) system (Fig. 37-1). The second and more familiar system is called *broadcast television*. This system uses high-power transmitters to radiate video and sound signals from antennas so that the signals can be intercepted by receiving antennas and conveyed by wires into standard home-type television receivers. Fig. 37-2 shows in block diagram form some of the main elements in a television broadcasting station.

The master control room is the nerve center of the television broadcasting station (Fig. 37-3). Here the various programs underway are monitored and programed for transmission either manually or automatically. The film and slide chain contains a TV camera, motion picture projector, slide projector, and multiplexer arranged so that motion pictures or slides can be televised for transmission. A video tape recorder also makes it possible to "store" TV programs and to retransmit them at a later date.

TELEVISION TRANSMISSION

Television broadcasting stations use high-frequency carrier waves. As a result, they are limited to straight-line transmission paths. In order to get long-distance transmission, "relay" or "repeater" towers have to be constructed. The relay towers receive the high-frequency radio signals, amplify them, and retransmit them to the next tower. Special telephone cables, called *coaxial cables,* are used for overland transmission and in conjunction with "microwave" relay towers.

Because of their great height above the earth, satellites are also used to relay the televi-

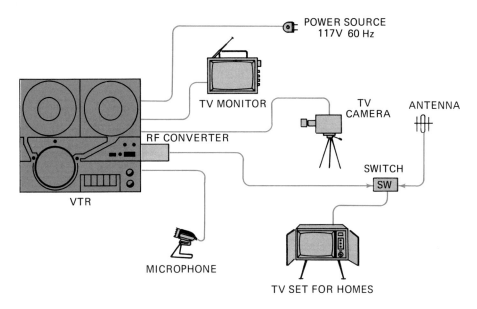

Fig. 37-1. Elements of a home-type closed-circuit television (CCTV) system. (Panasonic-Matsushita Electric Corp. of America)

Fig. 37-2. Main elements in a television broadcasting station

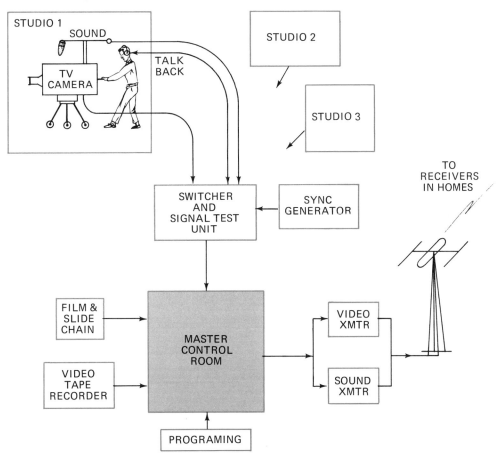

Fig. 37-3. Master control room. At extreme right is the automatic control system used for programing. *(Metromedia Television, WTTG-TV)*

sion signals from continent to continent. Sophisticated, digital television systems are used for transmitting pictures from satellites to ground stations.

TELEVISION CAMERAS

There are several types of television camera "pickup" tubes in general use today. Two of these are the *vidicon* tube and the *image orthicon* tube. Each tube has different characteristics. For example, the image orthicon tube has greater sensitivity, but the vidicon tube is more economical to operate, has good linear response, and requires less complex circuitry—thus permitting the construction of smaller, less expensive cameras.

Fig. 37-4A shows a block diagram of the basic circuits that develop the TV signal in a typical television camera. The schematic wiring diagrams that follow for this television camera show some of the variations used by different manufacturers in identifying components by symbol and letter. For example, the letter designation *TR* is used for a transistor instead of *Q*; and *D* is used for diode instead of *CR*. Generally, the variations in symbols and letter designations are slight compared with ANSI standards, and little difficulty should be encountered in identifying specific items. For clarity, the specifications

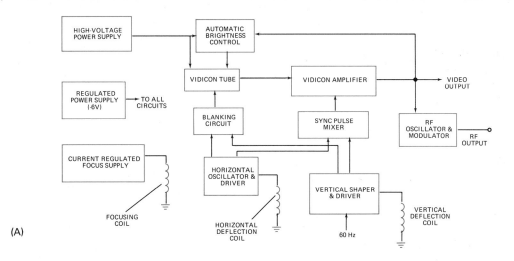
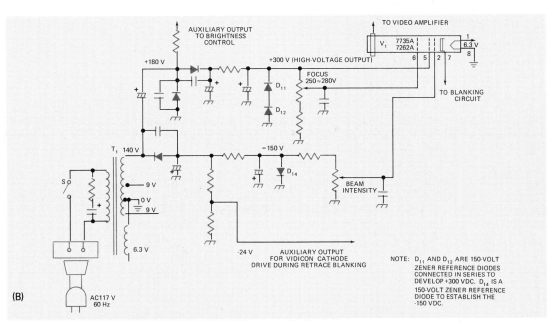
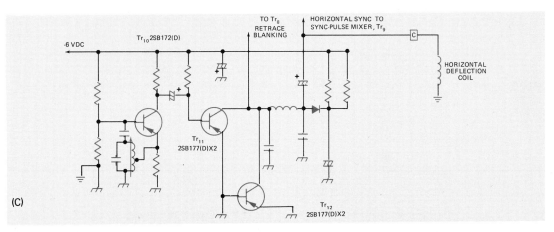

Fig. 37-4. Selected television camera diagrams: (A) block diagram, (B) high-voltage power supply circuit, (C) horizontal output circuit.

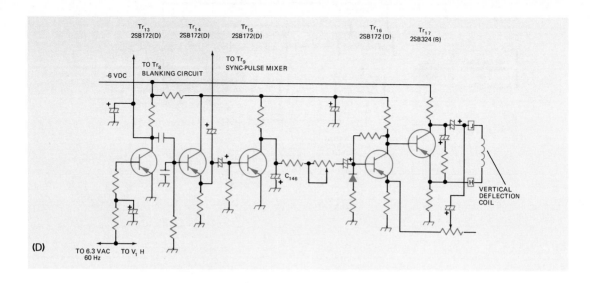

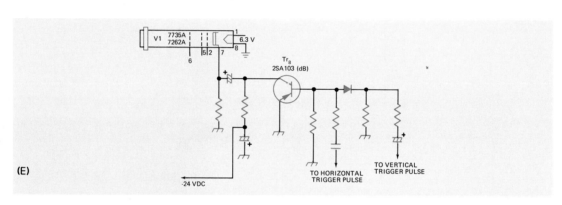

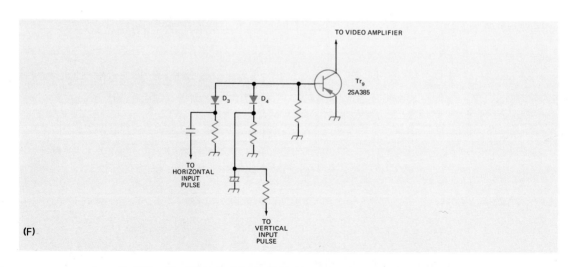

Fig. 37-4. (D) vertical output circuit, (E) blanking circuit, (F) sync pulse mixer circuit.

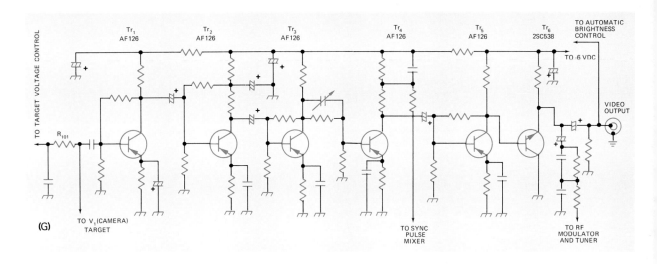

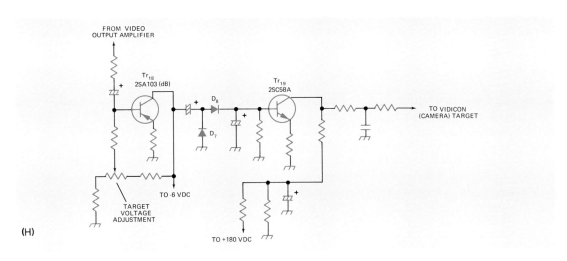

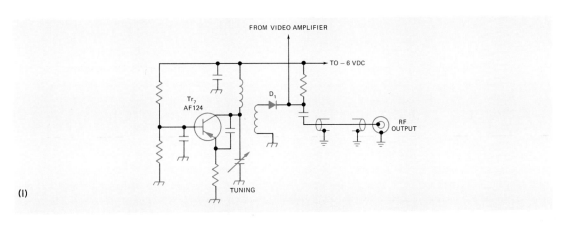

(G) video amplifier and r-f mixer circuit, (H) brightness-control circuit, (I) r-f oscillator and modulator circuit. (*Panasonic-Matsushita Electric Corp. of America*)

and letter symbols of most components have been omitted.

Power Supply. The high-voltage power supply (Fig. 37-4B) operates off of the 140-volt alternating-current winding of the power transformer T_1 and develops +300 volts direct current and −150 volts direct current for the vidicon. Because the operation of power supplies has been discussed in previous units, this unit will concentrate on circuits relating to generating video signals.

Horizontal Output. The horizontal oscillator-driver circuit (Fig. 37-4A) produces the sawtooth voltage for the horizontal deflection coil, which in turn causes the electron beam in the vidicon to scan horizontally. Transistor TR_{10} is the horizontal oscillator and TR_{11} is the horizontal output transistor (Fig. 37-4C). Transistor TR_{12} is connected as a diode and acts as the damper for the horizontal flyback. There are three outputs from the horizontal circuit; one to TR_8 for retrace blanking, another to TR_9 for horizontal synchronizing, and a third to the horizontal deflection coil.

Vertical Output. Fig. 37-4D shows the vertical shaper and driver circuit. This circuit takes some of the 6.3-volt alternating current from the filament winding of T_1, and then shapes and amplifies it to drive the vertical deflection coil. The vertical circuit also has two additional outputs—one to the blanking transistor TR_8 and another to the synchronization pulse mixer TR_9. Transistor TR_{13} is an overdriven amplifier whose output is a square wave. The incoming 60-hertz sine wave drives the stage into both cutoff and saturation. Transistor TR_{14} differentiates the square wave, amplifies the resulting trigger pulse, and couples it to TR_{15} which, with capacitor C_{146}, develops a sawtooth voltage. Transistors TR_{16} and TR_{17} amplify the sawtooth signal so that it can effectively drive the vertical deflection coil.

Blanking Circuit. Fig. 37-4E illustrates the blanking circuit, consisting of one transistor TR_8, which accepts the horizontal and vertical trigger pulses and applies them to the cathode of the vidicon. Since the trigger pulses occur during retrace time, and since TR_8 applies a positive pulse to the cathode, the vidicon electron beam is cut off during retrace time. If the blanking circuit were not used, annoying white lines would appear on the TV screen from right to left and from bottom to top.

Sync Pulse Mixer. Fig. 37-4F shows the synchronization pulse mixer, which consists of mixing diodes D_3 and D_4 and transistor TR_9. Diodes D_3 and D_4 accept and mix the horizontal and vertical sync pulses. Transistor TR_9 amplifies the combined signal and sends it to the video amplifier where the sync information is mixed with the video so that the TV receiver can correctly reconstruct the picture.

Video Amplifier and R-F Mixer. The video amplifier shown in Fig. 37-4G consists of transistors TR_1 through TR_6. The first three transistors are amplifiers that boost the small signal appearing across the vidicon target load resistor R_{101}. Transistor TR_4 is a mixer stage that adds the sync information from the sync pulse mixer to the picture signal. Transistor TR_5 amplifies the composite signal, while TR_6 is a buffer-output amplifier. There are three outputs from the video amplifier: one to the video output jack, one to the r-f modulator and tuner, and one to the automatic brightness-control circuit.

Brightness Control. Fig. 37-4H shows the automatic brightness-control circuit, which consists of transistors TR_{18} and TR_{19}, and diodes D_7 and D_8. This circuit monitors the peak signals coming from the video amplifier and adjusts the vidicon target voltage.

R-F Oscillator and Modulator. Fig. 37-4I shows the r-f oscillator and modulator. This circuit consists of transistor TR_7 and diode D_1. Its function is to develop the radio-frequency signal corresponding to channel 5 or 6, and modulate the radio frequency with the composite signal from the video amplifier. Transistor TR_7 is the r-f oscillator and diode D_7 is the modulator. The output of this circuit goes to the radio-frequency output jack on the camera. To display the video output signal on a regular TV receiver, it would be necessary to bypass the radio-frequency and intermediate-frequency stages. But with the radio-frequency output available on channel 5 or 6, it becomes a simple matter of tuning the TV receiver. The sound information is transmitted simultaneously on a separate frequency-modulation (FM) transmitter.

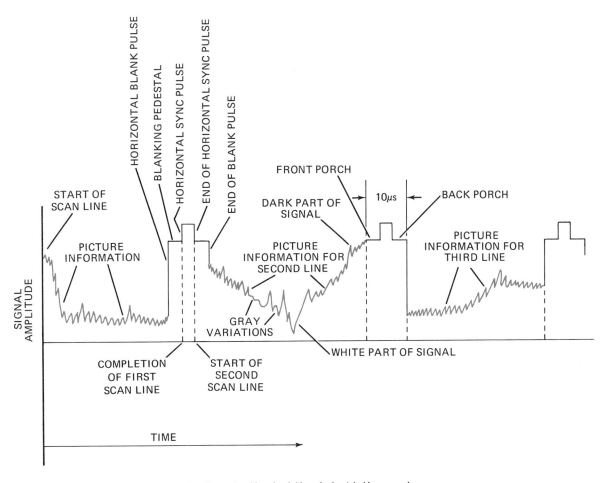

Fig. 37-5. Composite video signal (three horizontal video sweeps).

COMPOSITE VIDEO SIGNAL

Fig. 37-5 shows an artist's concept of a *composite video signal* during three horizontal sweeps, with video, blanking, and sync pulses shown. Notice that the composite video signal contains the picture or video signal and the blanking and sync pulses. The vertical sync pulses are not shown in this graphic view. Their purpose is to synchronize the vertical sweep current, which moves the horizontal scanning down a little each time and returns to the top of the frame after each field is scanned.

During a horizontal scan, the video signal varies, indicating the picture information from the televised scene. Each time the video signal senses a white area on the target, the video signal becomes more positive. When the video signal senses a dark area, it becomes more negative. Therefore, a video voltage can indicate light or dark picture elements. The variation between white and black is referred to as the *gray scale*. At the completion of the first line, a horizontal blanking pulse is transmitted with the corresponding sync pulse.

In the United States, the video signal and radio-frequency carrier use the *negative-modulation* technique. This means that the lowest radio-frequency output values correspond to the brightest elements of the televised scene. And, the highest peaks of radio-frequency output correspond to the darkest part of the televised scene. One of the major advantages of negative modulation is that a strong noise pulse caused by static will drive the radio-frequency signal into the blanking level, so the effect is not as noticeable as a "white" noise pulse.

TRANSMITTING THE TV SIGNAL

According to U.S. Federal Communications Commission (FCC) regulations, a television station must transmit both audio and video information on a channel that is no wider than 6 megahertz.

A TV station has two transmitters—one for the frequency-modulated aural or sound signals, the other for the amplitude-modulated picture or video signals. The frequency-modulated center frequency for transmitting sound is 4.5 megahertz higher than the carrier frequency of the video transmitter.

If an amplitude-modulated video signal were transmitted on double sideband, it would use a bandwidth of 8 megahertz; however, satisfactory results can be obtained if one sideband is suppressed slightly. It will be recalled from Unit 36 that the single-sideband transceiver suppressed one sideband completely. In a television signal, the lower sideband is not completely suppressed and a vestige (small amount) of it remains. Hence, the term *vestigial-sideband transmission*. By reducing the lower sideband and utilizing the full upper sideband, it is possible to transmit the video signals within the 6-megahertz limit, including the modulation frequencies needed for sound.

Fig. 37-6 shows a graphic view of a TV channel with the frequency allocations for the video and sound signals. Notice that the AM video carrier is 1.25 megahertz above the low end of the channel width. The frequency-modulated sound carrier is 4.5 megahertz above the video carrier, or 0.25 megahertz lower than the upper limit of the TV channel. Since the frequency deviation of the sound carrier is limited to ±25 kilohertz, there is ample band space for the sound frequencies without interfering with the video or extending beyond the 6-megahertz boundary for the TV channel. Notice that a small part (vestige) of the lower sideband of the video signal is also transmitted. This vestigial sideband has two main advantages: it reduces the bandwidth of the video signal, and it simplifies the receiving process.

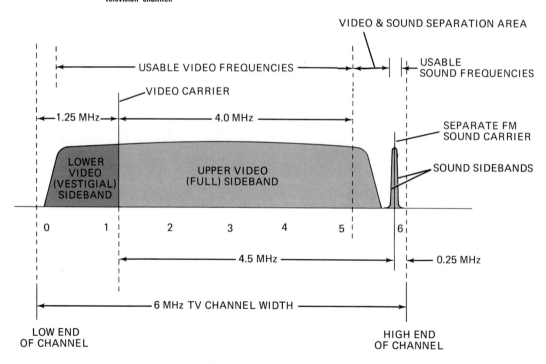

Fig. 37-6. Graphic view of the video and sound sidebands for black-and-white television channel.

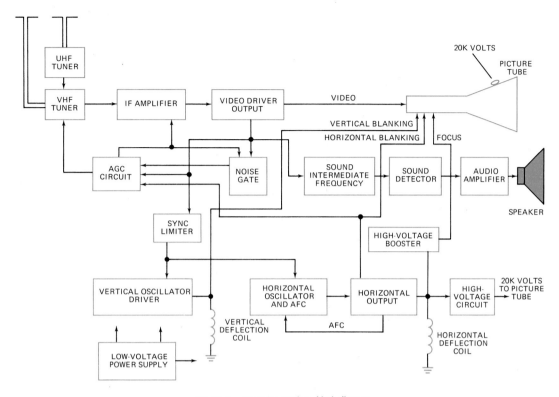

Fig. 37-7. Television receiver, block diagram.

There are 12 channels allocated by the FCC for the very-high frequency (VHF) range, numbers 2 through 13 (note that channel 1 was dropped because of a reallocation of frequencies). In the ultrahigh frequency (UHF) range, there are 70 channels. For example, VHF channel 4 has a range from 66 to 72 megahertz. Its amplitude-modulated video carrier is 67.25 and the frequency-modulated audio center frequency is 71.75. The UHF channel 28 has a frequency range of 554 to 560 megahertz. The amplitude-modulated video carrier frequency is 555.25 megahertz and the frequency-modulated audio center frequency is 559.75 megahertz. These frequency relationships correspond to the relationships shown in Fig. 37-6 of the video and sound frequencies.

TELEVISION RECEIVER CIRCUIT

Television receivers, like the television camera just described, have many similar components and electronic circuits. Fig. 37-7 is a block diagram of a TV receiver.

Tuner and R-F Amplifier. The tuner shown in Fig. 37-8A consists of an r-f amplifier and a local oscillator. The local oscillations mix with the incoming signal to produce a suitable signal for the i-f amplifier, which consists of transistors TR_1, TR_2, and TR_3. The first-stage TR_1 receives its base bias from the automatic gain control (AGC) circuit, and therefore has a variable gain depending on the strength of the incoming signal.

Video Amplifier. To the right of the i-f amplifier block (Fig. 37-7) is the video driver output circuit. This circuit, which is shown in Fig. 37-8B, consists of transistors TR_4 and TR_6 and diode X_1 (see also Fig. 37-8A), which detects the TV signals from the intermediate-frequency signal. The video driver TR_4 provides four auxiliary outputs, one to the noise-gate driver TR_5, one to the sync limiter TR_{10}, one to the AGC gate TR_9, and one

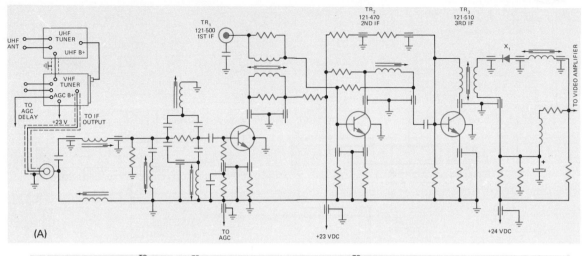

(A)

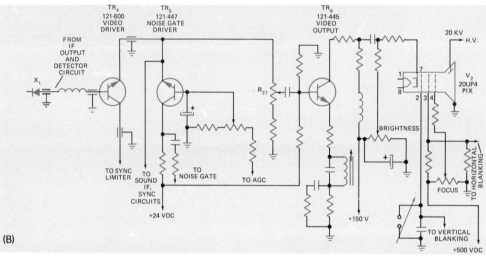

(B)

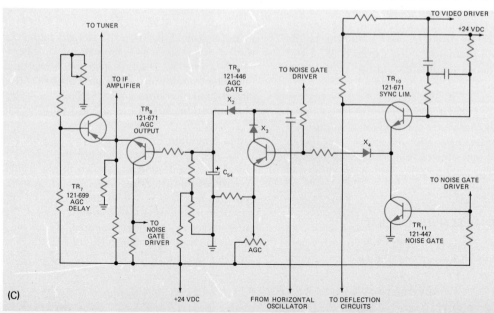

(C)

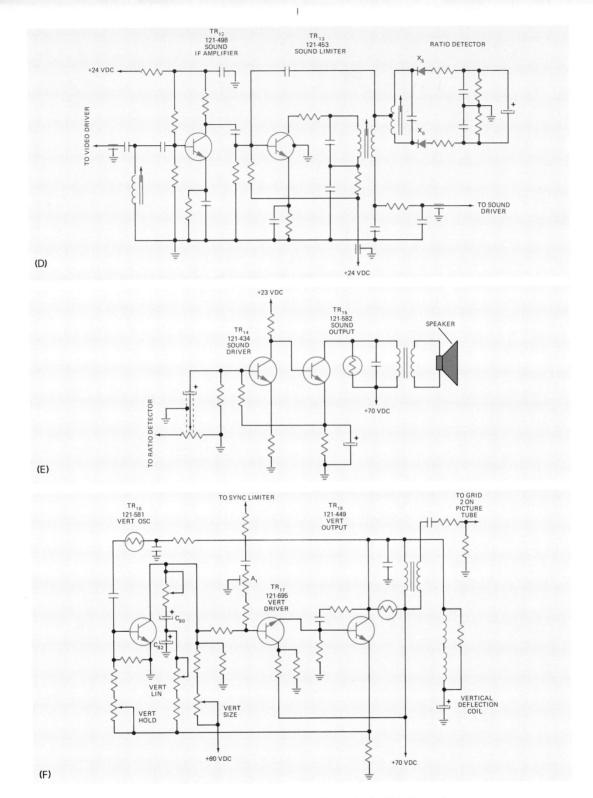

Fig. 37-8. Selected schematic wiring diagrams of a transistorized television receiver: (A) tuner and r-f amplifier circuit, (B) video amplifier circuit, (C) noise gate, automatic gain control (AGC), and sync limiter circuit, (D) sound detector and amplifier circuit, (E) sound driver and audio amplifier circuit, (F) vertical-output circuit.

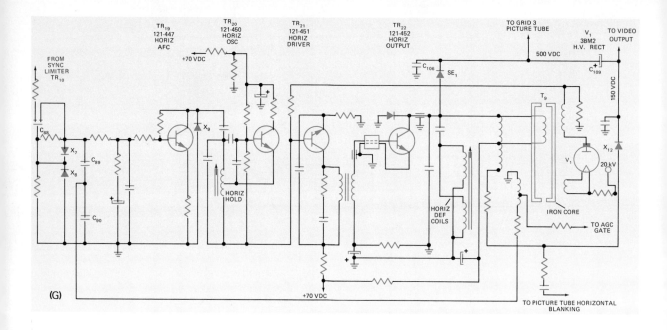

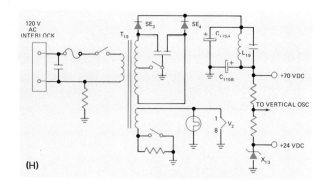

Fig. 37-8. (G) horizontal-output and high-voltage circuit, (H) low-voltage power-supply circuit. (*Zenith Radio Corp.*)

to the sound i-f amplifiers TR_{12}. The main output from the video driver is from contrast control R_{27} to the video output transistor TR_6. Transistor TR_6 amplifies the video signals and applies them to the cathode of the picture tube, where they modulate the beam to form the picture on the face of the tube.

Noise Gate. The noise gate circuit consists of transistors TR_5 and TR_{11} (Figs. 37-8B and C) that turn on the automatic gain control and sync circuits when a sync pulse is received. This improves the noise immunity of these circuits, since only noise pulses that are equal to the amplitude of the high-level sync pulses get through the gate.

Automatic Gain Control (AGC). Below the i-f amplifier shown in the block diagram of Fig. 37-7 is the AGC circuit, which gets its signals from the video circuit, the noise gate, and the horizontal oscillator. The AGC circuit consists of transistors TR_7, TR_8, and TR_9 and diodes X_2, X_3, and X_4 (Fig. 37-8C). The AGC gate transistor TR_9 conducts according to the level of the signals coming from the video amplifier and the noise gate. Diodes X_3 and X_2 are in the collector circuit of TR_9, with the horizontal oscillator pulse entering between them. The horizontal pulse allows TR_9 to conduct only during the horizontal sync time, the amount of conduction being dependent on the signal level at the base. During conduction, TR_9 charges the AGC capacitor C_{54}, which in turn controls the base current of TR_8. Transistor TR_8 supplies the AGC level to the first stage of the i-f amplifier, to the noise gate driver TR_5, to base, and finally to TR_7, the delay circuit for the tuner automatic gain control.

Sync Limiter. Below and to the right of the automatic gain control block (Fig. 37-7) is the sync limiter, which takes the sync pulses from the composite video signal. The stage consists of transistor TR_{10}, which gets its signals from the video driver TR_4 and the noise gate TR_{11} (see Fig. 37-8C). Transistor TR_{10} can conduct only when the noise gate is on, and is therefore somewhat immune to noise. The output of TR_{10} goes to the vertical oscillator and to the horizontal oscillator where the respective sync pulses are filtered out and used to synchronize the horizontal and vertical deflection circuits.

Sound Detector and Amplifier. To the right of the noise gate on the block diagram (Fig. 37-7) is the sound intermediate-frequency and detector circuit, which consists of transistors TR_{12} and TR_{13} and diodes X_5 and X_6 (Fig. 37-8D). Transistor TR_{12} gets its input from the video driver, and amplifies it before coupling the sound signal to the sound limiter TR_{13}. Limiter TR_{13} continues to amplify the signal, so that any residual amplitude interference is removed. Then the signal goes to the frequency-modulated ratio detector, diodes X_5 and X_6. The output of the ratio detector goes to the audio amplifier block TR_{14} and TR_{15}, where the signal is amplified to drive the loudspeaker (Fig. 37-8E).

Vertical Output. Below the sync limiter (Fig. 37-7) is the vertical output circuit consisting of transistors TR_{16}, TR_{17}, and TR_{18} (Fig. 37-8F). Sync information enters the vertical oscillator through integrator A_1, which separates the vertical sync pulses from the horizontal sync pulses. At TR_{17}, the sync signal triggers the vertical oscillator to operate at the correct frequency. Transistor TR_{16} is a feedback sawtooth oscillator that develops the sweep voltage across capacitors C_{80} and C_{82}. The sawtooth signal then goes to the vertical driver TR_{17} and the vertical output stage TR_{18}. A feedback circuit from the collector of TR_{18} returns some signal to the oscillator TR_{16} to ensure vertical sweep in the absence of sync information. The output of TR_{18} goes to the vertical deflection coils and to the first grid of the picture tube (for blanking during vertical retrace).

Horizontal Output and High Voltage. To the right of the vertical circuit (Fig. 37-7) is the horizontal oscillator and automatic frequency control. Farther to the right is the horizontal output. The horizontal automatic frequency control circuit consists of transistor TR_{19} and diodes X_7, X_8, and X_9 (Fig. 37-8G). The horizontal oscillator is transistor TR_{20}. The horizontal sync pulses come from the sync limiter TR_{10} and are separated from the vertical pulses by capacitor C_{88} and the network around diodes X_7 and X_8. The incoming sync pulses are compared with the horizontal sweep voltage between capacitors C_{89} and C_{90}. The comparison results in a direct-current voltage being applied to the base of TR_{19}, which shunts the tuned circuit of the horizontal oscillator TR_{20}. If the horizontal sawtooth gets out of synchronization with the horizontal sync pulses, TR_{19} will shunt the oscillator either more or less to correct the frequency. The output of the horizontal oscillator goes to the horizontal driver, transistor TR_{21}, which squares up the horizontal signal. It then goes to the horizontal output stage TR_{22}, which drives the horizontal deflection coils, the high-voltage booster circuit, and the flyback transformer T_9 for developing the 20,000 volts for the picture tube.

The rectifier SE_1 is the high-voltage booster diode that rectifies the high-peak voltage across the horizontal deflection coils. This voltage develops across capacitors C_{106} and C_{109} and supplies voltage to the second and third grid (focus) in the picture tube. The high-peak voltages developed across the deflection coil also appear across the primary of flyback transformer T_9. The primary voltage is stepped up many times and then applied to the plate of the high-voltage rectifier V_1, which develops the 20,000-volt direct current for the picture tube. Transformer T_9 has three auxiliary outputs. One provides the sawtooth signal for the AFC circuit (TR_{19}), one supplies a horizontal gating pulse to the AGC gate, TR_9, and another supplies signals to the second picture tube grid for horizontal retrace blanking. This signal is also rectified by diode X_{12}, which provides about 100 volts direct current for the video output stage TR_6.

Low-voltage Power Supply. The low-voltage power supply is shown in the lower left corner of the block diagram (Fig. 37-7). It consists of transformer T_{10}, rectifiers SE_3 and SE_4 and other components (Fig. 37-8H). The rectifiers SE_3 and SE_4 rectify the secondary alternating-current voltage and apply it to the filter network C_{115A} and C_{115B} and L_{19}. The output voltages, 70 volts and 24 volts direct current, the latter regulated by zener diode X_{13}, go to the various circuits in the receiver.

THE KINESCOPE

The basic purpose of the video (display) circuits is to develop an electronic image. The main part of the video circuit is the large picture tube called a *kinescope* on which the image appears (Fig. 37-9). The image is focused and displayed electronically on the wide face of the picture tube by circuits that adjust for brightness, focus, and contrast.

Located inside the picture tube (within a vacuum) is an electron gun and a luminescent surface called a *screen*. The tube is funnel-shaped and its neck contains the elements of the electron gun. A socket is provided on the end of the neck in order to make connections with the internal elements of the gun. The electron gun produces a narrow beam of electrons representing the video signal. The screen, located on the wide surface of the tube, is coated with a phosphor material, which glows whenever a beam of electrons strikes it.

The glow on the tube's screen remains for only a short time because the magnetic deflection coils surrounding the electron gun keep the electron beam sweeping across and down the screen. These movements are synchronized with the transmitted video signals by varying the magnetic strength of four deflection coils placed 90 degrees apart. The deflection coils are located in the deflection yoke mounted on the neck of the picture tube. As in the camera tube, the process of deflecting the electron beam is called *scanning*.

COLOR TELEVISION

Color television allows for a more pleasing picture representation than is possible with black-and-white (monochrome) TV sets. More variety is achieved with color than with the various shades of grey, and color improves the quality of the image on the picture tube.

The three primary colors, red, green, and blue, are used in color television. Practically all colors can be reproduced using the proper mixture of these three colors. Each of the primary colors has a corresponding chrominance signal representing it: blue, green, and red video signals. However, to make it possible for monochromic television receivers to receive this signal, all three video signals are combined in one radio frequency and then transmitted.

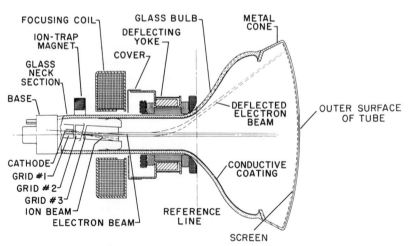

Fig. 37-9. Structure of a picture tube using coils to deflect the electron beam.

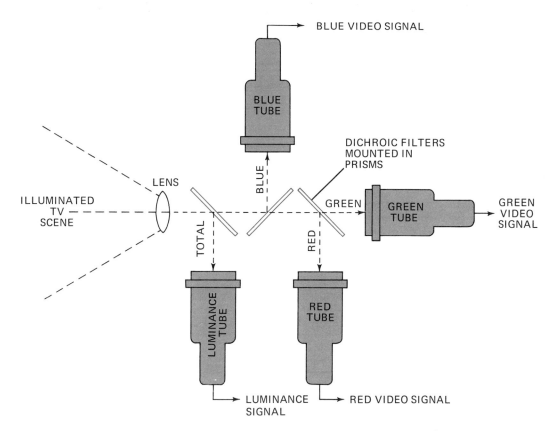

Fig. 37-10. Four-tube television color camera.

COLOR CAMERA

A color camera is, in many respects, similar to a regular black-and-white television camera, except that it has more pickup tubes. Fig. 37-10 shows a simplified view of a color camera that has four pickup tubes. Some color cameras use only three tubes: one tube for each color.

The four-tube camera can be thought of as producing two types of signals: luminance and color. The luminance tube can be used alone, or with the three color tubes. Through a simple change in the camera's optical system, all incoming light can be directed onto the luminance tube. Thus, during black-and-white transmission, the heaters of the color tubes can be turned off, thereby increasing their active life. When operating in this mode, the function of the luminance tube is the same as in any other black-and-white television camera.

During a color transmission, the incoming light is directed to all the dichroic filters. These filters, mounted in prisms, make it possible to split the light in such a way that each of the three color-pickup tubes "sees" only its respective color from the televised scene. The wavelength of light representing the color on the object is called the *hue*. It is the blueness, greenness, or redness of the scene. Thus, the blue tube develops a video signal that reflects the electrical image of the blue color on the light-sensitive surface of the blue tube. And the green and red tubes do likewise with their respective colors. A part of this light is also reflected to the luminance tube, which provides for the varying degrees of light, from black to brilliant white, that give the brightness to the televised scene.

In a color transmission using a four-tube color camera, four different video signals are developed—the red video signal (R), the green video signal (G), the blue video signal (B), and the luminance signal (Y). Actually, two luminance signals are available because the red, green, and blue signals can be recombined for transmission purposes by adding their color signals together.

One reason for recombining the color signals is to save space in the radio spectrum. Otherwise, it would be necessary to transmit three separate color signals—each with a frequency range of about 4.5 megahertz. This would be impractical, however, because the fixed limitation of 6 megahertz on the width of each television channel applies whether the transmission is in color or in black and white.

Another reason for recombining color signals is that many homes do not have color sets. By combining the black-and-white and color signals, however, both types of sets will receive the transmission, but only the color set will reproduce the transmission in color. This is because its circuitry can detect and separate the color signals and project them onto the picture tube. A monochrome or black-and-white television set will not respond or detect the color signals because it does not have the necessary color circuits or the correct picture tube. Thus, a viewer will see a color program in black and white.

Fig. 37-11 illustrates how light from an illuminated scene is split and recombined to form the complex signal generated by a color television transmitter. In this illustration, the light source is arbitrarily given a light value of 10. This value can be thought of as a voltage representing white light. When the filters split the light, a portion is picked up by each of the four tubes as shown, and their respective signals at any instant will add to the total voltage (if losses are not considered).

Each color tube develops a video signal, which then amplitude-modulates a chrominance subcarrier (red subcarrier, green subcarrier, and blue subcarrier). These subcarriers have a common frequency of 3.58 megahertz, and are multiplexed so that they differ in phase. The method of modulation used results in a suppressed carrier. Only the double sidebands remain. These sidebands contain the chrominance signals, which are then used to modulate the luminance video r-f carrier.

The method of modulation makes it possible to interweave the energy of the double sidebands of color within the regular bandwidth of the luminance video signal. This is possible because the luminance video signal generally does not use its full bandwidth space, but instead clusters about the horizontal sweep frequency. The chrominance sideband energy is designed

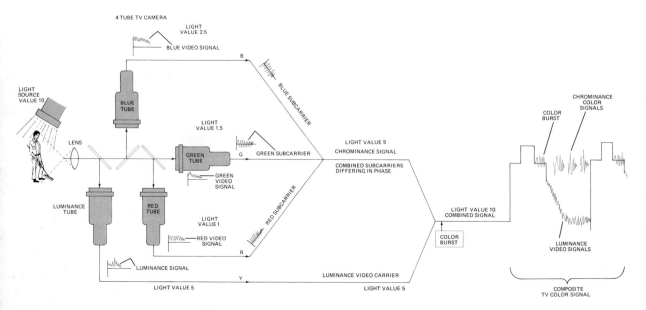

Fig. 37-11. Simplified process of developing a composite color television signal.

to fall within the sideband space "left open" by the luminance video signal. Therefore, these signals appear as sideband clusters of energy on each side of the clusters of energy developed by the luminance video signal. Thus, the chrominance and luminance signals are transmitted within the same frequency spectrum (Fig. 37-11).

COLOR BURST

Color burst is actually a series of short pulses (8 to 11 cycles) or bursts of a 3.58-megahertz subcarrier signal placed on the "back porch" portion of the horizontal blanking pulse (see Fig. 37-11). These pulses enable the television receiver's chrominance oscillator to develop a subcarrier frequency that is phase-locked with the transmitted subcarrier. It is, in effect, a reference signal that is transmitted as a part of the composite color television signal.

Within the receiver, color burst sets the proper phase of the subcarrier frequency so that it is in exact step with that of the transmitter's chrominance subcarrier. The chrominance signal that is 90 degrees out of phase with the luminance signal is labeled Q or *quadrature subcarrier signal*. The chrominance signal that is in phase is referred to as the *I signal*. The third chrominance signal is derived by the algebraic sum of the *I* and *Q* signals—thus, the three primary colors can be detected at any instant by the television receiver.

COLOR TELEVISION RECEIVER

The color receiver's large picture (display) tube or kinescope is similar, in many respects, to a conventional (black-and-white) television tube. There are several important differences, however. One type of color tube contains three electron guns that produce three separate beams of electrons. When these electron beams strike the fluorescent surface within the tube, they produce red, green, and blue light. The property of producing light below the temperature of incandescence by exciting certain materials with an electric energy beam is called *phosphorescence*. The materials used are called *phosphors*. The green phosphor light is produced from zinc, silicon dioxide, and manganese. The red phosphor light is produced from zinc phosphate and manganese. The blue phosphor light is produced from zinc sulfide, silver, and manganese oxide. The strength of the electron beam striking the various phosphors determines the amount of light output from the color kinescope's fluorescent surface.

Several different methods are used to reproduce the various color mixtures on the face of the tube. One of the most common involves the use of three electron beams. Each beam represents one color, and is directed at one of three phosphor dots located on the screen surface. The three phosphor dots (red, green, and blue) are arranged in a triangle as shown in Fig. 37-12. On a 15-inch kinescope there are

Fig. 37-12. Basic operating principles of a 21-inch shadow mask color picture tube. (RCA)

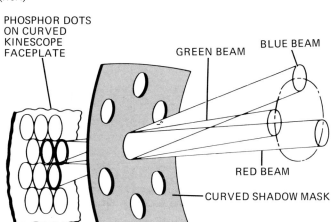

200,000 three-dot triangles, or 600,000 individual phosphor dots. The large number of dots that make up the fluorescent surface helps improve the quality of the total picture.

Fig. 37-13 shows that color receivers have much in common with monochrome receivers. A tuner is still required to select the desired channel by heterodyning it to the intermediate frequency and an i-f amplifier is included to boost the intermediate-frequency signal. Note also that the sound section is the same as that found in a monochrome receiver.

Among the additional stages found in a color receiver are the following: A chrominance amplifier, which is used to boost the color signal before it is sent on to the color demodulators. A burst amplifier, which, when keyed on at the proper times by the horizontal deflection circuit, amplifies the 3.58-megahertz color burst on the back porch of the horizontal blanking pulse and sends it on to the color automatic frequency control stage. The color afc stage compares the phase of the color burst with the phase of the local 3.58-megahertz oscillator and develops an error signal to correct the oscillator. One output of the oscillator directly feeds one of the color demodulators. The other output is shifted in phase by 90 degrees and then sent to the other color demodulator. The outputs of the color demodulators are sent to the red, green, and blue control grids in the kinescope. The cathode of the kinescope is driven by the luminance signal, which, with the chrominance signal, interacts in the kinescope to give a color picture.

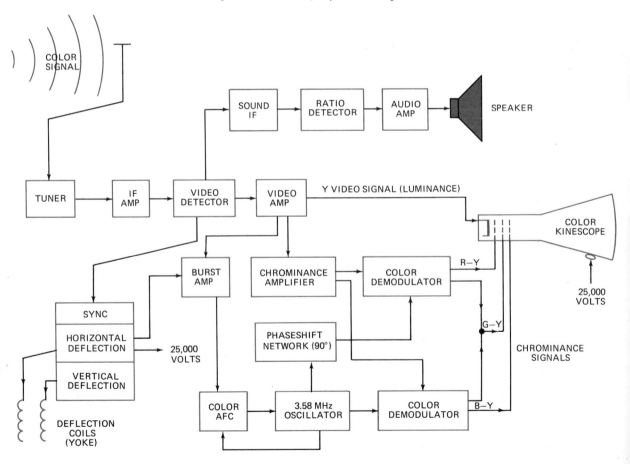

Fig. 37-13. Color TV set, simplified block diagram.

DIODE TUNING

A new system for tuning both VHF and UHF television channels makes use of a diode which acts as a voltage-sensitive variable capacitor. This type of diode is very commonly referred to as a *Varactor* or as a *Varicap*.

In a tuning circuit, the diode is reverse-biased by a variable voltage which may range from 0.5 to 28 volts. During the process of tuning, a channel input signal has the effect of increasing or decreasing the reverse bias applied to the diode. If the reverse bias is increased, the depletion region at the p-n junction of the diode widens and its capacitance decreases. Conversely, if the reverse bias is made to decrease, the depletion region becomes less wide and its capacitance increases. As a result of this action, the diode is able to produce resonance of the tuning circuit at any specified frequency.

For Review and Discussion

1. Describe a closed-circuit television system.
2. For what purpose are "relay" or "repeater" towers used in a transmission system?
3. Name the principal components of a composite video signal.
4. Define negative modulation.
5. Draw a block diagram of a typical black-and-white television receiver.
6. What is the function of a flyback transformer in a television receiver?
7. What is the function of the dichroic filters used in a color television camera?
8. Describe how the video signals are developed in a four-color camera tube.
9. What is the purpose of the "color burst"?

UNIT 38. Radar and Lasers

Radar was originally developed as an electronic aircraft defense system in England during World War II. Since then, it has evolved into an electronic system that is now used in many activities ranging from weather forecasting to the preparation of maps showing contours of the moon's surface.

The laser was originally developed in 1960. In this relatively new system, a combination of electronic and optical assemblies is utilized for the purpose of voice and picture communications as well as a number of other important activities.

RADAR

The word *radar* is derived from the phrase "ra(dio) d(etecting) a(nd) r(anging)." It is a system by means of which it is possible to detect the presence of objects and to determine their velocity, direction, and range (distance). In addition, some types of radar systems also make it possible to roughly analyze the composition (make-up) of the detected object.

Pulse-modulated Radar. Radar detection is most often accomplished by transmitting pulses of a

Fig. 38-1. Transmitted and echo signals used in radar.

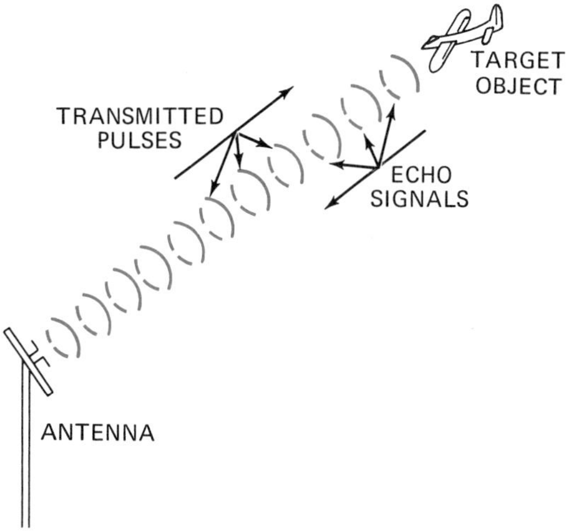

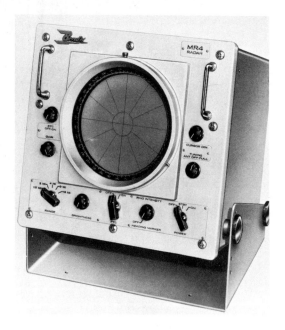

Fig. 38-2. Cathode-ray tube radar indicator screen and operating controls. *(Bendix Marine Products)*

relatively narrow beam of high-frequency electromagnetic energy over a region to be searched. A portion of this energy is then reflected from the object encountered, and forms what is referred to as the *echo signal* (Fig. 38-1). This signal then returns to the radar system, where it is received and is usually indicated upon the screen of a cathode-ray type of tube known as the *indicator* (Fig. 38-2). Since electromagnetic

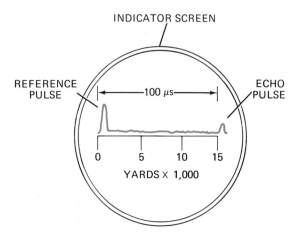

Fig. 38-3. A-scan radar presentation.

energy travels at the speed of light, the time period that elapses between the transmission of the signal and the reception of its echo can be used to determine the range of the target object.

In order to obtain accurate indications of range, the linear sweep of the cathode-ray tube must be timed precisely. The sweep trace then provides a time axis or *base* that makes it possible to convert a measured time interval into an indication of distance.

A-scan. A simple method of determining range by means of pulse-modulated radar involves the use of an A-scan presentation. In this basic type of presentation, the sweep trace of the cathode-ray tube indicator forms a horizontal baseline upon the screen of the tube. The transmitted pulse and the echo signal appear as vertical deflections or *blips* along the horizontal sweep trace. Since the horizontal trace and the transmitted pulse begin at exactly the same time, the distance between the blips is proportional to the range of the target object.

For example, assume that the entire linear sweep trace of the cathode-ray tube indicator is completed in 100 microseconds. Also assume that the transmitted or reference pulse and the echo pulse appear upon this trace as indicated in Fig. 38-3. The total time required for the pulse to be transmitted to the target object, reflected from it, and received is 100 microseconds. One-half of this time period, or 50 microseconds, is the time required for the pulse to reach the target. The velocity of the pulse through space is 186,000 miles per second, or 328 yards per microsecond. Hence, the target is 328 × 50, or 16,400 yards from the radar antenna. Although, in this case, the time axis (*trace*) is calibrated in yards, it could be calibrated in terms of land miles, nautical miles, or some other convenient unit of distance.

Transmitter. The basic pulse radar transmitter is essentially a blocking-type oscillator circuit, which may or may not be operated in conjunction with amplifier stages. The electron tube used in the oscillator stage may be one of several types that can be triggered into delivering powerful pulses of energy at high frequencies. Tubes such as the *magnetron* and the *reflex klystron* are commonly used for this purpose.

The number of pulses transmitted per sec-

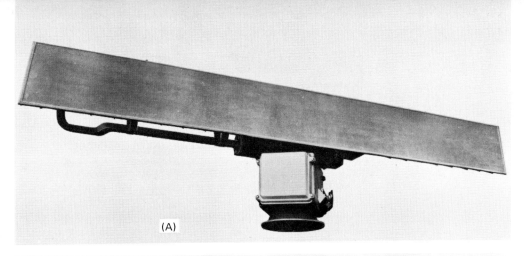
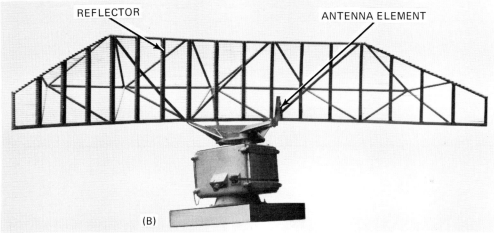

Fig. 38-4. Typical radar antennas: (A) slotted-waveguide antenna, (B) parabolic antenna. *(RCA)*

ond is referred to as the *pulse-repetition rate* (PRR). This must be precisely controlled at one rate. The average duration of each pulse, the pulse time or pulse width, is approximately 0.4 microsecond. Pulses of shorter duration are used for short-range detection, while pulses of longer duration are used for the detection of long-distance targets.

Receiver. The radar receiver circuit is a superheterodyne circuit designed to operate over the frequency range of its associated transmitter. The output stage of the receiver is a wideband video amplifier connected to the vertical deflection system of the cathode-ray tube indicator. The indicator is also connected to the horizontal synchronization and deflection (sweep) circuits, which are similar to those found in a typical television receiver. In the pulse-modulated radar system, the *synchronizer,* also known as the *timer,* provides the synchronizing pulses that are simultaneously applied to the transmitter and the indicator.

Antenna Systems. The most practical radar antenna system consists of a single antenna that is used for both transmitting and receiving. Such an antenna is operated in conjunction with an electronic transmit-receive (TR) switch that automatically transfers the antenna from the transmitter to the receiver and back again at the pulse-repetition rate. To provide the necessary shielding of pulse transmissions and to reduce signal loss, the transmission line connecting the antenna to the transmitter is a coaxial cable or waveguide.

The two most common types of antennas used with present-day radar systems are forms of the slotted waveguide and parabolic arrays (Fig. 38-4). The slotted-waveguide antenna is characterized by its light weight and outstanding performance.

The parabolic antenna is most commonly used with radar systems operating at higher frequencies. In this antenna, the reflector is formed so that its curvature is that of a mathematical parabola; hence, the name *parabolic*. The reflector, which may be constructed of either a solid sheet or a screen-like material, focuses the signal and acts as a shielding device.

Planned-Position Indicator (PPI) Radar. In the planned-position indicator (PPI) type of radar system, the antenna rotates continuously through 360 degrees. This rotation is synchronized with the deflection circuits of the indicator. As a result, the sweep trace appearing upon the screen of the indicator is a rotating line, beginning at the center of the screen and extending outward in a direction corresponding to the direction in which the antenna is pointing.

The echo signals returning to the receiving antenna of a planned-position indicator presentation do not produce *distortions* or *blips* upon the sweep trace. Instead, the echo signals cause certain areas of the screen that correspond to the location of the target object or objects to become brighter. Thus, a map-like presentation of all target objects appears upon the screen (Fig. 38-5). The position of the brightened

Fig. 38-5. Planned-position indicator (PPI) radar display of a harbor entrance. (*RCA*)

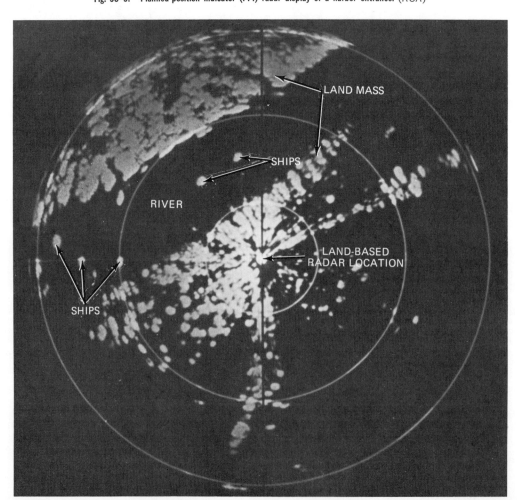

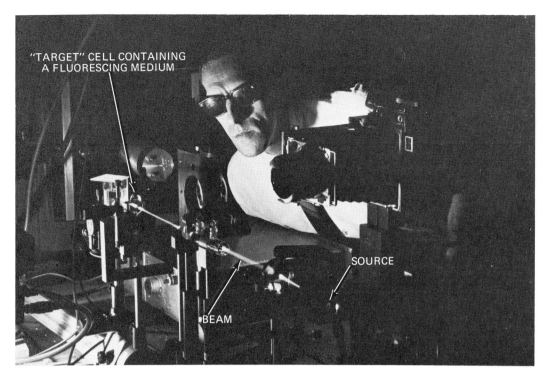

Fig. 38-6. **Laser beam.** (*Bell Telephone Laboratories*)

areas relative to the calibrated position of the antenna can then be determined.

A planned-position indicator presentation usually does not indicate the range of target objects directly. However, the range can be calibrated in terms of the distance from the center of the indicator screen to a brightened area. The direction-range feature of this radar system makes it extremely useful for navigational purposes and for providing storm-warning information.

LASERS

The word *laser* is derived from the phrase "*l*(ight) *a*(mplification by) *s*(timulated) *e*(mission of) *r*(adiation)." Basically, the laser is a system that produces a highly focused or concentrated beam of light (Fig. 38-6). However, unlike ordinary light, this beam does not diffuse or spread out appreciably as it travels away from the source. Because of this characteristic, it possesses a tremendous amount of energy that is utilized in various ways, including applications in the fields of communications, medical technology, and production.

Masers. Lasers were developed as an extension of a communications system known as the *maser* [*m*(icrowave) *a*(mplification by) *s*(timulated) *e*(mission of) *r*(adiation)]. In this system, the output is radio-frequency electromagnetic energy. Two of the most important maser applications are their use as extremely high-frequency amplifiers and as very stable oscillators or signal sources.

Incoherent and Coherent Light. All light is a form of electromagnetic radiation. Ordinary light, such as that produced by an incandescent lamp, consists of electromagnetic radiations of several different frequencies. Therefore, the visible light from the lamp actually consists of several different colors that appear to the eye as white. Such light is referred to as *incoherent* or *polychromatic*. Since incoherent light is a random, out-of-phase emission of energy, it diffuses rapidly, and thus cannot be precisely or accurately

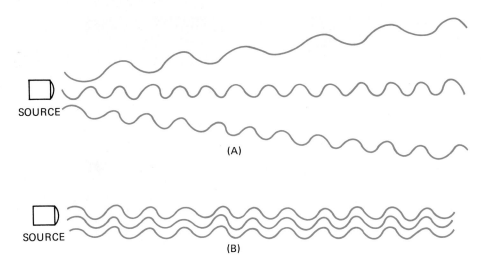

Fig. 38-7. Light radiation patterns: (A) incoherent light, (B) coherent light.

transmitted without a significant loss of energy (Fig. 38-7A).

The light generated by a laser is coherent light consisting of electromagnetic radiations that are in phase (Fig. 38-7B). Because of this characteristic, coherent light does not diffuse appreciably and can be transmitted over long distances without significant loss of energy.

TYPES OF LASERS

Three basic types of lasers have been developed and are in practical use. These lasers are designated in terms of the active materials within which coherent light is generated, namely, solid-state, gas, and semiconductor or injection lasers.

The active material in the original laser was ruby. In more recently developed solid-state lasers, the active materials are glass, synthetic crystals, or rare-earth-doped crystals.

The active materials most commonly used in gas lasers are carbon dioxide, argon, and a combination of helium and neon. Semiconductor lasers are also made from a number of materials, the most common of which is gallium arsenide. A fourth type of laser, which has not been fully developed, uses a liquid active material.

Ruby Lasers. The ruby laser provides a good example of how a solid-state laser operates. In all solid-state lasers, coherent light is generated as the result of a change in the energy levels of electrons within the atoms or ions of the active material when the material is exposed to an intense, incoherent light (Fig. 38-8).

The principal parts of a simple ruby laser assembly are shown in Fig. 38-9. The active material of the laser is a polished ruby rod made of crystalline aluminum oxide in which a small number of aluminum ions are replaced by chromium ions. It is the chromium ions that produce the laser effect. The ends of the rod are coated with a silver compound that acts as a light-reflecting surface or mirror. One end is heavily coated, the other end less heavily coated in order to form a semitransparent surface from which the beam of coherent light is emitted. These end surfaces and the rod itself form what is referred to as an *optical resonant cavity*.

The operation of this laser depends upon the fact that an incoherent light, produced by the xenon flash lamp, can be used to excite the chromium ions into higher, unstable energy levels. In solid-state laser terminology, the process whereby this occurs because of the absorption of incoherent light energy (or some other form of energy) is known as *pumping*. As a result

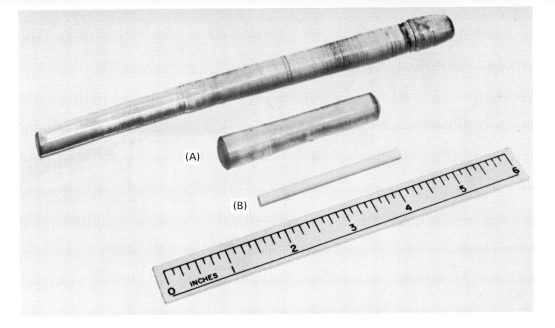

Fig. 38-8. Rod-shaped active material of a solid-state laser: (A) steps in the fabrication of a rod consisting of calcium tungstate that has been doped with enodymium, (B) final shape of the rod. (*Sperry Gyroscope Division, Sperry Rand Corp.*)

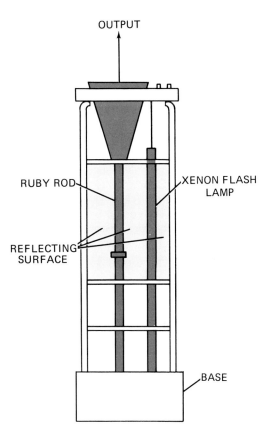

Fig. 38-9. Principal parts of a simple ruby laser.

of pumping, the chromium ions change their physical state by a transfer of electrons from one shell or orbit to another shell that represents a higher level of energy.

When the flash lamp is turned off, incoherent light energy is no longer directed toward the ruby rod and some of the electrons within the chromium ions return to their original, lower energy levels. Because of this action, the electrons release energy in the form of incoherent light, which resonates (surges back and forth through the rod) as it strikes the reflecting surfaces of the rod ends. As a result, more and more electrons are stimulated into releasing energy in the form of light that is in phase with the stimulating light. The light eventually becomes intense enough to escape from the semitransparent end of the rod as a powerful beam of visible, red coherent light. The entire cycle of events resulting in the generation of the beam of coherent light by the laser takes place within an extremely short time period of a few milliseconds.

The ruby laser and most other solid-state lasers are designed to produce an intermittent or pulsed output of coherent light. However,

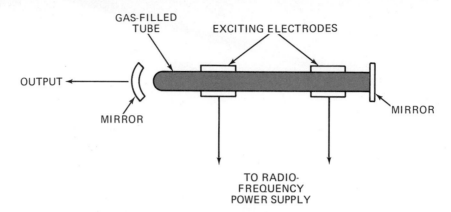

Fig. 38-10. The basic gas laser.

some solid-state lasers, such as the one that employs calcium tungstate doped with neodymium as the active material, are designed to produce a continuous output. These are referred to as *continuous wave* (CW) *lasers*.

Gas Lasers. In the gas laser, coherent light is generated within a glass tube containing a gas such as carbon dioxide or neon or a combination of gases such as helium and neon (Fig. 38–10). The gas is excited by the application of radio-frequency energy, which causes an electrical discharge through the tube in much the same way as the discharge through an ordinary neon lamp or tube. When this happens, atomic collisions occur within the gas, resulting in changes in the energy levels of atoms or ions. These changes produce a coherent light in a manner similar to the action of a solid-state laser.

Gas lasers are usually designed to produce a continuous wave output of one or more colors. One outstanding advantage of gas lasers is that they operate with a relatively high efficiency as compared with other types of lasers.

Semiconductor Lasers. The semiconductor laser is also commonly referred to as an *injection* or *diode laser*. In one laser of this type, the active material consists of a specially constructed gallium arsenide p-n junction diode that is operated in a forward-biased condition. The pumping energy for this laser is supplied by a direct-voltage source that can deliver a large amount of current to the diode. Under this condition, there is a vast recombination of electrons and holes within the depletion region of the junction. As a result of these recombinations, energy is released in the form of coherent light emitted from the junction region. Semiconductor lasers can be operated either continuously or in the form of pulses.

LASER APPLICATIONS

The rapid development of lasers has created a variety of possible applications to several areas of civilian, space, and military activities. Some of these applications are now a reality, while others are the subject of intense scientific and technological investigation.

Communications. The application of lasers to the fields of communications and radar appear to be most promising. Some experts feel that the laser will eventually replace most ordinary microwave systems now is use in the simultaneous point-to-point multichannel transmission of information. In this connection, the extremely high frequencies of laser carrier beams, 470 million megahertz or higher, have a distinct advantage over typical microwave carriers. This is because a laser beam, with its higher frequency, can be modulated to provide a much greater bandwidth. Hence, the beam can be modulated to provide greater numbers of single-sideband frequencies, each of which can be used as a carrier for transmitting telegraph, telephone, and/or television information.

The laser beam also offers the possibility of more efficient ground-to-space vehicle communication than is provided by systems now in use. This application was initially tested when astro-

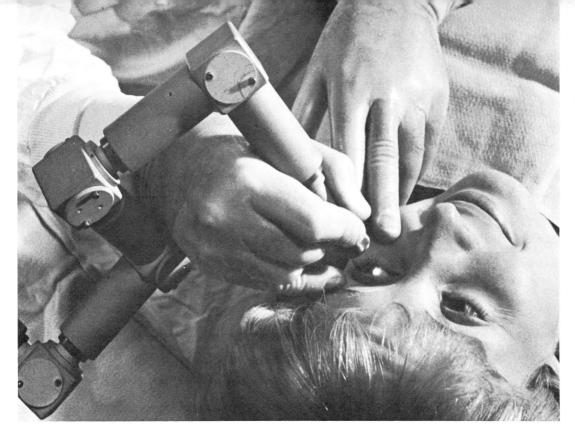

Fig. 38-11. Surgeon using a "laser knife" to repair a detached retina within the eye of a patient. (AT&T)

nauts aboard a Gemini spacecraft successfully communicated with a Hawaiian ground station by means of a laser beam.

Other Applications. In the medical field, the laser is being used to cut and attach internal bodily tissues that cannot be reached without first cutting through surface tissue (Fig. 38-11). The laser is also used in conjunction with microscopes to observe previously unknown processes of bodily cell behavior.

As a "machine tool," the laser beam has been found useful in cutting, shaping, drilling, and welding metal, minerals, plastics, and fabrics. The beam is also being used in industry as the medium by which precise measurements can be made. In the field of scientific research, the laser has opened up an entirely new method of examining various structures and processes. These and other applications are outstanding examples of how electronic and optical research and technology have been combined and put to use.

For Review and Discussion

1. What is the basic function of a radar system?
2. Explain the operation of a fundamental, pulse-modulated radar system.
3. For what purpose is a transmit-receive switch used in a radar system?
4. Name and describe the two basic types of antennas used with radar systems.
5. State the phrases from which the terms "radar," "laser," and "maser" are derived.
6. What is the basic function of a laser system?
7. Describe the difference between incoherent and coherent light.
8. Name the three general types of practical lasers.
9. Explain the operation of a simple ruby laser.
10. What is meant by "pumping" in a laser system?

SECTION 8

SENSING AND CONTROL SYSTEMS

The basic principles of operation relating to a number of sensor (transducer) devices have been presented in various units of this book. In this section, several of these devices will be discussed together with their application in present-day sensing and industrial control circuitry.

UNIT 39. Sensing and Control Circuits

A typical sensing and/or control system consists primarily of at least three principal unit groups: the sensing (transducer) unit, the control or receiving-transmitting unit, and the indicating or controlled unit (Fig. 39-1). The function of the control unit is to convert the response of the sensor into an output signal that may or may not be further amplified and that is capable of actuating a load such as an electromagnetic relay or an electronic switching circuit.

When the sensor is used in conjunction with a telemetry system, its response modulates the carrier signal of a radio-type transmitter. This signal is then converted into the desired monitor or control signal by the associated receiving and receiver output circuits.

PHOTOELECTRIC SENSING AND CONTROL

The photoelectric sensor, usually a photoconductive cell, is a widely used sensing device. The basic components of a popular type of photoelectric control system are shown in Fig. 39-2. In this system, the control unit contains the photoconductive cell sensing element, one or more sensor-output amplifier stages, and a load control relay or some other form of switching assembly. In other systems, the photoconductive cell is contained within a separate photocell receiver unit, the output of which is applied to the amplifier control unit.

Fig. 39-1. Basic units in a control system.

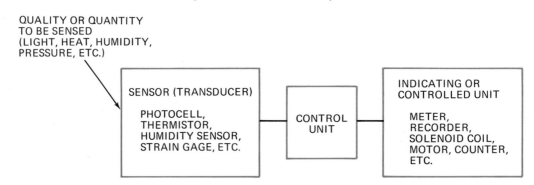

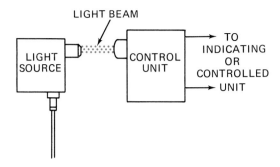

Fig. 39-2. Sensing elements of a photoelectric control system.

Operation. In the photoelectric sensing and/or control system, a beam of light is the medium through which the photocell is actuated. In most practical applications, the light beam is interrupted (modulated) by an object in accordance with the action to be sensed or controlled. The photocell then responds to the modulated beam by a variation in its resistance. Because the photocell is an integral part of the control circuit, the variation in its resistance affects the operation of the circuit in a manner that corresponds to the intensity of the light striking the cell.

In one type of on-off control unit, for example, the photocell is connected in series with the base-bias resistor network of a transistor amplifier stage (Fig. 39-3). When light strikes the photocell, its resistance decreases, thereby causing the forward bias on the transistor to increase. Hence, the transistor conducts a sufficient current to fully energize the relay coil in the collector (output) circuit. With the output (A and B or B and C) terminals of the relay connected in an appropriate load circuit, the manner in which the beam of light striking the photocell is modulated can then be made to control the action of the load or loads.

In burglar-alarm systems and in other systems where an invisible light beam is desirable or necessary, the lens of the light source used with a photoelectric control system is equipped with an infrared filter. When a single beam is used to guard the access to several areas, it is used in conjunction with mirrors placed in appropriate concealed locations.

Application. Several representative applications of photoelectric control processes are illustrated in Fig. 39-4. Because of their flexibility in application and stability of operation, these and many more types of photoelectric control systems are

Fig. 39-3. Photoelectric controller.

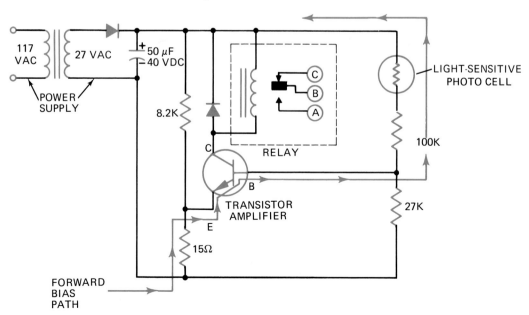

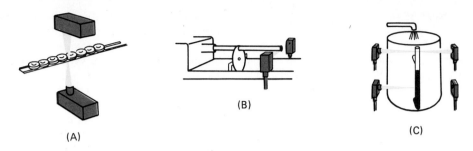

Fig. 39-4. Light beams supervise and control work: (A) counting small round objects on a moving conveyor, (B) shearing material or parts to size, (C) controlling the liquid level.

employed in all phases of industrial, commercial, and product-distribution activities.

THERMISTOR APPLICATIONS

In addition to their current-limiting function, thermistors serve as the sensing or otherwise active element in a wide variety of circuits. Several of the more common thermistor sensing applications are described in the following paragraphs.

Temperature Measurement. A simple thermistor circuit that can be used for measuring temperature is shown in Fig. 39-5A. As the temperature of the area surrounding the thermistor in this circuit changes, the resistance of the thermistor also changes, thus producing variations in the circuit current. Since the meter is in series with the circuit, its scale can be calibrated in terms of degrees of temperature.

Because thermistor resistance in such a circuit is relatively high (100 kilohms or more), any change in the resistance of the connecting wires that results because of changes in ambient temperature is negligible. Hence, the thermistor can be mounted some distance from the meter or other indicating device. The accuracy of this and similar circuit systems is primarily dependent upon the stability of the cell (or battery) supply voltage.

A more sensitive temperature-measuring circuit results when a thermistor is used as one "leg" of a bridge circuit (Fig. 39-5B). In this circuit, the bridge is unbalanced to a greater or lesser degree by changes in thermistor resist-

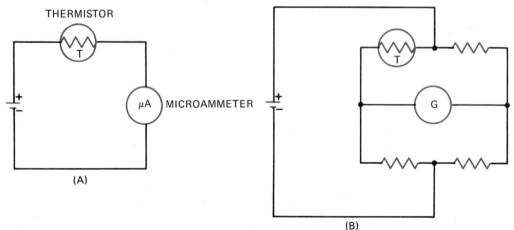

Fig. 39-5. Basic thermistor circuits.

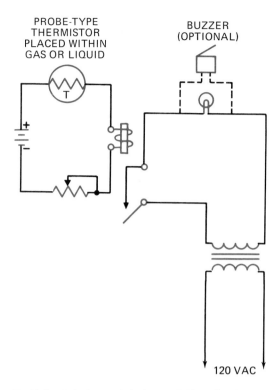

Fig. 39-6. A visual or aural circuit controlled by a thermistor.

ture at which the relay contacts will close to complete the load (lamp) portion of the circuit can be selected by adjusting the rheostat to the proper position. At temperatures below this point, the resistance of the thermistor reduces the current through the relay coil to a magnitude that causes the relay contacts to remain open.

Load Switching. A basic load-switching application of a thermistor is illustrated by Fig. 39-7. Under normal conditions with all the series-connected lamps operating, only a small current passes through each thermistor because of the relatively small voltage drop across each lamp. However, if any one of the lamps burns out, the total line voltage appears across the thermistor that is shunted across the defective lamp. The resulting surge of current through the thermistor quickly heats it. The resistance of the thermistor then sharply decreases, causing the voltage drop across the thermistor to correspond to the voltage drop that existed across the lamp while it was in operation. Hence, the circuit (minus one lamp) is restored to normal operation by the load-switching action of the thermistor. Thermistors are found also in time delay and liquid leveling circuits.

DETECTION OF NUCLEAR RADIATION

Nuclear radiation is the result of radioactivity, which is the property of certain types of atomic nuclei to disintegrate spontaneously. During this process, energy is emitted from the nuclei in the form of alpha and beta particles and gamma rays.

Alpha (α) *particles* are nuclei of helium atoms, each of which consists of two protons and two neutrons. *Beta* (β) *particles* are high-speed electrons. *Gamma* (γ) *rays* are radiations similar to visible light but having a much shorter wavelength than visible light.

ance. The extent of the unbalance is then indicated by the galvanometer, whose scale is calibrated in degrees. With the proper selection of galvanometer, such a circuit can be designed to operate satisfactorily over a wide range of temperature changes.

Temperature Signal. A circuit that is particularly useful for providing a visual and/or aural indication when the temperature of a given mass of gas or liquid has reached the desired point is shown in Fig. 39-6. In this circuit, the tempera-

Fig. 39-7. Load switching with a thermistor.

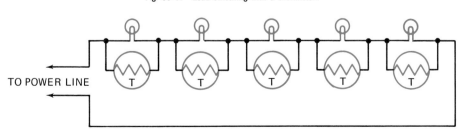

The most common natural radioactive elements are *radium* and *uranium*. Several artificial radioactive elements and compounds are formed as the result of various nuclear reactions and by the action of high-energy particle accelerators.

Nuclear radiations (alpha and beta particles and gamma rays) of significant intensity have pronounced physiological effects that can seriously damage body tissues. Since beta and gamma rays can penetrate most materials, it is important that means of detecting these radiations are available.

The detection and measurement of nuclear radiations is also important because of the rapidly increasing use of such radiations in industrial applications. These include: tracing and testing of materials, sterilization of products, measurement of liquid flow rate, and the analysis of various kinds of mixtures.

X-rays. X-rays, which have properties similar to those of gamma rays, have a wavelength which is between that of ultraviolet light and gamma rays. These rays are usually produced as the result of a sudden change in the velocity of a beam of electrons (a cathode ray). Because of their extremely high penetration property and their ability to affect radiographic film and the pick-up element of an image-intensifier tube, x-rays are used in several industrial applications, the most common of which is quality-control product inspection.

The Geiger Counter. The Geiger counter is an instrument designed to detect beta and gamma ray radiations electronically. Because this instrument is portable (battery-operated) and highly efficient as a beta detector, it is commonly used for locating radioactive ores and in civil defense applications.

The active detecting element in the typical Geiger counter is the gas-filled Geiger-Mueller (G-M) tube (Fig. 39-8A). This tube is most often contained within a cord-connected shielding probe equipped with "windows" that form the target areas through which the radiation to be detected is passed into the tube.

For example, when beta rays enter the tube, they collide with gas molecules, causing some of the gas to ionize. As a result of this process, electrons are dislodged from the gas molecules, thereby producing free electrons and positive gas ions within the tube. As the free electrons are attracted to the positively charged anode, they collide with other gas molecules, thus producing more free electrons and gas ions. The free electrons are then attracted to the anode, where they form the output (plate) current of the tube. The ions are attracted to the cathode, where they gain electrons and become neutralized gas molecules.

In one type of Geiger counter, the output signal of the Geiger-Mueller tube is passed through a resistance-capacitance coupling network to the first stage of an amplifier circuit

Fig. 39-8. Basic elements of a Geiger counter tube: (A) Geiger-Mueller tube, (B) resistance-capacitance coupling network for Geiger-Mueller tube.

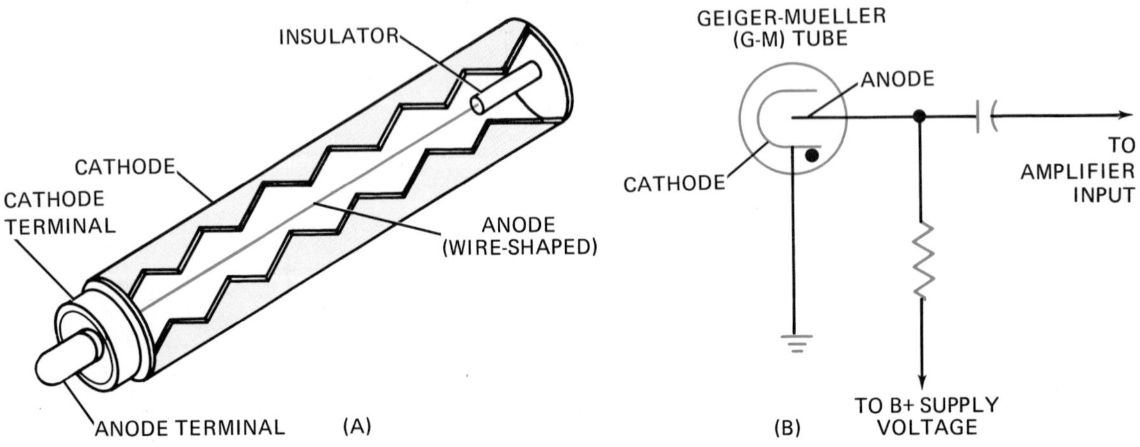

(Fig. 39-8B). The output of the amplifier is then connected to a meter and a headset, which provide visual and aural indications of radiation intensity.

The number of electron pulses passing through the Geiger-Mueller tube of the Geiger counter is proportional to the number of beta rays entering the tube. Hence, the intensity of radiation is indicated by the number of pulses of current present in the output circuit during any given period of time. In basic instruments, an increasing intensity (count) is indicated by an increased deflection of the meter pointer and by the higher frequency of the "clicks" produced by the headset. In more sophisticated instruments, the output-current pulse rate is indicated by a register that is calibrated to show the number of pulse counts occurring within a given period of time.

Radiography. In industrial radiography, x-rays are used to penetrate the product (object) under inspection. The shadow image of the product is then permanently recorded upon a radiation-sensitive film. When this film is developed, a photographic image (picture) of the test item is produced. Since the x-rays pass through different kinds and different thicknesses of materials with varying degrees of absorption, the image provides an extremely accurate indication of internal structures. For this reason, radiography is used for inspecting a wide variety of both metallic and nonmetallic products ranging from transistors to large tires (Fig. 39-9).

TIMERS

When it is necessary to provide energy pulses to a load at specified intervals of time and when it is necessary to energize or to deenergize a load after a given time period has elapsed, an electronic timer circuit or a clocklike time-switch mechanism driven by a synchronous motor is used.

Electronic Timers. The time-sensing element in most electronic timing circuits is some form of resistance-capacitance (RC) combination that provides a time interval that is dependent upon the resistance-capacitance time-constant period of the circuit. This method of time control is relatively simple and assures accurate timing

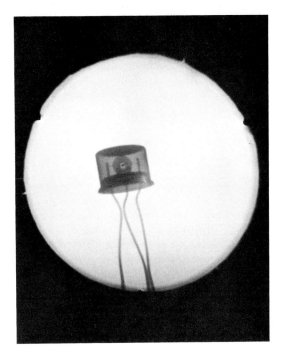

Fig. 39-9.   X-ray photograph of a transistor.

if all the associated components are of good quality.

An electronic timing circuit that functions as a relaxation-type oscillator is shown in Fig. 39-10. In this circuit, an adjustable resistance-capacitance time-constant period is established by resistor R_1 and capacitor C, which are connected directly across the battery.

When switch S_1 is turned on, the capacitor begins to charge toward the battery voltage at

Fig. 39-10.   An electronic timing circuit.

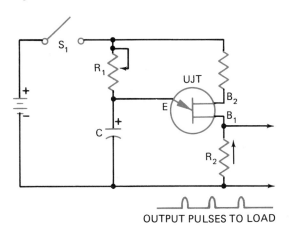

a rate determined by the $R_1 \times C$ time-constant period. When the voltage across C reaches a certain magnitude (determined by the operational characteristics of the unijunction transistor), it triggers the transistor into conductivity. As a result, the capacitor discharges through that portion of the circuit consisting of R_2 and the B_1 to E section of the transistor. This action causes a voltage drop (the signal pulse) to be developed across R_2 and applied to the load circuit.

As the capacitor continues to discharge, its voltage rapidly decreases to a magnitude that causes the p-n (emitter-to-B_1) diode junction of the transistor to be reverse-biased. At this time, the transistor ceases to conduct from B_1 to E, and the signal pulse voltage is zero. The cycle is then repeated at a frequency that is controlled primarily by the time constant. Hence, signal pulses having a predetermined time interval are applied to the load.

The resistance-capacitance circuit, operated in conjunction with associated switching devices such as diodes, silicon controlled rectifiers, and transistors, is used in a large number of different time-interval switching circuits, including pulse generators, flasher controls, and various types of industrial process control systems. The resistance-capacitance circuit is also used in conjunction with relays to provide a time-delay switching action of a relay that is necessary in many industrial and commercial operations. In time-delay circuits of this kind, the resistance-capacitance circuit is used alone or with a solid-state switching device that energizes the relay coil after being triggered into conductivity by the resistance-capacitance (timing) circuit.

ELECTRONIC SWITCHING

Although the electromagnetic relay is satisfactorily used in a wide variety of sensing and/or control circuit applications, it does have the disadvantage of operating at a speed that is lower than that required for the operation of specialized types of circuits. This is due to the

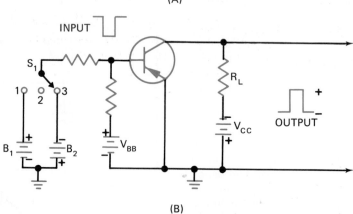

Fig. 39-11. A transistor used as a switching device.

mechanical switching action of the relay. The modern electronic switching circuit that most often utilizes a transistor or some other semiconductor device as the active switching element offers the advantage of increased speed of operation without the use of mechanical moving parts. For this reason, electronic switching is very commonly used in control circuits and in other circuit systems, including communications networks and computers.

A basic p-n-p transistor circuit that can function as an on/off switching stage is shown in Fig. 39-11. In this circuit, the control (input) signal is represented by batteries B_1 and B_2, either of which can be connected to the base input of the transistor by switch S_1.

With S_1 open (position 2), the emitter-base junction of the transistor is reverse-biased to the cutoff point by V_{BB}, the base supply voltage (Fig. 39-11A). In this condition, the current in the collector (output) circuit of the transistor is the small reverse current (I_{CBO}) that is present in the collector circuit when the transistor is biased to a nonconducting state. As a result, the transistor switch can be considered to be turned off.

When S_1 is closed to position 1, a positive voltage is applied to the base of the transistor. Since this voltage does not affect the reverse-bias condition of the base-emitter junction, the switch remains off.

The transistor is turned on by closing S_1 to position 3 (Fig. 39-11B). In this condition, a negative voltage is applied to the base of the transistor. The negative voltage overcomes the reverse bias applied to the base-emitter junction by V_{BB}, and the transistor is "switched" to the conducting state. At this time, a significant voltage drop (the output signal) appears across the output circuit.

From the preceding circuit description, it is evident that the p-n-p transistor switching stage can be turned off by a no-signal input to the base or by a positive-voltage pulse applied to the base. Conversely, the switch can be turned on by applying a negative-voltage pulse to the base. Hence, the transistor will act as an on/off switch, with the application of direct-voltage pulses to its input circuit or an alternating voltage to the input circuit.

With an alternating-voltage input signal, the switch will be off during the positive alternations of the input or control signal. The magnitude of the control voltage required to cause the switching action will be determined by the specific transistor used, the magnitude of V_{BB}, and the magnitude of V_{CC}.

In addition to p-n-p and n-p-n transistors, devices such as unijunction transistors, silicon controlled rectifiers, and diodes are commonly used to perform the switching function. As in the p-n-p or n-p-n transistor, the desirable switching characteristics of these devices result from the fact that they can be very rapidly changed from a nonconducting (off) state to a conducting (on) state by the application of a signal pulse of the correct polarity.

For Review and Discussion

1. Name the three principal units of a typical sensing and/or control system.
2. Briefly explain the operation of a basic photoelectric control system.
3. Draw the schematic diagram of a thermistor bridge circuit that can be used for the measurement of temperature.
4. Explain how a thermistor can be used as a load-switching device.
5. Define nuclear radiation.
6. Define alpha particles, beta particles, and gamma radiation.
7. Explain the operation of the Geiger-Mueller tube.

PROJECTS

INTRODUCTION

This section contains information for the construction of ten electronic projects of varying complexity and cost. All have been thoroughly tested and will work well when properly constructed.

The most important consideration in building projects like these is safety. *Poorly constructed electronic devices are dangerous and are not welcome in any home or school.* By following safe practices and the instructions of the teacher, the student will be able to build devices that will give him much satisfaction.

One major problem in building electronic projects is the difficulty in obtaining the proper parts. In some cases, substitutions may be made. Consultation with the teacher or a local parts distributor may prove helpful if a substitution must be made. For example, distributors usually have lists of transistor substitutions that will work in all but very critical applications.

Certain construction practices are very important when working with diodes and transistors. A heat sink should be used when soldering these components into a circuit. Good quality 60/40 solder and a low-wattage soldering pencil should be used when working with small components. When components are being soldered to a printed circuit, the foil should be cleaned with an ordinary pencil eraser. This will aid dramatically in producing a clean fast solder joint.

Semiconductor components may become damaged by heat during normal operation. To prevent this, the semiconductor component is usually mounted to a metal heat sink, which transfers the heat from the component to the air. When mounting such a device to a heat sink, it is usually necessary to isolate the component electrically from the heat sink.

When solid-state devices are purchased, the box containing them often includes a mica washer, nylon bushings, and fiber washers to isolate the component from its heat sink. Coating the mica washer on both sides with a *thin* layer of silicon grease will improve the heat transfer. Be sure to inspect the photographs in this project section carefully to see where heat sinks are used.

When beginners try their first electronic project it often refuses to work. They tend to hurry through the task, producing poor solder joints (or none at all), making wiring errors, burning the insulation off the wires, and making a general mess of things. It is important to *take the time necessary to be neat and careful*. When in doubt, the teacher should be consulted and earlier sections of this book dealing with the theory involved in the project should be reviewed. *A project should not be tried out until someone competent has checked the wiring*. Wiring errors often cause severe component damage and safety hazards. The important thing to remember is that anything worth doing is worth doing well.

LINE FILTER

Line filters are useful for eliminating radio-frequency (r-f) noise that interferes with radio and television reception. However, line filters will *not* eliminate all types of radio and television interference, because the noise often enters the set directly by radiation and not through the line. Because it is very difficult to predict just how the interference is getting into the set, a line filter must be tried to determine whether it will be effective in a specific case.

The filter can be connected between the radio or television and the outlet, or between the source of noise and the outlet. For example, if there is one electrical appliance that always seems to disturb reception, the line filter can be used between that appliance and the outlet. The motor-speed controller project (page 453) is the type of device that creates radio-frequency noise. If you build the speed controller, you should also consider building the line filter to use with it.

The line filter is actually a low-pass filter that transfers electrical energy at 60 hertz with little loss, but acts as a short circuit to electrical energy above 100 kilohertz. When wiring the line filter, connect the black wire to the gold screw in the plug, and be sure to connect the silver screw and white wire together. If your home is equipped with grounding-type outlets, also connect a green wire to a green terminal on the grounding plug. This green wire will run through the center of the circuit as shown on the schematic and printed circuit drawings. If a metal case is used, it should be grounded via this green wire.

CAUTION

Be very sure the green circuit is actually at ground potential. A neon test light should *not* indicate a trace of glow when connected across a good ground and the green wire. This ensures that the green circuit is not live as the result of some fault or mistake in the wiring of the home. Also check to see whether the neon test light glows when it is connected across the black and green wires. This shows that the green wire is a good ground.

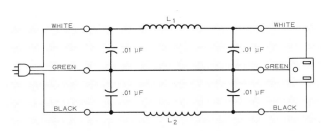

L_1 AND L_2 30 TURNS OF #16 MAGNET WIRE, 5/16" COIL DIAMETER
ALL CAPACITORS RATED AT 600 VDC

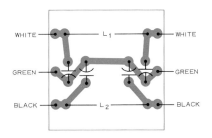

SAFETY FLASHER

Motorists and boating enthusiasts will find this safety flasher helpful to warn oncoming vehicles and summon assistance. It is designed to operate from a 12-volt battery and does not require its own energy source. It can be adapted for other uses by adding a 12-volt lantern battery.

The circuit is basically an astable multivibrator. Only one of the transistors (HEP-704) conducts the large current required by the lamp. Because this transistor becomes warm during extended operation, a heat sink should be provided. *Be sure to isolate the power transistor and lamp socket from the metal cabinet to prevent grounding it to the automobile body.*

The power connection may be made using alligator clips or by obtaining a plug to fit the lighter receptacle on the dashboard. A long power cord made from 18-gauge two-conductor wire will allow the flasher to be stationed at a distance from the disabled vehicle to provide ample warning to oncoming motorists.

The flasher uses two inexpensive automotive lenses to provide a warning signal in both directions. The cabinet can be metal with rubber feet or rubber suction cups attached to the bottom.

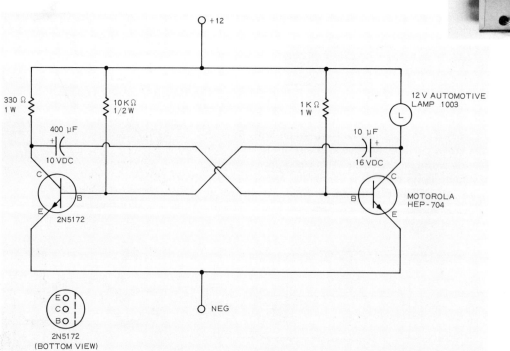

BATTERY CHARGER

The battery charger is a heavy-duty unit that provides a charging current of 10 amperes. Most modern automotive storage batteries have an ampere-hour rating of approximately 50. This charger will usually bring a completely dead battery up to full charge overnight.

The circuit uses a step-down transformer and bridge rectifier to provide 12 volts direct current. The transformer secondary is center-tapped; hence, the bridge rectifier could be connected across half the secondary to charge 6-volt batteries. A SPDT switch capable of carrying at least 10 amperes could be added to select 6 or 12 volts.

The transformer also has a tapped primary, which makes it convenient to control the charging rate. By switching in the correct tap, it is possible to obtain a low, medium, or high charging rate. The switch for this job carries only a few amperes but must be rated at 125 volts alternating current. When using the charger, always start with the rate switch in the *low* position. If the meter indicates less than 8 amperes and a faster charge is desired, try moving the switch to the *medium* position. If the meter moves beyond 10 amperes, move the rate switch back to *low*. The control may be moved from *medium* to *high* by following the same procedure.

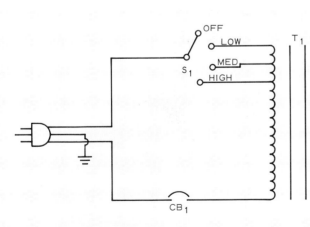

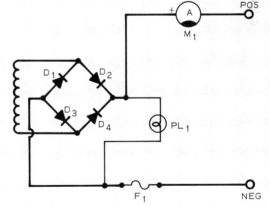

CB_1 - CIRCUIT BREAKER - 125 VAC - 1.5 AMPS
  (MALLORY CBB - 150)

D_1 THROUGH D_4 - SILICON DIODE - 15 AMPS AT 50 PIV
  MOTOROLA HEP - 151

F_1 - FUSE - 15 AMPS TYPE 3 AG

M_1 - METER - 0 TO 10 DC AMPS - EMICO RF 2 - 1/4 C

PL_1 - 12 V PILOT LAMP

S_1 - SINGLE POLE 4 POSITION - 125 VAC

T_1 - TRANSFORMER TAPPED PRI SEC 10, 11, 12, VCT
  AT 11 AMPS TRIAD F-29U

A fuse is used to protect the diodes in the bridge rectifier. A so-called "dead" battery usually has enough energy to destroy the diodes quickly if the wrong polarity is used when the unit is connected. Remember that the positive terminal of an automotive storage battery is slightly larger in diameter than the negative terminal. Whenever polarity markings cannot be found on the battery posts, check to see which post is larger.

Use #16 heavy-gauge wire in the heavy current parts of the circuit and for the output cables. The heavy-duty wiring is shown in bold type on the schematic.

CAUTION

Battery acid is highly corrosive and dangerous. Be very careful to keep it away from your skin and clothing. If you do come in contact with the acid, wash the area immediately with large quantities of cold water. *Never* cause sparks or flames near a charging battery because of the highly explosive gas that is produced. *Always* connect and disconnect the clips to the battery with the charger in the *off* position to avoid causing sparks.

TRANSISTOR TESTER

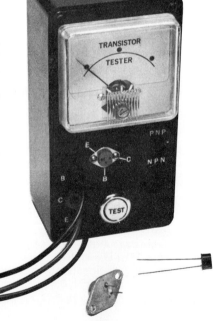

Transistors are generally much more reliable than vacuum tubes. They do occasionally become defective, however, and it is desirable to have some means of testing them. Most manufacturers of electronic equipment solder transistors into the circuit to save the cost of sockets and avoid the problems of intermittent connections. Removing transistors to test them is a poor procedure at best because heat damage may occur and much time is often wasted.

The transistor tester described here will test transistors both in circuit and out of circuit. It works by subjecting the transistor to a dynamic test to determine whether it is capable of amplifying. Occasionally, transistors are encountered that will pass such a test but are defective. When this is suspected, the transistor must be removed from the circuit and tested using a more elaborate instrument.

450

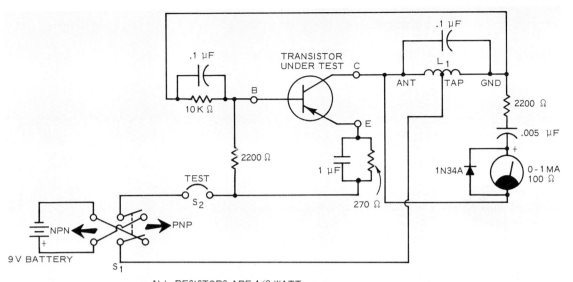

ALL RESISTORS ARE 1/2 WATT
L_1 SUPEREX VARI-LOOPSTICK VLT-950
S_1 DPDT SWITCH
S_2 MOMENTARY CONTACT PUSHBUTTON SWITCH

The circuit of the tester is a Hartley oscillator. The output from the oscillating transistor is rectified to drive a direct-current meter movement. When testing a transistor, an up-scale movement of the meter indicates that the transistor is capable of amplifying. How far the meter pointer moves up scale is not important. A defective transistor or an improperly connected transistor will produce no meter indication. Under some conditions, an improperly connected transistor will produce a slight reading on the meter. When a very small meter indication is obtained, the connections should be checked.

It is always good practice to know what type of transistor (p-n-p or n-p-n) is being tested. Also, you should know which lead is the base, which is the collector, and which is the emitter. However, this information is not always available and the tester can be used to identify unknown transistors according to polarity and lead configuration. A trial-and-error process is employed until a meter indication is obtained. At that point, the polarity switch will indicate p-n-p or n-p-n and the leads will determine emitter, base, and collector.

This tester is designed for checking bipolar transistors, field-effect transistors, and unijunction transistors. Other types of transistors will have to be tested using other equipment. Another limitation is sometimes encountered when testing in-circuit transistors that are heavily swamped by other devices. An example of this can be found in the power supply project (see page 457), where Q_2 is directly connected to Q_3 in what is known as the *Darlington connection*. Transistors connected in the Darlington configuration must be tested together as a single unit. Use the base lead of Q_2, the common collector point, and the emitter of Q_3. This will produce a simultaneous test on both transistors. Experience with the tester will suggest ways to overcome its limitations and properly interpret the results.

PHOTOELECTRIC CONTROLLER

Controls of this type are generally used to automatically turn on lights at night and then back off again the next morning. The project accomplishes this by using a cadmium-sulphide cell. This type of cell is light-sensitive and exhibits a marked difference in conductivity when exposed to light.

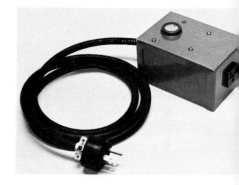

The light-sensitive cell is connected in series with the relay coil. When light is entering the cell, its resistance is low and sufficient current flows to hold the relay armature down. When night approaches, the cell resistance goes up and the relay current goes down. This will allow the spring to overcome the magnetic pull and the relay armature moves up, closing the normally closed (NC) contacts. The relay contacts then energize the lights.

The two resistors form a voltage-divider network and one of them is adjustable. This adjustment should be made carefully, using an insulated tool. Block the light entering the cell with your hand to simulate night and remove your hand to simulate day. Adjust the variable resistor to provide sure and positive operation of the relay under these two conditions. The output of the variable resistor is rectified and filtered because the relay coil requires direct current.

Follow good wiring practices when connecting the plug and receptacle. The black wire contacts the gold-colored screw, the white wire contacts the silver screw, and the green wire contacts the green screw.

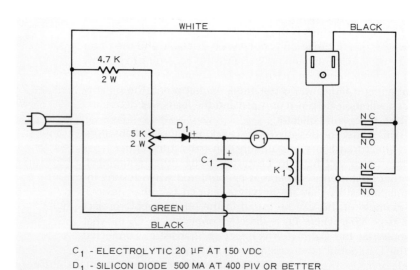

C_1 - ELECTROLYTIC 20 µF AT 150 VDC
D_1 - SILICON DIODE 500 MA AT 400 PIV OR BETTER
K_1 - RELAY POTTER AND BRUMFIELD LM11
       5,000 OHM COIL 5 AMP DPDT CONTACTS
P_1 - PHOTOCELL RCA TYPE 7163

MOTOR-SPEED CONTROLLER

This device will provide reliable speed control for power tools and appliances. Electric drills, blenders, and other devices offer much more versatility when speed control is available.

CAUTION

This speed control can be used only with universal-type motors. Be sure the motor has brushes and a commutator.

The heart of the speed control is the silicon controlled rectifier (SCR). It will conduct from cathode to anode when the proper signal is applied to the gate terminal. Like any rectifier, it will not conduct from anode to cathode.

Diode D_1 and the associated resistors form a variable direct-current power supply. As the slider on the 500-ohm potentiometer is moved in the direction of the arrow, more direct-current voltage is applied to the gate of the silicon controlled rectifier. The effect of this increased gate voltage is to enable the rectifier to turn on much earlier during the positive alternation. The earlier the moment of turn-on, the more energy will be delivered to the motor, which will run faster.

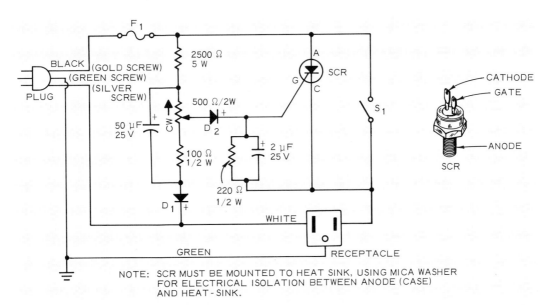

NOTE: SCR MUST BE MOUNTED TO HEAT SINK, USING MICA WASHER FOR ELECTRICAL ISOLATION BETWEEN ANODE (CASE) AND HEAT-SINK.

D_1-D_2 SILICON RECTIFIER, INTERNATIONAL RECTIFIER 8D-4
F_1 - 5 AMPERE FUSE OR CIRCUIT BREAKER
S_1 SPST SWITCH 5 AMPS @ 120 VAC (CLOSE FOR HIGH SPEED)
SCR SILICON CONTROLLED RECTIFIER, GENERAL ELECTRIC C20-B (C30-B FOR HEAVY-DUTY WORKSHOP USE.)

Because the silicon controlled rectifier can never turn on during the negative alternation, maximum speed will be approximately half the normal speed without the controller in the circuit. Switch S_1 is therefore provided to bypass the rectifier for full-speed operation.

The motor-speed controller also provides speed regulation. When the work load on the motor increases, this will tend to slow the motor down. As the motor speed drops, the induced electromotive force across the motor windings will decrease. The induced electromotive force is fed back to the gate circuit. The polarities of applied gate voltage and induced electromotive force are arranged to oppose one another. The result is that when the feedback drops, the applied gate voltage is able to turn on the silicon controlled rectifier sooner, which tends to make the motor run faster. The overall effect is that the motor speed tends to remain constant.

When using the speed controller, be very careful not to allow the motor to overheat. Most universal-type motors generate a lot of heat, and they are provided with built-in cooling fans to overcome this problem. When operating a motor at reduced speed, the fan does not work efficiently. Occasionally running the tool at full speed with no load is recommended to allow the cooling fan to quickly lower the temperature.

INTERCOM

This project uses a minimum number of parts. Nevertheless, it offers top performance because the integrated circuit amplifier exhibits all the necessary gain and audio power in one small container. The intercom uses battery power but can be converted to line operation if extended operation is contemplated.

The circuit is a basic audio amplifier with a few minor details added to make it useful for intercom operation. The master and remote speakers can be electrically interchanged by the DPDT talk-listen switch. The master station houses the switch and other circuitry, while the remote station contains only a speaker.

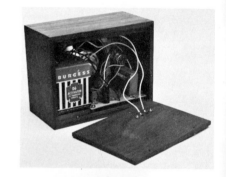

The loudspeakers operate as dynamic microphones. The low impedance of a speaker-microphone must be transformed to a high impedance to match the amplifier input. This impedance transformation is performed by transformer T_1. This transformer can be practically any audio output transformer from a junk vacuum tube-type radio or television. The leads that formerly connected the transformer to the speaker will again be connected to the speaker in the intercom circuit.

Transformer T_2 matches the output impedance of the amplifier to the speakers. This application is a bit more critical and the exact part or replacement should be obtained.

A 5,000-ohm potentiometer serves as a volume control. The sensitivity and volume of this intercom will be adequate for almost any application. Nearly any length of wire can be used to provide communication between widely separated rooms in the home or school. A more mellow tone can be obtained from the circuit by connecting a 0.01-microfarad capacitor across the primary of T_2.

Printed circuit wiring is advisable for this type of project. The RCA CA3020 integrated circuit exhibits a very high gain, and oscillations are likely to result if the wiring is needlessly long. Lead lengths are held to an absolute minimum with printed circuitry, and unwanted feedback paths can be more easily controlled.

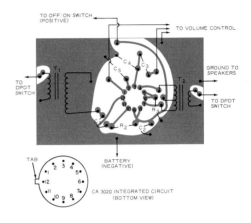

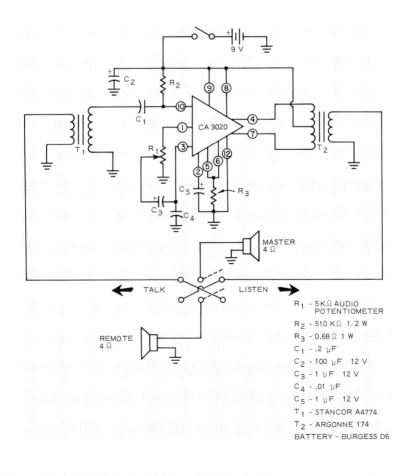

R_1 - 5 KΩ AUDIO POTENTIOMETER
R_2 - 510 KΩ 1/2 W
R_3 - 0.68 Ω 1 W
C_1 - .2 μF
C_2 - 100 μF 12 V
C_3 - 1 μF 12 V
C_4 - .01 μF
C_5 - 1 μF 12 V
T_1 - STANCOR A4774
T_2 - ARGONNE 174
BATTERY - BURGESS D6

TRANSISTOR POWER SUPPLY

A good power supply is essential for experimenting with electronic devices and circuits. Because most modern devices use transistors, a modern power supply should deliver low voltage at reasonable current levels and exhibit excellent ripple suppression and voltage regulation. The power supply described here meets these requirements.

This power supply will deliver from 0 to 12 volts at a current of 500 milliamperes, with transistor 2N3054 (Q_3) acting as a series regulator. If the output voltage tends to drop because of an increase in load current or because of a fluctuation in line, the 2N3054 transistor will produce more power to maintain a constant output voltage.

To visualize the regulating action, imagine that the output voltage suddenly drops. This will cause the voltage at the arm of R_7 to become less positive because R_7 is part of a divider network across the output of the supply. Thus, the base of Q_1 will become less positive. Transistor Q_1 will not conduct as much current from emitter to collector because some of its forward bias has been removed. Then the collector voltage at Q_1 will increase because the current through the 10K load resistor has dropped. This action makes Q_2 work harder, and Q_3 will also work harder because of the direct (Darlington) connection be-

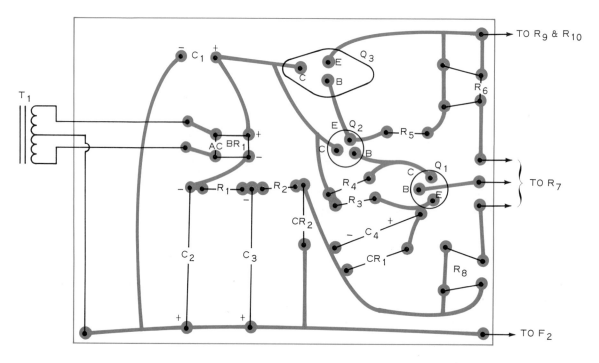

tween them. Finally, when Q_3 is conducting better, it will drop less voltage and will raise the output voltage to its original value. This action is nearly instantaneous and the output voltage seems to remain constant.

Resistor R_6 is adjusted for zero output when R_7 is turned all the way down, and R_8 is adjusted for 12 volts output when R_7 is turned all the way up. Interaction may require these adjustments to be repeated.

The meter function switch enables one meter movement to measure both voltage and current. This is desirable because meters are expensive. The 0.2-ohm R_{10} shunt will produce a range of 0 to 500 milliamperes, and the 15,000-ohm multiplier R_9 will produce a voltage range of 0 to 15 volts.

One drawback to the series regulator type of supply is that a short circuit at the output will usually destroy the regulator transistor. Thus, short circuits must be avoided when using this supply. The ½-ampere fuse in the output lead will provide some protection, but fuses often do not act quickly enough to save solid-state devices. Therefore, it is important to take the precaution of slowly advancing the voltage control R_7 from 0 while monitoring the current. A short circuit will be detected by the rapidly rising current and damage to the supply will be avoided. Current-limiting transistors are sometimes added to this type of power supply, but additional complexity and expense are incurred.

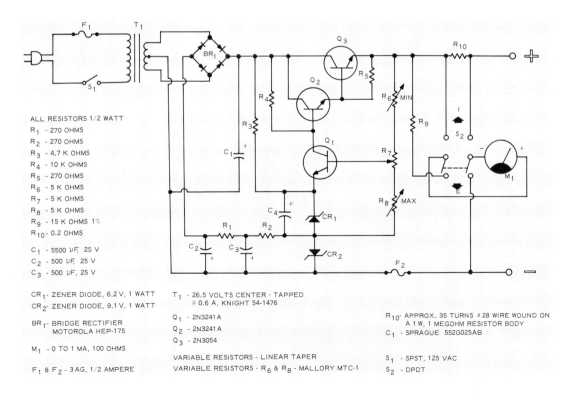

ALL RESISTORS 1/2 WATT
R_1 - 270 OHMS
R_2 - 270 OHMS
R_3 - 4.7 K OHMS
R_4 - 10 K OHMS
R_5 - 270 OHMS
R_6 - 5 K OHMS
R_7 - 5 K OHMS
R_8 - 5 K OHMS
R_9 - 15 K OHMS 1%
R_{10} - 0.2 OHMS

C_1 - 5500 µF, 25 V
C_2 - 500 µF, 25 V
C_3 - 500 µF, 25 V

CR_1 - ZENER DIODE, 6.2 V, 1 WATT
CR_2 - ZENER DIODE, 9.1 V, 1 WATT

BR_1 - BRIDGE RECTIFIER
    MOTOROLA HEP-175

M_1 - 0 TO 1 MA, 100 OHMS

F_1 & F_2 - 3 AG, 1/2 AMPERE

T_1 - 26.5 VOLTS CENTER - TAPPED
    @ 0.6 A, KNIGHT 54-1476

Q_1 - 2N3241A
Q_2 - 2N3241A
Q_3 - 2N3054

VARIABLE RESISTORS - LINEAR TAPER
VARIABLE RESISTORS - R_6 & R_8 - MALLORY MTC-1

R_{10} - APPROX. 35 TURNS #28 WIRE WOUND ON
    A 1 W, 1 MEGOHM RESISTOR BODY
C_1 - SPRAGUE 552G025AB

S_1 - SPST, 125 VAC
S_2 - DPDT

PORTABLE PUBLIC ADDRESS SYSTEM

This project represents one of the many applications of linear integrated circuits. The electrical efficiency and small size of integrated circuits makes them especially suitable for portable and battery-operated equipment.

The energy source for the amplifier consists of three 12-volt lantern batteries connected in series. Smaller batteries would not be appropriate because the current drain is high at large volume levels. Although this is not shown in the schematic diagram, it is always good practice to bypass the battery in such equipment with a large electrolytic filter capacitor. The bypass capacitor should be connected *after* the off-on switch to prevent discharging the batteries when the unit is not in use. A 1,500-microfarad electrolytic capacitor rated at 40 volts will do an effective job.

A more mellow and pleasing tone can be obtained by connecting a 0.01-microfarad ceramic-disk capacitor across the primary of the driver transformer. This capacitor will also limit the high-frequency response of the circuit.

A good strong briefcase or other enclosure should be used to house the amplifier because the three lantern batteries make the amplifier rather heavy. Also, at high volume, the speaker will tend to shake apart a flimsy enclosure.

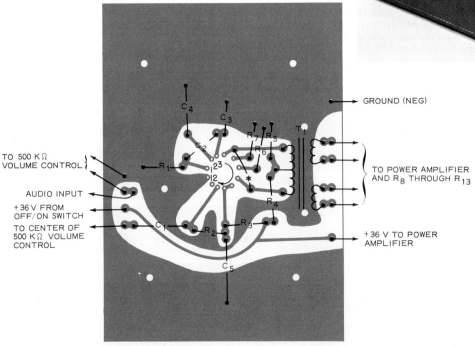

*OPTIONAL TONE CAPACITOR SEE TEXT

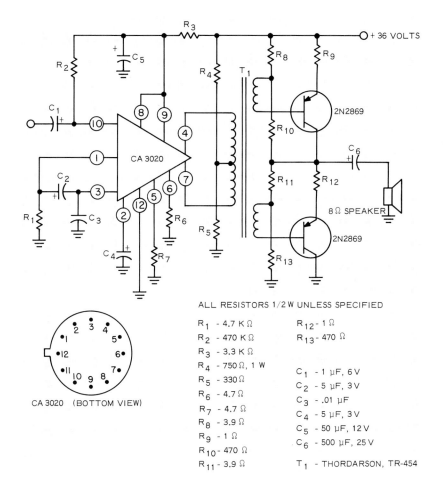

ALL RESISTORS 1/2 W UNLESS SPECIFIED

R_1 - 4.7 K Ω
R_2 - 470 K Ω
R_3 - 3.3 K Ω
R_4 - 750 Ω, 1 W
R_5 - 330 Ω
R_6 - 4.7 Ω
R_7 - 4.7 Ω
R_8 - 3.9 Ω
R_9 - 1 Ω
R_{10} - 470 Ω
R_{11} - 3.9 Ω
R_{12} - 1 Ω
R_{13} - 470 Ω

C_1 - 1 µF, 6 V
C_2 - 5 µF, 3 V
C_3 - .01 µF
C_4 - 5 µF, 3 V
C_5 - 50 µF, 12 V
C_6 - 500 µF, 25 V

T_1 - THORDARSON, TR-454

While practically any audio source will drive the amplifier, crystal or ceramic microphones are recommended for public address applications. A dynamic microphone can be used by connecting an impedance matching transformer at the input to the amplifier. Consult the intercom project to see an example of such an input transformer. In any case, the microphone cable should be long and shielded to prevent acoustical feedback. Battery-powered radios and record players may also be used with this amplifier to provide music and entertainment at picnics.

It is a good idea to use printed circuit wiring for the driver section of the amplifier. The RCA CA3020 integrated circuit exhibits very high gain, and instability is likely to result if the wiring is needlessly long.

This amplifier is capable of delivering at least 10 watts of audio power to the loudspeaker. This will prove to be enough volume for most applications. Do not try to use a small, inexpensive speaker in this project because it will not last very long. The speaker must be rated at 10 watts or more.

ENGINE ANALYZER

This instrument will be valuable to those who enjoy doing automotive maintenance and repair. It measures volts, dwell, and revolutions per minute (rpm), and provides a bright strobe light for observing and adjusting ignition timing.

The strobe light utilizes the charge on a 2-microfarad, 600-volt capacitor to produce an intense blue-white flash. A high-tension wire running to spark plug 1 provides the high-voltage pulse needed for initial ionization of the gas in the flash tube. After ionization is attained, the capacitor discharges through the tube. A simple filament transformer oscillator steps up the 12 volts from the automotive electrical system and provides the capacitor with over 400 volts direct current after rectification.

CAUTION

A charged capacitor can be dangerous. Do not come in contact with this circuit. If the oscillator circuit fails to start, try reversing the two transformer primary leads.

When switch S_2 is in the *volts* position, a 15-kilohm multiplier resistor is switched in series with the negative lead, making the meter read from 0 to 15 volts. This function is valuable for observing and adjusting the charging system of an automobile.

With S_2 in the rpm position, the signal from the distributor is amplified by the 2N109 transistor and coupled to the meter circuit through the 0.5-microfarad capacitor. The meter is calibrated from 0 to 1,500 rpm. This allows the same scale on the meter to be used for reading voltage and rpm. To calibrate the rpm function, a good-quality tachometer should be borrowed.

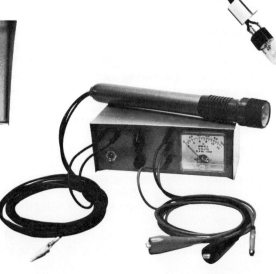

When S_2 is in the *dwell* position, the meter becomes an ohmmeter and the electrical system of the automobile serves as the energy source. The emitter-base junction of the 2N109 transistor serves as a voltage regulator to prevent voltage changes in the automobile from upsetting the accuracy of the reading.

To use the instrument to measure dwell, adjust the dwell calibrate control for a full-scale reading when the points are closed. Then start the engine, and the meter will indicate the percentage of dwell. To convert to dwell angle in degrees, use this formula:

Dwell angle, in degrees = percentage of dwell $\times \dfrac{360}{\text{number of cylinders}}$

For example, if the meter was being used on an 8-cylinder automobile and a reading of 60 percent was obtained, the dwell angle would be 60 percent, or $0.60 \times 360/8 = 27$ degrees.

If the dwell function is used on a positive-ground system, the meter will read 0 when the points are closed and 100 when the points are open. With the distributor points open, the calibrate control should be set for a meter reading of 100. Then the engine should be started and the reading on the meter noted. *The meter reading must now be subtracted from 100 to obtain the correct percentage of dwell.* Finally, the formula is used as before to convert the reading to dwell angle in degrees.

For example, if a positive-ground 6-cyclinder automobile is under test and a meter reading of 40 percent is obtained, the dwell angle may be found by subtracting 40 from 100 to get 60 (the true percentage of dwell). Hence, Dwell angle = $0.60 \times 360/6 = 36$ degrees.

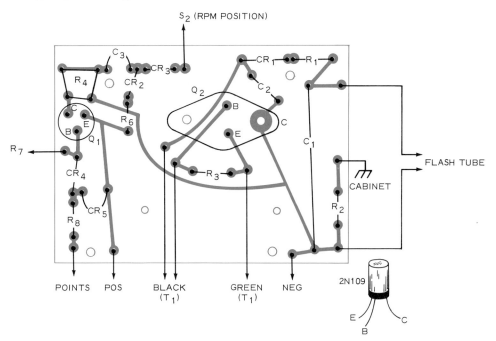

The engine analyzer is very flexible; it will make measurements on both positive- and negative-ground cars. The car battery can be either 6 or 12 volts. Four, six, or eight cylinders may be accommodated. However, two important items must be pointed out: the tachometer calibration will hold for one number of cylinders only; and the timing light will be less bright when used on 6-volt systems.

CAUTION

When using an instrument like this, *great care must be taken to keep hands away from the fan and associated belts and pulleys near the front of the engine. Caution must also be observed about the wires and the timing light itself as these could become tangled in the fan or belts and cause serious damage or injury.*

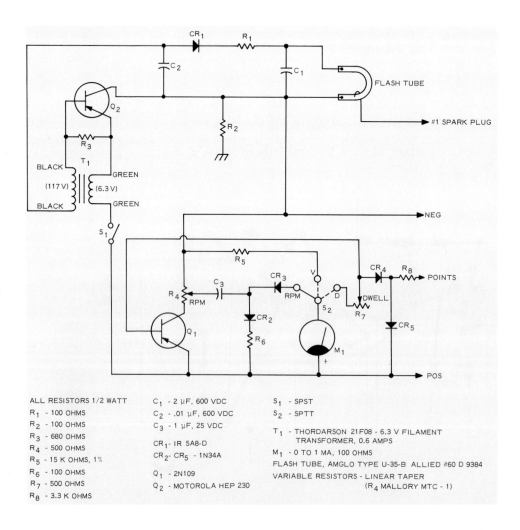

ALL RESISTORS 1/2 WATT
R_1 - 100 OHMS
R_2 - 100 OHMS
R_3 - 680 OHMS
R_4 - 500 OHMS
R_5 - 15 K OHMS, 1%
R_6 - 100 OHMS
R_7 - 500 OHMS
R_8 - 3.3 K OHMS

C_1 - 2 µF, 600 VDC
C_2 - .01 µF, 600 VDC
C_3 - 1 µF, 25 VDC

CR_1 - IR 5A8-D
CR_2 - CR_5 - 1N34A

Q_1 - 2N109
Q_2 - MOTOROLA HEP 230

S_1 - SPST
S_2 - SPTT

T_1 - THORDARSON 21F08 - 6.3 V FILAMENT TRANSFORMER, 0.6 AMPS
M_1 - 0 TO 1 MA, 100 OHMS
FLASH TUBE, AMGLO TYPE U-35-B ALLIED #60 D 9384
VARIABLE RESISTORS - LINEAR TAPER
(R_4 MALLORY MTC - 1)

APPENDICES

APPENDIX A. TRIGONOMETRIC FUNCTIONS

Trigonometry is the study of the relationships between the angles and the sides of right triangles. These relationships are very important in the study of electricity and electronics, particularly in the analysis of alternating-current circuits.

The *relationships* between the sides of a right triangle are referred to as *functions* and are expressed as *ratios between the sides*. The most commonly used trigonometric functions are known as the *sine* (sin), the *cosine* (cos), and the *tangent* (tan). The angle between two sides of a right triangle is then expressed in terms of a function as shown in Fig. A-1. With the mathematical equivalent of a function known, the angle can be determined by referring to a table of natural trigonometric functions (Table A-1). Conversely, with an angle and one side of a right triangle known, the remaining side can be determined by solving for the side in question. As an example, in Fig. A-1,

$$a = c(\sin \text{ of angle } A)$$

$$c = \frac{a}{\sin \text{ of angle } A}$$

$$b = c(\cos \text{ of angle } A)$$

Fig. A-1. Basic trigonometric functions.

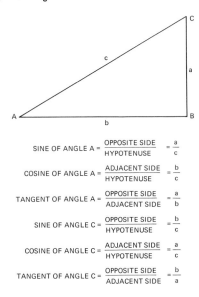

Table A-1. Natural trigonometric functions.

Angle	sin	cos	tan	Angle	sin	cos	tan
0°	0.0000	1.000	0.0000	45°	0.7071	0.7071	1.0000
1	.0175	.9998	.0175	46	.7193	.6947	1.0355
2	.0349	.9994	.0349	47	.7314	.6820	1.0724
3	.0523	.9986	.0524	48	.7431	.6691	1.1106
4	.0698	.9976	.0699	49	.7547	.6561	1.1504
5	.0872	.9962	.0875	50	.7660	.6428	1.1918
6	.1045	.9945	.1051	51	.7771	.6293	1.2349
7	.1219	.9925	.1228	52	.7880	.6157	1.2799
8	.1392	.9903	.1405	53	.7986	.6018	1.3270
9	.1564	.9877	.1584	54	.8090	.5878	1.3764
10	.1736	.9848	.1763	55	.8192	.5736	1.4281
11	.1908	.9816	.1944	56	.8290	.5592	1.4826
12	.2079	.9781	.2126	57	.8387	.5446	1.5399
13	.2250	.9744	.2309	58	.8480	.5299	1.6003
14	.2419	.9703	.2493	59	.8572	.5150	1.6643
15	.2588	.9659	.2679	60	.8660	.5000	1.7321
16	.2756	.9613	.2867	61	.8746	.4848	1.8040
17	.2924	.9563	.3057	62	.8829	.4695	1.8807
18	.3090	.9511	.3249	63	.8910	.4540	1.9626
19	.3256	.9455	.3443	64	.8988	.4384	2.0503
20	.3420	.9397	.3640	65	.9063	.4226	2.1445
21	.3584	.9336	.3839	66	.9135	.4067	2.2460
22	.3746	.9272	.4040	67	.9205	.3907	2.3559
23	.3907	.9205	.4245	68	.9272	.3746	2.4751
24	.4067	.9135	.4452	69	.9336	.3584	2.6051
25	.4226	.9063	.4663	70	.9397	.3420	2.7475
26	.4384	.8988	.4877	71	.9455	.3256	2.9042
27	.4540	.8910	.5095	72	.9511	.3090	3.0777
28	.4695	.8829	.5317	73	.9563	.2924	3.2709
29	.4848	.8746	.5543	74	.9613	.2756	3.4874
30	.5000	.8660	.5774	75	.9659	.2588	3.7321
31	.5150	.8572	.6009	76	.9703	.2419	4.0108
32	.5299	.8480	.6249	77	.9744	.2250	4.3315
33	.5446	.8387	.6494	78	.9781	.2079	4.7046
34	.5592	.8290	.6745	79	.9816	.1908	5.1446
35	.5736	.8192	.7002	80	.9848	.1736	5.6713
36	.5878	.8090	.7265	81	.9877	.1564	6.3138
37	.6018	.7986	.7536	82	.9903	.1392	7.1154
38	.6157	.7880	.7813	83	.9925	.1219	8.1443
39	.6293	.7771	.8098	84	.9945	.1045	9.5144
40	.6428	.7660	.8391	85	.9962	.0872	11.43
41	.6561	.7547	.8693	86	.9976	.0698	14.30
42	.6691	.7431	.9004	87	.9986	.0523	19.08
43	.6820	.7314	.9325	88	.9994	.0349	28.64
44	.6947	.7193	.9657	89	.9998	.0175	57.29
				90	1.0000	.0000	

APPENDIX B. LOGARITHMS

The *logarithm* (log) of a given number is the *power* or *exponent* to which another number (the *base*) must be raised to equal the given number. Common logarithms use the number 10 as the base, and the logarithm of a number to the base 10 is sometimes written as $\log_{10}$. However, the logarithm of any number is always assumed to be to the base 10 unless some other base is specifically indicated. Therefore, since

$$1 = 10^0 \qquad \log 1 = 0$$
$$10 = 10^1 \qquad \log 10 = 1$$
$$100 = 10^2 \qquad \log 100 = 2$$
$$1{,}000 = 10^3 \qquad \log 1{,}000 = 3$$

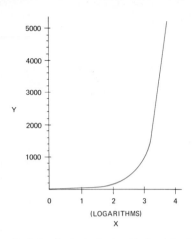

Fig. B-1. Graph showing logarithmic relationships.

Table B-1. Common logarithms of numbers 10 through 25.

N	0	1	2	3	4	5	6	7	8	9	N
10	0000	0043	0086	0128	0170	0212	0253	0294	0334	0374	10
11	0414	0453	0492	0531	0569	0607	0645	0682	0719	0755	11
12	0792	0828	0864	0899	0934	0969	1004	1038	1072	1106	12
13	1139	1173	1206	1239	1271	1303	1335	1367	1399	1430	13
14	1461	1492	1523	1553	1584	1614	1644	1673	1703	1732	14
15	1761	1790	1818	1847	1875	1903	1931	1959	1987	2014	15
16	2041	2068	2095	2122	2148	2175	2201	2227	2253	2279	16
17	2304	2330	2355	2380	2405	2430	2455	2480	2504	2529	17
18	2553	2577	2601	2625	2648	2672	2695	2718	2742	2765	18
19	2788	2810	2833	2856	2878	2900	2923	2945	2967	2989	19
20	3010	3032	3054	3075	3096	3118	3139	3160	3181	3201	20
21	3222	3243	3263	3284	3304	3324	3345	3365	3385	3404	21
22	3424	3444	3464	3483	3502	3522	3541	3560	3579	3598	22
23	3617	3636	3655	3674	3692	3711	3729	3747	3766	3784	23
24	3802	3820	3838	3856	3874	3892	3909	3927	3945	3962	24
25	3979	3997	4014	4031	4048	4065	4082	4099	4116	4133	25

Logarithms of numbers are divided into two parts: the *integral* or *whole number* part, known as the *characteristic,* and the *decimal* part, known as the *mantissa*.

The characteristic of the logarithm of any number greater than 1 is positive and is always 1 fewer than the number of digits to the left of the decimal point in the number. The mantissa of a logarithm is determined by referring to a table of common logarithms (Table B-1). For example:

$$\log 11 = 1.0414$$
$$\log 13.5 = 1.1303$$
$$\log 20.9 = 1.3201$$
$$\log 25 = 1.3979$$

Antilogarithms. The *antilogarithm* (*antilog*) of a given logarithm is the number from which the logarithm was derived. The process of finding an antilogarithm is essentially the reverse of that used to determine a logarithm.

Uses of Logarithms. Logarithms are used to simplify the operations of multiplication, division, raising to a power, and the extraction of square root. With logarithms, complicated problems involving these operations are reduced to problems requiring only simple addition, subtraction, multiplication, or division. Logarithmic terms are also found in some formulas. The formula used to compute the power output of microphones is an example of such a formula.

Logarithmic Relationships. Fig. B-1 is an example of the relationship between certain numbers and their logarithms. This graph shows that small increases along the X-axis of the graph result in larger and larger increases in the values of the numbers along the Y-axis. Therefore, the curve of the graph indicates a logarithmic relationship between the numerical values along the axes.

Logarithmic Taper. A logarithmic relationship between the amount of shaft rotation and the resulting variation in resistance is commonly used in the construction of potentiometers that are found in many audio circuits. These potentiometers are referred to as having a *logarithmic* (nonlinear) *taper*.

For example, assume that the X-axis of the graph of Fig. B-1 represents increments of shaft rotation. The Y-axis of the graph would then represent the variation in resistance that occurs as the result of a shaft adjustment. Thus, a given amount of rotation from one position of the shaft to another results in a much larger (or smaller) variation in resistance than is produced by a similar amount of rotation from a different position of the shaft.

APPENDIX C.   THE SLIDE RULE

The slide rule is an extremely versatile device that can be used to perform a number of mathematical computations with an accuracy that is sufficient to solve most practical problems (Fig. C-1). This appendix discusses the use of the slide rule in performing the operations of multiplication, division, squaring, and finding the square root. To perform these operations, the A, B, C, and D scales of the slide rule are used.

Principle of Operation. The slide rule operates on the principle of the logarithm. Segments on the scales of the slide rule represent logarithms but they are identified by the antilogarithms corresponding to these logarithms. For example, when the slide rule is used to multiply, two line segments represented by given lengths of the C and D scales are added. The sum of these line seg-

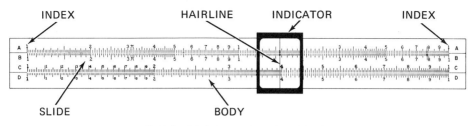

Fig. C-1. Principal parts of a typical slide rule.

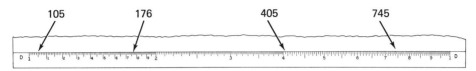

Fig. C-2. Examples of numbers read from the D scale of a slide rule.

ments is then represented by the antilogarithm of the sum of the two logarithms. According to a rule of logarithms, this sum is equal to the logarithm of the product.

Reading the Scales. Efficient operation of the slide rule requires the ability to read the scales quickly and accurately. Although the scales appear rather complicated, with practice, they can be mastered in a short time. As an aid in reading the scales, refer to Fig. C-2.

Placing the Decimal Point. Locating or placing the decimal point in the "answer" of a slide rule computation is one of the most difficult procedures of slide rule operation. To do this, a method of scientific notation that is presented in typical slide rule manuals is often used. However, the decimal point can also be accurately located in the answer by means of several rules that are not difficult to memorize if sufficient practice is done. This method is desirable when the mathematical background of the slide rule operator does not provide a thorough knowledge of exponents and scientific notation.

To place the decimal point by means of the "rules" method, it is necessary to assign what may be referred to as *point number* (PN) values to numbers. The PN value is dependent upon the location of the decimal point in a sequence of digits, as indicated by the following examples:

Number	PN Value
5	1 (odd)
37.02	2 (even)
469	3 (odd)
0.234	0 (even)
0.09	−1 (odd)
0.003	−2 (even)
0.000756	−3 (odd)

After the slide rule has been adjusted to "read" a given answer number, the decimal point in the answer must be located in accordance with the associated rule (rules are indicated below). For example, assume that the number read upon the scale of a slide rule is 647. Also assume that the PN value of the answer according to associated rule is 1. In this case, the decimal point in the number (647) must be moved two places to the left so that the answer is 6.47, a number which has a PN value of 1.

USING THE SLIDE RULE

The procedures for performing the operations of multiplication, division, squaring, and finding the square root are as follows:

To Multiply:
1. Place an index of the C scale over the multiplicand found on the D scale.
2. Place the hairline over the multiplier found on the C scale.
3. Read the answer (product) under the hairline on the D scale.

Rules for Placing the Decimal Point:
1. If the slide extends to the left, add the PN values of the numbers being multiplied. This sum will indicate the PN value of the product.
2. If the slide extends to the right, add the PN values of the numbers being multiplied and subtract 1 from this sum. The resulting difference will indicate the PN value of the product.

To Divide:
1. Place the hairline over the dividend found on the D scale.
2. Adjust the slide so that the divisor, found on the C scale, is directly below the hairline.
3. Read the answer (quotient) on the D scale directly below the index of the C scale.

Rules for Placing the Decimal Point:
1. If the slide extends to the left, subtract the PN value of the divisor from the PN value of the dividend. The resulting difference will indicate the PN value of the quotient.
2. If the slide extends to the right, subtract the PN value of the divisor from the PN value of the dividend and add 1. This sum will indicate the PN value of the quotient.

To Square a Number:
1. Place the hairline over the number found on the D scale.
2. Read the answer (square) under the hairline on the A scale.

Rules for Placing the Decimal Point:
1. If the square of a number is found in the right-hand section of the A scale, the PN value of the square is twice the PN value of the number being squared.
2. If the square of a number is found in the left-hand section of the A scale, the PN value of the square is twice the PN value of the number being squared minus 1.

To Find the Square Root of a Number:
1. If the PN value of the number is even, place the hairline over the number found in the right-hand section of the A scale. Read the answer (root) directly below the hairline on the D scale. Note that a PN value of zero is considered to be even.
2. If the PN value of the number is odd, place the hairline over the number found in the left-hand section of the A scale. Read the answer (root) directly below the hairline on the D scale.

Rules for Placing the Decimal Point:
1. If the PN value of the number is even, divide this PN value by 2. The resulting quotient will indicate the PN value of the root.
2. If the PN value of the number is odd, add 1 to the PN value and then divide by 2. The resulting quotient will indicate the PN value of the root.

APPENDIX D. PLANE VECTORS

A vector is a line segment used to represent a quantity that has both magnitude and direction (Fig. D-1A). The word "direction" is used here in a broad sense, since a vector can also be used to show the effect of one quantity upon a given condition as compared with the effect of another quantity upon the same condition.

Use of Vectors. In work dealing with electricity and electronics, vectors are often used to represent circuit conditions, including impedance and phase relationships. An example of a vector notation (a vector diagram) showing the relationship between the resistance R and the inductive reactance X_L in an alternating-current circuit is shown in Fig. D-1B. In this diagram, the R and X_L vectors are drawn at right angles to each other. The reason for this is that resistance alone has no effect upon the phase relationship between current and voltage, while inductive reactance alone causes a phase shift of 90 degrees between current and voltage.

Addition of Vectors. To use the diagram for finding the impedance of the circuit, the vectors must be added. This can be done by the *parallelogram* method. To use this method, a rectangle is drawn, using the equivalent of the vectors as the sides (Fig. D-1C). The diagonal of the rectangle (Z vector) then represents the magnitude and direction of the circuit impedance. The numerical value of this vector is found by using the *Pythagorean theorem*:

$$Z = \sqrt{R^2 + X_L^2} = 5 \text{ ohms}$$

The impedance Z vector (Fig. D-1C) is sometimes referred to as the *resultant*. It represents the single "force" that has the same effect upon the current of the circuit as the combined effects of the resistance and the inductive reactance.

Fig. D-1. Plane vectors: (A) vector line segment, (B) vector diagram, (C) addition of vectors.

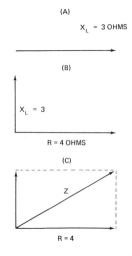

APPENDIX E. INTERNATIONAL SYSTEM OF UNITS

The International System of Units was officially adopted by the U.S. National Bureau of Standards

(NBS) in 1964. These units, their numerical equivalents, and their symbols are as follows:

Unit Prefix	Numerical Equivalent	Symbol
tera	10^{12}	T
giga	10^{9}	G
mega	10^{6}	M
kilo	10^{3}	k
hecto	10^{2}	h
deka	10	da
deci	10^{-1}	d
centi	10^{-2}	c
milli	10^{-3}	m
micro	10^{-6}	μ
nano	10^{-9}	n
pico	10^{-12}	p
femto	10^{-15}	f
atto	10^{-18}	a

APPENDIX F. THE GREEK ALPHABET

Letters of the Greek alphabet are often used as symbols to designate various electrical, magnetic, and mathematical quantities or conditions. The most commonly used letters of this alphabet and the quantity or condition they designate are as follows:

Name	Capital	Lower Case	Commonly Used to Designate
Alpha	A	α	Angles. Area. Coefficients
Beta	B	β	Angles. Flux density. Coefficients
Gamma	Γ	γ	Conductivity. Specific gravity
Delta	Δ	δ	Variation. Density
Epsilon	E	ϵ	Base of natural logarithms
Zeta	Z	ζ	Impedance. Coefficients. Coordinates
Eta	H	η	Hysteresis coefficient. Efficiency
Theta	Θ	θ	Temperature. Phase angle
Iota	I	ι	Current
Kappa	K	κ	Dielectric constant.
Lambda	Λ	λ	Wavelength
Mu	M	μ	Micro. Amplification factor. Permeability
Pi	Π	π	Ratio of circumference to diameter = 3.1416
Rho	P	ρ	Resistivity
Tau	T	τ	Time constant. Time phase displacement
Phi	Φ	ϕ	Magnetic flux. Angles
Psi	Ψ	ψ	Dielectric flux. Phase difference
Omega	Ω	ω	Capital, ohms. Lower case, angular velocity

APPENDIX G. N-P-N AND P-N-P TRANSISTOR CIRCUIT SYMBOLS

C_c	Collector-to-case capacitance
f_c	Cutoff frequency
G_{PE}	Large signal average power gain for common-emitter circuit
G_{pe}	Small signal average power gain for common-emitter circuit
I_B	Base current
I_C	Collector current
I_{CB}	Collector cutoff current
I_{CBO}	Collector cutoff current with emitter open
I_{CEO}	Collector cutoff current with base open
I_E	Emitter current
I_{EBO}	Emitter cutoff current with collector open
MAG	Maximum available amplifier gain
R_L	Load resistance
R_{oe}	Output resistance of common-emitter circuit
T_A	Ambient temperature
T_L	Lead soldering temperature
V_{BB}	Base supply voltage
V_{BC}	Base-to-collector voltage
V_{BE}	Base-to-emitter voltage
V_{CB}	Collector-to-base voltage
V_{CBO}	Collector-to-base voltage with emitter open
V_{CC}	Collector supply voltage
V_{CE}	Collector-to-emitter voltage
V_{CEO}	Collector-to-emitter voltage with base open
V_{EB}	Emitter-to-base voltage
V_{EBO}	Emitter-to-base voltage with collector open
V_{EE}	Emitter supply voltage

APPENDIX H. CHASSIS WIRING COLOR CODE

Electrodes and/or Terminals	Color
Transistor Emitters, Diode Cathodes, and Electron-Tube Cathodes	Yellow
Transistor Bases and Electron-Tube Control Grids	Green
Transistor Collectors and Electron-Tube Anodes (plates)	Blue
Grounds and Grounded Electrodes	Black
Heaters or Filaments	Brown
Screen Grids	Orange
Power Supply B+	Red
Power Supply B−	Violet
A-C Power Lines	Gray

Adapted from Military Standard (MIL-STD) 681A

*APPENDIX I. SELECTED FREQUENCY ALLOCATIONS

Frequency Band (kHz)	Nature of Service(s)
Below 10	Not allocated
19.95–20.05	Standard frequency
59–61	Standard frequency
535–1605	Standard AM broadcasting
1800–2000	Amateur
	Loran
2500	Standard frequency
3500–4000	Amateur
5000	Standard frequency
5950–6200	International broadcasting
7000–7300	Amateur
9500–9775	International broadcasting
10000	Standard frequency
11700–11975	International broadcasting
14000–14350	Amateur
15000	Standard frequency
15100–15450	International broadcasting
17700–17900	International broadcasting
20000	Standard frequency
21000–21450	Amateur
21450–21750	International broadcasting

Frequency Band (MHz)	Nature of Service(s)
25.0	Radio astronomy Standard frequency
25.6–26.1	International broadcasting
26.96–27.23	Citizens' radio
28–29.7	Amateur
50–54	Amateur
54–72	Television broadcasting: 55.25 MHz Video / 59.75 MHz Sound — Channel 2; 61.25 MHz Video / 65.75 MHz Sound — Channel 3; 67.25 MHz Video / 71.75 MHz Sound — Channel 4
76–88	Television broadcasting: 77.25 MHz Video / 81.75 MHz Sound — Channel 5; 83.25 MHz Video / 87.75 MHz Sound — Channel 6
88–108	FM broadcasting
136–137	Space telemetering and tracking
144–148	Amateur
174–216	Television broadcasting: 175.25 MHz Video / 179.75 MHz Sound — Channel 7; 181.25 MHz Video / 185.75 MHz Sound — Channel 8; 187.25 MHz Video / 191.75 MHz Sound — Channel 9; 193.25 MHz Video / 197.75 MHz Sound — Channel 10; 199.25 MHz Video / 203.75 MHz Sound — Channel 11; 205.25 MHz Video / 209.75 MHz Sound — Channel 12; 211.25 MHz Video / 215.75 MHz Sound — Channel 13
220–225	Amateur
420–450	Amateur
451–454	Public safety
462.5375–462.7375	Citizens' radio
467.5375–467.7375	Citizens' radio
470–890	Television broadcasting
1215–1300	Amateur
2300–2450	Amateur
3300–3500	Amateur
7250–7300	Communication satellite

*Adapted from Federal Communications Commission, *Rules and Regulations,* Volume II, May 1966; Part 2

Index

A-scan presentation, 430
Accelerator, particle, 4–5
Air gap, magnetic, 80–81
Alloys, ferromagnetic, 76
Alpha, 367
Alternating current, 20–21
Ammeter, 264–267
Ampere-turns, 85
Amplification, 244
Amplifier:
   audio, 229–231, 392–393
   audio frequency, 389
   bilateral, 408
   buffer, 245
   distortion in, 230–231
   high-fidelity, 230
   intermediate-frequency, 396–398
   low-fidelity, 230
   push-pull, 400–401
   radio-frequency, 382–383, 419–420
   speech, 377–379
   stereo, 238–239
   video, 419–422
Amplifier stage:
   class A, 217
   class AB, 217
   class B, 217
Amplitude modulation, 377
Angle modulation, 386
Angular velocity, 139
Anode, 24–25
Antenna, 370–374
   directional beam, 373
   parabolic, 432
   radar, 431–432
   slotted-waveguide, 432
Antilogarithm, 464
Artificial respiration, 72–73
Assembly techniques, 320
Atom, 4–6
Atomic quota, 7–8
Atomic solar system, 5–6
Atomic structure, 4–7
Atomic weight, 7
Audio amplifier, 392–393
Audio detector, 398–400
Automatic gain control, 420–422
Autotransformer, 112
AVC circuit, 398–400

Back voltage, 53
Bandpass, 168, 171

Bandwidth, 385
Base, tube, 223
Battery, testing of, 307
Beta, 467
Bias:
   feedback, 214–215
   single-source, 213–214
Bias stabilization, 214
Biasing, 212
Bilateral amplifier, 408
Binding energy, 7
Blanking circuit, 414–416
Breadboarding, 334
Brightness-control circuit, 415–416
Broadcast television, 410
Buffer amplifier, 245

Cable, coaxial, 410
Cadmium-sulphide cell, 452
Calculator, parallel resistor, 134
Camera:
   color, 425–427
   television, 412–416
Capacitance, 51
   distributed, 162–163
   effect of, 150–154
   factors affecting, 54
Capacitive reactance, 54, 154
Capacitor, 51–63
   ceramic, 57
   checking of, 303–305
   defective, 231
   discharging, 69
   electrolytic, 58–60
   film-type, 56
   fixed, 55
   glass, 57–58
   metallized, 56
   mica, 56–57
   multisection, 61
   nonpolarized, 60–61
   operation of, 52–54
   padder, 62
   paper, 56
   tantalum, 60
   trimmer, 62
   variable, 61–62
Capacitor checker, 294–295
Capacitor color code, 62, 63
Capacitors:
   in parallel, 161
   in series, 160

Carrier:
   majority, 200–201
   minority, 179, 200
   uncontrolled, 179
Cartridge, stereo, 238
Cathode, 24
   electron-tube, 222
   marking, 180–181
Cell:
   cadmium-sulphide, 452
   photoconductor, 32
   photovoltaic, 33
   testing of, 307
Centi, 467
Characteristic, 464
Charge, electric, 8–13
Charge storage, 53
Chassis, preparation of, 323–325
Choke:
   filter, 104–105
   radio-frequency, 105
Circuit:
   amplifier, 366
   AVC, 398–400
   blanking, 414–416
   bridge rectifier, 183
   brightness-control, 415–416
   capacitive, 150
   common-base, 209–211
   common-collector, 211
   common-emitter, 206–207
   complementary-symmetry, 235
   electric, 4
   emitter-follower, 217
   encoder, 367
   filter, 173, 181
   full-wave doubler, 185
   half-wave doubler, 185
   high-voltage power supply, 413–416
   horizontal output, 413–416
   hybrid, 364
   inductive, 135–136
   integrated, 357–364
   LC oscillator, 244
   magnetic, 80
   modulation, 415–416
   monolithic, 359–364
   oscillator, 366, 415–416
   parallel, 125, 149
   parallel *RC*, 159–160
   parallel *RL*, 144

Circuit (*Cont.*):
  parallel-tuned, 169
  phase-inverter, 234-235
  power supply, 366
  printed, 345
  pure *LC*, 169
  Q of, 167
  quadrupler, 186
  r-f mixer, 415-416
  radiation, 368
  rectifier, 180, 182
  resistive, 123
  resonance, 164
  sensor, 387-388
  series, 123-125, 148, 159
  series-parallel, 128
  series-tuned, 165-166
  symmetry, 235
  sync pulse mixer, 414-416
  transmitter, 366-369
  tuned, 164
  vertical output, 414-416
  voltage-doubler, 184-185
  voltage-tripler, 186
Circuitry:
  ac-dc, 67-69
  flexible, 355
Cleaning aids, 71
Closed-circuit television, 410
Coaxial cable, 410
Coefficient of coupling, 105
Coercive force, 86
Coil:
  checking of, 305-306
  deflection, 413, 422-423
Cold-cathode emission, 24
Collector characteristics, 208
Color burst, 427
Color camera, 425-427
Color coder, 47
Color television, 424-428
  receiver, 427-428
  transmission, 425-427
Colpitts oscillator, 245
Component failure, 301
Components:
  location of in chassis, 320-323
  mounting of, 325-331
  thin-film, 364
Composite video signal, 417
Conductor, 4, 21
Connection, Darlington, 451
Connections, loudspeaker, 239-241
Continuity test, 273, 344
Control:
  tone, 236

Control (*Cont.*):
  treble, 236
  volume, 236
Control grid, function of, 225
Converter, 396
  transistorized, 254-256
Copper loss, 108
Cord, three-conductor line, 65-66
Corona discharge, 12
Cosine, 463
Coulomb, 9
Coulomb's law, 78
Counter electromotive force, 53
Coupling, 231
  capacitance, 231
  direct, 232
  interstage, 231
  resistance, 234
  signal, 245
  transformer, 231, 232, 234
Crossover network, 242
Crosstalk, 83
Crystalline structure, 26-27
Current, 14
  alternating, 20-21
  direct, 19
  electric, 16
  ionic:
    in gases, 17
    in liquid, 16
  leakage, 179
  negative electronic, 208
  parallel circuit, 126
  polarity, 208
  reverse, 179
  series circuit, 124

Darlington connection, 451
Deci, 467
Deflection, electrostatic, 279
Deflection coil, 413, 422-423
Delta, 467
Demagnetization, 89
Designation, tube type, 224
Detector, 388-389
  audio, 398-400
  oscillating, 290-291
  ratio, 421-423
  sound, 423
  superregenerative, 390-392
Device, control, 4
Diagram:
  block, 317
  connection, 319-320
  function, 317
  pictorial, 315

Diagram (*Cont.*):
  printed circuit, 347
  schematic, 316
  single-line, 317
  wiring, 319-320
Dichroic filter, 425
Dielectric, 52
Dielectric constant, 55
Dielectric field, 9-10
Diode:
  crystal, 192-193
  forward bias, 176-178
  junction, 176
  reverse bias, 179
  testing of, 300
  tunnel, 219-220
  zener, 187
Diode rectifier, 179
Diode tuning, 429
Dipoles, 74
Direct current, 19
  pulsating, 19
  "pure," 19
  varying, 19
Discharge:
  corona, 12
  electric, 11
Distortion in amplifier, 230
Domains, magnetic, 74
Donor impurity, 29
Doping, 27
Drafting practices, 314-315
Driver, relay, 393
Dynode, 24-25

Eddy currents, 107-108
Electric charge, 8-13
Electrical impulse, 19
Electrical power, 37
Electrical shock, 64-65
Electricity:
  definition of, 2
  static, 11-13
Electrodes, 196
Electrolysis, 17
Electromagnet, 84-85
  advantages of, 88
Electromotive force, 14
Electron, 5
  free, 7
  velocity of, 19
Electron emission, 22-24
Electronic current, 18-19
Electronic drafting, 309-320
Electronics, definition of, 2
Electrostatic field, 9-10
Electrostatic induction, 10-11

Elements, ferromagnetic, 75
Emission:
  cold-cathode, 24
  electron, 22–24
  field, 24
  photoelectric, 23–24
  secondary, 24
  thermionic, 23
Energy:
  binding, 7
  electric, 2–4
  kinetic, 2
  law of conversion, 2
  potential, 2
  source of, 4
Energy levels, 6–7

Farad, 54
Feedback, 244
Ferrites, 76
Fidelity, 230
Field:
  dielectric, 9–10
  electrostatic, 9–10
  flux, 76
  magnetic, 76
Field emission, 24
Filter:
  band-reject, 175
  bandpass, 174
  dichroic, 425
  high-pass, 174
  low-pass, 173
  mechanical, 408–409
Flux density, 76
Flux lines, 82, 90
Flyback transformer, 423
FM transmitter, 384
Force:
  electromotive, 14
  magnetomotive, 76, 85
Formula:
  conductance, 127
  power, 38
  power gain, 208
  voltage gain, 208
Formulas for Ohm's law, 36–38
Free electrons, 7
Frequency, 20
  resonant, 164
Frequency distortion, 230
Frequency-divider network, 242
Frequency measurements, 282
Functions, 463

Gauss, 76
Geiger counter, 442–443

Generator:
  audio signal, 287–288
  marker, 289–290
  r-f signal, 288–289
  signal, 286–290
  sweep, 289
  three-phase, 97–98
  Van de Graaff, 12
Germanium:
  n-type, 27–29
  p-type, 29–30
Gilbert, 76
  conversion to, 85
Grounding, safety practice, 65
Gun, soldering, 336

Hartley oscillator, 244–245
Headset, 120
Heat sink, 339
Henry, 102–103
Hertz, 20
High-voltage power supply circuit, 413–416
Holes, 18, 29
  properties of, 31
Horizontal output, 422–423
  circuit, 413–416
Hysteresis:
  versus frequency, 87
  magnetic, 86
Hysteresis loop, 87–88
Hysteresis loss, 86

Image orthicon tube, 412
Impedance, 54, 102, 141–142, 157, 160, 170
  input and output, 210
Impedance matching, 113
Implosion, 70
Induced voltage, magnitude of, 90
Inductance:
  effect of:
    in ac circuit, 138
    in dc circuit, 136
  factors affecting, 103
Induction:
  electromagnetic, 89–90
  electrostatic, 10–11
  magnetic, 79
  mutual, 105–106
Inductive reactance, 102, 138, 142
Inductor, 103–105
  quality factor of, 104
  variable, 104

Inductors:
  in parallel, 147
  in series, 146
Insulation:
  reinforced, 67
  removal of, 333
Insulator, 22, 329
Integrated circuit, 357–364
Ions, 8
Iron, soldering, 335
Isotopes, 7

Jacks, 329
Junction:
  depletion region, 31–32
  p-n, 31–32
  potential barrier, 31–32

Kilo, 467
Kilohertz, 20
Kilohm, 15
Kinescope, 424

Lambda, 467
Laminate, copper-clad, 346
Lamp:
  incandescent, 2–3
  neon, 17
Laser, 433
  continuous wave, 436
  gas, 436
  "pumping," 434
  ruby, 434–436
  semiconductor, 436
  solid-state, 434–436
  types of, 434–436
Laser applications, 436–437
Law of attraction, 78
Law of repulsion, 78
Layout, chassis, 323–324
Left-hand rule, 85
Light:
  coherent, 434
  incoherent, 433
Limiter, sync, 420–423
Lissajous figure, 284
Load, 4
Load line, 216–217
Logarithm, 463–464
Loudspeaker, 119–120, 239–242
  coaxial, 242
  electrodynamic, 120
  phasing of, 240
  rating of, 120
  testing of, 306–307

Loudspeaker connections, 239–240
Low voltage, danger of, 64
Lugs, 333

Magnet, permanent, 74
Magnetic fields, 82–84
Magnetic phono cartridge, 117–118
Magnetic shield, 81
Magnetism, 74
   residual, 80
Magnetizing, 88–89
Magnetostriction, 81
Mantissa, 464
Manual, tube, 224
Maser, 433
Materials:
   diamagnetic, 75
   ferromagnetic, 75
   paramagnetic, 75
Matter, 4
Maxwell, 76
Measurement, frequency, 282
Mechanical filter, 408–409
Mega, 467
Megahertz, 20
Megohm, 15
Meter:
   absorption type, 291
   care of, 261
   D'Arsonval, 258–263
   grid dip, 290
   reading of, 262–263
   relative strength, 291
Meter action, 261
Meter protection, 274–275
Meter scales, 262
Meter shunts, 264, 266–267
Metric system, 14
Micro, 467
Microphone:
   directional, 118–119
   dynamic, 118
   power output of, 119
Microphonics, 230–231
Milli, 467
Modulation, 377–386
   amplitude, 377
   angle, 386
   low-power, 383–384
   negative, 417
   pulse, 386
Modulation process, 379–380
Modulator, 377–379
   reactance tube, 384–385
   ring-balanced, 405–407

Modulator circuit, 415–416
Modules, printed, 356
Molecule, 4
Monolithic circuit, 359–364
Moving magnetic field, 83
Mu, 467
Multimeters, 273–274
Multiplier, 267
Multivibrator, 248–249
   astable, 249–250
   bistable, 251–252
   driven, 250
   flip-flop, 251–252
   free-running, 249–250
   monostable, 250

N-type germanium, 27–29
Nano, 467
Negative modulation, 417
Neon lamp oscillator, 248
Network:
   crossover, 242
   voltage-divider, 215
Neutrons, 6–7
Noise gate, 420–422
Nonsinusoidal oscillator, 248
Nuclear radiation, 441–443
Nucleus, 5–6

Ohm, 15, 102
Ohm's law, 36–41, 143, 158
   formulas for, 36–38
   limitations of, 38
   memory aids for, 39
Ohm's law calculator, 39–41
Ohm's law chart, 39–40
Ohmmeter, 269–275
   range of, 271
   use of, 271–273
Omega, 467
Oscillator, 244–253
   blocking, 253
   Colpitts, 245
   crystal-controlled, 246
   feedback, 244
   Hartley, 244–245
   neon lamp, 248
   nonsinusoidal, 248
   RC, 247–248
   tuned-circuit, 244
Oscillatory action, basic, 172
Oscillatory response, 172
Oscilloscope, 278–285
Oscilloscope probes, 286
Overloading, 231

P-n junction, 31–32
P-type germanium, 29–30
Padder capacitor, 62
Pads, T and L, 236
Particles:
   alpha, 441
   beta, 441
Permeability, 79
Phase angle, 141, 145, 156, 159
Phase shift, 140, 155
Phosphorescence, 427
Phosphors, 427
Photocathode, 24–25
Photoconductivity, 32
Photoelectric emission, 23–24
Photoelectric sensor, 438–440
Photoengraving process, 348–349
Phototube, 23–24
   multiplier, 24–25
Pi, 467
Pico, 467
Pictorial diagram, 315
Piezoelectric effect, 246
Plasma, 17–18
Point number (PN) value, 465
Polarity of induced magnetism, 80
Poles, magnetic, 78
Potential, 13
   difference in, 13–14
   zero, 13
Potentiometer, 44–45, 330
   checking of, 305
   volume-control, 236
Power, 131
   apparent, 147–148
   electrical, 37
   in inductive circuits, 147
   true, 148
Power factor, 148
   correction of, 149
PPI radar, 432–433
Preamplifier, 237
Printed circuit, 345
   soldering, 350–351
   types of: die-stamped, 356
      painted, 356
      plated, 356
   working with, 351–355
Printed circuit diagram, 347
Probes, oscilloscope, 286
Protons, 5–6
Pulse modulation, 386
Puncture of dielectric, 55
Push-pull amplifier, 400–401
   operation of, 233–234

Q of a circuit, 167
Quantum, 7

Radar:
    definition of, 429
    PPI, 432–433
    pulse-modulated, 429–430
Radar antenna, 431–432
Radar receiver, 431
Radar transmitter, 430–431
Radiation, nuclear, 441–443
    gamma, 441
Radiator, 370–374
Radiography, 443
Ratio detector, 421–423
RC oscillator, 247–248
Receiver:
    color television, 427–428
    five-tube, 401–403
    radar, 431
    single-channel RC, 390–393
    superheterodyne, 394–403
Recombination, 7
Record, stereo, 238
Recording, tape, 121–122
Rectification, 20–21, 176
    three-phase, 188
Rectifier, 176
    selenium, 193–194
    silicon controlled, 191–192, 453–454
Relay driver, 393
Reluctance, definition of, 79
Remanence, magnetic, 80
Reproducer, 389–390
Reproducer system, 401
Rescue of shock victims, 72–73
Residual magnetism, 80
Resistance, 14–15, 21
    checking of, 302–303
    effect of, 167, 173
    parallel circuit, 126
    series circuit, 124
    series-parallel circuit, 128–131
    standard values, 47
Resistance box, 276–277
Resistor, 42–46
    adjustable, 43–44
    bleeder, 70, 186–187
    carbon, 42
    defective, 231
    emitter, 215
    glass-tin oxide, 43
    metal-glaze, 42–43
    physical size of, 48
    power, 43
    power rating of, 48

Resistor (*Cont.*):
    safety factor, 49
    trimmer, 46
    wire-wound, 43
Resistor color code, 46–47
Resonance, conditions of, 171
R-F amplifier, 419–420
R-F mixer circuit, 415–416
Rheostat, 44–45, 330
    checking of, 305
Ring-balanced modulator, 405–407
Rounding of numbers, 134

Safety factor of capacitor, 55
Safety rules, 65
    general, 71–72
Schematic diagram, 316
Schmitt trigger, 253
Secondary emission, 24
Selectivity:
    definition of, 166
    in parallel-tuned circuit, 171
    in series-tuned circuit, 166
Self-inductance, 101–102
Semiconductor, 26–34
    intrinsic, 27
Sensor, photoelectric, 438–440
Sensor circuit, 387–388
Shield, magnetic, 81
Shock victims, rescue of, 72
Shunts, meter, 264, 266–267
Signal, composite video, 417
Signal coupling, 245
Signal generators, 286–290
Signal tracer, 291
Silicon controlled rectifier, 191–192, 453–454
Silkscreen process, 249
Sine, 463
Sine waveform, 20
    generation of, 91–93
    mathematical basis of, 93–95
Single-channel RC receiver, 390–393
Single-sideband generation, 385–386
Single-sideband transceiver, 405–409
Sinusoidal voltage, 91–97
Skin effect, 22
Slide rule, use of, 465–466
Socket, tube, 223
Soldering:
    dip, 351
    wave, 351
Soldering gun, 336

Soldering iron, 335
Soldering semiconductors, 338–339
Soldering techniques, 335–339
Solenoid, 84
Sound detector, 423
Speaker systems, 241
Speech amplifier, 377–379
Spin, electron, 74
Stability, magnetic, 80
Stabilization:
    diode bias, 216
    thermistor, 215–216
Static electricity, 11–13
Stereo amplifier, 238–239
Stereo cartridge, 238
Stereo magnetic tape, 238
Stereo records, 238
Stereo system, 237–238
Structure:
    atomic, 4
    crystalline, 26–27
    lattice, 26–27
Superconductivity, 22
Superheterodyne receiver, 394–403
Superregenerative detector, 390–392
Suppressor grid, 24
Surface barrier, 22
Switch, transmit-receive, 431
Switches, interlock, 71
Switching, electronic, 444–445
Symbol:
    for n-p-n transistor, 200
    for p-n-p transistor, 201
Symbols, electrical, 309–311
    rules for drawing, 311–313
Sync limiter, 420–423
Sync pulse mixer, 414–416
System:
    atomic solar, 5–6
    reproducer, 401
    television, 410

Tangent, 463
Tape, magnetic stereo, 238
Tape recording, 121–122
Taper, linear, 45–46
Techniques:
    assembly, 320
    wiring, 320
Television:
    broadcast, 410
    closed-circuit, 410
    color, 424–428
Television camera, 412–416

Television receiver circuit, 419–424
　color, 427–428
Television systems, 410
Television transmission, 410–412
Television transmitter, 418–424
Temperature, Curie, 80
Temperature coefficient, 21, 55
Terminal strips, 329
Tester:
　diode, 292
　transistor, 292
　tube, 293
Thermionic emission, 23
Thermistor, 49–50
　applications, 440–441
Time constant, 136–137, 151
Timer, electronic, 443–444
Tolerance, 45
Tower:
　relay, 410
　repeater, 410
Transceiver, single-sideband, 405–409
Transformer, 106–116
　characteristic of, 108–112
　checking of, 306
　core construction, 107
　flyback, 423
　interstage, 115–116
　isolation, 112–113
　microphone, 115
　output, 115
　power, 112
　rectifier, 182
　step-down, 114–115
　voltage ratio, 108
Transformer coupling, 234
Transistor, 194–195, 198
　current control, 205–206
　cutoff frequency, 208
　defective, 230
　fabrication of, 195
　field-effect, 201–203, 218–219
　high-frequency, 218
　junction, 195
　methods of construction, 198
　n-p-n, 199, 206
　p-n-p, 200, 207, 210
　testing of, 298–300
　unijunction, 203–204
Transmission:
　color television, 425–427
　television, 410–412
　vestigial sideband, 418
Transmitter:
　FM, 384

Transmitter (Cont.):
　low-power AM, 375–376
　radar, 430–431
　television, 418–419
Transmitter circuit, 366–369
Treble control, 236
Trigger, Schmitt, 253
Trigonometry, 463
Trimmer capacitor, 62
Troubleshooting, 296–307
Tubes:
　cathode-ray, 70
　defective, 230
　diode, 221
　electron, 222
　　testing of, 297–298
　electron-ray, 286
　image orthicon, 412
　indicator, 228
　multiunit, 226
　pentode, 226
　rectifier, 224
　shielding, 226–227
　substitution test, 293
　tester, 293
　tetrode, 226
　thyratron, 227
　triode, 224–225
　vidicon, 412
Tuner:
　broadcast band, 388, 395
　television, 419
Tuning, 167
　diode, 429

Units, conversion of, 15

Vacuum tube voltmeter, 273–274
Van de Graff generator, 12
Varactor, 429
Variable capacitor, 61–62
Varicap, 429
Vectors, use of, 466
Vertical output circuit, 414–416
Video amplifier, 419–422
Video tape recorder, 410–411
Vidicon tube, 412
Visual inspection, 296
Volt-ohm-milliammeter, 273–274
Voltage, 14
　avalanche, 180
　bias, 225
　breakdown, 22
　checking for, 66–67
　checking of, 301–302

Voltage (Cont.):
　high induced, 138
　induced, 106
　negative, 188
　parallel circuit, 125–126
　parallel RL circuit, 144
　peak-reverse, 180
　series circuit, 124
　sinusoidal, 91–97
　three-phase, 97–98
Voltage characteristics, 137–138
Voltage divider, 188
Voltage drop, 124
　polarity of, 125
Voltage gain, 208
Voltages, comparison of, 165, 282
Voltmeter, 267–269
　digital, 269
　loading effect, 268
　vacuum tube, 273–274
Volume control, 236

Wave:
　ground, 374
　sky, 374
　standing, 369
Waveform:
　current, 91
　nonsinusoidal, 99–100
　peaked, 99
　sawtooth, 100
　sine, 20
　sinusoidal, 91–97
　square, 100
　voltage, 91
Waveform observation, 281
Wavetrap, 173
Weight, atomic, 7
Wheatstone bridge, 275–276
Wire, 4, 331–334
　bus, 334
　cable, 332
　color coding, 334
　grounding, 66
　shielded, 332
　testing, 344
Wire connecting, 341–342
Wire harness, 342
Wire routing, 341–342
Wiring from a diagram, 339–341
Wiring techniques, 320, 331–334
Working voltage, 55

X-rays, 442

Zero potential, 13